DIFFERENTIAL GEOMETRY FOR PHYSICISTS

ADVANCED SERIES ON THEORETICAL PHYSICAL SCIENCE
A Collaboration between World Scientific and Institute of Theoretical Physics

Series Editors: Dai Yuan-Ben, Hao Bai-Lin, Su Zhao-Bin
(Institute of Theoretical Physics Academia Sinica)

Vol. 1: Yang-Baxter Equation and Quantum Enveloping Algebras
(Zhong-Qi Ma)

Vol. 6: Differential Geometry for Physicists
(Bo-Yu Hou & Bo-Yuan Hou)

Forthcoming:
Geometric Methods in Physics *(Wu Yong-Shi)*
Special Relativity and its Experimental Foundation *(Zhang Yuan-Zhong)*
Liquid Crystal Models of Biomembranes *(Ouyang Zhong-Can, Xie Yu-Zhang and Liu Ji-Xing)*
Group Theory for Condensed Matter Physics *(Tao Rui-Bao)*

Advanced Series on Theoretical Physical Science
Volume 6

DIFFERENTIAL GEOMETRY FOR PHYSICISTS

Bo-Yu Hou
Northwest University
Xian, China

Bo-Yuan Hou
Academia Sinica
Beijing, China

World Scientific
Singapore • New Jersey • London • Hong Kong

Published by

World Scientific Publishing Co. Pte. Ltd.
5 Toh Tuck Link, Singapore 596224
USA office: 27 Warren Street, Suite 401-402, Hackensack, NJ 07601
UK office: 57 Shelton Street, Covent Garden, London WC2H 9HE

British Library Cataloguing-in-Publication Data
A catalogue record for this book is available from the British Library.

Advanced Series on Theoretical Physical Science — Vol. 6
DIFFERENTIAL GEOMETRY FOR PHYSICISTS

ISBN-13 978-981-02-3105-7
ISBN-10 981-02-3105-9

Contents

Preface

The developments of physics and mathematics are closely interlinked. In certain respects, mathematics is an important tool for physics, while many significant mathematics problems originate from physics. The relationship between modern theoretical physics and differential geometry is more profound. Physics not only gives new ideas and motivations to modern differential geometry, but also provides powerful alternative methods for geometry. The proof of Atiyah-Singer Index theorem in terms of supersymmetric quantum mechanics[1]; Donaldson's analysis of differential structure of $\mathbf{R}^4$ in terms of the solutions of sourceless non-Abelian gauge field equations[2]; and the calculation of Donaldson invariants by Seiberg-Witten in terms of supersymmetric Abelian gauge field with spin[3] are only three of the most prominent examples. On the other hand, in the last few decades, there have been extensive applications of topology and geometry in physics, such as, Riemannian geometry in general relativity[4], Poisson-Lie algebra and symplectic geometry in integrable systems[5], complex geometry and algebraic geometry in conformal field theory and string theory[6], fibre bundle differential geometry in the global analysis of Yang-Mills theory[7] and quantum anomaly[8], etc.. The global and topological analysis gives an impetus to inaugurate many physical idea. All fundamental interactions are gauge interactions, the gauge field potential and gauge field strength identified with the connection and curvature on fibre bundle [9], respectively. In modern quantum field theory and statistical physics, global and nonlinear analysis plays a critical role. The concepts of connection, curvature, characteristic class in differential geometry are showing up in various branches of theoretical physics, and no longer limited to gravity and Yang-Mills theory. For instance, Berry phase[10] arises from non-trivial topological properties of parametrical space.

In light of the present situation, this book is written in the hope that it will acquaint our readers and graduate students with some of the basic ideas and methods of modern differential geometry and their applications in physics.

The book is divided into fourteen chapters with another eighteen appendices as introduction to prerequisite topological and algebraic knowledge. The first seven chapters focus on local analysis. This part could be used as an introductional text book for graduate students of theoretical physics. Chapters 8 - 10 discuss geometry on fibre bundle which could be further reference to researchers. The last four chapters deal with the Atiyah-Singer index theorem and its generalization, quantum anomaly, cohomology field theory, and noncommutative geometry, which may give the reader a glimpse of the frontier of current research in theoretical physics.

The first three chapters expound the differential geometry without a metric, in these chapters, three kinds of important differential operators: exterior differential, Lie derivative, and covariant derivative, are introduced. To combine them together, attention is given to some basic concepts of manifold, mapping, connection and integrability. In chapter one differentiable manifold and tensor field are introduced, and we put emphasis upon the point that tensor fields are geometric entities, independent of the choice of coordinate systems; we emphasize further the coordinate independence of equations. In chapter two, mapping between manifolds and its induced mapping between tensor fields; Lie group manifolds and its invariant vector fields are introduced. In chapter three, we take into consideration additional structures on the manifold: connection and curvature on manifold, though these do not necessarily involve a metric, and analysis of the problem in terms of moving frames are introduced. These are intended to give the readers a visual picture of fibre bundle.

After introducing the connection on manifold, in chapters 4 - 6 we introduce metric structure, symplectic structure, and complex structure on manifold, respectively. In these three chapters, the applications to general relativity, analytical mechanics, symmetric space and integrable systems are also discussed.

Chapter 7 is a critical chapter, from which we start the global analysis on manifold. In chapters 7 and 8, homotopy and homology, which are the fundamental concepts in algebric topology, are introduced, so are the basic structure and classification of fibre bundles. In chapters 9 and 10, the connection and characteristic class on the fiber bundle are introduced, and we proceed with the global topological analysis in gauge theory. Fibre bundle is the best language for studying the relationship between the global and local properties of manifold.

In chapters 11 - 13, Atiyah-Singer Index theorem, Atiyah-Patodi-Singer Index theorem and family index theorem are introduced briefly. With quantum anomaly as an example, we treat the global analysis in quantum field theory and take account various topological constructions. The last chapter is about non-commutative geometry, quantum group, and q-gauge theory, which are some of the popular topics in current theoretical physics.

At the end of each chapter, a notations and formulae are attached for its summary and reference.

There are many books on differential geometry, and we list some of these which we are more familiar as general references at the end of this book. On the one hand, it is necessary to have an extensive mathematical knowledge, no matter how difficult it is to grasp them immediately for physicists, on the other hand, to the best of our knowledge, these books we known do not go beyond A-S index theorem, and the extension of A-S index theorem (family index theory, index theorem for open and infinity manifold), and their application on quantum field theory are little mentioned. The materials in these areas can only be found in certain articles. Our purpose is to introduce differential geometry, especially the fundamental concept, methods, and results of differential geometry. As for various mathematical terms and definitions, only the most necessary are included, as we never mean to be all-embracing. The straightforward and intuitive approach, which is more acceptable to physicists while not affecting the final mathematical results, has been adopted. References are cited for the problems discussed, and they are by no means complete, and we apologize to those authors for not mentioning their works. For reader's convenience, the notations and terminology in this book are consistent with the general convention.

Acknowledgement: It is pleasure to thank Profs. Y .S. Duan, M. L. Ge, H. Y. Guo, Z. Q. Ma, X. C. Song, P. Wang, S. K. Wang, K. Wu, L. C. Xu, L. N. Zhang, for our pleasant and effective cooperations, to thank profs. B. L. Hao and C. Y. Zhu for their support and help. We would like to thank D. Mi, B. Wu, M. Xie, F. Z. Yang, Z. X. Yang, Q. R. Zheng, for their translation and arrangement of the manuscript, in particular, Dr. Z. X. Yang's revision and translation in English of the whole book. Without their help, this book is difficult to be completed. This book is supported in part by the National Natural Science Foundation of China.

Chapter 1

Differentiable Manifolds and Differential Forms

In Euclidean geometry, two figures are said to be identical if they overlap completely after an Euclidean movement which does not change the Euclidean distance between two points. All of the Euclidean movements form a group. The Euclidean geometry studies the invariant properties of figures under the action of Euclidean movements. From this point of view Klein gave a definition of geometry in 1871 as follows: Given a set E (called space) and a group G which acts on the space E, the geometry studies the invariant properties of the space E on which the transformation group G acts. Differential geometry studies the invariant properties of differentiable manifolds under the action of diffeomorphism transformations. The major research objects of differential geometry are differentiable manifolds and various tensor fields on them.

1.1 Manifold

Many problems in physics are related to the continuous spaces, such as the usual space-time in kinetics and dynamics, curved space-time in general relativity, phase space in statistical physics, interior space and related base space(ordinary space-time) in gauge field theory, etc.. These spaces have some common properties: they are all continuous spaces and have a definite dimension. To study them, the concept of manifold is introduced. Manifold is the generalization of point, line, surface and other higher dimensional continuous spaces which we are familiar with.

Definition 1.1 *Manifold: A real n-dimensional manifold M is a space which is like $\mathbf{R}^n$ locally. More precisely, a real n-dimensional manifold is such a Hausdorff space in which any point has a neighborhood homeomorphic to $\mathbf{R}^n$.*

In the following, we briefly explain some mathematical term in this definition.
1). Real n-dimensional linear space $\mathbf{R}^n$.

$\mathbf{R}^n$ is an n-dimensional linear space over real field. Its element is called a point or a vector, which can be represented by an n-tuple real numbers as:

$$x = (x^1, x^2, \cdots, x^n) \in \mathbf{R}^n,$$

$x^i (i = 1, \cdots, n)$ are called the components of vector x, or the coordinates of point x. The addition operation of vectors in $\mathbf{R}^n$ is defined as addition of the corresponding real components,

$$x + y = (x^1 + y^1, x^2 + y^2, \cdots, x^n + y^n) \in \mathbf{R}^n.$$

Multiplication of vector x and a real number a is defined as

$$a\, x = (ax^1, ax^2, \cdots, ax^n) \in \mathbf{R}^n.$$

It can be easily checked that the addition defined above has the properties: There exists a zero element

$$0 = (0, 0, \cdots, 0) \in \mathbf{R}^n,$$

$$x + 0 = 0 + x = x.$$

For each $x \in \mathbf{R}^n$, there is an inverse $\overline{x}$

$$\overline{x} \equiv -x = (-x^1, -x^2, \cdots, -x^n) \in \mathbf{R}^n,$$

$$x + \overline{x} = 0.$$

The addition is also associative and commutative:

$$x + (y + z) = (x + y) + z.$$

$$x + y = y + x.$$

Operations endowed with the above four properties are called Abelian.

The multiplication defined above is associative, i.e.,

$$a(bx) = (ab)x.$$

and is distributive with respect to addition

$$a(x + y) = ax + ay.$$

$$(a + b)x = ax + bx.$$

Set $\{x\}$ which has the above properties is an n-dimensional real linear space $\mathbf{R}^n$. A real n-dimensional vector space $\mathbf{R}^n$ can be defined as follows

Definition 1.2 *Linear space: An n-dimensional linear space* $\mathbf{R}^n$ *on real field is such a space that each element can be represented by a real n-tuple, on which an Abelian addition and a multiplication with a number,satisfying associativity and distributivity, are defined.*

2). Open set and continuous mapping.

Distance functions are usually introduced in order to analyse the continuity of spaces and mappings on them. As we know, continuity depends only on neighborhood. Changing the definition of distance function(i.e., changing the metric) does not influence continuity. Therefore we usually lay aside distance functions and study the more abstract topological spaces rather than metric spaces. To study neighborhood instead of distance, the concept of open set is introduced. Open set is a fundamental concept in describing the topological properties of spaces. Its precise definition is given in Appendix B. The following is an intuitive usual definition which commonly used:

Definition 1.3 open set: *An open set* U *is a subset of space* S*, in which the neighborhood of each point of* U *is completely in* U*, where the neighborhood can be defined by any choice of distance functions, of which the definition of open set is independently.*

For example, on real axis $\mathbf{R}^1$ (one-dimensional linear space $\mathbf{R}^1$), an open interval (a, b) which does not contain two end points is an open set. But a closed interval $[a, b]$ which contains two end points is not an open set, because the neighborhoods of end points a and b are not entirely within the set $[a, b]$.

In definition 1.3, the "neighborhood" of a point x is a subset, where the distance from any point in the subset to this point x is less than a given real number. A distance function is needed to define a "neighborhood". But it should be emphasized that "neighborhood" can be defined by any distance function so it is independent of the choice of distance functions. For example, for any two points in $\mathbf{R}^n$, there are two types of distance functions:

$$d_1(x, y) = [\sum_{i=1}^{n}(x^i - y^i)^2]^{\frac{1}{2}},$$

$$d_2(x, y) = \max |x^i - y^i|.$$

The "neighborhood" defined by d_1 is a disk while the 'neighborhood' defined by d_2 is a square. The shape of the "neighborhood" depends on the definition of distance functions. But we know that any disk contains a square and any square contains a disk. Therefore, if U is an open set with respect to distance function d_1, then U is also an open set with respect to d_2. Namely the definition of open set is independent of the choice of distance functions.

A continuous function $f(x)$ on a manifold is actually a continuous map from the manifold to a number $f : M \to \mathbf{R}$. To define a continuous function, a conventional trick is to use $\varepsilon-\delta$ description which makes use of distance functions. For a topological space with no distance function given, continuous mapping can be defined through 'open set'as follows:

Definition 1.4 *Continuous mapping : For given two manifolds M and N, the mapping $f : M \to N$ is continuous if the inverse image of any open set in N is an open set in M.*

Closely related to open set, there are some other concepts, such as closed set, neighborhood, etc..

Definition 1.5 *Closed set: The complementary set of an open set A in set S is a closed set. A closed set contains all of its limit points.*

Definition 1.6 *Neighborhood: a neighborhood of a point x in space S is an open set of S which contains x.*

3). Homeomorphic mappings and topological properties of manifolds

The geometric properties of spaces are usually related to invariance under the action of transformation groups so geometry can be classified according to the properties of these groups. For instance, nonsingular linear transformations transform straight lines to straight lines and preserve the affine properties of spaces. The corresponding geometry is called affine geometry. The orthogonal transformations further preserve the Euclidean classification of graphs and the corresponding geometry, which studies the invariance of spaces under the orthogonal transformations, is called Euclidean geometry. Now we have to look at more general spaces where the concept of straight lines is absent, and study mapping more general than linear mappings. The first one we consider is the continuous mappings mentioned above. Continuous mappings do not destroy the nearest neighborhood property but they could alter the dimension of manifolds. One famous example is the Peano curve (in 1890) which maps a one-dimensional real interval I continuously onto a two-dimensional region I^2 where the mapping f is defined as the limit of a set of mappings $f_n : I \to I^2$, $f = \lim_{n\to\infty} f_n$. Figure 1.1 gives the first three steps $f_1(I)$, $f_2(I)$ and $f_3(I)$ and shows how to draw it. The n-th step f_n can be sketched similarly. After the n-th step, each point on I^2 is within a distance of $(\frac{1}{2})^n$ from $f_n(I)$. As $n \to \infty$, we get a continuous onto mapping which maps a one-dimensional interval I onto a two-dimensional region I^2.

Notice that the above mapping is not a one-to-one mapping. It only guarantees that each point on a two-dimensional region I^2 is in a certain neighborhood of the image $f(I)$. This example shows that continuous onto mappings do not necessarily preserve the dimension of a manifold. Further restrictions on continuous mappings are needed if we want to keep the dimension of spaces invariant.

On the other hand, although we know that one-to-one correspondence always keeps different sets distinct, this alone does not preserve the dimension of spaces. For

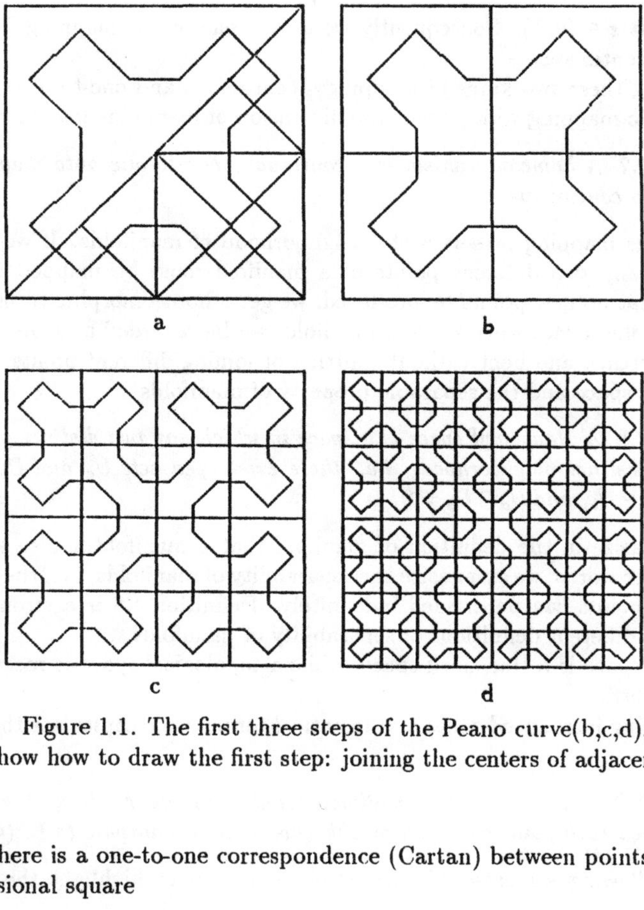

Figure 1.1. The first three steps of the Peano curve(b,c,d), and (a) show how to draw the first step: joining the centers of adjacent triangles.

instance, there is a one-to-one correspondence (Cartan) between points (x, y) of the two-dimensional square

$$I^2 = \{(x, y) \in \mathbf{R}^2 | 0 < x, y < 1\},$$

and points z of an open line segment

$$I = (0, 1) = \{z \in R | 0 < z < 1\}.$$

Here x and y are real decimal numbers ($0 < x, y < 1$)

$$x = 0.x_1x_2\cdots,$$

$$y = 0.y_1y_2\cdots,$$

and a real decimal number z can be constructed from x, y as

$$z = 0.x_1y_1x_2y_2\cdots.$$

It is clear that $z \in (0,1)$. Consequently we have a one-to-one mapping from I^2 to I which is not continuous.

Combining these two kinds of mappings (continuous and one-to-one) we get the homeomorphic mapping which keeps the dimension of manifolds invariant.

Definition 1.7 *A homeomorphism is a continuous one-to-one onto mapping whose inverse is also continuous.*

Continuous mapping preserves the neighborhood of manifolds. If we further requiring that any two different points of a manifold must be mapped to different points and that no new points be produced, we get a homeomorphic transformation. A homeomorphic transformation of a manifold can be regarded as a piece of rubber arbitrary stretched and bent without splitting or joining different points.

4). Hausdorff space and the separable property of manifolds

Definition 1.8 *A Hausdorff space is a space in which any two distinct points a and b may have distinct neighborhoods, i.e., there exist open sets U_a and U_b, such that $a \in U_a$ and $b \in U_b$, and $U_a \cap U_b = \emptyset$.*

We emphasize in the definition of manifold that a manifold is also a Hausdorff space. In other words, we emphasize the separability of manifolds, i.e., the line joining two distinct points can be divided indefinitely. Definition 1.8 is a stronger version among many different definitions of separability of manifolds.

A metric space is a Hausdorff space, while a topological space is metrizable only if it is Hausdorff.

After having introduced these mathematical terms, we recapitulate the definition of manifolds.

Definition 1.9 *A real (complex) n-dimensional manifold M is such a Hausdorff space, in which each point has open neighborhoods homeomorphic to $\mathbf{R}^n(\mathbf{C}^n)$.*

In the following we present some simple examples to highlight the concept of manifold.

Example 1) The circle S^1 is a one-dimensional manifold.

Example 2) The one-dimensional figures in Figure 1.2 are not manifolds because the neighborhoods at knots are not homeomorphic to open sets in $\mathbf{R}^1$.

Example 3) The unit sphere S^2 in three-dimensional Euclidean space $\mathbf{E}^3$ is a two-dimensional manifold.

Example 4) The cone surface in $\mathbf{E}^3$

$$M = \{(x,y,z) \in E^3 | x^2 - y^2 - z^2 = 0\}$$

is not a manifold because open neighborhood at the origin point is not homeomorphic to an open set in $\mathbf{R}^2$. If we take the half cone surface

$$M = \{(x,y,z) \in \mathbf{E}^3 | x^2 - y^2 - z^2 = 0, \quad x \geq 0\}.$$

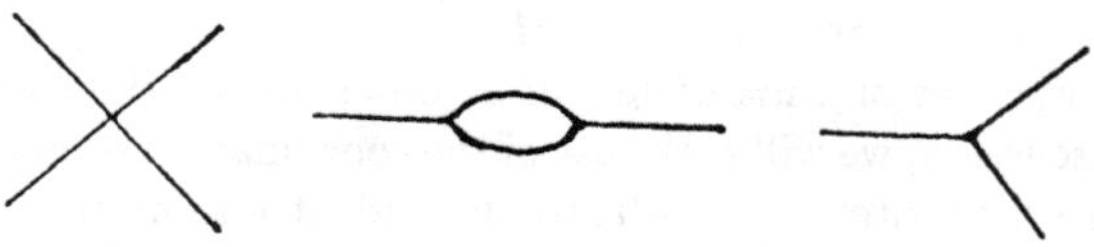

Figure 1.2. Examples of one-dimensional figures which are not manifolds.

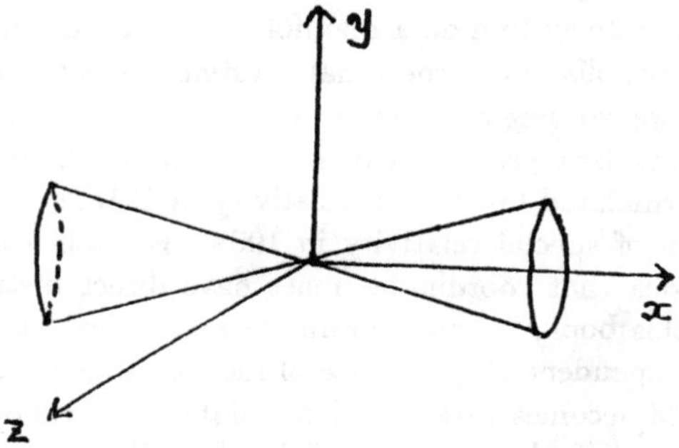

Figure 1.3. The cone surface in $\mathbf{E}^3$.

The half cone surface is a two-dimensional manifold. After we have studied differentiable structure of manifold, it is easy to show that the half cone surface is not a smooth manifold (i.e., it is not differentiable.).

1.2 Differentiable manifold

In order to carry out differentiation of functions and tensor fields on a manifold M, it is useful to introduce a coordinate system on the manifold. A local coordinate system is established through a homeomorphic mapping from an open set in M to an open set in $\mathbf{R}^n$. From its very definition, a manifold is locally like $\mathbf{R}^n$. For any point in M there is an open set U containing this point which is homeomorphic to an open set V in $\mathbf{R}^n$. Let φ be the corresponding homeomorphism

$$\varphi: \quad U \to V = \varphi(U) \in \mathbf{R}^n.$$

For the point $p \in U$ in M, the n-tuple $(x^1, x^2, \cdots, x^n)$ of $\varphi(p)$ in $\mathbf{R}^n$ is called the coordinate of the point p. Through this homeomorphism φ we obtain the coordinate for each point in the open set U in M. The pair (U, φ) is called the coordinate chart of M or the local coordinate system on M.

The local properties of a manifold can be described by the coordinate system. In the following discussion, we will make use of the coordinate system but not constrained by it. Sometimes we write out coordinate dependent formulations explicitly, but the more important properties are the invariant ones under coordinates transformations. The local coordinate itself of a manifold M has no geometric significance, and our attention is mostly focused on those properties which are invariant under coordinate transformations.

For a given coordinate system on a manifold, we often use its metric properties. However, due to we establish local coordinate systems on a manifold in terms of the homeomorphism, so we can free ourselves from the constraints of the metric and can study directly the invariant properties of manifold under homeomorphism. When A.Einstein finally formulated his general relativity in 1915, seven years had elapsed since his formulation of special relativity in 1908. Einstein explained that giving up the traditional idea that coordinates must have direct metric meaning was by no means easy. In this book, we use coordinate systems which can be transformed arbitrarily and is independent of the choice of metric. When a metric is introduced such that the manifold becomes a Riemannian manifold, we will point it out explicitly.

A n-dimensional manifold M is like $\mathbf{R}^n$ locally. However, in general it is not possible to find a single chart to cover the whole manifold M. For example, the n-dimensional sphere S^n needs at least two open sets to cover it. If the whole manifold M needs many open sets $\{U_\alpha\}$ to cover, then we express this as

$$\bigcup_\alpha U_\alpha = M.$$

The collection of open sets $\{U_\alpha\}$ is called the open covering of M. The family of all coordinate charts

$$\mathcal{A} = \{(U_\alpha, \varphi_\alpha)\}$$

is called the atlas of M. If two open sets U_α and U_β in the collection of open sets $\{U_\alpha\}$ overlap, i.e., $U_\alpha \bigcap U_\beta \neq 0$, as illustrated in Figure 1.4, they have to satisfy a compatibility condition between these two charts $(U_\alpha, \varphi_\alpha)$ and (U_β, φ_β). φ_α and φ_β homeomorphically map the intersection region $U_\alpha \bigcap U_\beta$ into two non-empty open sets in $\mathbf{R}^n$ and the mapping between these two open sets is equivalent to a coordinate transformation on the manifold:

$$\begin{aligned} f = \varphi_\beta \cdot \varphi_\alpha^{-1} : \varphi_\alpha(U_\alpha \textstyle\bigcap U_\beta) &\rightarrow \varphi_\beta(U_\alpha \textstyle\bigcap U_\beta) \\ \{x^i\} &\rightarrow \{y^j = f^j(x)\} \\ g = \varphi_\alpha \cdot \varphi_\beta^{-1} : \varphi_\beta(U_\alpha \textstyle\bigcap U_\beta) &\rightarrow \varphi_\alpha(U_\alpha \textstyle\bigcap U_\beta) \\ \{y^i\} &\rightarrow \{x^j = g^j(y)\} \end{aligned}$$

here f and g are n real continuous functions on open sets of $\mathbf{R}^n$ respectively. In order to carry out the differential operation on the manifold, we have to require the

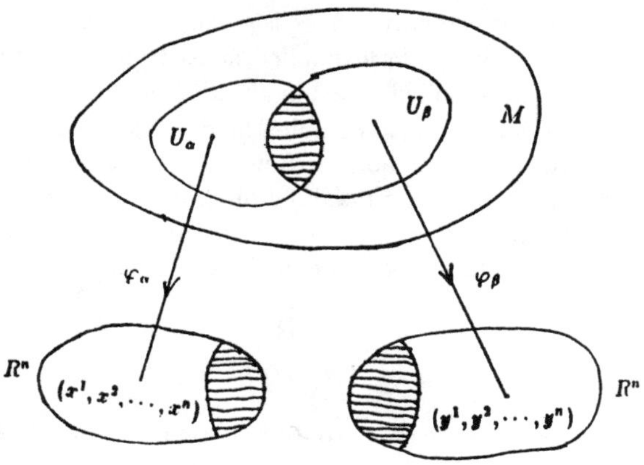

Figure 1.4. Two overlapping charts on manifold M.

coordinate transformations f and g to be differentiable. If these real functions have k-th order partial derivatives, i.e., the mapping f belongs to C^k classes, we say the two coordinate charts are C^k compatible.

Next we introduce the important concept of differentiable structure on a manifold. Intuitively, differentiable structure of a manifold measures its "smoothness" and shows how different open sets in an open covering of the manifold are patched together.

Definition 1.10 *Atlas $\mathcal{A} = (U_\alpha, \varphi_\alpha), (\alpha \in I)$ of a manifold M is called the C^k differential structure(or complete C^k compatible atlas.) if:*

1). $\{U_\alpha\}$ is an open covering of M.

2). Any two charts in $\mathcal{A}$ are C^k compatible.

3). $\mathcal{A}$ is the maximum atlas which has the above two properties.

The atlas satisfying the first two conditions is called a base of C^k differentiable structure or simply C^k atlas.

Two atlases $\mathcal{A} = \{(U_\alpha, \varphi_\alpha)\}$ and $\mathcal{A}' = \{(U_\beta, \varphi_\beta)\}$, are equivalent if the union $\{(U_\alpha, \varphi_\alpha), (\varphi_\beta, \varphi_\beta)\}$ is still a C^k atlas. Each C^k compatible atlas is contained in a unique complete C^k compatible atlas. So when constructing a differentiable structure, it is sufficient to point out one C^k compatible atlas. For example, $\mathcal{A} = \{(U_\alpha, \varphi_\alpha)\}$ is a C^k atlas of M, then the differentiable structure can be obtained as:

$$\mathcal{A} = \{(V, \psi) || (V, \psi) \text{ are } C^k - \text{compatible with all the charts in} \{(U_\alpha, \varphi_\alpha)\}\}.$$

A manifold is called a C^k compatible differential manifold, i.e., it has a C^k atlas equivalent class. A manifold is called a smooth manifold or differentiable manifold if it is endowed with C^∞ differential structure.

When a C^ω-differentiable structure on M is given, i.e., the mappings $\varphi_\beta \cdot \varphi_\alpha^{-1}$ are real analytic functions, which can be expanded at each point on the manifold by a convergent Taylor series, M is called a real analytic manifold.

If we change all real field $\mathbf{R}^n$ into complex field $\mathbf{C}^n$ and each mapping $\varphi_\beta \cdot \varphi_\alpha^{-1}$ into complex analytic function(holomorphic function) in the previous definitions, the manifold is called a complex analytic manifold or simply called complex manifold.

For a differentiable manifold, we can discuss the differentiable functions on it. As shown in Figure 1.5, f is a real function defined on M, i.e.,

$$\begin{aligned} f: M &\rightarrow \mathbf{R} \\ p \in M &\rightarrow f(p) \in \mathbf{R} \end{aligned}$$

suppose $(U_\alpha, \varphi_\alpha)$ is an admissible chart containing point p, then $F = f \cdot \varphi_\alpha^{-1}$ is a real

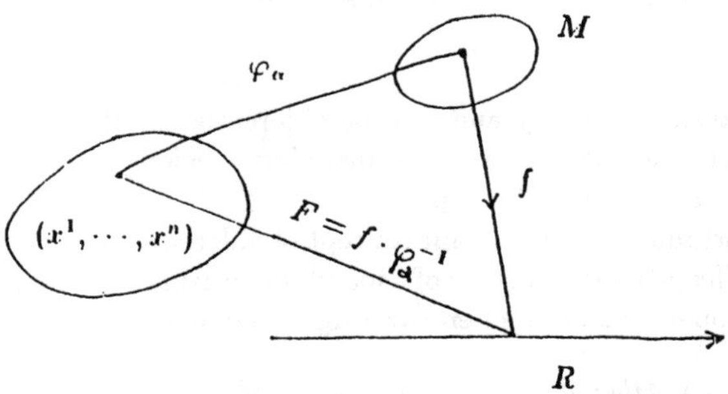

Figure 1.5. The real function f on manifold M.

function defined on an open set of $\mathbf{R}^n$. Function f is called differentiable at point p if F is differentiable at point $x = \varphi_\alpha(p)$. It can be proved that the differentiability of f is independent of the choice of the admissible atlas. If there is another chart (U_β, φ_β) which containing point p, write

$$f \cdot \varphi_\beta^{-1} = (f \cdot \varphi_\alpha^{-1}) \cdot (\varphi_\alpha \cdot \varphi_\beta^{-1})$$

as $\varphi_\alpha \cdot \varphi_\beta^{-1}$ belongs to C^∞ class(smooth), $f \cdot \varphi_\beta^{-1}$ is differentiable if and only if $f \cdot \varphi_\alpha^{-1}$ is differentiable.

Similarly, we can discuss the differentiability of continuous mappings between differentiable manifolds. For example, let f be a continuous mapping from an m-dimensional differentiable manifold M to an n-dimensional differentiable manifold N:

$$\begin{aligned} f: \ M &\rightarrow N \\ p &\rightarrow q = f(p) \end{aligned}$$

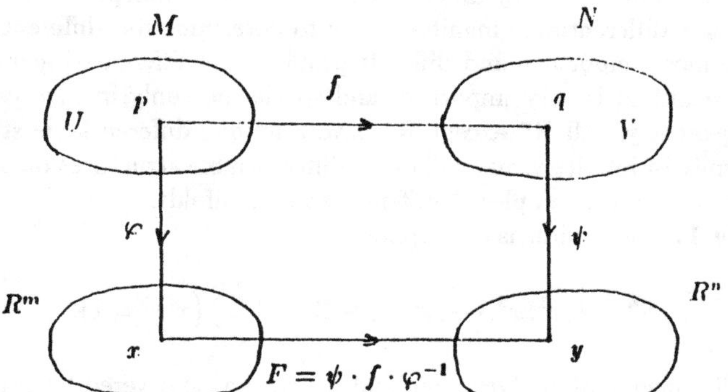

Figure 1.6. The mapping between manifolds.

Suppose (U, φ) containing point p and (V, ψ) containing point q, are charts on M and N respectively, the differentiability of

$$F = \psi \circ f \circ \varphi^{-1} : x \to y = F(x)$$

determines the differentiability of mapping f. For simplicity, from now on, we usually do not distinguish f from F, points p and q from their coordinates x and y. Functions $q = f(p)$ and $y = F(x)$ are understood to mean the same.

Homeomorphism between manifolds M and N with the same dimension is called a diffeomorphism if we further require the homeomorphism be differentiable. If these two manifolds are diffeomorphic, it must be homeomorphism. But the inverse is not true in general.

Topology is the most natural mathematical structure in studying continuity. All topological spaces can be classified by homeomorphism. The differentiable structure of manifolds is the most natural mathematical structure in studying differentiability. Manifolds which are homeomorphic can be further classified by diffeomorphism. Manifolds belonging to different diffeomorphism classes have different differentiable structures. That is to say, manifolds which are homeomorphism may have different differentiable structures. For instance[11], in 1956, Milnor discovered that for seven dimensional sphere S^7 (see Appendix H), in addition to the differential structure of unit sphere in $\mathbf{R}^8$, there exists another differentiable structure: Milnor's exotic sphere Σ^7. These two spheres are homeomorphic but not diffeomorphic. Mathematicians have proved that for low dimensional ($d \leq 3$) manifolds, there is a unique differentiable structure. But for higher dimensional manifolds (including $\mathbf{R}^4$), there could be different differentiable structures. Since homeomorphic manifolds may have inequivalent differentiable structures, we say differentiable structure is independent of topological structure. For a differentiable manifold, how to determine its differential structure is one of the most important and difficult problems in differential geometry. Four-dimensional manifold is very important and special not only in physics but also in differential geometry. All $\mathbf{R}^n$ except $\mathbf{R}^4$ have a unique differentiable structure, but there is a family of infinitely many distinct differentiable structures on $\mathbf{R}^4$.

Next, we give some examples of differentiable manifolds.

Example 1) The n-dimensional sphere S^n.

$$S^n = \{(x^1, x^2, \cdots, x^{n+1}) \in \mathbf{R}^{n+1} \mid \sum_{i=1}^{n+1} \left(x^i\right)^2 = 1\}.$$

This is a well known real analytic manifold which can be covered by two open sets:

Northern hemisphere: $U_+ = S^n - (0, 0, \cdots, -1)$.

Southern hemisphere: $U_- = S^n - (0, 0, \cdots, 1)$.

The corresponding diffeomorphic mappings can be chosen as stereographic projection from north pole and south pole respectively. It is easy to prove that the two charts are real analytic compatible. Taking the two-dimensional sphere S^2 as an example, we have the explicit expressions.

$$S^2 = \{(x, y, z) \in \mathbf{R}^3 \mid x^2 + y^2 + z^2 = 1\}.$$

The atlas consists of the following two charts:

$$U_+ = \{(x, y, z) \in S^2; z \neq -1\}, \quad \varphi_+ : (x, y, z) \to \left(\frac{x}{1+z}, \frac{y}{1+z}\right)$$

$$U_- = \{(x, y, z) \in S^2; z \neq 1\}, \quad \varphi_- : (x, y, z) \to \left(\frac{x}{1-z}, \frac{y}{1-z}\right)$$

It is easy to see that they are C^∞ compatible. Hence S^2 is a smooth manifold.(cf. exercise 1 of this chapter).

Lower dimensional spheres S^n $(n \leq 6)$ have only one differentiable structure. Higher dimensional sphere, however, have many different differentiable structures in general. For instance, S^7 has 28, S^9 has 8 and S^{11} has 992 different differentiable structures.

Example 2) Lie group manifold.

A Lie group is a group which is also given the structure of a smooth manifold. We can use real parameters in the definition of the group to represent the coordinates of a point on a Lie group manifold. Some simple group manifolds are homeomorphic to simple topological manifolds. For instance, the one-dimensional unitary group

$$U(1) \simeq S^1, \text{ the group element can be expressed as } \{e^{i\theta}\},\ 0 \leq \theta < 2\pi.$$

Another example is $SU(2) \simeq S^3$, with the element expressed as

$$u = \begin{pmatrix} a & b \\ -\bar{b} & \bar{a} \end{pmatrix} = \begin{pmatrix} x^1 + ix^2 & x^3 + ix^4 \\ -x^3 + ix^4 & x^1 - ix^2 \end{pmatrix},$$

satisfying the condition

$$\det u = |a|^2 + |b|^2 = \sum_{i=1}^{4} \left(x^i\right)^2 = 1,$$

i.e., $SU(2)$ group is a three parameters group, and the parameter space is diffeomorphic to S^3.

Example 3) Real projective space RP(n).

The real n-dimensional projective space $RP(n)$ is the set of straight lines through the origin in $\mathbf{R}^{n+1}$ space. Since each straight line through the origin intersects with the sphere S^n at two points. $RP(n)$ is equivalent to manifold $S^n/\mathbf{Z}_2$, in which the antipodal points on sphere S^n are taken to be one point. For example,

$$RP(3) \simeq S^3/\mathbf{Z}_2 \simeq SU(2)/\mathbf{Z}_2 \simeq SO(3),$$

where $\mathbf{Z}_2 \simeq S^0$ is a group consisting of two elements.

Due to the existence of this natural covering mappings $\varphi : S^n \to RP(n)$, where φ is a C^∞ mapping and the inverse of φ is locally C^∞, the C^∞ differentiable structure of S^n can induce a C^∞ differentiable structure of $RP(n)$. $RP(n)$ is also an n-dimensional differentiable manifold.

Example 4) Complex projective space CP(n).

The complex n-dimensional projective space $CP(n)$ is the set of all complex straight lines through the origin in $\mathbf{C}^{n+1}$. The points Z and Z' in the following

$$Z = \left(z^0, z^1, \cdots, z^n\right) = cZ',$$

where $\{z^i\}$ are complex numbers, at least one of which is non-zero and c is any non-zero complex number, here after Z= $\{z^i\}$ is taken as one point in $CP(n)$. The $n+1$ complex numbers, is called the homogeneous coordinate of this point. When the homogeneous coordinate is multiplied by an arbitrary complex number c, it still represents the same point, so $CP(n)$ is still an n-dimensional complex manifold.

In terms of homogeneous coordinates, the open sets U_k in $CP(n)$ are taken as $U_k = \{Z \in \mathbf{C}^{n+1} \mid z^k \neq 0\}$. It is easy to prove that

$$\bigcup_{k=0}^{n} U_k = CP(n),$$

i.e., the collection $\{U_k\}$ is an open covering of $CP(n)$, here U_k is an n-dimensional complex manifold homeomorphic to an open set in $\mathbf{C}^n$. Notice that for the interior points of U_k, since their $z^k \neq 0$, we may define

$$\zeta^i_{(k)} \equiv z^i/z^k,$$

and obtain

$$\zeta_{(k)} = \left(\zeta^0_{(k)}, \zeta^1_{(k)}, \cdots, \zeta^n_{(k)}\right) \text{ (neglecting } \zeta^k_{(k)} = 1).$$

It is called the Euclidean coordinate of an interior points in U_k, which has n complex numbers and completely determines the points in U_k. The homeomorphism is

$$\varphi_k : Z \in U_k \to \zeta_{(k)} \in \mathbf{C}^n.$$

Therefore the set $\{U_{k,}\varphi_k\}$ constitutes an atlas of the manifold $CP\,(n)$. Within the intersection, $U_k \bigcap U_j \neq 0$, $z^k \neq 0$, $z^j \neq 0$, the compatibility condition demands that the mapping f

$$f = \varphi_k \varphi_j^{-1} : \zeta^i_{(j)} \to \zeta^i_{(k)} = z^i / z^k = \zeta^i_{(j)} \left(\zeta^k_{(j)}\right)^{-1}$$

be a holomorphic analytic function. Hence, $CP\,(n)$ is a complex analytic manifold.

In the following we will discuss $CP\,(1)$ in detail. The homogeneous coordinate of $CP\,(1)$ is

$$Z = (z^1, z^2) = (cz^1, cz^2)\ (z^1, z^2 \text{ do not vanish simultaneously.}).$$

$CP\,(1)$ can be covered by two open sets:

U_1 is the region where $z^1 \neq 0$, the Euclidean coordinate is

$$\zeta = z^2 / z^1 = u + iv.$$

U_2 is the region where $z^2 \neq 0$, the Euclidean coordinate is

$$\zeta' = z^1 / z^2.$$

At the intersection area (Both z^1 and z^2 are non-zero.),

$$\zeta' = 1/\zeta = \frac{u}{u^2 + v^2} - i \frac{v}{u^2 + v^2} (\text{ Either u or v is non-zero.}).$$

The compatibility condition, i.e., the mapping between U_1 and U_2

$$\begin{aligned} f : U_1 &\to U_2 \\ \zeta &\to f\,(\zeta) = 1/\zeta \\ (u, v) &\to \left(\frac{u}{u^2 + v^2}, \frac{-v}{u^2 + v^2}\right) \end{aligned}$$

is a complex analytic mapping. Therefore $CP\,(1)$ is a complex analytic manifold. Let $(x^2 + y^2 + z^2 = 1)$

$$u = \frac{x}{1 + z}, \quad v = \frac{y}{1 + z},$$

we have

$$\frac{u}{u^2 + v^2} = \frac{x}{1 - z}, \quad \frac{-v}{u^2 + v^2} = \frac{-y}{1 - z}.$$

Compare with example 1), we see that $CP\,(1)$ and S^2 have the same differentiable structure and are diffeomorphic to each other.

$CP(n)$ is the complex line bundle through the origin of $\mathbf{C}^{n+1}$, it can be taken as points of sphere S^{2n+1} in $\mathbf{C}^{n+1}$, but there is one arbitrary phase factor $U(1)$, as $\{z^k\} \sim \{e^{i\alpha}z^k\}$. Hence,

$$CP(n) \simeq S^{2n+1}/U(1) \simeq \mathbf{C}^n \bigcup CP(n-1),$$

$$CP(1) \simeq S^3/U(1) \simeq S^2 \simeq \mathbf{C}^1 \bigcup \{\infty\}.$$

Example 5) The product manifold.

The product of two differentiable manifolds $M \times N$ is still a differentiable manifold. Suppose that the atlases of M and N are $\{U_\alpha, \varphi_\alpha\}$ and $\{V_\beta, \psi_\beta\}$ respectively, for the product manifold $M \times N$, we have the following smooth projections:

$$\Pi_1 : M \times N \to M,$$

$$\Pi_2 : M \times N \to N.$$

Therefore $\{U_\alpha \times V_\beta; \varphi_\alpha \cdot \Pi_1 + \psi_\beta \cdot \Pi_2\}$ is the atlas of the manifold $M \times N$, and determines the differentiable structure of the product manifold. For example,

$$\text{cylinder } = S^1 \times \mathbf{R},$$

$$\text{torus } T^2 = S^1 \times S^1.$$

Finally, we introduce the notion of compactness (Please refer to Appendix B for details.).

Definition 1.11 *Manifold M is compact if any open covering of M has a finite subcovering.*

Compactness is a topological property. Homeomorphism preserves compactness. A manifold homeomorphic to a closed subset with supremum in Euclidean space is compact. We can use this property to decide the compactness of a manifold. In the example above, $S^n, RP(n), CP(n), T^2$ are compact manifolds. The cylinder is not a compact manifold.

1.3 Tangent space and tangent vector field

In the preceding two sections, we have studied the properties of differentiable manifold: a manifold locally like $\mathbf{R}^n$ and is differentiable, so we can use analytic tools to study its properties. An important analytic method is to study its infinitesimal part. Its advantage lies in the fact that the problem can be linearized in this way. For instance:

The properties of a function in the neighborhood of a point can be represented by the derivative of the function at that point.

The local properties of a Lie group in the neighborhood of identity can be represented by the Lie algebra of the group.

The behavior of a curve in the neighborhood of a point can be represented by the tangent vector at the point.

Similarly, for a differentiable manifold M, the properties at neighborhood of each point of M can be approached through a linear space (tangent space). A tangent space consists of tangent vectors. First of all we look at how a tangent vector can be defined at a point p of M along with a given direction, in a way that it is independent of the choice of coordinates.

Usually, a vector in linear space $\mathbf{R}^n$ can be expressed by a straight line along with the direction which connects two points in the space. However, for differentiable manifold, we can not even define a vector in this way, as we no longer have straight lines in general. How can we generalize the concept of tangent vector to a manifold?

For a differentiable manifold, we may have the definition of smooth functions in a way which is coordinate free. We know that the sum of differentiable functions (sum pointwise) on a manifold is still a differentiable function. A differentiable function multiplies by a real number is still a differentiable function. The product of two differentiable functions (product pointwise) is also a differentiable function. Therefore, we have an algebra $\mathcal{F}(M)$ over real field consisting of the set of all differentiable functions on M. Now we study the set of differentiable functions defined in the neighborhood of a point p on M , denoted by $\mathcal{F}_p(M)$. Tangent vectors at p on M can be defined by linear differential operators that act on the set $\mathcal{F}_p(M)$. This is a natural generalization of vectors in $\mathbf{R}^n$. To show this, let us first have a close look at the linear differential operators that act on function f of $\mathbf{R}^n$.

In $\mathbf{R}^n$, we have Taylor expansion for function $f(x)$, when the vector $\triangle x$ is sufficiently small,

$$\begin{aligned} f(x+\triangle x) &= f(x)+\sum_{i=1}^{n}(\triangle x)^i\frac{\partial}{\partial x^i}f + \text{infinitesimal} \\ &\equiv f(x)+\delta f + \text{infinitesimal}. \end{aligned}$$

Here we remember that point x in $\mathbf{R}^n$ can also be called a vector x. For given coordinate system, vector x can be represented by n real numbers $\{x^i\}$ which are coordinate dependent. However, the vector x itself is independent of the choice of the frames. In the same way, vector $\triangle x$ is a vector through point x and is independent of the choice of the coordinate even though its component $(\triangle x)^i$ is not. Combining $(\triangle x)^i$ and $\partial_i = \frac{\partial}{\partial x^i}$, we get a direction derivative δ which acts on differentiable function f

$$\delta = \sum_{i=1}^{n}(\triangle x)^i\partial_i.$$

δ is the linear differential operator which acts on function f and is the directional derivative along the displacement direction $\triangle x$. The operator δ itself is independent of the choice of coordinates and is a linear combination of directional derivatives $\{\partial_i\}$ along with the coordinate axes. The coefficients $(\Delta x)^i$ can be seen as coordinate components of the directional derivative δ which are just the components of the displacement vector Δx. We can use the coordinate free directional derivative to represent the displacement vector Δx through point x.

For a differentiable manifold M, tangent vectors along a given direction for each point on M can be defined similarly. One may draw a smooth curve $x(t)$ through a point p on M. Through curve $x(t)$, one can select a definite direction at point p. The smooth curve $x(t)$ is equivalent to a mapping from line fragment $(-\epsilon, \epsilon)$ (parameterize by t) onto M

$$\begin{aligned} x : (-\epsilon, \epsilon) &\to M \\ t &\to x(t) \in M \end{aligned}$$

and $t = 0$ is mapped to p : $x(0) = p$. For a differentiable function $f \in \mathcal{F}_p(M)$ given on

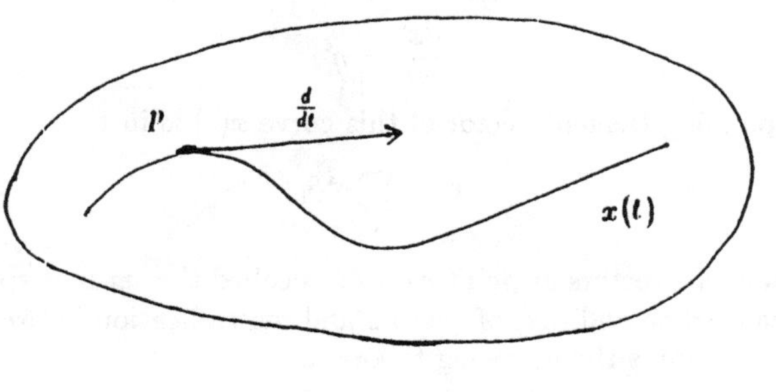

Figure 1.7. Tangent $X = \frac{d}{dt}$ given by curve $x(t)$.

manifold M, when we constrained on smooth curve $x(t)$, one obtains a differentiable function $f(x(t))$ with respect to parameter t. The rate of change of the function along the curve at point p can be represented as

$$X_p f = \frac{d}{dt} f(x(t)) \mid_{t=0} .$$

The linear differential operator $X = \frac{d}{dt}$, which acts on functions on a manifold, can be taken as the tangent vector which is tangent to the curve $x(t)$. Tangent vectors thus defined are generalizations of the directional derivative on $\mathbf{R}^n$.

In the neighborhood of point p, there is a local coordinate $x = (x^1, x^2, \cdots, x^n)$. The equation above can be written as

$$X_p f = \sum_{i=1}^{n} \frac{dx^i}{dt}\Big|_p \frac{\partial}{\partial x^i} f\Big|_p .$$

If the curve through point p is chosen along the coordinate direction x^j, i.e., $x^j = t$, one gets the tangent vector $\partial_j = \frac{\partial}{\partial x^j}$ along the coordinate curve x^j . From the equation above we see that the tangent vector X_p at point p along any given direction is a linear combination of $\{\partial_j\}$. The set $\{\partial_j, j = 1, 2, \cdots, n\}$ is called the coordinate base vectors of the tangent vector X_p , which is determined by the local coordinate frame. The coefficients $\frac{dx^i}{dt}|_p$ are the components of the tangent vector X_p. Here we emphasize that the tangent vector $X_p = \frac{d}{dt}|_{t=0}$ is independent of the choice of the coordinate frame but the components $\frac{dx^j}{dt}|_p$ are not.

Curves and vectors are closely related. On the one hand, making use of the curve $x(t)$ through point p, one can get the tangent vector $X_p = \frac{d}{dt} = \frac{dx^i}{dt}\frac{\partial}{\partial x^i}$ along this curve. On the other hand, it can be proved that for a given linear combination

$$Y = \sum_{i=1}^{n} \eta^i(x)\partial_i, \; \eta^i(x) \in \mathcal{F}(M),$$

it is always possible to define an integral curve $x(t)$ as solution to the differential equations

$$\frac{dx^i}{dt} = \eta^i,$$

and the corresponding tangent vector of this curve $x(t)$ is just

$$Y = \sum_i \eta^i \partial_i.$$

The set of all tangent vectors at point p on M is called the tangent space $T_p(M)$ at p, in which one can define addition of vectors and multiplication between a vector and a real numbers. There is the following theorem:

Theorem 1.1 *For an n-dimensional manifold M, the tangent space $T_p(M)$ at a point p is an n-dimensional linear space over real field. When a local coordinate frame is chosen, the bases $\{\partial_i, i = 1, 2, \cdots, n\}$ are linearly independent and complete.*

Proof: If the bases $\{\partial_i\}$ are not linearly independent, i.e., there exist n numbers a^i, and not all of them vanish, such that

$$\sum_{i=1}^{n} a^i \partial_i = 0.$$

Now let it acts on the coordinate axes x^j through point p , one has

$$\sum_{i=1}^{n} a^i \frac{\partial x^j}{\partial x^i} = a^j = 0, \; j = 1, 2, \cdots, n.$$

So we meet a contradiction. Therefore the bases $\{\partial_i\}$ are linearly independent. In addition, any tangent vector through point p can be expanded by the bases $\{\partial_i\}$. Hence this bases is complete . Q. E. D.

The union of all tangent spaces at all point on manifold M,

$$T(M) = \bigcup_p T_p(M)$$

is called the tangent bundle of M.

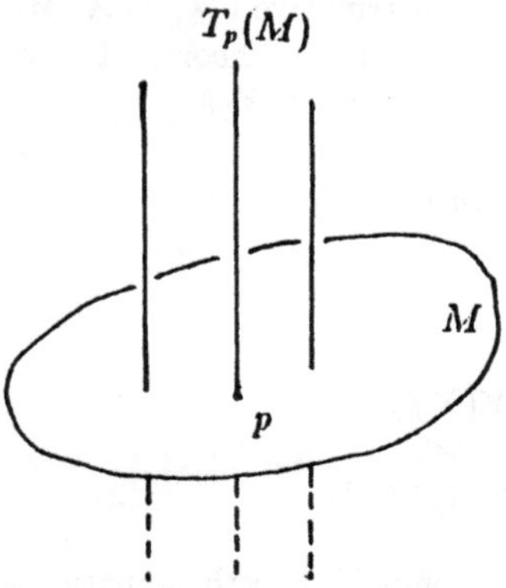

Figure 1.8. Sketch of tangent bundle.

The tangent bundle $T(M)$ is a $2n$-dimensional manifold, which is locally a direct product manifold but globally is not so in general. The tengent vectors at the same point can be added up but tangent vectors at different points are not related at present. Tangent bundle $T(M)$ is illustrated in Figure 1.8. At each point p on M, there is an n-dimensional tangent space $T_p(M)$, which is represented by a straight line through p. To show that tangent spaces at different points are not related, the straight lines at different points are drawn parallel.

In the following, we introduce vector field $X(x)$. A vector field $X(x)$ on a manifold is an assignment of a vector X_p to each point p of M. The vector field $X(x)$ is also called a section of the tangent bundle $T(M)$, since for each point p on manifold M, only one tangent vector $X_p \in T_p(M)$ is selected. When a local frame is chosen, the vector field can be written as

$$X(x) = \sum_{j=1}^{n} \xi^j(x)\partial_j.$$

Vector field $X(x)$ is called continuous and differentiable if the functions $\{\xi^j(x)\}$ are continuous and differentiable.

The differentiable vector field $X(x)$ on manifold M can be taken as a linear differential operator which acts on differentiable functions $f \in \mathcal{F}(M)$ on the manifold. It has the following two properties:

1).It is a linear operator

$$X(af + bg) = aXf + bXg, \quad (a, b \in R, \text{ and } f, g \in \mathcal{F}(M))$$

2).It obeys the Leibniz rule

$$X(fg) = fXg + gXf.$$

Satisfying these two criteria can be taken as the definition of differential vector fields on a manifold. To verify the differentiability of vector field X, it is sufficient to prove that for any differentiable function $f \in \mathcal{F}(M)$, Xf is also a differentiable function on M.

Let $\mathcal{X}(M)$ be the space of all differentiable vector fields on M, i.e., the set of sections of the tangent bundle. The addition of vector fields can be defined and it obeys the rules of Abelian group. The product between vector fields and functions is also defined, which obeys the associative law and distributive law, so $\mathcal{X}(M)$ is an infinite dimensional linear space over real number, and from Theorem 1.1, it is an rank-n $\mathcal{F}(M)$-module. In addition, for any two vector fields in $\mathcal{X}(M)$

$$X = \sum_j \xi^j(x)\partial_j, \; Y = \sum_j \eta^j(x)\partial_j,$$

a Lie bracket $[X,Y]$ can be defined

$$\begin{aligned} [X,Y]f &= X(Yf) - Y(Xf) \\ &= (\xi^j \frac{\partial \eta^k}{\partial x^j} - \eta^j \frac{\partial \xi^k}{\partial x^j})\partial_k f, \end{aligned}$$

where the repeated indices are summed up, i.e., we omit $\sum_{j,k}$ in the equation above. From now on, we will use this summation convention. The equation above is valid for any real function $f \in \mathcal{F}(M)$ on M. It seems that the commutator of two linear operators should be a second order differential operator. However, the calculation above shows that the second order parts cancel out and only first order parts are left. Therefore the commutator of two vector fields is still a vector field, i.e.,

$$[X,Y] \in \mathcal{X}(M).$$

We see that the set $\mathcal{X}(M)$ is a space of vector fields with Lie algebra structure. For any three vector fields $X, Y, Z \in \mathcal{X}(M)$, the Jacobi identity is satisfied

$$[[X,Y],Z] + [[Y,Z],X] + [[Z,X],Y] = 0.$$

Notice that the vector field $X \in \mathcal{X}(M)$ can be multiplied by a number and also can be multiplied by any differentiable function $f \in \mathcal{F}(M)$ on M such that $fX \in \mathcal{X}(M)$. Hence, the set of vector fields $\mathcal{X}(M)$ is a module over the algebra $\mathcal{F}(M)$ of differentiable functions on M.

1.4 Cotangent vector field

For an n-dimensional vector space V, we may use the set of linear homeomorphic mappings from V to the real field **R**,

$$\mathrm{Hom}(V, R) : V \to \mathbf{R},$$

i.e., the set of linear functions on vector space V to construct the dual vector space V^* of V. Since addition of linear functions is still a linear function and multiplication between a linear function and a number is also a linear function, the set of all linear functions forms a linear space V^*. V^* is also an n-dimensional space and is called the dual space of V. The space V and V^* are dual to each other, $(V^*)^* = V$ (Please refer to Appendix D.). We now look at linear mappings that preserve the linear structure of $\mathcal{X}(M)$,

$$\Theta(M) \equiv \mathrm{Hom}(\mathcal{X}(M), \mathcal{F}(M)) : \mathcal{X}(M) \to \mathcal{F}(M),$$

where any element $\sigma \in \Theta(M)$ is called the cotangent vector field, which maps any tangent vector field X to a function

$$\sigma : X \to \sigma(X) \equiv <\sigma, X> \in \mathcal{F}(M).$$

Here σ is a linear mapping and has the following properties

$$\sigma(X+Y) = \sigma(X) + \sigma(Y) \quad (X, Y \in \mathcal{X}(M)),$$

$$\sigma(fX) = f\sigma(X) \quad (f \in \mathcal{F}(M)).$$

It is easy to prove that linear combination $f\sigma + g\tau$ of any two cotangent vector fields σ and τ, which have the above two properties, still has those two properties. The set $\{\sigma\}$ of all cotangent vector fields which satisfy the above two properties forms a linear space. This linear space is called a linear form $\Theta(M)$, which is also a $\mathcal{F}$-module.

To define a cotangent vector field dual to $\mathcal{X}(M)$, we may start from any given function $f \in \mathcal{F}(M)$, and define a linear mapping df as

$$\begin{aligned} df : \mathcal{X}(M) &\to \mathcal{F}(M) \\ X &\to <df, X> = Xf \in \mathcal{F}(M) \end{aligned}$$

The cotangent vector df is called a total differentiation of function f, which maps an arbitrary element in $\mathcal{X}(M)$ to a real function. To choose a local frame in the neighborhood of a point, we notice that the natural bases of tangent space is $\{\partial_i\}$. If $f = x^j$, then

$$<dx^j, \partial_i> = \frac{\partial}{\partial x^i} x^j = \delta_i^j. \tag{1.1}$$

We can choose $\{dx^i\}$ to be the bases of covector field $\Theta(M)$, dual to the bases $\{\partial_j\}$ of tangent field $\mathcal{X}(M)$. It can be proved that $\{dx^i\}$ are linearly independent and complete. The total differentiation for a function f is

$$df = \frac{\partial f}{\partial x^i} dx^i.$$

It should be emphasized that any cotangent vector (for instance dx^j) is determined by its action on the whole tangent field $\mathcal{X}(M)$. As shown from Eq. 1.4, it is determined by its action on all bases $\{\partial_i\}(i = 1, \cdots, n)$.

When a local coordinate $x = (x^1, x^2, \cdots, x^n)$ in the neighborhood of p is given, the cotangent vector field $\sigma(x) \in \Theta(M)$ can be expressed as

$$\sigma(x) = \sigma_i(x)dx^i, \quad \sigma_i(x) \in \mathcal{F}(M).$$

Here $\sigma(x)$ is a differentiable cotangent vector field if function $\sigma_i(x)$ is differentiable. The differentiable cotangent vector field $\sigma(x)$ is called a Pfaff form or a linear differential form. $\sigma(x)$ is dual to the vector field $X(x) = \xi^j(x)\partial_i$, in the sense that it takes any tangent vector $X(x)$ to $\mathcal{F}(M)$.

$$\begin{aligned} < \sigma(x), X(x) > &= \sigma_i(x)\xi^j(x) < dx^i, \partial_j > \\ &= \sigma_i(x)\xi^i(x) \in \mathcal{F}(M). \end{aligned}$$

Let $\Theta(M)$ be the set of all cotangent vector fields on M, it is an n-rank $\mathcal{F}(M)$-module also.

In addition, on every point p on M the cotangent space $T_p^*(M)$ is the dual space of the tangent space $T_p(M)$. Both spaces are n-dimensional linear spaces, which are isomorphic from the point of view of the structure of algebra. The union set of all cotangent vector spaces at each point on M

$$T^*(M) = \bigcup_p T_p^*(M) \subset M \times \mathbf{R}^n$$

is called cotangent bundle of M. The cotangent bundle is a $2n$-dimensional manifold. The cotangent field σ is a section of the cotangent bundle $T^*(M)$, i.e., following certain rules, there is a cotangent vector $\sigma_p \in T_p^*(M)$ at each point p.

However, on a manifold, the global properties of cotangent vector fields are completely different from those of tangent vector fields. This will become more clear when we discuss properties of tensor fields on a manifold (see Chapter 2).

Both tangent vector fields and cotangent vector fields on a manifold are geometric objects which are invariant under the transformation of coordinates. However, their explicit components in a given local coordinates, will change with the transformation of the frame. In the following, we will deal with the laws, governing transformation of the components of tangent vector fields and cotangent vector fields, under coordinate transformations.

For a general coordinate transformation

$$x^i \to y^i(x),$$

if it is a diffeomorphism, the Jacobi determinant is non-zero:

$$\det(\frac{\partial y^i}{\partial x^k}) \neq 0,$$

and there is an inverse transformation

$$y^i \to x^i(y).$$

For a given local coordinate system, tangent vector field $X(x)$ and cotangent vector field $\sigma(x)$ are written as

$$X(x) = \xi^i(x)\frac{\partial}{\partial x^i} = \xi'^i(y)\frac{\partial}{\partial y^i},$$

$$\sigma(x) = \sigma_i(x)dx^i = \sigma_i'(y)dy^i.$$

As a result of the transformation of the bases $\{\partial_i\}$ and $\{dx^i\}$

$$dy^i = \frac{\partial y^i}{\partial x^k}dx^k,$$

$$\frac{\partial}{\partial y^i} = \frac{\partial x^k}{\partial y^i}\frac{\partial}{\partial x^k},$$

the components of tangent vector field $\{\xi^i(x)\}$ and cotangent vector field $\{\sigma_i(x)\}$ transform respectly as

$$\xi'^i = \frac{\partial y^i}{\partial x^k}\xi^k,$$

$$\sigma_i' = \frac{\partial x^k}{\partial y^i}\sigma_k.$$

The bases $\{\partial_i\}$ and $\{dx^i\}$ are called natural frames. The components $\{\sigma_k\}$ transformed as the vector field base $\{\partial_k\}$, are called the covariant component; The components $\{\xi^k\}$ transformed as the covector field base dx^k, are called the contravariant component. Now we give an example in mechanics to show the applications of cotangent and tangent vector fields in physics. The orbit parameter of the movement of a single particle is time t, the tangent vector along the orbit is

$$V = \frac{d}{dt} = \frac{dx^i}{dt}\partial_i,$$

i.e., the velocity vector. When a local frame is given, the components of the velocity vector is $\frac{dx^i}{dt}$.

On the other hand, when dealing with the dynamical properties of the movement of a particle, an action S is introduced, which is a function on a manifold, and

$$dS = p_i dx^i,$$

i.e., the components of momentum are components of the cotangent vector field.

When one studies dynamics of particles, the Lagrangian is a function of coordinates and velocities, i.e., a function on tangent bundle. The Hamiltonian is a function of coordinates and momenta, i.e., a function on cotangent bundle.

1.5 Tensor product, exterior product and various higher order tensor fields

Let $X, Y \in \mathcal{X}(M)$ be vector fields, the bilinear mapping $K \in \mathrm{Hom}(\mathcal{X} \times \mathcal{X}; \mathcal{F})$,

$$K: \ X, Y \to K(X,Y) \in \mathcal{F}(M)$$

satisfies

$$K(fX,Y) = fK(X,Y) = K(X,fY),$$
$$K(X+Y,Z) = K(X,Z) + K(Y,Z).$$

The set of bilinear mappings $\mathrm{Hom}(\mathcal{X} \times \mathcal{X}; \mathcal{F})$ can be given a vector space structure and it is also an $\mathcal{F}$-module, i.e., let $K, H \in Hom(\mathcal{X} \times \mathcal{X}; \mathcal{F})$, then $(K + H)$ and fK are all in $Hom(\mathcal{X} \times \mathcal{X}, \mathcal{F})$

$$\begin{aligned} (K+H)(X,Y) &= K(X,Y) + H(X,Y), \\ (fK)(X,Y) &= fK(X,Y). \end{aligned}$$

This $\mathcal{F}$-module is called the tensor product of co-vector fields, denoted as

$$\overset{2}{\otimes}\, \Theta \equiv \Theta \otimes \Theta = Hom(\mathcal{X} \times \mathcal{X}; \mathcal{F})$$

which is an n^2-rank $\mathcal{F}$-module. When the local coordinate system in an open set U on a manifold is given, two vector fields X, Y could be written as $X = \xi^i(x)\partial_i, Y = \eta^j(x)\partial_j$, from the equations above we have

$$K(X,Y) = \xi^i \eta^j K(\partial_i, \partial_j).$$

Let $K_{ij} = K(\partial_i, \partial_j)$, K_{ij} are called components of tensor K in the given coordinate system. Thus the tensor field K can also be written as

$$K = K_{ij} dx^i \otimes dx^j,$$

$$< K; X, Y >= K_{ij}\xi^k \eta^l < dx^i \otimes dx^j; \partial_k, \partial_l >= K_{ij}\xi^k\eta^l \delta^i_k \delta^j_l = K_{ij}\xi^i \eta^j,$$

here $dx^i \otimes dx^j$ is the natural frame bases of the tensor field $\overset{2}{\otimes}\, \Theta$, $\overset{2}{\otimes}\, \Theta(M)$ is an n^2-rank $\mathcal{F}$-module.

Next we study the exterior product of covector fields, which is an skewsymmetric linear mapping, called 2-form and constitute a space $\Lambda^2(M)$. The bases of the 2-form space $\Lambda^2(M)$ are

$$\begin{aligned} dx^i \wedge dx^j &= (dx^i \otimes dx^j - dx^j \otimes dx^i) \\ &= \delta^{ij}_{kl} dx^k \otimes dx^l, \ (i \neq j), \end{aligned}$$

where

$$\delta^{ij}_{kl} = \delta^i_k \delta^j_l - \delta^i_l \delta^j_k = \begin{vmatrix} \delta^i_k & \delta^i_l \\ \delta^j_k & \delta^j_l \end{vmatrix}$$

is called generalized Kronecker δ symbol. The space has $\binom{n}{2} = \frac{1}{2}n(n-1)$ bases and is a $\frac{1}{2}n(n-1)$ rank $\mathcal{F}$-module. Any 2-form $\alpha_2 \in \Lambda^2(M)$ can be written as

$$\alpha_2 = \frac{1}{2} f_{ij}(x) dx^i \wedge dx^j, \; f_{ij} = -f_{ji},$$

where summation convention for repeated indices is used. More explicitly,

$$\alpha_2 = \frac{1}{2} \sum_{i \neq j} f_{ij}(x) dx^i \wedge dx^j = \sum_{i<j} f_{ij}(x) dx^i \wedge dx^j.$$

Similarly, we can consider the multi-tensor product $\mathcal{T}_r(M) \equiv \overset{r}{\otimes} \Theta(M)$ of co-vector fields

$$\overset{r}{\otimes} \Theta(M) = \Theta(M) \otimes \Theta(M) \otimes \cdots \otimes \Theta(M).$$

The bases are

$$dx^{i_1} \otimes \cdots \otimes dx^{i_r}, \quad i_1, \cdots, i_r = 1, 2, \cdots, n.$$

There are n^r bases. Any element $K \in \overset{r}{\otimes} \Theta(M)$ can be written as

$$K = K_{i_1 \cdots i_r} dx^{i_1} \otimes \cdots \otimes dx^{i_r},$$

where the summation convention for repeated indices is used. K is called an r-order covariant tensor field. The coefficients $K_{i_1 \cdots i_r}$ are called components of tensor K. Addition among tensors can be defined as the addition of their corresponding components which obeys the rule of Abelian group. Multiplication between tensor and function $f \in \mathcal{F}(M)$ can also be defined, which is distributive and associative. Hence $\mathcal{T}_r(M)$ is an n^r rank $\mathcal{F}$-module and called order-r covariant tensor field. Similarly, for the linear space $\Lambda^r(M)$, the bases are

$$\begin{aligned} dx^{i_1} \wedge \cdots \wedge dx^{i_r} &= \sum_{\sigma \in S_r} \mathrm{Sgn}(\sigma)(dx^{i_{\sigma_1}} \otimes \cdots \otimes dx^{i_{\sigma_r}}) \\ &= \delta^{i_1 \cdots i_r}_{k_1 \cdots k_r} dx^{k_1} \otimes \cdots \otimes dx^{k_r}, \end{aligned}$$

where

$$\delta^{i_1 \cdots i_r}_{k_1 \cdots k_r} = \begin{vmatrix} \delta^{i_1}_{k_1} & \cdots & \delta^{i_1}_{k_r} \\ \cdot & \cdots & \cdot \\ \cdot & \cdots & \cdot \\ \delta^{i_r}_{k_1} & \cdots & \delta^{i_r}_{k_r} \end{vmatrix}$$

$$= \begin{cases} +1, & \text{when subscripts is a even permutation of the superscripts,} \\ -1, & \text{when subscripts is a odd permutation of the superscripts,} \\ 0, & \text{otherwise,} \end{cases}$$

is the generalized Kronecker δ symbol and is completely skewsymmetric with respect to both subscripts and superscripts. For details of δ symbols, please refer to Appendix I.

$\Lambda^r(M)$ is an $\begin{pmatrix} n \\ r \end{pmatrix} = \frac{n!}{r!(n-r)!}$ rank $\mathcal{F}$-module.

Any r-form $\alpha_r \in \Lambda^r(M)$ can be written as

$$\alpha_r = \frac{1}{r!} f_{i_1 \cdots i_r}(x) dx^{i_1} \wedge \cdots \wedge dx^{i_r},$$

where function $f_{i_1 \cdots i_r}(x)$ is completely skewsymmetric with respect to its subscripts, and $\Lambda^1(M) = \Theta(M)$ is the set of cotangent vector fields on M.

Furthermore we can introduce $\Lambda^*(M)$, which is the direct sum of all orders of differential forms:

$$\begin{aligned} \Lambda^* &= \Lambda^0 \oplus \Lambda^1 \oplus \cdots \oplus \Lambda^n \\ &\equiv \oplus \sum_{i=0}^{n} \Lambda^i, \end{aligned}$$

where zero-forms are just the differentiable functions on the manifold and one-forms are the cotangent vector fields. Λ^* is a 2^n rank $\mathcal{F}(M)$-module, where one defines the wedge product among differential forms as

$$\alpha_p \wedge \beta_q = \frac{(p+q)!}{p!q!} \sum_{\sigma} (-1)^{\deg_\sigma} (\alpha_p \otimes \beta_q)^\sigma = (-1)^{pq} \beta_q \wedge \alpha_p,$$

where the summation runs over all permutations of $p+q$ indices. The result is a $(p+q)$-forms. Wedge product operation is associative, distributive and skew commutative:

1). Associative:

$$(\alpha \wedge \beta) \wedge \gamma = \alpha \wedge (\beta \wedge \gamma),$$

2). Distributive:

$$(\alpha + \beta) \wedge \gamma = \alpha \wedge \gamma + \beta \wedge \gamma,$$

3). Skew commutative:

$$\alpha \wedge \beta = (-1)^{pq} \beta \wedge \alpha, \quad \alpha \in \Lambda^p, \quad \beta \in \Lambda^q.$$

The 2^n-rank $\mathcal{F}$-module Λ^* with wedge product operations constitutes a Cartan algebra and is simply called exterior algebra.

Because of skew commutativity, wedge product of two same 1-forms is zero identically. There is a theorem:

Theorem 1.2 $\theta^1, \theta^2, \cdots, \theta^r \in \Lambda^1 = \Theta(M)$ *are linearly independent if and only if:*

$$\theta^1 \wedge \theta^2 \wedge \cdots \wedge \theta^r \neq 0.$$

Theorem 1.3 (Cartan Lemma) *For two set of one forms* $\{\theta^i, \omega^i, i = 1, \cdots, r\} \in \Lambda^1(M)$, *if*

$$\sum_{i=1}^{r} \theta^i \wedge \omega^i = 0$$

and $\{\theta^i\}$ *are linearly independent, then each* ω^i *can be written as a linear combination of* $\{\theta^i\}$

$$\omega^i = \sum_{j=1}^{r} c^i_j \theta^j,$$

where the coefficient functions satisfy

$$c^i_j = c^j_i.$$

Proof: Since $\theta^1, \theta^2, \cdots, \theta^r \in \Lambda^1$ are linearly independent, they can be chosen as part of the bases of Λ^1 : $\theta^1, \theta^2, \cdots, \theta^r, \theta^{r+1}, \cdots, \theta^n$, then ω^i can be expressed as

$$\omega^i = \sum_{j=1}^{r} c^i_j \theta^j + \sum_{a=r+1}^{n} c^i_a \theta^a.$$

Put this into the first formula in Theorem 1.3, we have

$$0 = \sum_{i,j=1}^{r} c^i_j \theta^i \wedge \theta^j + \sum_{i=1}^{r} \sum_{a=r+1}^{n} c^i_a \theta^i \wedge \theta^a.$$

As $\{\theta^\alpha \wedge \theta^\beta, 1 \leq \alpha \leq \beta \leq n\}$ are the linearly independent bases of Λ^2, so $c^i_a = 0, c^i_j = c^j_i$, the lemma is proved.

$\Lambda^r(M)$ is a subspace of order-r covariant tensor field $\overset{r}{\otimes}\Theta(M)$, i.e., $\Lambda^r(M) \subset T_r(M)$. Hence there is a contraction operation between differential forms and tangent vector fields, called the evaluation of differential forms. When a vector field X is contracted with a differential form, only the first index of the form is contracted. Because of the complete skewsymmetry of the coeffients of differential forms, this ensures that X is contracted with each index in the differential form. For instance, let $X = \xi^i \partial_i, Y = \eta^i \partial_i$, then

$$\begin{aligned} < dx^i \wedge dx^j, X > &= < (dx^i \otimes dx^j - dx^j \otimes dx^i), X > \\ &= < dx^i, X > dx^j - < dx^j, X > dx^i \\ &= \xi^i dx^j - \xi^j dx^i. \end{aligned}$$

$$< dx^i \wedge dx^j; X, Y >= \xi^i \eta^j - \xi^j \eta^i.$$

The evaluation of differential forms is closely related to the matrix determinant. For example,

$$\begin{aligned} < dx^i \wedge dx^j; \partial_k, \partial_l > &= \begin{vmatrix} < dx^i, \partial_k > & < dx^i, \partial_l > \\ < dx^j, \partial_k > & < dx^j, \partial_l > \end{vmatrix} \\ &= \delta^{ij}_{kl}. \end{aligned}$$

Consequently

$$< dx^i \wedge dx^j; X, Y >= \xi^k \eta^l \delta^{ij}_{kl}.$$

This evaluation of differential forms can be generalized. It can be proved that

$$\begin{aligned} < dx^{i_1} \wedge \cdots \wedge dx^{i_r}; \partial_{j_1}, \cdots, \partial_{j_r} > &= \det(< dx^{i_\alpha}, \partial_{j_\beta} >) \\ &= \delta^{i_1 \cdots i_r}_{j_1 \cdots j_r}, \end{aligned}$$

$$< dx^1 \wedge \cdots \wedge dx^n; X_{(1)}, \cdots, X_{(n)} >= \det(< dx^i, X_{(j)} >).$$

The contraction between differential forms and vector fields X can be written as the action of the interior operator i_X on differential forms, defined as

$$i_X \alpha \equiv X \rfloor \alpha \equiv < \alpha, X > . \tag{1.2}$$

The interior product operator i_X is an antiderivative one, i.e.,

$$i_X(\alpha_p \wedge \beta_q) = (i_X \alpha_p) \wedge \beta_q + (-1)^p \alpha_p \wedge (i_x \beta_q). \tag{1.3}$$

For instance, let

$$\alpha = f_i dx^i, \beta = g_k dx^k, X = \xi^i \partial_i,$$

$$\begin{aligned} \alpha \wedge \beta &= f_i g_k dx^i \wedge dx^k \\ &= \frac{1}{2}(f_i g_k - f_k g_i) dx^i \wedge dx^k, \\ i_X(\alpha \wedge \beta) &= (f_i g_k - f_k g_i)\xi^i dx^k \\ &= (i_X \alpha) \wedge \beta - \alpha \wedge (i_X \beta). \end{aligned}$$

Contraction of differential forms with vector fields is independent of the choice of frames. One often uses a given coordinate system to calculate the interior product. However, the results are independent of the choice of the frames.

The Lie bracket $[X, Y]$ for two tangent vector fields on a manifold is still a tangent vector field, i.e., is a linear differential operator which acts on $\mathcal{F}(M)$. Therefore the set of all tangent vector fields constitutes a Lie algebra. But there is no such operation for cotangent vector fields. Through the completely skewsymmetric wedge product, higher order differential forms can be obtained. The r-form is a skewsymmetric r linear functional on the sections of the tangent bundle $T(M)$, which is a mapping from a r-dimensional parallelogram on $T(M)$ to a real function. For instance, a 2-form

$$\alpha_2 = \frac{1}{2} f_{ij} dx^i \wedge dx^j,$$

takes two tangent vector fields X, Y into a real function

$$\begin{aligned} < \alpha_2; X, Y > &= \frac{1}{2} f_{ij}(\xi^i \eta^j - \xi^j \eta^i) \\ &= f_{ij}\xi^i \eta^j \ (\text{ as } f_{ij} = -f_{ji}). \end{aligned}$$

When $X = Y$, the right hand side of the equation is zero, ensuring α_2 as a linear functional on the section of tangent bundle $T(M)$. It is the skewsymmetry that makes α_2 a linear functional on the section of $T(M)$. A symmetric bilinear functional has no such property. This example shows that differential forms $\Lambda^*(M)$ are closely related to the structure of all higher order linear mapping spaces on the sections of $T(M)$. And consequently the differential forms are the most important tensor fields in the study of manifold.

In the following, we deal simply with the linear space which is made up of multi-tensor products. For instance, order-r contravariant tensor space $T^r(M)$ is a $r-$linear tensor field space formed by the tensor product of r tangent fields. The following three definitions are equivalent compeletely.

1). $T^r(M)$ is a r-multiplicative tensor product of $\mathcal{X}(M)$,

$$T^r(M) = \mathcal{X}(M) \otimes \mathcal{X}(M) \otimes \cdots \otimes \mathcal{X}(M).$$

The basis are

$$\{\partial_{i_1} \otimes \partial_{i_2} \otimes \cdots \otimes \partial_{i_r}\}, \ i_1, i_2, \cdots, i_r = 1, 2, \cdots, n.$$

There are n^r bases. Any element $K \in T^r(M)$ can be written as

$$K = K^{i_1 \cdots i_r} \partial_{i_1} \otimes \cdots \otimes \partial_{i_r}, \quad K^{i_1 \cdots i_r} \in \mathcal{F}(M).$$

2). Tensor can be defined by the transformation properties of its components under a change of coordinate systems. For instance, the bases of the tangent field are transformed as

$$\partial'_i = \frac{\partial x^j}{\partial x'^i}\frac{\partial}{\partial x^j} \equiv L_i^j e_j, \qquad detL = \frac{\partial(x^1, \cdots, x^n)}{\partial(x'^1, \cdots, x'^n)} \neq 0.$$

Since tensor K is a geometric object which is independent of the choice of frames, the components of K are transformed as

$$K'^{i_1 \cdots i_r} = (L^{-1})^{i_1}_{j_1} \cdots (L^{-1})^{i_r}_{j_r} K^{j_1 \cdots j_r}.$$

The set of n^r real numbers which are transformed like this is called a order-r contravariant tensor K. The n^r real functions $\{K^{i_1 \cdots i_r}\}$ are called its components. This is the well known definition of tensor by coordinate transformations.

3). $\mathcal{T}^r(M)$ is the set of r-linear functions on the cotangent tensor field

$$\mathcal{T}^r(p) = \mathrm{Hom}(\Theta(M), \Theta(M), \cdots, \Theta(M); \mathcal{F}(M)),$$

where element K is a r-linear mapping

$$\begin{aligned} K : \Theta(M), \Theta(M), \cdots, \Theta(M) &\rightarrow \mathcal{F}(M) \\ \sigma\ ,\ \ \tau\ ,\ \ \cdots,\ \ \omega &\rightarrow K(\sigma, \tau, \cdots, \omega) \in \mathcal{F}(M) \end{aligned}$$

where $\sigma, \tau, \cdots, \omega \in \Theta(M)$ are any cotangent vector fields on M. This mapping is a r-linear and satisfies:

$$\begin{aligned} K(f\sigma, \tau, \cdots, \omega) &= K(\sigma, f\tau, \cdots, \omega) = \cdots \\ &= fK(\sigma, \tau, \cdots, \omega), \quad f \in \mathcal{F}(M). \end{aligned}$$

$$K(\sigma, \tau, \cdots, \omega) + K(\rho, \tau, \cdots, \omega) = K(\sigma + \rho, \tau, \cdots, \omega).$$

The equations above are also true for other arguments, i.e., K is a linear mapping which preserves the linear structure with respect to all its arguments. Therefore, tensor K is completely determined if its action on the bases of cotangent vector fields is given. For a given bases $\{dx^i, i = 1, \cdots, n\}$ of cotangent vector fields,

$$K(dx^{i_1}, dx^{i_2}, \cdots, dx^{i_r}) \equiv K^{i_1 i_2 \cdots i_r} \in \mathcal{F}(M),$$

are called the components of the order-r contravariant tensor K. Thus tensor K can be written as

$$K = K^{i_1 i_2 \cdots i_r} \partial_{i_1} \otimes \partial_{i_2} \cdots \otimes \partial_{i_r}.$$

The third definition of $\mathcal{T}^r(M)$ tensor is independent of the choice of coordinate systems. Subsequently we mainly use this definition.

An order-s covariant tensor field $\mathcal{T}_s(M)$ can be defined similarly as

$$\begin{aligned} \mathcal{T}_s(M) &= \Theta(M) \otimes \Theta(M) \otimes \cdots \otimes \Theta(M) \\ &= \mathrm{Hom}(\mathcal{X}(M), \mathcal{X}(M), \cdots, \mathcal{X}(M); \mathcal{F}(M)). \end{aligned}$$

For tensor fields of the same type, addition can be defined as the addition of the corresponding components at each point. Multiplication of tensors by differentiable function on the manifold can be defined as multiplication of the components at each point. Hence, the set of all tensor fields of the same type constitutes an $\mathcal{F}(M)$–module.

Tensor products among tensor fields can be introduced to get higher order tensor fields. For example, let

$$K \in \mathcal{T}_s^r(M), \quad H \in \mathcal{T}_v^u(M),$$

then

$$K \otimes H \in \mathcal{T}_{s+v}^{r+u}(M).$$

$\mathcal{T}_s^r(M)$ tensor field can be defined as the direct product of order-r contravariant tensor field and order-s covariant tensor field $\mathcal{T}_s^r(M) = \mathcal{T}^r(M) \otimes \mathcal{T}_s(M)$, which is an n^{r+s}-rank $\mathcal{F}$-module.

For the (r, s) type of tensor fields $\mathcal{T}_s^r(M)$ with $r, s \geq 1$, contraction operation of tensor fields can be defined, which reduces the order of the tensor fields by two. These operations among tensor fields are independent of the choice of the frames.

$$\begin{aligned} C : \mathcal{T}_s^r(M) &\rightarrow \mathcal{T}^{r-1,s-1}(M) \\ (CK)^{i_2 \cdots i_r}_{j_2 \cdots j_s} &= K^{ii_1 \cdots i_r}_{ij_1 \cdots j_s}, \end{aligned}$$

i.e., summation for a pair of upper and lower indices.

Direct sum can be defined for different types of tensor fields. Direct sum for all types of tensors is

$$\mathcal{T}(M) = \oplus \sum_{r,s=0}^{\infty} \mathcal{T}_s^r,$$

where the element is a formal sum of all types of tensors. $\mathcal{T}(M)$ constitutes an infinite rank $\mathcal{F}$-module, in which there exists a tensor product so that $\mathcal{T}(M)$ is an associative algebra and is called tensor algebra.

In addition, the tensor field space can be taken as a linear space which is of multi-linear mappings on vector or covector field. Some different linear mappings constitute isomorphic tensor field spaces. For instance, $\mathcal{T}_1^1(M)$ type tensor

$$\begin{aligned} \mathcal{T}_1^1(M) &= \mathcal{X}(M) \otimes \Theta(M) \\ &= \mathrm{Hom}(\Theta(M), \mathcal{X}(M); \mathcal{F}(M)) \\ &= \mathrm{Hom}(\mathcal{X}(M), \mathcal{X}(M)) \equiv End(\mathcal{X}(M)) \\ &= \mathrm{Hom}(\Theta(M), \Theta(M)) \equiv End(\Theta(M)), \end{aligned}$$

where any $K \in \mathcal{T}_1^1(M)$ can be written as

$$K = K_j^i \partial_i \otimes dx^j. \tag{1.4}$$

For the set of endomorphisms of tangent field space $\mathcal{X}(M)$, the element

$$K \in \mathrm{Hom}(\mathcal{X}(M); \mathcal{X}(M)) \equiv End(\mathcal{X}(M)),$$

$$\begin{aligned} K : \mathcal{X}(M) &\rightarrow \mathcal{X}(M) \\ \partial_j &\rightarrow K(\partial_j) = K_j^i \partial_i \in \mathcal{X}(M) \end{aligned}$$

so that the linear transformation operator K in vector field space $\mathcal{X}(M)$ can be represented by Eq. 1.4, i.e., the endomorphism K in a vector field space is a $\mathcal{T}_1^1(M)$ type tensor field and contraction is just the trace operation of the endomorphism operator. Both direct product and contraction of tensor fields are independent of the choice of coordinate systems.

1.6 Exterior differentiation

In this section, we first study an important linear differential operator: the exterior differentiation d, which is a differential operator within Cartan exterior algebra $\Lambda^*(M)$, and acts on differential forms:

$$d : \Lambda^r(M) \to \Lambda^{r+1}(M).$$

The exterior differential operator d is a generalization of the conventional total differential operator which acts on function

$$df = f_{,i}\, dx^i, \ (f_{,i} \equiv \frac{\partial f}{\partial x^i}). \tag{1.5}$$

For a p-form

$$\alpha_p = \frac{1}{p!} f_{i_1 \cdots i_p} dx^{i_1} \wedge \cdots \wedge dx^{i_p},$$

its exterior differentiation is defined as

$$\begin{aligned} d\alpha_p &= \frac{1}{p!} df_{i_1 \cdots i_p} \wedge dx^{i_1} \wedge \cdots \wedge dx^{i_p} \\ &= \frac{1}{p!} f_{i_1 \cdots i_p, k}\, dx^k \wedge dx^{i_1} \wedge \cdots \wedge dx^{i_p} \\ &= \frac{1}{p!(p+1)!} \delta^{k i_1 \cdots i_p}_{j_1 j_2 \cdots j_{p+1}} f_{i_1 \cdots i_p, k}\, dx^{j_1} \wedge \cdots \wedge dx^{j_{p+1}}. \end{aligned} \tag{1.6}$$

Apply d to Eq. 1.5 once more, we have

$$d(df) = f_{,ij}\, dx^j \wedge dx^i = 0,$$

because $f_{,ij} = f_{,ji}$ which is symmetric but $dx^j \wedge dx^i$ is skewsymmetric. In the same way, apply d to Eq. 1.6 once more gives zero. In general

$$d^2 = 0.$$

Making use of Leibniz rule for usual differentiation of functions, it can be proved that

$$\begin{aligned} d(\alpha_p \wedge \beta_q) &= \frac{1}{p!q!} d(f_{i_1 \cdots i_p} h_{k_1 \cdots k_p}) \wedge dx^{i_1} \wedge \cdots \wedge dx^{i_p} \wedge dx^{k_1} \wedge \cdots \wedge dx^{k_q} \\ &= (d\alpha_p) \wedge \beta_q + (-1)^p \alpha_p \wedge d\beta_q. \end{aligned}$$

In conclusion, the exterior differentiation has the following features:
1). It is a first order linear differential operator, $d : \Lambda^r \to \Lambda^{r+1}$ and

$$d(a\sigma + b\tau) = ad\sigma + bd\tau \ (\sigma, \tau \in \Lambda^*, a, b \in \mathcal{R}).$$

2). It is a skew differential operator, i.e.,

$$d(\alpha_p \wedge \beta_q) = (d\alpha_p) \wedge \beta_q + (-1)^p \alpha_p \wedge d\beta_q.$$

3). Its action to a functions is

$$df = f_{,i}\, dx^i.$$

4).

$$d^2 = 0.$$

These four features can be taken as the definition of the exterior differential operator d.

A differential form α is said to be closed if

$$d\alpha = 0.$$

A r-form α is exact if and only if it can be represented globally as an exterior differentiation of a $(r-1)$-form β,

$$\alpha = d\beta.$$

An exact form is also closed but the inverse is not true in general. A closed form can be represented as an exact form only locally. In general a closed form can not be represented globally as an exact form. This problem will be studied in the context of homology, which is closely related to the global topology of manifolds. For more information, please refer to Chapter Seven.

For a smooth function $f(x)$ on a smooth manifold, when acted upon by the exterior differential operator, a one form $df = f_{,i}\, dx^i$ is obtained. For the exterior algebra Λ^* which consists of all differential forms on a manifold, the exterior differential operation is well defined so that differential forms of one more order can be obtained. However, we can not define exterior differential operation on tangent vector fields, due to tangent vector fields at different points are not related yet. Only when additional structures are introduced, then covariant derivative and Lie derivative can be defined. These will be the subjects of the next two chapters. Since the action of exterior differential operator on differential forms does not require additional structure on manifold, they play an important role in the study of manifolds. E. Cartan obtained many results on partial differential equation problems and on local geometric problems of manifolds with the help of exterior differential operator. Next we take vector analysis in three-dimensional Euclidean space as an example, to show the application of the exterior differential operation. Gradient, curl and divergence of vectors all can be expressed in term of exterior differentiation.

Example 1. Vector analysis in three-dimensional Euclidean space $\mathbf{E}^3$.

In $\mathbf{E}^3$, a generic vector field $\vec{A}(x) = (A_1(x), A_2(x), A_3(x))$ can be expressed as a 1-form

$$A(x) = A_i(x)dx^i.$$

An axial vector $\vec{B}(x) = (B_1(x), B_2(x), B_3(x))$ can be represented by a 2-form

$$\begin{aligned} B(x) &= B_1(x)dx^2 \wedge dx^3 + B_2(x)dx^3 \wedge dx^1 + B_3(x)dx^1 \wedge dx^2 \\ &= \frac{1}{2}\epsilon_{ijk}B_i(x)dx^j \wedge dx^k. \end{aligned}$$

Apply the exterior differential operator on a zero-form $f(x)$ yields the gradient of the function

$$df = f_{,i}\, dx^i \equiv \nabla f \cdot d\vec{x}. \tag{1.7}$$

Apply the exterior differential operator on the one-form $A = A_i dx^i \equiv \vec{A} \cdot d\vec{x}$ gives the curl of vector field $\vec{A}$

$$dA = A_{i,j}\, dx^j \wedge dx^i = \frac{1}{2}\epsilon_{ijk}(\nabla \times \vec{A})_i dx^j \wedge dx^k. \tag{1.8}$$

Apply the exterior differential operation on the 2-form $B = \frac{1}{2}\epsilon_{ijk}B_i dx^j \wedge dx^k$ gives the divergence of the vector field

$$dB = \frac{1}{2}\epsilon_{ijk}B_{i,l}\, dx^l \wedge dx^j \wedge dx^k = \nabla \cdot \vec{B} dx \wedge dy \wedge dz. \tag{1.9}$$

Apply the exterior differential operator to Eq. 1.7, $d^2 f = 0$, which is simply $\nabla \times \nabla f = 0$. Apply the exterior differential operator to Eq. 1.8, $d^2 A = 0$, which is just simply $\nabla \cdot \nabla \times \vec{A} = 0$. In the three-dimensional space, the 3-form is the highest order form, the exterior differential operator on Eq. 1.9 is identically zero, so it does not provide any new information.

In the case of vector analysis in three-dimensional space, the use of differential forms greatly simplifies the calculations. Here we take the Dirac magnetic monopole as an example.

Example 2. Dirac's monopole problem.

In 1931[12], Dirac proposed the possible existence of the magnetic monopole by considering the duality between electricity and magnetism. He found that if point magnetic charge exists then not only the magnetic field is singular at the point of the magnetic charge, but also there must be a singular string for the magnetic field when represented by gauge potential. Magnetic charge has to be quantized if this singular string is not be observed. Now we know that the existence of the singular string is due to the nontrivial topology of S^2 surface enclosing the magnetic charge.

To study the static monopole problem, the static magnetic field $\vec{B}$ can be written in terms of a vector potential $\vec{A}$,

$$\vec{B} = \nabla \times \vec{A}. \tag{1.10}$$

If the vector potential is expressed by a one form

$$A = A_i dx^i,$$

the exterior differentiation of A,

$$dA = \frac{1}{2}\epsilon_{ijk}B_i dx^j \wedge dx^k \equiv B, \tag{1.11}$$

is just the Eq. 1.10. For a given field B, the potential A may be altered by adding an arbitrary exact form, i.e., there is a gauge transformation

$$A \to A' = A + d\phi, \quad \phi \in \mathcal{F}(E^3).$$

For the static monopole which has spherical symmetry, since the topology properties of S^2 is nontrivial, B can not be expressed as an exact form globally, i.e., Eq. 1.11 can not be satisfied everywhere on S^2. To cover all of S^2 at least two open sets are needed: U_+ is the region which excludes the south pole on S^2; U_- is the region which excludes the north pole on S^2. For regions $U_\pm$, there are gauge potential $A^\pm$ correspondingly. In three-dimensional Cartesian coordinates, the gauge potentials are

$$A^\pm = \frac{q}{r}\frac{1}{z \pm r}(xdy - ydx) \equiv A_i^\pm dx^i, \tag{1.12}$$

where $r = \sqrt{x^2 + y^2 + z^2}$ and

$$A_x^\pm = -\frac{q}{r}\frac{y}{z \pm r}, \quad A_y^\pm = \frac{q}{r}\frac{x}{z \pm r}, \quad A_z^\pm = 0.$$

Therefore the gauge field strength $F = \frac{1}{2}\epsilon_{ijk}F_i dx^j \wedge dx^k$ can be written as

$$F_x = \frac{\partial A_z^\pm}{\partial y} - \frac{\partial A_y^\pm}{\partial z} = -\frac{\partial}{\partial z}\left(\frac{q}{r}\frac{x}{z \pm r}\right) = q\frac{x}{r^3}.$$

Similarly,

$$F_y = q\frac{y}{r^3}, \qquad F_z = q\frac{z}{r^3}.$$

Eq. 1.12 is indeed the gauge potential around a magnetic charge. The field strength has a unified expression on all of S^2, but the gauge potential has not, one has to use different gauge potentials for different open coverings.

Due to the spherical symmetry of this problem, it is more convenient to work in spherical coordinates. In the local moving frame

$$\sigma^r \equiv dr, \qquad \sigma^\theta \equiv rd\theta, \qquad \sigma^\varphi \equiv r\sin\theta d\varphi.$$

Gauge potentials for the south and the north regions are

$$A^\pm = q(\pm 1 - \cos\theta)d\varphi = A_\varphi^\pm \sigma^\varphi,$$

where

$$A_\varphi^\pm = q\frac{\pm 1 - \cos\theta}{r\sin\theta}, \quad A_r^\pm = 0, \quad A_\theta^\pm = 0. \tag{1.13}$$

The field strength is

$$\begin{aligned} F &= dA^{\pm} = q \sin\theta d\theta \wedge d\varphi \\ &= \frac{q}{r^2}\sigma^{\theta} \wedge \sigma^{\varphi}. \end{aligned}$$

Only radial field strength $\frac{q}{r^2}$ is present, and the magnetic charge of the point magnetic monopole is q.

The field strength F is expressed in different regions by the exterior differentiation of $A^{\pm}$ respectively, and $F = dA^{\pm}$ is a closed form($dF = 0$). Because of the nontrivial topology of S^2, F can not be expressed as an exact form on all of S^2. $A^{\pm}$ have singular points on S^2. F can be written as $F = dA^{\pm}$ only for certain regions. Each open covering $U_{\pm}$ is topologically trivial so that closed forms can be expressed as exact forms (Poincaré lemma, see chapter 7). However, for the topologically nontrivial manifolds (for example S^2), closed forms are not exact forms in general.

When the field strength F is given, there is a gauge transformation for gauge potential A, i.e., there is an arbitrary exact form $d\psi$ and that

$$A \rightarrow A' = A + d\psi,$$

with $F = dA = dA'$, i.e., the field strength F is gauge invariant. In the overlapping region $U_+ \cap U_-$, i.e., in the neighborhood of the equator, the field strength F can be expressed as $dA^+ = dA^- = F$, i.e., there exists a gauge transformation between A^+ and A^-. This gauge transformation along the equator is a function of φ, i.e., the following condition must be satisfied

$$e^{i\psi_-} = e^{in\varphi}e^{i\psi_+} \equiv Ue^{i\psi_+},$$

where n is an integer so that the gauge transformation at the equator is single valued. That is to say,

$$A_+ = A_- + U^{-1}idU = A_- + nd\varphi.$$

In addition, from Eq. 1.13, we have

$$A_+ = A_- + 2qd\varphi.$$

Comparing these two equations, it is clear that the magnetic charge must be quantized, i.e.,

$$q = \frac{n}{2}.$$

The gauge potential can be uniquely defined for each region on S^2 respectively. The above equation is just the quantization condition of the magnetic charge.

1.7 Orientation and Stokes formula

An important property of manifolds is their orientability. For instance, consider the two-dimensional surface in three-dimensional Euclidean space. A sphere S^2 and a torus T^2 both have the inner side and outer side. There are two normal lines at each point on the surface: the inner normal line and the outer normal . Surfaces, which have two sides, globally are called orientable surfaces.

On the other hand, if the two sides can not be distinguished , the surface is not orientable and is called a single-sided surface. The simplest one-sided surface is the Móbius band. As shown in Figure 1.9, it is a rectangle twisted by 180°. When points

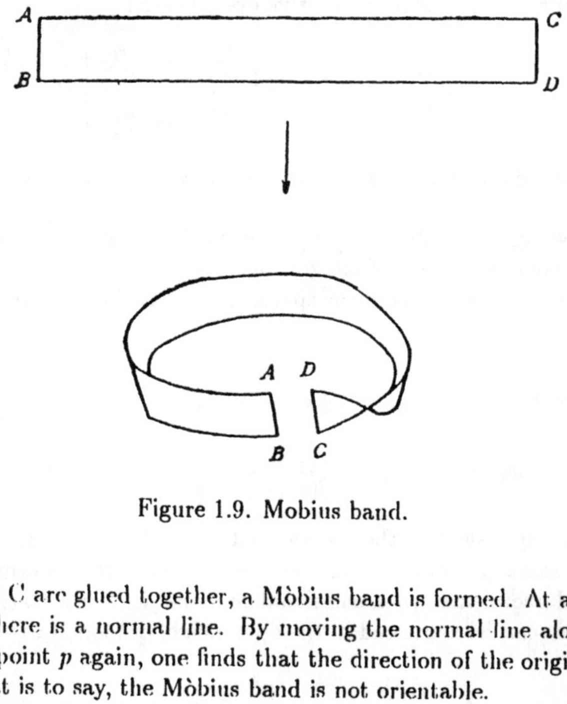

Figure 1.9. Mobius band.

A and D, B and C are glued together, a Mòbius band is formed. At a point p on the Mòbius band, there is a normal line. By moving the normal line along the surface, until it reaches point p again, one finds that the direction of the original normal line is reversed. That is to say, the Mòbius band is not orientable.

To determine whether a higher dimensional manifold is orientable or not, one needs a mathematical object which has plus and minus signs. When the direction is changed, the sign of this mathematical object is changed. In an n-dimensional manifold, the set of n-forms is a one-dimensional linear space. For two n-forms ω_1 and ω_2 which are non-zero everywhere on connected manifold, there exists a continuous function f such that $\omega_1 = f\omega_2$, where f is positive definite (negative definite) everywhere. When f is positive definite everywhere, ω_1 and ω_2 are called equivalent. The orientation of a manifold M is a choice of the equivalent classes of n-forms. And the following definition is adopted.

Definition 1.12 *A n-dimensional manifold is called orientable if there exists a continuous n-form ω which is non-zero everywhere on the manifold.*

Connected orientable manifold has two possible orientations. For an orientable manifold, there exists a coordinate atlas $\{(U_\alpha, \varphi_\alpha\}$ which is consistent with the differential structure of the manifold, such that for transformation functions (when $U_\alpha \cap U_\beta \neq 0$,)

$$\begin{aligned} f = \varphi_\beta \cdot \varphi_\alpha^{-1} : \varphi_\alpha(U_\alpha \cap U_\beta) &\rightarrow \varphi_\beta(U_\alpha \cap U_\beta) \\ \{x^i\} &\rightarrow \{y^i = f^i(x)\} \end{aligned}$$

the corresponding Jacobi matrix is positive everywhere :

$$J = \frac{\partial(y^1, \cdots, y^n)}{\partial(x^1, \cdots, x^n)} = \begin{vmatrix} \frac{\partial y^1}{\partial x^1} & \cdots & \frac{\partial y^1}{\partial x^n} \\ \cdots & \cdots & \cdots \\ \frac{\partial y^n}{\partial x^1} & \cdots & \frac{\partial y^n}{\partial x^n} \end{vmatrix}.$$

A manifold is called an orientable manifold if it has a coordinate atlas satisfying this condition.

In the following, we study integration on manifold. First we need to know the measure or the volume element of integration. On an n-dimensional manifold, set Λ^n of n forms is a one-dimensional linear space, for a given local frame, the bases of Λ^n are as

$$dx^1 \wedge dx^2 \wedge \cdots \wedge dx^n.$$

Under a coordinate transformation , $\{x^i\} \rightarrow \{y^i(x)\}$,

$$dy^1 \wedge \cdots \wedge dy^n = \frac{\partial(y^1, \cdots, y^n)}{\partial(x^1, \cdots, x^n)} dx^1 \wedge \cdots \wedge dx^n.$$

This equation is equivalent to the transformation of the integration volume element for an n-dimensional integral. Therefore, we can take the n-form $\tau(x)$ as the n-dimensional volume element(n-dimensional measure with orientation). In a given local frame, $\tau(x)$ can be expressed in the open neighborhood U as

$$\tau(x) = dx^1 \wedge dx^2 \wedge \cdots \wedge dx^n.$$

Using this measure, integration of function $f(x) \in \mathcal{F}(M)$ over any n-dimensional closed submanifold V within U is

$$\int_V f(x)\tau(x) = \int_V f(x) dx^1 dx^2 \cdots dx^n, \quad V \subset U, \tag{1.14}$$

where the right hand side is the usual n-dimensional integration in $\mathbf{R}^n$ and the left hand side is the integration expressed by differential forms. It is clear that the result of integration is independent of the coordinate chart chosen.

When the region of integration V is in the intersection $U_1 \cap U_2 \neq 0$ of two open coverings U_1 and U_2 on manifold M, the definition of the partition of unity is needed. The partition of unity ρ_α is non-zero only on U_α

$$\rho_\alpha(x) = 0, \quad \text{if } x \notin U_\alpha,$$

and in the region of intersection, $\sum_\alpha \rho_\alpha = 1$, as shown in Figure 1.10. Therefore,

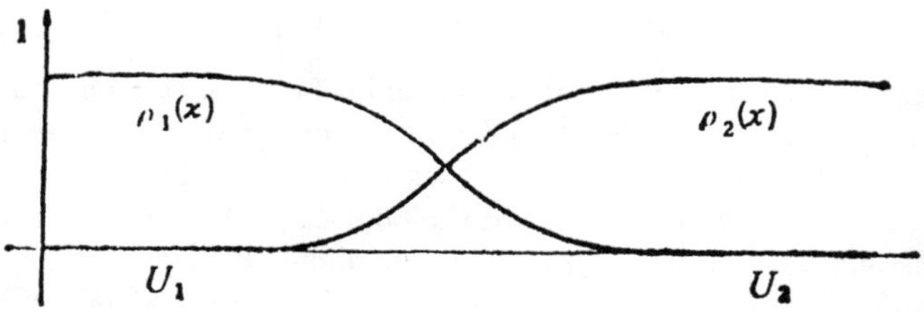

Figure 1.10. Partition of unity ρ_α.

$$\begin{aligned}\int_V f(x)\tau(x) &= \int_V \sum_\alpha \rho_\alpha(x) f(x)\tau(x)\\ &= \int_V \rho_1(y)f(y)\tau(y) + \int_V \rho_2(z)f(z)\tau(z),\end{aligned}$$

where $y \in U_1$ and $z \in U_2$ are local coordinate for each chart. The integration can be reduced to integration on a closed submanifold within each coordinate neighborhood U_α.

When there are many elements in the open covering $\{U_\alpha\}$ of manifold M, to ensure that integration on the manifold is well defined, the partition of unity of the whole manifold is introduced. There exists a set of smooth functions $\{\rho_\alpha\}$ on M, with

1). $0 \leq \rho_\alpha \leq 1$,

2). when $x \notin U_\alpha$, $\rho_\alpha(x) = 0$,

3). $\sum_\alpha \rho_\alpha(x) = 1$.

the set $\{\rho_\alpha\}$ is called the partition of unity with respect to covering $\{U_\alpha\}$. As all differentiable manifolds are paracompact(refer to Appendix B), i.e., for an arbitrary point p on manifold M, there is a neighborhood U such that the set $\{U \cap U_\alpha\}$ is a finite set, summation $\sum_\alpha \rho_\alpha(x)$ is well defined since only finite terms in the sum is non-zero. Hence, the integral of $f(x)$ on the whole of M is

$$\int_M f\tau = \sum_\alpha \int_{U_\alpha} \rho_\alpha(x_\alpha) f(x_\alpha) dx_\alpha^1 \cdots dx_\alpha^n.$$

It is easy to prove that for different choices of the partition of unity, the result of integration is the same, i.e., $\int_M f\tau$ as defined above is independent of the patition used. Eq. 1.14 represents integration over an n-dimensional closed submanifold of M, where V is a manifold with boundary. Here the concept of the manifold with boundary is needed.

The definition of manifolds with boundary is similar to the definition 1.1. There are two kinds of charts: one of the open neighborhoods are homeomorphic to open sets of $\mathbf{R}^n$; the other kind of open neighborhoods homeomorphic to open sets of $\mathbf{R}^n_+$, where

$$\mathbf{R}^n_+ = \{(x^1, x^2, \cdots, x^n) \in \mathbf{R}^n | x^n \geq 0\}.$$

$\mathbf{R}^n_+$ is the region with $x^n \geq 0$ in an n-dimensional linear space $\mathbf{R}^n$, where $x^n = 0$ is an $(n-1)$-dimensional linear space $\mathbf{R}^{n-1}$, called the boundary of $\mathbf{R}^n_+$. When performing a homeomorphic mapping to a manifold V with boundary, the set of points whose image are on the boundary of $\mathbf{R}^n_+$ is called the boundary ∂V of V. ∂V is an $(n-1)$-dimensional manifold if V is an n-dimensional manifold.

When manifold M is orientable with boundary, the induced orientation of the boundary ∂M can be obtained from the orientation of M such that ∂M is an $(n-1)$-dimensional orientable manifold. When the standard orientation $dx^1 \wedge \cdots \wedge dx^n$ is used for $\mathbf{R}^n_+$, the induced orientation of $\partial \mathbf{R}^n_+$ is defined as the equivalence class of $(-1)^n dx^1 \wedge \cdots \wedge dx^{n-1}$, the factor $(-1)^n$ has been introduced in order to remove a particular sign in the proof of Stokes theorem (see Appendix G).

For the integration on manifold M with boundary, there exists the famous Stokes formula:

$$\int_M d\omega = \int_{\partial M} \omega,$$

where $d\omega$ is an exact n-form. If $\partial M = 0$, integration on the right hand side is zero. For general proof of the Stokes formula, please refer to Appendix G. Here we present some examples to illustrate the Stokes formula.

1). $M = [a, b]$ is a closed interval of $\mathbf{R}^1$, ∂M are two points a and b,

$$\int_M df = \int_a^b \frac{\partial f}{\partial x} dx = f(b) - f(a).$$

2). M is a closed region on $\mathbf{R}^2$,

$$\begin{aligned}
\int_M d(\vec{A} \cdot d\vec{l}) &= \int_M d(A_1 dx^1 + A_2 dx^2) \\
&= \int_M \left(\frac{\partial A_1}{\partial x^2} - \frac{\partial A_2}{\partial x^1}\right) dx^1 \wedge dx^2 \\
&= \int_M (\text{rot}\vec{A}) \cdot d\vec{s} \\
&= \int_{\partial M} \vec{A} \cdot d\vec{l}.
\end{aligned}$$

This is just the usual Stokes theorem.

3). M is a closed region on $\mathbf{R}^3$, then

$$\begin{aligned}\int_M d(\vec{E}\cdot d\vec{s}) &= \int_M d(\frac{1}{2}\epsilon_{ijn}E_i dx^j \wedge dx^n) \\ &= \int_M \nabla\cdot\vec{E}dV \\ &= \int_{\partial M} \vec{E}\cdot d\vec{s},\end{aligned}$$

which is just the Gauss theorem.

The Stokes formula has a wide variety of applications both in physics and in differential geometry. Noticing the bilinear property of the integrand differential form and the region of integration in Eq. 1.14, a notation can be introduced to represent the integration:

$$\int_M \omega \equiv < M,\omega > \in R,$$

which satisfies the following:

$$\begin{aligned}< M, a\omega > &= < aM,\omega > = a < M,\omega > ,\ a\in R, \\ < M,\omega+\tau > &= < M,\omega > + < M,\tau >, \\ < M_1 + M_2,\omega > &= < M_1,\omega > \ + \ < M_2,\omega > .\end{aligned}$$

In this case, the Stokes formula can be written as

$$< M, d\omega > \ = \ < \partial M,\omega > .$$

It shows the duality between the exterior differential operator d and the boundary operator ∂ of a manifold. The exterior differential operator d is a local analytical operator acting on differential forms. The boundary operator ∂ of a manifold is a global topological operator acting on the whole region of integration. The Stokes formula links the region of integration and differential forms together and joins local properties and global properties of the manifold together by the exterior differential operator d and the boundary operator ∂ on manifold. This duality partly explains the importance of Stokes' theorem in algebraic topology (see Chapter 7).

Notations and Formulae

$\mathcal{F} = \mathcal{F}(M)$ denote the ring of C^∞ functions on M.
$\mathcal{X} = \mathcal{X}(M)$ denote the $\mathcal{F}$-module of all C^∞ vector fields on M.
$\Lambda^k = \Lambda^k(M)$ denote the $\mathcal{F}$-module of all k-forms on M.
$\Theta = \Theta(M) \equiv \Lambda^1(M)$ denote the $\mathcal{F}$-module of all 1-forms on M.
$\Lambda^* = \Lambda^*(M) = \Lambda^0 \oplus \Lambda^1 \oplus \cdots \oplus \Lambda^n$.

- A vector field $X \in \mathcal{X}$ is a line differential operator on $\mathcal{F}$, i.e., it satisfy:
  1. R-linear:
  $$X\,(af+bg) = aXf + bXg \quad (a,b \in R, \text{ and } f,g \in \mathcal{F}(M)).$$
  2. Derivation:
  $$X\,(fg) = fXg + gXf \quad (f,g \in \mathcal{F}(M)).$$
- A covector field $\sigma \in \Theta \equiv \mathrm{Hom}(\mathcal{X},\mathcal{F})$ is a $\mathcal{F}$-linear map $\sigma : \mathcal{X} \to \mathcal{F}$ which satisfies
  $$\sigma\,(fX+gY) = f\sigma(X) + g\sigma(Y) \quad (\text{for } f,g \in \mathcal{F} \text{ and } X,Y \in \mathcal{X}).$$
  Usualy we denote $\sigma(X)$ by $<\sigma, X> \equiv \sigma(X) \in \mathcal{F}(M)$,
  $$<df, X> = Xf \quad \in \mathcal{F}(M).$$
  Chosen local coordinate $\{x^i\}_1^n$, the bases of Θ are $\{dx^i\}_1^n$, the bases of $\mathcal{X}$ are $\{\partial_i\}_1^n$,
  $$<dx^i, \partial_j> = \delta^i_j,$$
  $$<\sigma, X> = \sigma_i \xi^i, \quad for \quad \sigma = \sigma_i(x)dx^i, \quad X = \xi^j(x)\partial_j.$$
- The tensor product of covector field $K = K_{ij}dx^i \otimes dy^j \in \Theta \otimes \Theta = Hom(\mathcal{X} \times \mathcal{X}; \mathcal{F})$ is a $\mathcal{F}$-bilinear map $K : \mathcal{X} \times \mathcal{X} \to \mathcal{F}$.
- A 2-form $\alpha_2 \in \Lambda^2(M)$ is a skew symmetric $\mathcal{F}$-bilinear map $\alpha_2 : \mathcal{X} \times \mathcal{X} \to \mathcal{F}$
  $$\alpha_2 = \frac{1}{2} f_{ij} dx^i \wedge dx^j, f_{ij} = -f_{ji},$$
  $$dx^i \wedge dx^j = \delta^{ij}_{kl} dx^k \otimes dx^l \ (i \neq j),$$
  $$<dx^{i_1} \wedge dx^{i_2}; \partial_{k_1}, \partial_{k_2}> = det(<dx^i, \partial_k>) = \delta^{i_1 i_2}_{k_1 k_2}.$$
- A k-form $\alpha_k \in \Lambda^k$ is a skewsymmetric $\mathcal{F}$-multilinear map $\alpha_k : \mathcal{X} \times \cdots \times \mathcal{X} \to \mathcal{F}$.
  $$\alpha_k = \frac{1}{k!} f_{i_1 \cdots i_k}(x) dx^{i_1} \wedge \cdots \wedge dx^{i_k},$$
  where $f_{i_1 \cdots i_k}$ is skew symmetric w.r.t., its subscript,
  $$dx^{i_1} \wedge \cdots \wedge dx^{i_r} = \delta^{i_1 \cdots i_r}_{k_1 \cdots k_r} dx^{k_1} \otimes \cdots \otimes dx^{k_r},$$
  $$<dx^1 \wedge \cdots \wedge dx^n; X_{(1)}, \cdots, X_{(n)}> = \det(<dx^i, X_{(j)}>).$$

- The exterior differentiation is a map $d : \Lambda^k \to \Lambda^{k+1}$ satisfy
  (1) $\mathcal{R}$-liear,
  (2) skew differential operator,

$$d(\alpha_k \wedge \beta_l) = d\alpha_k \wedge \beta_l + (-1)^k \alpha_k \wedge d\beta_l,$$

  (3) $df = f_{,i} dx^i$,
  (4) $d^2 = 0$.

- The contraction between vector field X and k-form α can be witten as

$$\begin{aligned} i_X\alpha \;\equiv\; X\rfloor\alpha &= \;<\alpha, X> \;\in \Lambda^{k-1}, \\ i_X\alpha(X^{(1)}, \cdots, X^{(k-1)}) &= \alpha(X, X^{(1)}, \cdots, X^{(k-1)}). \end{aligned}$$

  Thus, intrior product $i_X : \Lambda^k \to \Lambda^{k-1}$ is an $\mathcal{F}$-linear antiderivation satisfying
  (1) $\mathcal{F}$-linear: $i_{fX+gY} = fi_X + gi_Y, \quad f, g \in \mathcal{F}, \quad$ and $X, Y \in \mathcal{X}$,
  (2) antiderivation: $i_X(\alpha_p \wedge \beta_q) = (i_X\alpha_p) \wedge \beta_q + (-1)^p \alpha_p \wedge (i_x\beta_q)$,
  (3) $i_X\sigma =< \sigma, X >, \quad \sigma \in \Lambda^1(M)$,
  (4) $i_Xi_X = 0, \quad$ and $i_X 1 = 0$.

- The Stokes formula:

$$\int_M d\omega = \int_{\partial M} \omega.$$

Exercises

1. Prove that the two-dimensional surface S^2 in E^3 is an analytic manifold, i.e., the transition function at the overlapping region is an analytic function.
2. Show that $CP(1)$ and S^2 are diffeomorphic, i.e., they have identical differential structure.
3. Analyse the differential structure of $RP(3)$, is $RP(3)$ orientable?
4. For given tangent field $X = \xi^i \partial_i$ and cotangent field $\sigma = \sigma_i dx^i$,

$$< \sigma, \; X >= \sigma_i \xi^i \; \in \mathcal{F}(M)$$

is it coordinate dependent?
5. For differentiable vector fields X and Y, show that
   (1) XY in general is not a differentiable vector field (Unless one of them is zero.).
   (2) $[X, \; Y]$ is a differentiable vector field.
6. Let cotangent field bases $\{\theta^i\}$ on manifold M be the dual bases of tangent field bases $\{e_i\}$, namely,

$$< \theta^i, \; e_j >= \delta^i_j,$$

prove that

$$<\theta^{i_1}\wedge\cdots\wedge\theta^{i_r};e_{j_1},\cdots,e_{j_r}>=det(<\theta^{i_\alpha},\ e_{j_\beta}>)=\delta^{i_1\cdots i_r}_{j_1\cdots j_r}.$$

7. For vector field $X=\xi^i e_i$ and r-form

$$\alpha_r=\frac{1}{r!}f_{i_1\cdots i_r}\theta^{i_1}\wedge\cdots\wedge\theta^{i_r},$$

show that

$$i_X\alpha_r=X|\alpha_r=\frac{1}{(r-1)!}f_{i_1\cdots i_r}\xi^{i_1}\theta^{i_2}\wedge\cdots\wedge\theta^{i_r}.$$

8. Prove

$$i_X(\alpha_p\wedge\beta_q)=(i_X\alpha_p)\wedge\beta_q+(-1)^p\alpha_p\wedge(i_X\beta_q),$$
$$d(\alpha_p\wedge\beta_q)=d\alpha_p\wedge\beta_q+(-1)^p\alpha_p\wedge d\beta_q.$$

9. For differential form in three-dimensional Euclidean space

$$\alpha=x^4dx+y^3dx\wedge dy+xyzdx\wedge dz,$$

find $d\alpha$.
10.For $\alpha=\frac{1}{2}f_{ij}dx^i\wedge dx^j,\beta=g_idx^i$, the exterior product

$$\gamma=\alpha\wedge\beta=\frac{1}{3!}h_{ijk}dx^i\wedge dx^j\wedge dx^k.$$

Find h_{ijk} in terms of $f_{ij},\ g_i$.
11.For gauge potential $A=A_idx^i$, field strength $F=dA=\frac{1}{2}F_{ij}dx^i\wedge dx^j$, find the explicit expressions of F_{ij} in terms of A_i.
12.Analyse the monopole in conventional Cartesian coordinates. In this case, gauge potential 1-form at the northern and southern parts are

$$A_\pm=\frac{q}{r}\frac{1}{z\pm r}(xdy-ydx).$$

Find the components of gauge potential A and field strength F.

Chapter 2

Transformation of Manifold, Manifolds with Given Vector Fields and Lie Group Manifold

Continuous mapping is one of the most important tools in the study of manifolds. In this chapter we look at mappings between manifolds and the induced mappings between tensor fields, with the emphasis on problems associated with transformation groups of manifolds. §2.2, §2.3 study the integrable conditions of vector fields and differential equations, and Frobenius Theorem. In §2.4 we introduce the Lie derivative operator acting on any tensor field on the manifold, it has a wide range of applications in analytic mechanics and in general relativity. In the last two sections we look at Lie group manifold and its invariant vector fields, orbit space and homogeneous space, which are very important in the study of manifolds.

2.1 Continuous mapping between manifolds and its induced mapping

Let f be a smooth mapping from one differentiable manifold M to another one N

$$\begin{aligned} f : M &\rightarrow N \\ p &\rightarrow q = f(p), \quad p \in M, \; q \in N. \end{aligned}$$

The induced linear mapping from the tangent space $T_p(M)$ at point p to the tangent space $T_q(N)$ at point q is called the tangent mapping f_*:

$$f_* : T_p(M) \rightarrow T_q(N)$$

i.e., the mapping takes the tangent vector X at point p on curve $x(t)$ on M, to a tangent vector f_*X at point q on curve $f(x(t))$ on N. For a given local coordinate system, it is written as

$$f_* : X = \xi^i(x)\frac{\partial}{\partial x^i}|_p \to f_*X = \xi^i(x(y))\frac{\partial y^k}{\partial x^i}\frac{\partial}{\partial y^k}|_{q=f(p)}.$$

The tangent mapping f_* maps the tangent space on M to the corresponding tangent space on N, and determines the properties of f at the neighborhood of point p. Hence, sometimes $df \equiv f_*$ is used. The tangent mapping has the following properties:

1). Linearity,

$$f_*(X+Y) = f_*X + f_*Y, \quad X, Y \in \mathcal{X}(M).$$

2). Commutes with Lie bracket,

$$[f_*X, f_*Y] = f_*[X, Y].$$

3). for two consecutive mappings,

$$(g \cdot f)_* = g_* \cdot f_*.$$

The mapping between two linear spaces can induce the dual mapping between their corresponding dual spaces. The dual mapping reverses the direction of the mapping and is usually called a pull back mapping. Mapping f between manifolds can induce a pull back mapping f^* between differential forms

$$\begin{aligned} f^* : T_p^*(M) &\leftarrow T_q^*(N) \\ f^*\omega &\leftarrow \omega \end{aligned}$$

i.e., mapping f^* pulls differential forms at point q on N back to differential forms at point p on M. In a given coordinate system, the mapping can be written out explicitly.

For the 0-forms which are functions on the manifold, function $\varphi(y)$ on manifold N can be pulled back to function $\psi(x)$ on manifold M by this pull back mapping:

$$f^*\varphi(y) = \varphi(f(x)) \equiv \psi(x),$$

as illustrated in the following diagram,

$$\begin{array}{ccccc} & M & \xrightarrow{f} & N & \\ \psi & \downarrow & \xleftarrow{f^*} & \downarrow & \varphi \\ & R & & R & \end{array}$$

where $y = f(x)$ and $\psi = f^*\varphi = \varphi \cdot f$.

From a given 1-form on N,

$$\omega_i(y)dy^i \in \Lambda^1(N),$$

the corresponding one form on M can be obtained as:

$$f^*\omega_i(y)dy^i = \omega_i(y(x))\frac{\partial y^i}{\partial x^k}dx^k \in \Lambda^1(M).$$

Similarly for a r-form ω_r

$$\omega_r = \frac{1}{r!}f_{i_1\cdots i_r}(y)dy^{i_1}\wedge\cdots\wedge dy^{i_r} \in \Lambda^r(N),$$

its pull back is:

$$f^*\omega_r = \frac{1}{r!}f_{i_1\cdots i_r}(y(x))\frac{\partial(y^{i_1},\cdots,y^{i_r})}{\partial(x^{k_1},\cdots,x^{k_r})}dx^{k_1}\wedge\cdots\wedge dx^{k_r} \in \Lambda^r(M).$$

The pull back mapping has the following properties:
1). Linearity,

$$f^*(\omega+\tau) = f^*\omega + f^*\tau.$$

2). Commutes with the exterior product,

$$f^*(\omega\wedge\tau) = f^*\omega\wedge f^*\tau.$$

3). Commutes with the exterior differential operator,

$$f^*d\omega = df^*\omega.$$

4). For two consecutive pull back mappings,

$$(g\cdot f)^* = f^*\cdot g^*.$$

As mentioned above, mapping f between manifolds can induce mappings between tensor fields. There are two types of induced mappings: One is the tangent mapping f_*, which pushes tangent vector fields forward. Another is the mapping f^*, which pulls differential forms back. f_* and f^* are dual to each other and satisfy

$$< f^*\omega, X >_p = < \omega, f_*X >_{q=f(p)},$$

where the left hand side and the right hand side are functions on different manifolds:

$$< f^*\omega, X > \in \mathcal{F}(M), \text{ and } < \omega, f_*X > \in \mathcal{F}(N),$$

but both sides have the same value at corresponding points(p and $q = f(p)$). In general, for contraction of r forms and r tangent vector fields, there is a similar relation

$$< f^*\omega; X_1,\cdots,X_r >_p = < \omega; f_*X_1,\cdots,f_*X_r >_q,$$

where $\omega\in\Lambda^r(N)$, $X_1,\cdots,X_r \in \mathcal{X}(M)$.

2.2 Integral submanifold and Frobenius theorem

Let f be a smooth mapping from a k-dimensional manifold M to an n-dimensional manifold N

$$f: \ M \rightarrow N$$
$$p \quad \rightarrow q = f(p) \in N, \quad \text{for } p \in M,$$

the corresponding tangent mapping is

$$f_* : T_p(M) \rightarrow f_* T_p(M) \subset T_q(N). \tag{2.1}$$

Mapping f is called a rank-r mapping at point p if the linear space $f_* T_p(M)$ is r-dimensional. Given a local coordinate, the coordinate in the neighborhood of point p is $x = \{x^i; i = 1, \cdots, k\}$ and that of point q is y=$\{y^i; j = 1, \cdots, n\}$. Suppose the rank of the Jacobi matrix of mapping f at point p is r.

Mapping $f_*|_p$ is subjective if rank $r = n \leq k$.

Mapping $f_*|_p$ is injective if rank $r = k \leq n$.

Suppose $n = k$, i.e., M and N are smooth manifolds of equal dimension, if at point p, the rank r of $f_*|_p$ is $r = n = k$, then mapping f is a one to one diffeomorphism at the neighborhood of point p.

In short, if there exists a smooth mapping f from manifold M to manifold N, the charácteristics of f at the neighborhood of a point $(p \rightarrow q = f(p))$ determine the corresponding characteristics of the mapping between tangent spaces, and vice versa. And the rank of mapping f at point p is equal to the rank of the tangent mapping $f_*(= df)$ at point p, which is equal to the dimension of the space $f_*(T_p(M))$. For injective mapping, the rank of f is equal to the dimension of M. For subjective mapping, the rank of f is equal to the dimension of target space N.

Mapping f is called an immersion if $f_*|_p$ is injective at any point on M. Immersion mapping f is called an embedding if f itself is a $1-1$ mapping. In this case, manifold M and its image $f(M)$ are homeomorphic and (M, f) is called an embedded submanifold of N. Or simply M is a submanifold of N.

Figure 2.1 Two mappings from segment I to R^2 . a). an injective but not an immersion; b). an immersion but not an embedding.

For an embedded submanifold (M, f), because the mapping f is $1-1$ injective, the differential structure of M can be transplanted to that of the image $f(M)$, so that M and $f(M)$ are diffeomorphism. In addition, as $f(M) \subset N$, $f(M)$ has a topology induced from that of N. In general, the topology of $f(M)$ induced from

N is less refined than that brought to $f(M)$ by f. When the embedding topology is equivalent to the induced topology from N, the embedded submanifold is called a regular submanifold. For regular submanifold, M is diffeomorphism to $f(M)$. For simplicity, we usually do not make a distinction between M and $f(M)$.

Let M be a closed subset of N and if for any point $p \in M$, the coordinate chart (U, φ) containing point p in the complete coordinate atlas of N satisfies

$$M \cap U = \{x \in U | \varphi^i(x) = 0, i = k+1, k+2, \cdots, n\},$$

then M is called a k-dimensional closed submanifold of N. For instance, the $(n-1)$-dimensional sphere S^{n-1} in $\mathbf{R}^n$ is the closed submanifold of $\mathbf{R}^n$. Closed submanifold is a regular submanifold. Any compact smooth manifold M can be smoothly embedded in $\mathbf{R}^n$ for sufficiently large n. This is the reason why it is so important to study submanifolds in $\mathbf{R}^n$. There is an embedding theorem (see [VII] B.A.Dubrovin et al. Modern Geometry, Part II , P.90).

Theorem 2.1 (Whitney) *Any smooth connected closed n-dimensional manifold M can be smoothly embedded in $\mathbf{R}^{2n+1}$ and smoothly immersed in $\mathbf{R}^{2n}$.*

Let M be an embedded submanifold in N, the tangent space $T_p(M)$ of the submanifold M is a subspace of the tangent space $T_p(N)$ of manifold N:

$$T_p(M) \subset T_p(N) \qquad (M \subset N).$$

However, the inverse is not true, i.e., not only the vector in $T_p(N)$ is not a vector in $T_p(M)$ in general, but also it has not a unique image in $T_p(M)$, because for a generic manifold orthogonal projections are not well defined.

For submanifold M, the element σ in the cotangent space $T_p^*(M)$ is a homomorphism mapping $\sigma \in$Hom$(T_p(M), \mathbf{R})$, i.e.,

$$\sigma : T_p(M) \to \mathbf{R}.$$

But an element in cotangent space $T_p^*(M)$ can not be related to a definite element in cotangent space $T_p^*(N)$, as elements σ in $T_p^*(M)$ are not well defined for vectors not in $T_p(M)$. However, to the cotangent vector τ in $T_p^*(N)$,

$$\tau \in Hom(T_p(N), \mathbf{R}) \Longrightarrow f^*\tau \in Hom(T_p(M), R).$$

i.e., $T_p(M) \subset T_p(N)$ and the mapping from $T_p(M)$ to $\mathbf{R}$ can be determined, i.e., cotangent vectors in $T_p^*(M)$ are determined by cotangent vectors in $T_p^*(N)$. That is to say, when a k-dimensional manifold M is embedded in an n-dimensional manifold N, the tangent space is pushed forward and the cotangent space is pulled back. And there is an embedding f such that

$$\begin{array}{llll} f: & M & \longrightarrow & N \\ f_*: & X & \longrightarrow & f_*X \quad (X \in \mathcal{X}(M)) \\ f^*: & f^*\omega & \longleftarrow & \omega \quad\ \ (\omega \in \Lambda^*(N)). \end{array}$$

Although tangent space and cotangent space are dual to each other, they play different roles in the case of the embedding of manifolds. In this section, we use tangent vector field to study submanifolds. For simplicity, we consider the one-dimensional case first.

As is known, for a given curve on manifold N a tangent vector on each point of the curve is given. On the other hand, for a given vector field X starting from an arbitrary point p on N, can we find a curve such that the tangent vector at each point on the curve is just the vector field X at that point? Before answering this question, we have to look at the tangent vector field X at the neighborhood of any point p. If $X_p = 0$ at point $p \in N$, p is called a critical point of the tangent vector field X. Critical points of differentiable tangent vector fields on a manifold are closely related to the topology of the whole manifold, which will be one of the main topics in the last several chapters of this book. Although tangent fields near critical points are very complicated, they are fairly simple at the neighborhood of non-critical points called generic points. Let $p \in N$ be a generic point of the tangent vector field X. It can be proved that one and only one smooth curve through p can be found such that the tangent vectors on the curve are just the vector fields X at the points on the curve. This curve is called the integral curve of the vector field X. In the local coordinate system at the neighborhood of p, vector field X can be written as

$$X(x) = \xi^i(x)\frac{\partial}{\partial x^i}.$$

For the integral curve $x(t)$ (where t is a real parameter) through p

$$x : t \longrightarrow x(t)$$

and $x(0)$ is just the coordinate of p. For the tangent vector of this curve to be the given tangent vector field, it is equivalent to solving the following differential equation

$$\frac{dx^i(t)}{dt} = \xi^i(x(t)), \quad i = 1, \cdots, n\ , \tag{2.2}$$

with the initial condition $x^i(0)$ =the coordinates of p. This is a system of first order differential equations which has a unique solution at the neighborhood of p. The integral curve is a one-dimensional submanifold of the manifold N which has the following properties(for a given vector field):

1). Different integral curves do not intersect as a result of the uniqueness of solutions of Eq. 2.2.

2). There is one integral curve for each point (except the critical point of the vector field) on manifold N. The set of all integral curves forms itself an $(n-1)$-dimensional manifold.

3). At the neighborhood of point p, there exists a coordinate system such that the coordinate axis is just the integral curve. In other words, the coordinate system can be chosen such that the vector field is simply $X = \frac{\partial}{\partial x}$.

In conclusion, the maximum integral curves of vector fields are the immersion submanifolds of manifold N and forms a partition of N, where the dimension of the submanifold is one or zero. The later case occurs only at critical zero point of the vector fields.

In what follows, the argument above will be generalized to the case of integral manifold of k vector fields. One says there is a k-dimensional smooth distribution S^k on the manifold N if there is a k-dimensional subspace S_p^k of $T_p(N)$, at each point p of manifold N, which varies smoothly with respect to p.

If in the neighborhood U of any point $p \in N$, there are k differentiable tangent vector fields $X_1, X_2, \cdots, X_k$ which are linearly independent everywhere such that $S^k(x)$ is spanned by $X_1(x), X_2(x), \cdots, X_k(x)$, then the set $\{ X_1, X_2, \cdots, X_k\}$ is called the base of this k-dimensional distribution $S^k(x)$. In which case $S^k(x)$ is called a differentiable k-dimensional distribution on N, written as

$$S^k(x) = \{X_1(x), X_2(x), \cdots, X_k(x)\}.$$

If there exist $(n-k)$ linearly independent functions $\{f^a, a = 1, 2, \cdots, n-k\}$ on N satisfying

$$< df^a, X_i >= X_i(f^a) = 0, \quad i = 1, 2, \cdots, k, \tag{2.3}$$

these $(n-k)$ functions $\{f^a\}$ are called the first integrals of the k-dimensional distribution S^k and there is an integral manifold

$$M = \{x | f^a(x) = \text{const}, \quad a = 1, 2, \cdots, n-k\}. \tag{2.4}$$

We have $df = 0$ when restricted to the submanifold M as is clear in Eq. 2.3, although $df \neq 0$ on the whole manifold N. M is called the integral submanifold determined by the distribution S^k. This integral submanifold is a connected submanifold of N. For an arbitrary point p (the coordinates are $x^1_{(0)}, x^2_{(0)}, \cdots, x^n_{(0)}$) on N, there exists a k-dimensional integral submanifold satisfying:

$$f^a(x^1, \cdots, x^n)|_0 = f^a(x^1_{(0)}, \cdots, x^n_{(0)}), \qquad a = 1, \cdots, n-k. \tag{2.5}$$

For a k-dimensional integral submanifold M, a local coordinate system can be chosen such that at the neighborhood of a point, M is spanned by the following base vectors:

$$\{\frac{\partial}{\partial x^i}, \quad i = 1, 2, \cdots, k\}. \tag{2.6}$$

For a k-dimensional distribution on the manifold N, is there a k-dimensional integral manifold M (Eq. 2.4) for a generic point p on N such that Eq. 2.5 and Eq. 2.6 are satisfied? For $k = 1$ the answer is definitely positive. For $k > 1$, the answer is yes only if additional conditions are satisfied. This is the well known Frobenius theorem. We now try to formulate this additional condition. If at the neighborhood of a point

p Eq. 2.5 and Eq. 2.6 are satisfied, the bases of the distribution S^k can be expressed as

$$X_\alpha = \sum_{i=1}^{n} \xi_\alpha^i(x)\partial_i \quad (\alpha = 1, \cdots, k).$$

It is easy to prove that

$$[X_\alpha, X_\beta] = \sum_{\gamma=1}^{k} C_{\alpha\beta}^\gamma(x) X_\gamma \subset S^k, \quad \forall\, X_\alpha,\ X_\beta \subset\ S^k. \tag{2.7}$$

The distribution S^k is called involutive if it satisfies the Eq. 2.7

For the coordinate bases in Eq. 2.6, we have:

$$[\partial_i, \partial_j] = 0 \qquad (i, j, = 1, \cdots, k).$$

This equation is not invariant under arbitrary(nonsingular) coordinate transformations, although Eq. 2.7 is. A distribution S^k is integrable if in some neighborhood of each point of N there is a coordinate system x^i such that $x^j = const, j = k+1, ..., n$ define an integral manifold of S^k. The integrable condition of distribution S^k is that S^k should be involutive (i.e., Eq.2.7 is satisfied.). This is the Frobenius theorem.

Theorem 2.2 (Frobenius) *If a k-dimensional distribution $S^k = \{X_1, \cdots, X_k\}$ on manifold N satisfies involutive condition*

$$[X_\alpha, X_\beta] \subset S^k, \quad \forall\, X_\alpha,\ X_\beta \subset\ S^k,$$

then at the neighborhood of each point p on N, there exists a coordinate system such that

1).$\{\frac{\partial}{\partial x^i}, i = 1, 2, \cdots, k\}$ form the local bases of S^k.

2).$\{x^j$ =const, $j = k+1, \cdots, n\}$(for example $f^a = x^{k+a} = x^j$) defines an integral manifold of S^k. And any integral function can be expressed as the function of these $x^j (j = k+1, \cdots, n)$.

3).Through each point $p \in N$, there is a unique k-dimensional maximum integral manifold M. Any integral manifold through p is an open submanifold of M.

4).The set of maximum integral manifold for all points on N constitutes a family of manifolds, called a foliation. The foliation itself is an $(n-k)$-dimensional manifold (quotient manifold N/S^k).

We have pointed out that the constitution of the integral manifold is local. In general, the global integral manifold can not be found.

2.3 Integrability of differential equations and Frobenius theorem in terms of differential forms

The existence of integral manifold can be analyzed by differential forms. Due to the simplicity of differential forms and the exterior derivative operation, it is convenient to study the existence of integral manifold by differential forms. Partial differential equations can usually be written as a system of differential forms. The integrability of differential equations can be treated in the same way.

The straightforward problem is to find the integral manifold (M, f) satisfying the Pfaff equation on an n-dimensional manifold N

$$\omega = \omega_i(x)dx^i = 0,$$

i.e., f is an embedding

$$f : M \to N$$

with:

$$f^*\omega = 0.$$

There are two questions to ask:
1). Is there a coordinate system $\{y^i\}$ such that $\omega = dy^1$? In general, the answer is negative. But if

$$d\omega = 0, \tag{2.8}$$

there exists a function $\varphi(x)$ at the neighborhood U of point p such that

$$\omega = d\varphi, \tag{2.9}$$

and φ =const is an integral manifold of $\omega = 0$, which is called the hyper surface on N. Eq. 2.8 and Eq. 2.9 are special cases of the following theorem:

Theorem 2.3 (Poincare) *For a r-form $(r \geq 1)$ α, if $d\alpha = 0$, then at the neighborhood of a point there exists a $(r-1)$-form β such that $\alpha = d\beta$.*

2). Can we find two functions u and v such that

$$\omega = udv,$$

where v =const is the integral surface and u^{-1} is called the integral factor. It can be proved that the answer is yes if and only if:

$$\omega \wedge d\omega = 0. \tag{2.10}$$

This is called the integrable condition of the equation $\omega = 0$, which is also called the Frobenius condition. Next we give two examples.

Example 1. On R^2, for the 1-form

$$\omega = P(x,y)dx + Q(x,y)dy,$$

then

$$d\omega = 0 \quad \Longrightarrow \quad \frac{\partial P}{\partial y} = \frac{\partial Q}{\partial x}.$$

This is the integrable condition of $\omega = 0$, i.e., there exists a function $\varphi(x, y)$ such that $\omega = d\varphi$.

Example 2. If $d\omega \neq 0$, but on R^2 the Eq. 2.10 is always satisfied, i.e., $\omega \wedge d\omega = 0$. Then for the 1-form ω on the two-dimensional manifold there exists a function $\mu(x,y)$ such that

$$\mu(x,y)(P(x,y)dx + Q(x,y)dy) = d\varphi(x,y),$$

where

$$\mu(x,y) = \frac{1}{P}\frac{\partial\varphi}{\partial x} = \frac{1}{Q}\frac{\partial\varphi}{\partial y}$$

is the integral factor.

For the higher-dimensional case, the integral factor may not exist. In the following, we will study these cases.

Let $S^k(x)$ is a k-dimensional involutive distribution on manifold N. The bases of $S^k(x)$ are $\{X_1, \cdots, X_k\}$. As these bases are independent, they can be expanded to the bases of the tangent space at point p on N: $\{X_1, \cdots, X_k, X_{k+1}, \cdots, X_n\}$. The dual bases in cotangent space are $\{\theta^1, \cdots, \theta^k, \theta^{k+1}, \cdots \theta^n\}$, and we have: $<\theta^i, X_j>= \delta^i_j (i,j = 1, 2, \cdots, n)$. The set of 1-form equations $\{\theta^{k+1} = 0, \cdots, \theta^n = 0\}$ is a system of Pfaff equations, and $S_{n-k} = \{\theta^{k+1}, \cdots, \theta^n\}$ is called the $(n-k)$-dimensional co-distribution, which satisfying

$$<\theta^a, X_b>= 0 \quad (a = k+1, \cdots, n \text{ and } b = 1, \cdots, k).$$

We now try to find out the integral manifold determined by the $(n-k)$-dimensional co-distribution, i.e., look for functions g^a_b and f^a such that

$$\theta^a = \sum_b g^a_b df^b \qquad (a, b = k+1, \cdots, n),$$

then $M = \{x | f^b(x) = \text{const}, b = k+1, \cdots, n\}$ is the integral manifold of the co-distribution.

Theorem 2.4 *If the distribution $S^k(x)$ is involutive, then the elements $\{\theta^a\}$ of its dual co-distribution $S_{n-k}(x)$ satisfy*

$$d\theta^a = \sum_{b=k+1}^{n} \tau^a_b \wedge \theta^b \quad (a = k+1, \cdots, n), \tag{2.11}$$

where $\tau^a_b \in \Lambda^1(N)$ are 1-forms.

Proof: As

$$<d\theta^a; X_i, X_j>= X_i <\theta^a; X_j> - X_j <\theta^a; X_i> - <\theta^a; [X_i, X_j]>,$$

which will be proved in the next section (see Eq. 2.34), we have $<d\theta^a; X_i, X_j>= 0$, if $[X_i, X_j] \in S^k$ for any $X_i, X_j \in S^k$ and θ^a are elements of the doal co-distribution. Therefore, $d\theta^a$ also belong to the co-distribution as Eq. 2.11.

The bases set $\{\theta^\alpha\}_{k+1}^n$ of the co-distribution are linearly independent, so,

$$\Sigma = \theta^{k+1} \wedge \theta^{k+2} \wedge \cdots \wedge \theta^n \neq 0.$$

This Pfaff system is completely integrable iff

$$d\theta^a \wedge \Sigma = 0 \quad (\alpha = k+1, \cdots, n).$$

This is called the Frobenius local integrable condition of co-distribution,which is a result of S^k being involutive. An equivalent condition is $d\Sigma = \tau \wedge \Sigma$, where $\tau \in \Lambda^1(N)$.

In general, if θ^a are elements of co-distribution, by defining

$$\omega = \sum_{a=k+1}^{n} \tau^a \wedge \theta^a, \quad \text{where} \quad \tau^a \in \Lambda^*(N), \tag{2.12}$$

then we have

$$\omega(X_1, \cdots, X_k) = 0, \quad \text{where} \quad X_1, \cdots, X_k \in S^k.$$

Hence a subalgebra I of the exterior algebra $\Lambda^*(N)$ can be defined

$$I = \{\omega | \omega(X_1, \cdots, X_k) = 0, \text{ where } X_1, \cdots, X_k \in S^k\}.$$

It is easy to check that $I \subset \Lambda^*(N)$ is an ideal of $\Lambda^*(N)$. That is, if $\omega \in I$, then

$$\omega \wedge \tau \in I, \qquad \forall \tau \in \Lambda^*(N).$$

I is called the ideal produced by the Pfaff equations $\{\theta^a = 0\}$. If furthermore

$$d\theta^a \in I, \tag{2.13}$$

I is called a closed ideal of $\Lambda^*(N)$. The condition Eq. 2.13 has two equivalent formulations

1). Let $\Sigma = \theta^{k+1} \wedge \theta^{k+2} \wedge \cdots \wedge \theta^n$, then

$$d\theta^a \wedge \Sigma = 0.$$

2). There is

$$d\Sigma = \tau \wedge \Sigma, \quad \tau \in \Lambda^1(N). \tag{2.14}$$

The Frobenius theorem states that if the k-dimensional distribution S^k is involutive, then there is a unique k-dimensional maximum integral manifold for each point $p \in N$. In addition, the $(n-k)$-dimensional ideal that maps the involutive S^k to zero must be a closed ideal. Therefore the Frobenius theorem can be recapitulated as

Theorem 2.5 (Frobenius) *Pfaff equations $\{\theta^a = 0\}$ are completely integrable, if at the neighborhood of a generic point the ideal generated by the co-distribution $\{\theta^a\}_{k+1}^n$ is a closed ideal.*

In this case, there exists an integral manifold determined by the $(n-k)$-dimensional co-distribution at the neighborhood of generic points, i.e., we have functions g^a_b and f^a such that

$$\theta^a = \sum_b g^a_b df^b \qquad (a, b = k+1, \cdots, n).$$

Then these functions $\{f^a = \text{const}\}$ are solutions of $\{\theta^a = 0\}$. Equations $\{f^a = \text{const}\}$ determine the k-dimensional integral manifold M on N :

$$M = \{x | f^b(x) = \text{const}, b = k+1, \cdots, n\}.$$

M is also called the solution manifold.

To understand the significance of the Frobenius theorem, we give some simple examples.

Example 1. The Pfaff equation in three dimensional vector space R^3,

$$\omega = P(x,y,z)dx + Q(x,y,z)dy + R(x,y,z)dz = 0. \tag{2.15}$$

Without losing generality, take $R(x,y,z) \neq 0$. The above equation can be written as

$$dz = -\frac{P}{R}dx - \frac{Q}{R}dy.$$

This equation has solution $z = z(x,y,c)$ only if

$$\frac{\partial^2 z}{\partial x \partial y} = \frac{\partial^2 z}{\partial y \partial x}.$$

which is,

$$P(\frac{\partial R}{\partial y} - \frac{\partial Q}{\partial z}) + Q(\frac{\partial P}{\partial z} - \frac{\partial R}{\partial x}) + R(\frac{\partial Q}{\partial x} - \frac{\partial P}{\partial y}) = 0. \tag{2.16}$$

This is the integrable condition of Eq. 2.15. If the above equation is satisfied, there exist surfaces on R^3 defined by

$$f(x,y,z) = \text{const}, \tag{2.17}$$

and $df = 0$ and $\omega = 0$ are equivalent ($\omega = gdf$). The surfaces defined by Eq. 2.17 are called integral surfaces of the Pfaff equation.

Integrable condition Eq. 2.16 can be written in the form of Eq. 2.14

$$d\omega = \theta \wedge \omega, \tag{2.18}$$

where θ is another Pfaff form

$$\theta = a(x,y,z)dx + b(x,y,z)dy + c(x,y,z)dz.$$

From Eq. 2.18, it can be proved that

$$\frac{\partial R}{\partial y}-\frac{\partial Q}{\partial z}=bR-cQ,\quad \frac{\partial P}{\partial z}-\frac{\partial R}{\partial x}=cP-aR,\quad \frac{\partial Q}{\partial x}-\frac{\partial P}{\partial y}=aQ-bP.$$

consequently Eq. 2.16 is satisfied. From the analysis above we can see that the Frobenius condition is, in general, the integrable condition of differential equations.

To study integrable condition for the system of non-homogeneous partial differential equations, differential forms and the Frobenius Theorem are very convenient. In the following, we apply the Frobenius theorem to a system of differential equations by the prolongation method developed by Estabrook and Wahlquist[13].

Example 2. Let functions $f(x,y)$ and $g(x,y)$ satisfy the following equations

$$\begin{aligned}&\tfrac{\partial f}{\partial x}+P_1f+Q_1g=R_1,\\&\tfrac{\partial f}{\partial y}+P_2f+Q_2g=R_2,\\&\tfrac{\partial g}{\partial x}+U_1f+V_1g=W_1,\\&\tfrac{\partial g}{\partial y}+U_2f+V_2g=W_2,\end{aligned}\tag{2.19}$$

where $P_i,Q_i,R_i,U_i,V_i,W_i(i=1,2)$ are the given functions of x,y. To see if Eq. 2.19 are integrable or not, we rewrite it in terms of differential forms, let

$$\begin{aligned}\alpha&=df+Pf+Qg-R,\\\beta&=dg+Uf+Vg-W,\end{aligned}$$

where $P=P_1dx+P_2dy$, and Q,R,U,V,W take similar forms as P.

Through prolongation, manifold N is four-dimensional and consists of (x,y,f,g). On the solution manifold, $\alpha=\beta=0$. If solutions exist, (α,β) must be a closed ideal. That is to say,

$$\begin{aligned}d\alpha\wedge\alpha\wedge\beta=0,\\d\beta\wedge\alpha\wedge\beta=0.\end{aligned}\tag{2.20}$$

As $P,Q,\cdots,W$ are 1-forms on the two-dimensional manifold, so

$$Q\wedge dP=0,\cdots \text{ and similar equations.}$$

From Eq. 2.20, it is easy to prove the integrable condition of Eq. 2.19 is

$$\begin{aligned}&\tfrac{\partial P_2}{\partial x}-\tfrac{\partial P_1}{\partial y}+Q_1U_2-Q_2U_1=0,\\&\tfrac{\partial Q_1}{\partial y}-\tfrac{\partial Q_2}{\partial x}+Q_2V_1+P_2Q_1-Q_1V_2-P_1Q_2=0,\\&\tfrac{\partial R_2}{\partial x}-\tfrac{\partial R_1}{\partial y}+Q_1W_2-Q_2W_1+P_1R_2-P_2R_1=0,\\&\tfrac{\partial U_2}{\partial x}-\tfrac{\partial U_1}{\partial y}+U_1P_2-U_2P_1+V_1U_2-V_2U_1=0,\\&\tfrac{\partial V_2}{\partial x}-\tfrac{\partial V_1}{\partial y}+U_1Q_2-U_2Q_1=0,\\&\tfrac{\partial W_2}{\partial x}-\tfrac{\partial W_1}{\partial y}+U_1R_2-U_2R_1+V_1W_2-V_2W_1=0.\end{aligned}$$

2.4 The flow of vector fields, one parameter local Lie transformation groups and Lie derivative

For functions on the manifold, differentiation along a given curve can be defined. However, there is no such a natural way to define differentiation along a curve for vector fields, because vectors at different points on the manifold belong to different vector spaces.

Exterior differential operator is a generalization of the total differentiation of functions on manifold. However, exterior differential operator d acts only on differential forms. The key issue is to define a differential operator which can act on any kind of tensor fields on manifold.

For any tensor field on manifold, in general, there are two approaches to deal with differentiation or derivative of tensor fields with respect to the base manifold:

1). For a given vector field X on the manifold, define a one parameter Lie transformation group such that tensor fields on the same integral curve can be compared subsequently, so Lie derivative of tensor field K, $L_X K$ can be defined.

2). Connection coefficients are introduced at each point on the manifold, such that a relationship between tangent spaces at and near a given point is established. And covariant derivative ∇K of tensor field K can be defined such that ∇K is still a tensor field.

We first study Lie derivative in this section and leave covariant derivative to the next chapter.

Given a differentiable vector field X on manifold M , there is a diffeomorphism transformation on the manifold M

$$\begin{aligned} \varphi_t : M &\longrightarrow M \\ p &\longrightarrow q = \varphi_t(p) \end{aligned}$$

such that $\varphi_t(p)$ is the integral curve of vector field X starting from point p. In other words, vector X_q is tangent to $\varphi_t(p)$, where point p is assumed to be a generic point, i.,e., $X_p \neq 0$. In a local coordinate system at the neighborhood of p, vector field can be expressed as

$$X = \xi^i(x)\partial_i.$$

Mapping φ_t of the integral curve through p is

$$\varphi_t : x \longrightarrow y(x,t)$$

where x and y are coordinates of points p and q respectively. The point p corresponds to $t = 0$:

$$y^i(x,0) = x^i.$$

$y(x,t)$ is the integral curve of the vector field X through point p and is a solution of the following ordinary differential equations(ODE) (The coordinate x of p is taken as a fixed parameter.)

$$\frac{dy^i}{dt} = \xi^i(y), \qquad i = 1, 2, \cdots, n.$$

From the fundamental theory of ODE, the solution which satisfies the following initial condition

$$y^i(x, 0) = x^i$$

is unique. Except at the critical point of vector field X, there is only one integral curve through each point p on the manifold. For a sufficiently small parameter t, φ_t gives a one to one mapping on M and has:

$$\varphi_{t+s}(p) = \varphi_t(\varphi_s(p)), \qquad \forall t, s \in R.$$

The set of φ_t forms a one parameter Lie transformation group which is Abelian, and the one parameter transformation is called the flow.

The analysis above shows that given a vector field X on manifold M, a one parameter Lie transformation group is also given, and vice versa. Given a one parameter Lie transformation group φ_t on M, a vector field X on the manifold M can be induced in the following way: $\varphi_t(p)$ is the curve through a point p on M, X_p is chosen as the tangent vector of the curve at point p. At each point p on M, a definite tangent vector can be chosen so that a tangent vector field $X \in \mathcal{X}(M)$ is established on M.

Given a vector field X on manifold M, the diffeomorphism transformation φ_t on M is determined. Mapping between manifolds induces the mapping between tensor fields on the manifolds such that differential forms ω are pulled back as $\varphi_t^*\omega$ and tangent vector fields Y are pushed forward as $\varphi_{t*}Y$. Hence, given a vector field X, Lie derivative L_X of any tensor K is defined as:

$$L_X K|_p = \lim_{t \to 0} \frac{1}{|t|}(\varphi_t^* K - K)|_p, \tag{2.21}$$

or

$$L_X K|_p = \lim_{t \to 0} \frac{1}{|t|}(\varphi_{t*} K - K)|_p. \tag{2.22}$$

For differential forms or covariant tensors, Eq. 2.21 is used, i.e., Lie derivative is obtained by comparing the tensor at point x with the tensor at point y which is pulled back to x. For contravariant tensor fields, because the tensor at point y must be pushed along the integral curve to point x, Eq. 2.22 should be applied. Lie derivative so defined keeps the type of tensor fields invariant and commutes with the contraction of tensor fields, it is a kind of directional derivative and is independent of the choice of coordinates on the manifold.

Lie derivative has been applied extensively in general relativity. General relativity studies invariant properties of physics systems under generalized coordinate transformations, i.e., under diffeomorphism transformation of space-time manifold. Given a vector field on the space-time manifold, a one parameter diffeomorphism transformation of the space-time manifold is determined, and Lie derivative of various tensor fields on the space-time manifold can be defined. In the following, we give some examples of Lie derivatives on certain tensor fields.

Example 1. For a scalar field $f(x) \in \Lambda^0(M) = \mathcal{F}(M)$, let $X = \xi^i(x)\partial_i$ be a given vector field, $y(x,t)$ be its integral curve through x, $\frac{dy^i}{dt}|_p = \xi^i(x)$.

$$\varphi_t^* f = f(y(x,t)),$$

$$\begin{aligned} L_X f|_p &= \lim_{t\to 0} \frac{f(y(x,t)) - f(y(x,0))}{t} \\ &= \frac{\partial f}{\partial y^i}\frac{dy^i}{dt}|_{t=0} \\ &= \frac{d}{dt} f|_{t=0} = Xf|_p, \end{aligned}$$

here Xf is called the directional derivative of function f along tangent vector X, or Lie derivative with respect to X, which is written as $L_X f = Xf \in \mathcal{F}(M)$.

Example 2. For one form $\omega = \omega_i(y)dy^i \in \Lambda^1(M)$,

$$\varphi^*\omega = \omega_i(y(x,t))\frac{\partial y^i}{\partial x^j}dx^j.$$

Hence,

$$\begin{aligned} L_X\omega|_p &= \left[\frac{\partial \omega_i}{\partial y^k}\frac{dy^k}{dt}\frac{\partial y^i}{\partial x^j}dx^j + \omega_i\frac{\partial}{\partial x^j}(\frac{dy^i}{dt})dx^j\right]|_p \\ &= (\omega_{j,k}\xi^k + \omega_i\xi^i_{,j})dx^j|_p, \end{aligned}$$

or

$$L_X\omega_j \equiv (L_X\omega)_j = \omega_{j,k}\xi^k + \omega_i\xi^i_{,j}.$$

Example 3. For a vector field $Y = \eta^i(y)\frac{\partial}{\partial y^i}$, to get its Lie derivative, vector at point y must be pushed forward to x along the integral curve and compared with vector at x. For the push-forward mapping $\varphi_* : y \to x = x(y,\lambda)$ must be considered. In this case, $\frac{dx^i}{d\lambda}|_p = -\xi^i(p)$, but $\frac{dy^i}{d\lambda}|_p = \xi^i(p)$. Therefore,

$$\varphi_* Y|_p = \eta^i(y(x,\lambda))\frac{\partial x^j(y,\lambda)}{\partial y^i}\frac{\partial}{\partial x^j}|_p,$$

$$\begin{aligned} L_X Y|_p &= [\frac{\partial \eta^i}{\partial y^k}\xi^k\delta^j_i\frac{\partial}{\partial x^j} + \eta^i\frac{\partial}{\partial y^i}(\frac{dx^j}{d\lambda})\frac{\partial}{\partial x^j}]|_p \\ &= (\eta^j_{,k}\xi^k - \eta^i\xi^j_{,i})\frac{\partial}{\partial x^j}|_p = [X,Y]|_p, \end{aligned}$$

or

$$(L_X Y)^j \equiv L_X \eta^j = \eta^j_{,k}\xi^k - \eta^i \xi^j_{,i}.$$

Similarly, the Lie derivative for a general tensor field $K \in \mathcal{T}^r_s(M)$ is

$$(L_X K)^{i_1\cdots i_r}_{k_1\cdots k_s} = \xi^j K^{i_1\cdots i_r}_{k_1\cdots k_s,j} + K^{i_1\cdots i_r}_{jk_2\cdots k_s}\xi^j_{,k_1} + \cdots - K^{ji_2\cdots i_r}_{k_1\cdots k_s}\xi^{i_1}_{,j} - \cdots.$$

The first term on the right hand side is $XK^{i_1\cdots i_r}_{k_1\cdots k_s}$, then, each covariant tensor index contributes a term with plus sign, and each contravariant tensor index contributes a term with minus sign. In conclusion, Lie derivative has the following properties:

1). L_X is a linear operator and does not change the type of tensor field.

$$L_X : \mathcal{T}^r_s(M) \to \mathcal{T}^r_s(M),$$

$$L_X(aK + bK') = aL_X K + bL_X K', \qquad K, K' \in \mathcal{T}^r_s(M), \quad a, b \in \mathbf{R}.$$

For example,

$$\begin{aligned} L_X(aY + bZ) &= [X, aY + bZ] = a[X, Y] + b[X, Z] \\ &= aL_X Y + bL_X Z, \quad X, Y, Z \in \mathcal{X}(\mathcal{M}). \end{aligned}$$

2). L_X is a derivative and satisfies the Leibniz rule.

$$L_X(K \otimes K') = (L_X K) \otimes K' + K \otimes (L_X K'), \qquad \forall K, K' \in \mathcal{T}(M),$$

$$L_X(T^r_s(\omega_1, \cdots, \omega_r; X_1, \cdots, X_s))$$

$$= (L_X T^r_s)(\omega_1, \cdots, \omega_r; X_1, \cdots, X_s) + T^r_s(L_X\omega_1, \cdots, \omega_r; X_1, \cdots, X_s) + \cdots.$$

Example (i) There exists the following Jacobi identity among three vector fields:

$$L_X[Y, Z] = [L_X Y, Z] + [Y, L_X Z].$$

Example (ii) For $\omega, \sigma \in \Lambda^*(M)$, $\quad f \in \mathcal{F}(M)$,

$$L_X(f\omega) = fL_X\omega + (L_X f)\omega, \tag{2.23}$$

$$L_X(\omega \wedge \sigma) = (L_X\omega) \wedge \sigma + \omega \wedge L_X\sigma.$$

Example (iii)

$$L_X(<\omega, Y>) = <L_X\omega, Y> + <\omega, L_X Y>.$$

3). $L_X f = Xf, \qquad f \in \mathcal{F}(M)$.

4). $L_X Y = [X, Y], \qquad X, Y \in \mathcal{X}(M)$.

The first two properties show that L_X is the derivative acting on tensor field algebra $\mathcal{T}(M)$, such that the rule of Lie derivative for higher order tensor fields can be induced from that for the lower order tensor fields. The last two properties give the rule of Lie derivative on the lower order tensor fields (function f and tangent vector field Y), and Lie derivative on all orders of tensor fields in $\mathcal{T}(M)$ can be obtained subsequently. These four properties can also be used to define Lie derivative.

It is easy to prove that Lie derivative satisfies:

$$[L_X, L_Y] = L_{[X,Y]}, \qquad X, Y \in \mathcal{X}(M),$$

$$L_{aX+bY} = aL_X + bL_Y, \qquad a, b \in \mathbf{R}, \tag{2.24}$$

and commutes with the 'pull back' mapping, i.e., for smooth mapping φ between manifolds M, N

$$\begin{aligned} \varphi : M &\rightarrow N \\ X &\rightarrow \varphi_* X \\ \varphi^* \omega &\leftarrow \omega \end{aligned}$$

and Lie derivative satisfies

$$L_X \varphi^* \omega = L_{\varphi_* X} \omega.$$

Lie derivative L_X is a differential operator acting on all tensor fields including differential forms, while exterior differential operator d acts on differential forms only. Differential forms $\Lambda^*(M)$ are the most important tensor fields and make up the exterior algebra, so we shall consider the derivative algebra on exterior algebra in the following. The derivative D of the exterior algebra, obeying the Leibniz rule, is a linear mapping between differential forms on manifold M,

$$D : \Lambda^s(M) \rightarrow \Lambda^{s+k}(M), \tag{2.25}$$

satisfying

$$D(\omega \wedge \sigma) = (D\omega) \wedge \sigma + \omega \wedge (D\sigma), \tag{2.26}$$

or

$$D(\omega \wedge \sigma) = D(\omega) \wedge \sigma + (-1)^r \omega \wedge (D\sigma), \tag{2.27}$$

where ω, σ are differential forms and $\omega \in \Lambda^r(M)$ is a r-form. The linear mapping satisfying Eq. 2.25 and Eq. 2.26 is called the k-derivative on differential forms. And the linear mapping satisfying Eq. 2.25 and Eq. 2.27 is called the k-skew-derivative on differential forms. The exterior differential is a 1-skew-derivative. Lie derivative is a 0-derivative. For example, for $\omega \in \Lambda^p(M)$, $\sigma \in \Lambda^q(M)$,

$$d(\omega \wedge \sigma) = d\omega \wedge \sigma + (-1)^p \omega \wedge d\sigma \in \Lambda^{p+q+1}(M),$$

but

$$L_X(\omega \wedge \sigma) = L_X\omega \wedge \sigma + \omega \wedge L_X\sigma \in \Lambda^{p+q}(M).$$

Derivatives and skew derivatives on the exterior algebra of differential forms have the following properties:

1). The action of these operators on any differential form is determined completely by how these operators act on $\mathcal{F}(M)$ and $\Lambda^1(M)$.

2). If D, D' are k, k'-derivatives respectively, then $[D, D']$ is a $(k+k')$- derivative.

3). If D is a k-derivative and D' is a k'-skew-derivative, then $[D, D']$ is a $(k+k')$-skew-derivative.

4). If D, D' are k, k'-skew-derivatives respectively, then $DD' + D'D$ is a $(k+k')$-derivative.

From the last three properties, new derivative can be obtained from two other derivatives defined on differential forms. As we know, Lie derivative on tangent vector fields is very simple, i.e., the Lie bracket. However, Lie derivative acts on differential forms in a very complicated way. To get around this problem, interior product between differential forms and vector fields will be useful. The interior contractor i_X, introduced in the previous chapter is

$$i_X\omega \equiv X\rfloor\omega.$$

It is easy to prove that the interior contractor has the following properties:

1). It is a (-1)-order linear operator.

$$i_X : \Lambda^r \to \Lambda^{r-1}.$$

2). It is an odd operator.

$$i_X(a_p \wedge \beta_q) = (i_X a_p) \wedge \beta_q + (-1)^p \alpha_p \wedge (i_X \beta_q).$$

3). For 1-forms, $\omega \in \Lambda^1(M)$.

$$i_X\omega = \omega(X) \equiv <\omega, X> .$$

For instance,

$$i_X df =< df, X >= Xf, \tag{2.28}$$

$$(i_X\omega_r)(X_1, \cdots, X_{r-1}) = \omega_r(X, X_1, \cdots, X_{r-1}).$$

4). For functions, $f \in \mathcal{F}(M)$.

$$i_X f = 0. \tag{2.29}$$

From the first two properties we see that the interior contractor i_X is a (-1)-skew-derivative. Due to the derivative operator properties of i_X, Eq. 2.28 and Eq. 2.29 together completely determine i_X. The above four properties can also be used to define the interior contractor i_X. Furthermore, the interior contractor has following additional properties:

5). As differential forms are completely skewsymmetric covariant tensor fields, contraction of differential forms with two identical tangent vector fields must be zero:

$$i_X^2 = 0.$$

6).For $f, g \in \mathcal{F}(M)$ and $X, Y \in \mathcal{X}(M)$.

$$i_{fX+gY} = fi_X + gi_Y. \tag{2.30}$$

Close attention has to be paid to the comparison between Eq. 2.30 and Eq. 2.24. Although both i_X and L_X depend on vector field X, they have different properties. The interior contractor i_X depends only on the vector field at the point i_X acts on. Lie derivative L_X depends on all vector fields X at the neighborhood of the point L_X acts on, such that

$$i_{fX} = fi_X.$$

However, in general, $L_{fX} \neq fL_X$, instead

$$L_{fX}\omega = fL_X\omega + df \wedge i_X\omega.$$

For L_X, i_X and exterior differential operator d acting on differential forms, only two of them are basic operators. There exist the following relations among them:

1). Lie derivative L_X commutes with exterior differential operator d.

$$[L_X, d] = 0. \tag{2.31}$$

From the properties of general derivatives, the 0-derivative which commutes with exterior differential operator d must be a Lie derivative L_X with respect to a tangent vector field X. Hence,

2). The Lie derivative can be written as

$$L_X = d \cdot i_X + i_X \cdot d. \tag{2.32}$$

Both d and i_X are skew derivatives. The right hand side of Eq. 2.32 is a 0-derivative, which acts on differential forms and must be equal to Lie derivative L_X.

3).

$$[L_X, i_Y] = i_{[X,Y]}. \tag{2.33}$$

Eq.2.31-2.33 are called Cartan formulae. They are very useful in treating differential forms. The proof of these formulae can be carried out by their action on any function f and 1− form $\omega \in \Lambda^1(M)$.

Using the Cartan formulae, it is easy to prove that for $\omega \in \Lambda^1(M)$,

$$(d\omega)(X, Y) = L_X\omega(Y) - L_Y\omega(X) - \omega([X, Y]). \tag{2.34}$$

And further, when $\omega_r \in \Lambda^r(M)$,

$$\begin{aligned}(d\omega_r)(X_0, X_1, \cdots, X_r) = \quad & \textstyle\sum_{i=0}^{r}(-1)^i L_{X_i}(\omega_r(X_0, \cdots, \hat{X}_i, \cdots, X_r)) \\ & + \sum_{i<j} (-1)^{i+j}\omega_r([X_i, X_j], X_0, \cdots, \hat{X}_i, \cdots, \hat{X}_j, \cdots, X_r),\end{aligned} \tag{2.35}$$

where $\hat{X}_i$ represents the deletion of the variable. For instance, when ω is a 2-form,

$$\begin{aligned}(d\omega)(X,Y,Z) = \ & L_X(\omega(Y,Z)) + L_Y(\omega(Z,X)) + L_Z(\omega(X,Y)) \\ & -\omega([X,Y],Z) - \omega([Y,Z],X) - \omega([Z,X],Y).\end{aligned}$$

In conclusion, there are three derivatives which act on differential forms $\Lambda^*(M)$, they are L_X, d and i_X, which are 0, +1, −1-derivatives respectively. Only two of these three are basic derivatives. Except for the commutation relation $[L_X, d] = 0$, they do not commute in general. For instance,

$$[L_X, L_Y] = L_{[X,Y]}, \tag{2.36}$$

$$[L_X, i_Y] = i_{[X,Y]}. \tag{2.37}$$

2.5 Lie group, Lie algebra and exponential map

An n-dimensional Lie group itself is not only an n-dimensional real analytic manifold but also has group structure, i.e., there exists mapping:

$$\begin{aligned} G \times G \ &\rightarrow \ G \\ x\ ,\ y \ &\rightarrow \ xy. \end{aligned}$$

This mapping has the properties of group operations, i.e., for each group element there is a mapping taking this element to its inverse:

$$\begin{aligned} G \ &\rightarrow \ G \\ x \ &\rightarrow \ x^{-1}. \end{aligned}$$

These two mappings are differentiable and continuous with respect to group parameters. In general, we can define a Lie group as follows.

Definition 2.1 *A Lie group is a set that is a group in the usual algebraic sense and is also a differentiable manifold, such that the following mapping is analytic:*

$$\begin{aligned} G \times G \ &\rightarrow \ G \\ x\ ,\ y \ &\rightarrow \ xy^{-1}. \end{aligned}$$

As mentioned above, a Lie group is an analytic manifold with a group structure such that the group operation is analytic. Since a Lie group G has the characteristics of a differentiable manifold, we can talk about its dimension, compactness, connectivity, etc.. On the other hand, as a Lie group G has the group structure, we can analyze its solvability, nilpotent property, subgroup, coset and quotient group etc.. The two aspects are closely related.

As group operations on Lie groups are analytic, analytic tools are often used to study the structure of Lie groups. Now we give some examples of Lie groups.

Example 1) One-dimensional connected Lie group must be abelian. There are only two cases:

(1). The additive group of real number $\mathbf{R}$, which is simply connected but not compact.

(2). $U(1)$ group, which is a unit circle in complex plane with group multiplication being multiplication of complex numbers,

$$U(1) = \{\mathbf{z} \in \mathbf{C} \mid |z| = 1\} \simeq S^1 \simeq \mathbf{R}/\mathbf{Z} \simeq SO(2),$$

it is compact and connected, but not simply connected. In general, one can prove that any connected abelian group is a Cartesian product of these two kinds of groups.

Example 2) The general linear transformation group of a real n-dimensional linear space:

$$GL(n,\mathbf{R}) = \{A \in M(n,\mathbf{R}) \mid \det A \neq 0\},$$

is an n^2-dimensional manifold, which is neither compact nor connected. It consists of two open subsets of $\mathbf{R}^{n^2}$, where the connected part containing the identity element e (identity matrix) is

$$GL^+(n,\mathbf{R}) = \{A \in M(n,\mathbf{R}) \mid \det A > 0\}.$$

Example 3) Real orthogonal group $O(n,\mathbf{R})$ is a subgroup of $GL(n,\mathbf{R})$:

$$O(n,\mathbf{R}) = \{O \in GL(n,\mathbf{R}) \mid O\ O^t = 1\}.$$

Proposition: $O(n,\mathbf{R})$ is a closed submanifold of $GL(n,\mathbf{R})$.

In the following, we analyze the condition $O\ O^t = 1$, which is satisfied by each element of $O(n,\mathbf{R})$. It is equivalent to say that matrix elements A_{ik} satisfy:

$$\sum_{k=1}^{n} A_{ik}A_{jk} = \delta_{ij}. \tag{2.38}$$

There are $\frac{1}{2}n(n+1)$ constraints for n^2 matrix elements. Hence $O(n,\mathbf{R})$ group is a $\frac{1}{2}n(n-1)$-dimensional submanifold. Taking $i=j$ in Eq. 2.38 and summing up, we have

$$\sum_{i,k}(A_{ik})^2 = n.$$

Therefore $O(n,\mathbf{R})$ is a bounded closed subset of $M(n,\mathbf{R}) \cong \mathbf{R}^{n^2}$, and $O(n,\mathbf{R})$ is a compact manifold, actually it is the maximum compact submanifold of $GL(n,\mathbf{R})$. A compact manifold without boundary be called the closed manifold, so $O(n,\mathbf{R})$ is a closed submanifold of $GL(n,\mathbf{R})$.

But $O(n, \mathbf{R})$ is not connected. The connected part $SO(n, \mathbf{R})$, which contains the identity element e, is the maximum compact subgroup of $GL^+(n, \mathbf{R})$, it is a $\frac{1}{2}n(n-1)$-dimensional connected but not simply connected manifold (see Chapter 8 for detail).

Example 4) The general linear transformation group of an n-dimensional complex linear space

$$GL(n, \mathbf{C}) = \{A \in M(n, \mathbf{C}) \mid \det A \neq 0\},$$

in which the maximum compact simply connected Lie subgroup is

$$SU(n) = \{U \in M(n, \mathbf{C}) \mid U^+ U = 1, \det U = 1\}.$$

It is an $(n^2 - 1)$-dimensional real Lie group, which is compact and simply connected. Let us take SU(2) as an example. The group element is

$$g = \begin{pmatrix} z_1 & z_2 \\ -\bar{z}_2 & \bar{z}_1 \end{pmatrix} \in SU(2),$$

which satisfying

$$\det g = |z_1|^2 + |z_2|^2 = 1.$$

Let $z_1 = t + iz$, $z_2 = x + iy$, then the equation above can be written as

$$t^2 + z^2 + x^2 + y^2 = 1.$$

From this we see that $SU(2)$ group manifold is equivalent to a unit sphere S^3 in four-dimensional Euclidean space. Therefore it is a compact and simply connected manifold.

$SU(2)$ group is just the angular momentum group in quantum mechanics, which plays a very important role in physics. In addition, it is the simplest and most significant non-Abelian Lie group. Any other non-Abelian Lie group has $SU(2)$ as its subgroup, which is the basic building block to construct other non-Abelian Lie groups.

Other higher rank groups $SU(n)$ are also important in physics. For example, $SU(3)$ group plays the central role in the classification of fundamental particles and is the primary symmetry group of strong interactions in QCD. In the theory of Lie groups, any classical group can be embedded in a higher rank $U(n)$ group as a subgroup. This property can be used to analyze various classical Lie groups.

Let us now look at the local structure of Lie group. A Lie group manifold $G = \{x\}$ is an n-dimensional differentiable C^∞ manifold. A neighborhood of a point x in G locally is like $\mathbf{R}^n$. The tangent space $T_x(G)$ of G at a point x is $\mathbf{R}^n$. As mentioned in Chapter one, for general differentiable manifold, there exists a vector field $X(x)$ on manifold G. The collection of all vector fields, $\mathcal{X}(G)$ forms an infinite dimensional Lie algebra, i.e., if we choose moving frame bases $e_a (a = 1 \cdots, n)$ as the bases of vector fields, the Lie bracket:

$$[e_a, e_b] = f_{ab}^c(x) e_c,$$

here the structure functions $f^c_{ab}(x)$ are not constants but depend on x.

In addition, a Lie group manifold has a group structure. A group G can act on itself. For any $g \in G$, two kinds of differentiable homomorphisms(left, right) on the group manifold can be defined:

$$\begin{aligned} L_g: G &\rightarrow G \\ x &\rightarrow L_g(x) \equiv gx, \\ R_g: G &\rightarrow G \\ x &\rightarrow R_g(x) \equiv xg^{-1}. \end{aligned}$$

It is easy to prove that $\{L_g\}$ and $\{R_g\}$ commute:

$$L_{g_1} R_{g_2} = R_{g_2} L_{g_1}.$$

A transformation on a group manifold can lead to a transformation of tensor fields on this group manifold. One of the most important tensors on group manifold is the left(right) invariant tensor filed. Here we study the left invariant vector fields on G only. The right invariant vector field can be considered similarly.

A vector filed is a section on a vector bundle, i.e., in the tangent space $T_p(M)$ for each point $p \in M$, a vector X(p) is selected with certain rules. Starting with any tangent vector $A = A^a e_a$, which is a vector of tangent space $T_e(G)$ at identity e on the group manifold G, and using all left translation transformations L_{g*}, a vector field on the group manifold G can be obtained:

$$\begin{aligned} L_{g*}: \ T_e(G) &\rightarrow T_g(G) \\ A &\rightarrow L_{g*}A = X(g), \quad \forall\, g \in G, \end{aligned}$$

i.e., as g runs over group G from a given tangent vector A at identity, a vector field $X \in \mathcal{X}(G)$ on group G can be derived. It is easy to prove that this vector field is left invariant, its set denoted as $\mathcal{X}_l(G)$, i.e.,

$$L_{g*}X(x) = X(x) = L_{g*}A \in \mathcal{X}_l(G)$$

A set $\mathcal{X}_l(G)$ of all left invariant vector fields on G is an n-dimensional vector space which is isomorphic to $T_e(G)$. There are n linearly independent left invariant vector fields on group manifold G.

The tangent space $T_g(G)$ at any point $g \in G$ on group manifold can be related to the tangent space at identity by left translation transformations

$$L_{g*}: \ T_e(G) \rightarrow T_g(G).$$

The tangent bundle $T(G)$ on manifold G can be obtained as g goes over G. Since there are n linearly independent left invariant vector fields on G, i.e., there are n

global sections of tangent bundle $T(G)$, the tangent bundle of a Lie group is trivial and is a direct product of G and $\mathbf{R}^n$

$$T(G) \sim G \times \mathbf{R}^n.$$

Therefore any Lie group is parallelizable. For tangent field vectors, Lie bracket can be defined as

$$[X, Y] \in \mathcal{X}(G).$$

Because the tangent mapping L_{g*} commutes with Lie brackets, we get

$$L_{g*}[X,Y] = [L_{g*}X, L_{g*}Y] = [X,Y] \in \mathcal{X}_l(G), \quad \text{for} X, Y \in \mathcal{X}_l(G). \tag{2.39}$$

The set $\mathcal{X}_l(G)$ of left invariant vector fields forms a Lie algebra $\mathcal{X}_l(G)$, which is an n-dimensional subalgebra of tangent field Lie algebra $\mathcal{X}(G)$. Note that $\mathcal{X}(G)$ is of infinite dimension, but $\mathcal{X}_l(G)$ is n-dimensional, $dim(X_L) = dim(G) = n$. If we take a bases as $X_i (i = 1, \cdots, n)$. The bases X_i of Lie algebra $\mathcal{X}_l(G)$ consist of n linearly independent left invariant vector fields X_i, satisfying

$$[X_i, X_j] = C_{ij}^k \, X_k, \tag{2.40}$$

where C_{ij}^k are constants and are called structure constants of Lie algebra $\mathcal{X}_l(G)$. It can be proved that these structure constants are skewsymmetric

$$C_{ij}^k = -C_{ji}^k,$$

and satisfy Jacobi identity:

$$C_{ij}^k C_{mk}^l + C_{jm}^k C_{ik}^l + C_{mi}^k C_{jk}^l = 0. \tag{2.41}$$

The structure constants C_{jm}^k of Lie algebra $\mathcal{X}_l(G)$ are also called structure constants of Lie group G and completely determine the local properties of a Lie group.

Similarly, the right invariant vector field $X^R(x)$ can be defined:

$$R_{g*}X^R(x) = X^R(x) \in \mathcal{X}_r(G). \tag{2.42}$$

The set $\mathcal{X}_r(G)$ of right invariant vector fields can forms a Lie algebra. It can be proved that $\mathcal{X}_r(G)$ is algebraically isomorphic to $\mathcal{X}_l(G)$. Both $\mathcal{X}_r(G)$ and $\mathcal{X}_l(G)$ are called Lie algebra $L(G)$ of Lie group G,

$$L(G) \sim \mathcal{X}_l(G) \sim \mathcal{X}_r(G).$$

As there is an inverse to each group element, there exists an anti-isomorphism on Lie group G:

$$\begin{aligned} \gamma : G &\rightarrow G \\ g &\rightarrow g^{-1}. \end{aligned}$$

The tangent mapping γ_* at e is

$$\gamma_* \mid_e = -id_{T_eG}, \tag{2.43}$$

i.e., the identity operation of tangent space T_eG at the unit element e of group G is equivalent to $-\gamma_* \mid_e$. On the tangent bundle of group manifold there is a tangent mapping

$$\begin{aligned} -\gamma_* : \mathcal{X}_l(G) &\rightarrow \mathcal{X}_r(G) \\ X^L &\rightarrow -\gamma_* X^L \in \mathcal{X}_r(G) \end{aligned}$$

$$[X^R, Y^R] = [-\gamma_* X^L, -\gamma_* Y^L] = \gamma_*[X^L, Y^L].$$

Therefore the Lie algebras, which are formed by left and right invariant vector fields on group G, are isomorphic and are simply called Lie algebra $L(G)$ of Lie group G.

As we know in section two, once a vector field $X(x)$ on manifold M is given, the flow mapping φ_X can be defined as:

$$\begin{aligned} \varphi_X : \mathbf{R} \times G &\rightarrow G \\ t\ ,\ x &\rightarrow \varphi_X(t, x). \end{aligned}$$

The image of this mapping is also called the integral curve of vector field $X(x)$. The left invariant vector field $X(x) = L_{g_*}A \in \mathcal{X}_l(G)$ on Lie group manifold is complete. The flow $\varphi_A(t, x)$ which is determined by field X can be written as

$$\varphi_A(t, g) = g\ \varphi_A(t, e).$$

The flow $\varphi_A(t, e)$ satisfies:

$$\varphi_A(t_1 + t_2, e) = \varphi_A(t_1, e)\varphi_A(t_2, e),$$

and

$$\varphi_{\lambda A}(t, e) = \varphi_A(\lambda t, e).$$

The flow $\varphi_A(t, x)$ determined by left invariant vector field can be written as

$$\varphi_A(t, x) = x\ \exp(tA),$$

i.e., any left invariant vector field in Lie algebra $\mathcal{X}_l(G)$ can generate a one parameter subgroup of group G. An exponential map from Lie algebra $\mathcal{X}_l$ to group G can be defined as:

$$\begin{aligned} \exp : \mathcal{X}_l &\rightarrow G \\ A &\rightarrow \exp(A) \equiv \varphi_A(1, e). \end{aligned}$$

The exponential map represents the relation between Lie algebra and Lie group. The mapping is a generalization of the exponential mapping of the ordinary matrix.

Note: 1. The flow φ_A, generated by a left-invariant vector field, commutes with the left translation, from Eq.2.44 and 2.39 it is easy to see, $\varphi_A(t,x) = R_{e^{tA}}(x)$, the left-invariant flow represents infinitesimal right translation.

2. It can be proved that the integral curve of any left invariant vector field is also that of a right invariant vector field. Namely, when a vector field

$$Y = Ad_{g^{-1}}X,$$

i.e.,

$$e^{tY} = g^{-1}e^{tX}g,$$

then the integral curve along $X^L \in \mathcal{X}_l(G)$ through point g is the same as that along $Y^R \in \mathcal{X}_r(G)$ through point g.

In studying the structures of Lie groups and their Lie algebras, their representations are very useful. The representations are homomorphisms between a Lie group and a matrix group or between a Lie algebra and a matrix algebra. As we are quite familiar with the operation of matrices, the use of matrices will be very helpful. A set $\{\tau_a\}(a = 1, \cdots, n)$ of $m \times m$ matrices satisfying Eq. 2.40

$$[\tau_a, \tau_b] = C_{ab}^c \tau_c,$$

is called an m-dimensional representation of Lie algebra $L(G)$. Their exponentialization will give an m-dimensional representation $g = (g_{ij}), (i,j = 1, \cdots, m)$ of the corresponding Lie group G.

The vector field $X \in \mathcal{X}(M)$ on an n-manifold can be defined as differential operators acting on functions on the manifold. Similarly, left invariant vector fields on a group manifold can be defined as differential operators acting on functions on the group. And we can express the corresponding bases X_a of the left-invariant vector fields on the Lie group G as

$$X_a f(g) = \frac{d}{dt} f(ge^{t\tau_a}) \mid_{t=0},$$

or

$$X_a = \frac{d}{dt}|_{t=0} = \sum_{i,j=1}^{m} \frac{dg_{ij}}{dt}\frac{\partial}{\partial g_{ij}}|_{t=0} = \mathrm{Tr}(\frac{dg}{dt}\frac{\partial}{\partial g^T})|_{t=0} = \mathrm{Tr}(g\tau_a \frac{\partial}{\partial g^T}). \tag{2.44}$$

Similar discussions can be applied to cotangent vector fields on group manifold. If ω satisfies

$$L_a^* \omega = \omega, \qquad \forall a \in G,$$

ω is called a left invariant form. For example, the Maurer-Cartan form

$$\theta = g^{-1}dg,$$

satisfies

$$L_a^* \theta = (ag)^{-1} d(ag) = \theta,$$

i.e., θ is invariant under left translation by any constant group element. When we represent g by its corresponding m-dimensional matrix, left invariant forms $g^{-1}dg$ are 1-forms which take values on the Lie algebra $L(G)$ and can be expanded by τ_a as

$$\theta(x) = g(x)^{-1}dg(x) = \theta^a(x)\tau_a, \tag{2.45}$$

where $\theta^a(x)$ are n independent 1-forms on Lie group G.

It is easy to prove via Eq. 2.45 that

$$d\theta + \theta \bigwedge \theta = 0.$$

From Eq. 2.40, we see that θ^a obeys the Maurer-Cartan equation

$$d\theta^a + \frac{1}{2}C^a_{bc}\theta^b \bigwedge \theta^c = 0.$$

The identity $d^2\theta^a = 0$ is equivalent to the Jacobi identity Eq 2.41. The set of left invariant forms θ form an n-dimensional linear space. And the dual space of this linear space is a left invariant vector field space. The bases of the left invariant vector field space can be chosen to be the dual bases X_a of $\{\theta^a\}$ satisfying

$$< \theta^a, X_b >= \delta^a_b, \quad [X_a, X_b] = C^c_{ab}X_c,$$

where

$$X_a = Tr(g\tau_a \frac{\partial}{\partial g^t}) \in \mathcal{X}_l$$

or in terms of the representation matrix $g = (g_{ik})$ of group G,

$$X_a = \sum_{ijk} g_{ik}(\tau_a)_{kj}\frac{\partial}{\partial g_{ij}}, \tag{2.46}$$

$\{X_a\}$ are left invariant vector fields on G. Any other left invariant vector field is a linear combinations of $\{X_a\}$ with coefficients in $\mathbf{R}$. From Eq. 2.46, $\{X_a\}$ form the Lie algebra $L(G)$. $\{\theta^a\}$ and $\{X_a\}$ constitute frame fields and co-frame fields which are dual to each other on Lie group G.

Similarly, the right invariant Maurer-Cartan forms can be defined

$$\sigma = dg\ g^{-1} = \tau_a\sigma^a.$$

There is the relation

$$d\sigma - \sigma \bigwedge \sigma = 0,$$

or in terms of components

$$d\sigma^a - \frac{1}{2}C^a_{bc}\sigma^b\sigma^c = 0.$$

$\{\sigma^a\}$ constitute the right invariant co-frame fields on $L(G)$. The dual frame fields $\{Y_a\}$ are:

$$Y_a = Tr(\tau_a g \frac{\partial}{\partial g^t}) \in \mathcal{X}_r(G)$$

which satisfy

$$< \sigma^a, Y_b >= \delta^a_b, \quad [Y_a, Y_b] = -C^c_{ab} Y_c,$$

i.e., all structure constants are negative of that of left invariant vector fields. The left and right invariant vector fields commute with each other

$$[X_a, Y_b] = 0.$$

The properties of $\{X_a\}$ and $\{Y_a\}$ are very much like the relation between base space fixed generators and rigid body fixed generators of a top in quantum mechanics.

2.6 Lie transformation groups, orbit and the space of orbits

As we know, the set of all isometry on a Riemann manifold is a Lie group, and the primary object of Riemannian geometry is to study invariance of a Riemannian manifold under the isometry group. The set of all diffeomorphism transformations on a differentiable manifold M is the diffeomorphism group $Diff(M)$ of the manifold M, and differential geometry studies invariance under diffeomorphism, or the action of the group $Diff(M)$, with the invariance independent of the choice of coordinates on manifolds.

In the following, we study the actions of a Lie group G on a differentiable manifold M. The realization of the Lie group G is the Lie transformation group on M. The realization of a groups is also called a representations of the group.

Definition 2.2 *The left actions of a Lie group G on a manifold M are differentiable mappings and defined as follows:*

$$\phi: \quad G \times M \to M$$

$$g, x \to \phi(g, x)$$

such that

$$\phi(g_1, \phi(g_2, x)) = \phi(g_1 g_2, x),$$

$$\phi(e, x) = x,$$

where e is the identity element of G and g_1, g_2 are arbitrary elements of G.

Given any $g \in G$, on M there exists a diffeomorphism ϕ_g:

$$\begin{aligned} \phi_g: \quad M &\rightarrow M, \\ x &\rightarrow \phi(g, x). \end{aligned}$$

The set $\{\phi_g, g \in G\}$ satisfies:

$$\begin{aligned} \phi_{g_1} \cdot \phi_{g_2} &= \phi_{g_1 g_2}, \\ \phi_e &= id_M, \end{aligned}$$

i.e., mapping $g \rightarrow \phi_g$ is a homomorphism from G to the Lie transformation group of M, which is called the realization of group G on M or simply the representation.

Similarly, the right actions of a group G on a manifold M can be defined. The right action ψ:

$$\begin{aligned} \psi: \quad M \times G &\rightarrow M \\ x\ ,\ g &\rightarrow \psi(x, g) \end{aligned}$$

satisfies:

$$\begin{aligned} \psi(\psi(x, g_1), g_2) &= \psi(x, g_1 g_2), \\ \psi(x, e) &= x. \end{aligned}$$

It is easy to show that the diffeomorphism $\{\psi_g\}$ obtained from the right actions of group G on M

$$\begin{aligned} \psi_g: M &\rightarrow M \\ x &\rightarrow \psi(x, g) \end{aligned}$$

is not a representation of group G. However mapping $g \rightarrow \psi_{g^{-1}}$ is a representation of group G. For simplicity, the left and right actions of a group G on a manifold M are written as

$$\phi_g\, x \equiv gx,$$

$$\psi_g\, x \equiv xg,$$

respectively. In the following, we will use the above notations and mainly study the left action ϕ_g. The right action ψ_g can be studied similarly.

For Lie group G acting on manifold M, there are two special cases that call for special attention.

1). If only the identity e of G makes all points on M invariant, the action of G on manifold M is called effective, or G acts effectively on M.

2). If no other element than identity e of G has fixed point on M, i.e., every point on M is moved away from itself, the action of G on manifold M is called free, or G acts freely on M.

For effective action, for any element g aside from the identity of G, there must exist a point $x \in M$ such that $gx \neq x$.

For free action, all points $x \in M$ satisfy $gx \neq x$ for any element g aside from the identity of G.

Obviously, the condition in case 2) is stronger than that of in case 1). If the action is free, it must be effective. If the action is effective, it is not necessarily free. For example, consider $U(1)$ act on two-dimensional sphere S^2, the action is effective but not free on S^2, since the north and the south poles are fixed points. However, $U(1)$ has a free action on the equator of S^2.

The actions of G on M give a homomorphic mapping from G to $Diff(M)$, its kernel is

$$K = \{g \in G \mid gx = x, \forall x \in M\}.$$

The action of group is called effective if the homomorphism kernel K contains the identity element only. The kernel K is an invariant subgroup of G and represents the part in G which can not be realized by the action of group G on M. The action of the quotient group G/K on M is effective.

If the action of group G on M is effective, the elements which keep a given point x_0 on M fixed constitute a closed subgroup G_{x_0} of group G:

$$G_{x_0} = \{g \in G \mid gx_0 = x_0 \in M\}.$$

G_{x_0} is called the isotropic subgroup or stable subgroup of group G at point x_0. For a given point x_0 on M, the set of all points which can be derived by the action of all elements of group G on point x_0 is called an orbit O_{x_0} of G through point x_0:

$$O_{x_0} = \{x \in M \mid x = gx_0, \forall g \in G\}.$$

It is a submanifold of M.

If two orbits intersect, they must coincide completely. Therefore all orbits form a partition of M. If each orbit is considered as a point, the resulting manifold is called the space of orbits, or simply orbit space, and denoted as M/G.

Example 1. $\mathbf{R}/\mathbf{Z} = S^1$, where the orbit of real number x contains all $x + \mathbf{Z}$. The orbit space (a paste space where all points on the same orbit are pasted together) is the circle S^1.

Example 2. $RP(n) = S^n/\mathbf{Z}_2$, where $\mathbf{Z}_2$ is the parity transformation which acts on S^n. Each orbit is a pair of antipodean points. The space of orbits is the projective space $RP(n)$.

Next we introduce the so called transitive action of G on M and the G-homogeneous space.

Definition 2.3 *The action of a group G on a manifold M is transitive, if for any two points x, y on M, there exists a group element $g \in G$ such that $y = gx$. And manifold M is called the homogeneous manifold of G.*

Given a transformation group which acts on a manifold M, for any two points of the orbit O_x which are derived from $x \in M$ by the action of G,

$$x_1, x_2 \in O_x = G \cdot x,$$

there is a group element g satisfying $gx_1 = x_2$ (For instance, $x_1 = g_1x$, $x_2 = g_2x$, then $g = g_2g_1^{-1}$.), so, the action of group G on orbit O_x is transitive. Every orbit O_x of M related to G is a homogeneous space of G, but the action of G on M may be either transitive ($O_x = M$) or intransitive ($O_x \neq M$).

It can be proved that the isotropic groups G_x of points on the same orbit are conjugate to each other:

$$G_{gx} = gG_xg^{-1}.$$

So G_x is often called the isotropic subgroup of the group G.

If a manifold M is a homogeneous space under the action of G, the orbit derived by the action of G on any point of M is M itself. For the stable subgroup G_x on a point $x \in M$, there exists a one to one bijective mapping φ_x:

$$\varphi_x : G/G_x \to M.$$

If M is a connected locally compact manifold and G is a locally compact Lie group, it can be proved that φ_x is a diffeomorphism, i.e.,

$$G/G_x \sim M.$$

In this case, M is called a G-homogeneous manifold.

For a left action of G on M

$$\begin{aligned} \Phi : G \times M &\to M \\ g, x &\to \Phi(g, x) \in M. \end{aligned}$$

A mapping can be defined for any given point $p \in M$

$$\begin{aligned} \Phi_p : G &\to M \\ g &\to \Phi_p(g) = \Phi(g, p) \in M. \end{aligned}$$

Φ_p induces the push-forward map $\mu \equiv \Phi_{p*}$

$$\begin{aligned} \mu : L(G) &\to \mathcal{X}(M) \\ X &\to X_M \\ X_M f(x) &= \frac{d}{dt} f(e^{-tX}x)|_{t=0}, \quad \forall f \in \mathcal{F}(M). \end{aligned}$$

The target vector field $X_M(x)$ is called the fundamental vector field on M with respect to $X \in L(G)$. It can be shown that the choice of base point p in the orbit O_p dose not effect μ, in terms of which one can obtain a Lie algebra of the fundamental vector fields on M, which is the homomorphic image of the Lie algebra of group G. When M is the group manifold G itself, the fundamental vector fields are just the right invariant vector fields on group G. In general, when H is a closed subgroup of Lie group G, the coset manifold G/H relative to group G is a homogeneous space. There is a unique analytic manifold structure on the coset space G/H, such that G is a transformation group on space G/H. For any group element $g \in G$, there exists a correspondingly transformation of the coset space G/H.

$$\begin{aligned} g: \ G/H &\rightarrow G/H \\ g_1 H &\rightarrow g g_1 H. \end{aligned}$$

The action of group G on G/H is effective, real analytic and G/H is a real analytic manifold.

Any Lie group G can be acted on by itself. For example the left and right translation transformations L_g and R_g of a Lie group G act freely and transitively on itself. The Lie group G itself is a homogeneous space. Because of this, at any point of G the local properties are the same, so to study the local structure of G it is sufficient enough to analyze the neighborhood of the identity element of group G, i.e., to study the Lie algebra L(G).

Homogeneous manifold with certain symmetry groups are often considered in theoretic physics. In the following, some examples are given:

Example 1. $CP(n)$ is the set of complex lines through the origin of $\mathbf{C}^{n+1}$. The group $SU(n+1)$ which maps a complex line to another complex line is a transitive group on $CP(n)$. The isotropic subgroup, which does not change a given complex line, is $U(n)$. Hence

$$CP(n) \sim SU(n+1)/U(n).$$

For instance, when $n = 1$,

$$CP(1) \sim SU(2)/U(2) \sim S^2.$$

Example 2. $RP(n)$ is the set of real lines through the origin of $\mathbf{R}^{n+1}$. The linear transformation group $SO(n+1)$ is a transitive group on $RP(n)$. The isotropic subgroup is $O(n)$. Hence

$$RP(n) \sim SO(n+1)/O(n).$$

Example 3. Grassmannian manifold.

Both real and complex projective spaces are special cases of the Grassmannian manifold. In an real n-dimensional space $\mathbf{R}^n$, the set of all k-dimensional subspaces through the origin is called a Grassmannian manifold $G_r(n,k)$, it is a $k(n-k)$-dimensional differentiable manifold which can be represented as a coset manifold:

$$G_r(n,k) \simeq O(n)\,/O(n-k) \times O(k)\,.$$

$$RP(n) = G_r(n+1,1) \simeq O(n+1)/O(n) \times O(1).$$

Similarly, the complex Grassmann manifold $G_r^c(n,k)$ is the set of all k-dimensional complex subspaces through the origin of $\mathbf{C}^n$. It can be represented as:

$$G_r^c(n,k) \simeq SU(n)/SU(n-k) \times U(k),$$

$$CP(n) = G_r^c(n+1,1) \simeq SU(n+1)/SU(n) \times U(1),$$

$$CP(1) \simeq SU(2)/U(1) \simeq S^2.$$

In $\mathbf{C}^n$, the complex manifold determined by the set of complex homogeneous polynomials

$$P_i(z^0, z^1, \cdots, z^n) = 0 \; (1 \le i \le n)$$

is called an algebraic manifold, which is one of the major research topics in algebraic geometry. The complex Grassmannian manifold is an algebraic manifold.

Similarly, we can deal with the quaternion Grassmannian manifold

$$G_r^h(n,k) \simeq SP(n)/SP(n-k) \times SP(k).$$

An example is the quaternion projective space $HP(n)$:

$$HP(n) \simeq SP(n+1)/SP(n) \times SP(1).$$

Example 4. The Stiefel manifold $V_{n,j}(n \ge j)$, which is the set of all j-dimensional orthogonal frames in an n-dimensional Euclidean space.

$$\begin{aligned} V_{n,j} &\simeq O(n)/O(n-j) \;\; (n \ge j) \\ &\simeq SO(n)/SO(n-j) \;\; (n > j). \end{aligned}$$

For example:

$$V_{n,n} \simeq O(n), \qquad V_{n,n-1} \simeq SO(n).$$

The n-dimensional sphere S^n is a very simple manifold which is also a special Stiefel manifold

$$S^n \simeq V_{n+1,1} \simeq SO(n+1)/SO(n).$$

For instance,

$$S^0 \simeq V_{1,1} \simeq Z^2.$$

$$S^1 \simeq V_{2,1} \simeq SO(2) \simeq U(1).$$

$$S^{2n-1} \simeq SU(n)/SU(n-1).$$

$$S^{4n-1} \simeq SP(n)/SP(n-1).$$

In the table below, we list some Grassmannian manifolds and projective spaces for reference.

Grassmann manifolds	Projective spaces
$G_r(n,j) = O(n)/O(j) \times O(n-j)$	$RP(n) = G_r(n+1,1)$ $= SO(n+1)/O(n)$
$G_r^c(n,j) = U(n)/U(j) \times U(n-j)$	$CP(n) = G_r^c(n+1,1)$ $= SU(n+1)/SU(n) \times U(1)$
$G_r^h(n,j) = SP(n)/SP(j) \times SP(n-j)$	$HP(n) = G_r^h(n+1,1)$ $= SP(n+1)/SP(n) \times SP(1)$

Notations and Formulae

- The tangent map $f_* : \mathcal{X}(M) \to \mathcal{X}(M)$ satisfy: (for $X, Y \in \mathcal{X}(M)$)

  1). Linearity,
  $$f_*(X+Y) = f_*X + f_*Y.$$
  2). Commutes with Lie bracket,
  $$[f_*X, f_*Y] = f_*[X,Y].$$
  3). For consecutive mappings,
  $$(f \cdot g)_* = f_* \cdot g_*.$$

- The pull back mapping $f^* : \Lambda^*(M) \to \Lambda^*(M)$ satisfying: (for $\omega, \tau \in \Lambda^*(M)$)

  1). Linearity,
  $$f^*(\omega + \tau) = f^*\omega + f^*\tau.$$
  2). Commutes with the exterior product,
  $$f^*(\omega \wedge \tau) = f^*\omega \wedge f^*\tau.$$
  3). Commutes with the exterior differential operator,
  $$f^* d\omega = d f^*\omega.$$
  4). For consecutive mappings,
  $$(f \cdot g)^* = g^* \cdot f^*.$$

- Integrable condition (Frobenius theorem). The following two approaches are equivalent:

  (1) The tangent field approach: The k-dim distribution $S^k = \{X^{(1)}, \cdots, X^{(k)}\}$ is integrable iff this distribution S^k is involutive, i.e., $[X^{(\alpha)}, X^{(\beta)}] \in S^k$
  (2) The differential forms approach: The Pfaff forms $\{\theta^{(a)}\}_1^k$ is integrable iff the ideal generated by $\{\theta^{(a)}\}_1^k$ is a closed ideal, i.e., let $\Sigma = \theta^{(1)} \wedge \cdots \wedge \theta^{(k)}$,
  $$d\Sigma = \tau \wedge \Sigma, \qquad \tau \in \Lambda^1(M).$$

- The Lie derivative L_X is a R-linear mapping between the same order tensors satisfying: (for $K, K' \in T^r_s(M)$)

(1) R-linear

$$\begin{aligned} L_X(aK+bK') &= aL_XK + bL_XK', \\ L_{aX+bY} &= aL_X + bL_Y, \quad a,b \in \mathbf{R}. \end{aligned}$$

(2) Leibniz rule:

$$\begin{aligned} L_X[Y,Z] &= [L_XY,Z]+[Y,L_XZ], \\ L_X(f\omega) &= fL_X\omega + (L_Xf)\omega, \\ L_X(\omega\wedge\sigma) &= (L_X\omega)\wedge\sigma + \omega\wedge(L_X\sigma), \\ L_X(<\omega,Y>) &= <L_X\omega,Y> + <\omega,L_XY>. \end{aligned}$$

(3) $\qquad L_Xf = Xf.$
(4) $\qquad L_XY = [X,Y].$
L_X also satisfy (for $\omega_r \in \Lambda^r(M)$):

$$\begin{aligned} (d\omega_r)(X_0,X_1,\cdots,X_r) &= \sum_{i=0}^{r}(-1)^i L_{X_i}(\omega_r(X_0,\cdots,\hat{X}_i,\cdots,X_r)) \\ &+ \sum_{i<j}(-1)^{i+j}\omega_r([X_i,X_j],X_0,\cdots,\hat{X}_i,\cdots,\hat{X}_j,\cdots,X_r). \end{aligned}$$

$$\begin{aligned} [L_X,L_Y] &= L_{[X,Y]}, \quad [L_X,i_Y]=i_{[X,Y]}, \\ [L_X,d] &= 0, \qquad L_X = d\cdot i_X + i_X\cdot d. \end{aligned}$$

- The left-invariant vector field $\{X_a\}$ on group manifold G

$$\begin{aligned} X_af(g) = \frac{d}{dt}f(ge^{t\tau_a})\mid_{t=0}, \quad & X_a = \mathrm{Tr}(g\tau_a\tfrac{\partial}{\partial g^T}) \ \in \mathcal{X}_l(G), \\ [X_a,X_b] = C^c_{ab}X_c, \quad & [\tau_a,\tau_b] = C^c_{ab}\tau_c. \end{aligned}$$

- The right-invariant vector field $\{Y_a\}$ on group manifold G

$$\begin{aligned} Y_af(g) = \frac{d}{dt}f(e^{t\tau_a}g)\mid_{t=0}, \quad & Y_a = \mathrm{Tr}(\tau_ag\tfrac{\partial}{\partial g^T}) \ \in \mathcal{X}_r(G), \\ [Y_a,Y_b] = -C^c_{ab}Y_c, \quad & [X_a,Y_b] = 0. \end{aligned}$$

- The left-invariant 1-form $\{\theta^a\}$:

$$g^{-1}dg = \theta^a \tau_a, \qquad d\theta^a + \frac{1}{2}C^c_{ab}\theta^b \wedge \theta^c = 0,$$

$$< \theta^a, X_b >= \delta^a_b, \qquad X_b \in \mathcal{X}_l(G).$$

- The right-invariant 1-form $\{\sigma^a\}$:

$$dgg^{-1} = \tau_a \sigma^a, \qquad d\sigma^a - \frac{1}{2}C^a_{bc}\sigma^b \wedge \sigma^c = 0,$$

$$< \sigma^a, Y_b >= \delta^a_b, \qquad Y_b \in \mathcal{X}_r(G).$$

Exercises

1. Prove that if there is a coordinate transformation at the neighborhood of point p on manifold M

$$f:\ x \to y = f(x)$$

then for any differential form $\alpha_r \in \Lambda^r(M)$

$$d(f^*\alpha_r) = f^* d\alpha_r.$$

2. Find the integrable conditions for the partial differential equations:

$$\frac{\partial y^k}{\partial x^i} = \psi^k_i(x, y) \quad (i = 1, \cdots, n;\ k = 1, \cdots, m).$$

3. Given $X = \xi^i \partial_i$, $\omega = \omega_i dx^i$, $f \in \mathcal{F}(M)$,
find $L_X \omega$ and $L_{fX}\omega$.

4. For vector fields X, Y on S^2 in spherical coordinates,

$$X = \phi\frac{\partial}{\partial\theta}, \quad Y = \frac{\partial}{\partial\theta} + \phi\frac{\partial}{\partial\phi},$$

find $L_X Y$ and $L_{fX} Y$.

5. For any smooth function on manifold, $f \in \mathcal{F}(M)$, prove that Lie derivative L_X and exterior differential operator d commute, namely,

$$L_X f = Xf, \quad L_X df = d(Xf).$$

6. Prove the following equations and specify their validity.

$$L_X = d \cdot i_X + i_X \cdot d,$$

$$[L_X,\ L_Y] = L_{[X,Y]}, \quad [L_X,\ i_Y] = i_{[X,Y]}.$$

7. For $\omega \in \Lambda^1(M)$, $X, Y \in \mathcal{X}(M)$, prove

$$d\omega(X,Y) = L_X\omega(Y) - L_Y\omega(X) - \omega([X,Y]).$$

8. k-form $\alpha_k \in \Lambda^k(M)$ contracts with k vector fields $X, Y, \cdots, Z$ yields a function on manifold M

$$\alpha_k(X,Y,\cdots,Z) \in \mathcal{F}(M).$$

show that

$$\begin{aligned} L_W\alpha_k(X,Y,\cdots,Z) &= W(\alpha_k(X,Y,\cdots,Z)) \\ &\quad -\alpha_k([W,X],Y,\cdots,Z) - \cdots \\ &\quad -\alpha_k(X,Y,\cdots,[W,Z]). \end{aligned}$$

Chapter 3

Affine Connection and Covariant Differentiation

3.1 Moving frame approach to tensor field

Tangent vector fields on manifold are sections of tangent bundle. Tangent bundle of n-dimensional manifold is a $2n$-dimensional manifold, which is locally a Cartesian product of M and $\mathbf{R}^n$

$$T(M) \sim M \times \mathbf{R}^n$$

Note that the tangent bundle $T(M)$ is the space of pairs (p, v_p) for all $p \in M$ and all $v_p \in T_P(M)$, coordinate systems appropriate to a general $2n$-dimensional manifold are not all appropriate to $T(M)$, which has a fibre bundle structure. A vector field $X(p)$ is a smooth section of $T(M)$, i.e., a smooth assignment of a tangent vector X_p for each point: $p \in M$. For a given coordinate system on the manifold M, the vector field $X(x)$ can be written as

$$X(x) = \xi^i(x)\partial_i,$$

where $\{\partial_i\}(i = 1, \cdots, n)$ are the natural frame of $T(M)$ associated to the coordinate system. Sometimes moving frames are more convenient than natural frames. The moving frame $e_a(x)$ are linear combination of the natural frame ∂_i

$$e_a(x) = e_a^i(x)\partial_i \quad (a = 1, \cdots, n),$$

where $e_a^i(x) \in \mathcal{F}(M), \{e_a^i(x)\}$ constitute a non-singular $n \otimes n$ matrix, i.e., the determinant of the matrix is non-vanishing at the neighborhood of point p

$$det(e_a^i(x)) \neq 0.$$

$\{e_a(x)\}(a = 1, \cdots, n)$ is a set of n linearly independent vector fields, and is called the moving frame. Any vector field $X(x)$ can be written as

$$X(x) = \xi^i(x)\partial_i = \xi^a(x)e_a(x) \quad (i, a = 1, \cdots, n),$$

where $\xi^a(x) \in \mathcal{F}(M)$ are the components of $X(x)$ in the moving frame. Since vector fields are geometric objects on the manifold which are independent of the choice of the coordinate system. Let $\{e'_a(x)\}(a = 1, \cdots, n)$ be another frame of $T(M)$, which is a set of n linearly independent vector fields and can be expressed as a linear combination of the former frame as

$$e'_a(x) = L_a{}^b(x)e_b(x) \quad (a = 1, \cdots, n),$$

and

$$det(L_a{}^b(x)) \neq 0.$$

The space of pairs $(x, L^b_a(x))$ is called frame bundle and is denoted by $L(M)$ with the fibre $L^b_a(x) \in GL(n, R)$. A bundle whose fibre is a Lie group manifold is called a principal bundle. The frame bundle $L(M)$ is a principal bundle, for there is a bijection between the set of all frames in $T_x(M)$ and $GL(n, R)$.

In the moving frame, the bases of co-vector field can be chosen as

$$\theta^a(x) = \theta^a{}_i(x)dx^i \quad (a = 1, \cdots, n),$$

where matrix $(\theta^a_i(x))$ is the inverse of $(e_a{}^i(x))$ such that the duality between vector fields and co-vector fields is invariant

$$< \theta^a(x), e_b(x) >= \theta^a{}_i(x)e_b{}^j(x) \ < dx^i, \partial_j >= \theta^a{}_i(x)e_b{}^i(x) = \delta^a_b,$$

and components of vector field $X = \xi^i(x)\partial_i$ and co-vector field $\sigma = \sigma_i(x)dx^i$ are

$$X = \xi^a(x)e_a(x) = \xi^i(x)\partial_i, \quad \xi^a(x) =< \theta^a(x), X >= \theta^a{}_j(x)\xi^j(x).$$
$$\sigma = \sigma_a(x)\theta^a(x) = \sigma_i(x)dx^i, \quad \sigma_a(x) =< \sigma, e_a(x) >= e_a{}^i(x)\sigma_i(x).$$

The natural frame have properties:

$$[\partial_i, \partial_j] = 0, \quad d(dx^i) = 0,$$

but the bases $\{e_a(x)\}$ do not commute each other.

Now we consider the transformation of moving frame,

$$e'_a(x) = L^b_a(x)e_b(x), \quad \theta'^a(x) = (L^{-1})^a{}_b(x)\theta^b(x), \quad < \theta'^a(x), e'_b(x) >= \delta^a_b.$$

Components of the vector field in this frame are

$$\xi'^a(x) = (L^{-1})^a{}_b(x)\xi^b(x),$$

and the components of the co-vector field are

$$\sigma'_a(x) = L_a{}^b(x)\sigma_b(x).$$

In this book, subscripts denote the components which are transformed by $L = (L_a{}^b)$, and are called covariant indices. Superscripts denote the components which are transformed by $L^{-1} = ((L^{-1})^a{}_b)$, and are called the contravariant indices.

Vector fields $\mathcal{X}(M)$ and co-vector fields $\Theta(M)$ are dual to each other n-rank $\mathcal{F}$-module. The dual bases are $\{e_a(x)\}$ and $\{\theta^a(x)\}$ satisfy:

$$< \theta^a(x), e_b(x) >= \delta^a_b, \qquad [e_a(x), e_b(x)] = f^c_{ab}(x)e_c(x).$$

Then

$$d\theta^a(x) = -\frac{1}{2}f^a_{bc}(x)\theta^b(x) \wedge \theta^c(x),$$

here $f^c_{ab}(x)$ be called the structure functions with respect to the moving frame $\{e_a(x)\}$ and $\{\theta^a(x)\}$.

3.2 Affine connection and covariant differentiation

Exterior differential operator d is a generalization of the ordinary total differential operator and acts on differential forms. For all differential forms, exterior differential operator is well defined and no additional structures of the manifold are needed.

Lie derivative operator L_X is defined on various tensor fields including all orders of differential forms. However, we need a given tangent vector field to define a Lie derivative operator, and that the Lie derivative at point p depends on the given tangent vector field at the neighborhood of p.

Covariant differential introduced in the following is also defined on various tensor fields but with one additional structure: affine connection. Vectors at different points cannot be compared directly, as they are in different tangent spaces. A connection can be given by defining what is called parallel transport of these tangent spaces. With this structure one can compare two tangent spaces at two infinitesimally separated points. So differential structures can be defined on the tensor bundle by the connections. Connections on the tangent bundle $T(M)$ are called affine connections of M. The manifold with a given affine connection is called an affine connection manifold, which we should considered in this chapter.

The modern treatment of connections will be given after the study of fiber bundles. The modern theory of connections usually starts with connection on the principal bundle. In this chapter, the classical definition of connection is used to analyze the local properties of connection on the tangent bundle of the manifold. After introducing the fiber bundles in Chapter 8, connection of the principal bundle and associated bundle will be studied in detail.

When a linear connection is given on the tangent bundle $T(M)$ of manifold M, namely, there is a given continuous linear mapping between sections of the tensor bundle of M:

$$\begin{array}{rlcl} \nabla: & T^r_s(M) & \to & T^r_{s+1}(M) \\ & K \in T^r_s(M) & \to & \nabla K \in T^r_{s+1}(M) \end{array}$$

∇K is called the absolute covariant differentiation of tensor field K or simply the covariant differentiation. Mapping ∇ satisfies:

1). Linearity :

$$\nabla(aK + bK') = a\nabla K + b\nabla K', \qquad a, b \in \mathbf{R}.$$

2). Leibniz rule:

$$\nabla(K \otimes K') = (\nabla K) \otimes K' + K \otimes \nabla K', \quad K, K' \in \mathcal{T}(M),$$

$$\nabla < \omega, X >=< \nabla\omega, X > + < \omega, \nabla X > . \tag{3.1}$$

3). For function $f \in \mathcal{F}(M)$,

$$\nabla f = df. \tag{3.2}$$

4). For tangent vector field $X \in \mathcal{X}(M)$, when the local frame of tangent vector fields is chosen,

$$X = \xi^a(x) e_a(x),$$

$$\nabla X = d\xi^a \otimes e_a + \xi^a \nabla e_a. \tag{3.3}$$

From Eq. 3.3, ∇X can be calculated if covariant differentiation ∇e_a of the bases $e_a(x)$ of tangent vector fields is given. Through the properties above, covariant differentiation of any tensor field can be carried out. ∇e_a denote the infinitesimal variance of bases e_a at the neighborhood of a point and can be expressed as

$$\nabla e_a(x) = \Gamma^b_a(x) e_b(x) \qquad (a = 1, 2, \cdots, n), \tag{3.4}$$

where

$$\Gamma^b_a(x) =< \theta^b, \nabla e_a >= \Gamma^b_{ia}(x) dx^i \qquad (a, b = 1, 2, \cdots, n).$$

are n^2 1-forms called connection 1-forms, and $\Gamma^b_{ia}(x)$ are n^3 continuous functions on M called connection coefficients. Repeated indices are summed over in the above two equations as is the convention.

In Eq. 3.3, in order that covariant differentiation of a tensor field is still a tensor field, connection term must be added because differentiation to the components of a tensor field is not components of a tensor field. From Eq. 3.4 and Eq. 3.3, we have

$$\nabla X = (d\xi^a + \xi^b \Gamma^a{}_b) e_a. \tag{3.5}$$

Using Eq. 3.1, connections on the cotangent bundle can be induced from connections on the tangent bundle. As the dual bases of $\{e_a\}$, there are $\{\theta^a\}, < \theta^a, e_b >= \delta^a_b$, which is easy to prove

$$\nabla \theta^a(x) = -\Gamma^a{}_b(x) \theta^b(x).$$

Hence, covariant differentiation of cotangent field $\omega(x) = \omega_a(x)\theta^a(x)$ is

$$\nabla\omega = (d\omega_a - \Gamma^b_a \omega_b)\theta^a. \tag{3.6}$$

Similarly, covariant differentiation of any (r, s)-type tensor fields can be carried out yielding $(r, s+1)$-type tensor fields.

We often need covariant derivative of a tensor field along a given direction on the manifold. Suppose that a curve c passes point p on manifold M and the tangent vector at p is X, we define the covariant derivative of tensor field $K \in \mathcal{T}^r_s(M)$ along X as:

$$\nabla_X K = < \nabla K, X > \in \mathcal{T}^r_s(M), \tag{3.7}$$

where ∇K is a (r, s+1)-type tensor field. After ∇K contracts with tangent vector X, $\nabla_X K$ is a (r, s)-type tensor field, so K and $\nabla_X K$ are of the same type. $\nabla_X K$ is called the directional covariant derivative. From Eq. 3.7, directional covariant derivatives along different directions can be linearly added together:

$$\nabla_{fX+gY} K = f\nabla_X K + g\nabla_Y K, \qquad \forall\ f, g \in \mathcal{F}(M). \tag{3.8}$$

When a local coordinate system is given, directional covariant derivative along the natural bases ∂_i is:

$$\nabla_i K = < \nabla K, \partial_i >,$$
$$\nabla K = dx^i \nabla_i K.$$

For instance, from Eq. 3.5

$$\nabla_i X = (\xi^a_{,i} + \Gamma^a_{ib}\xi^b)e_a \equiv \xi^a_{;i}e_a.$$

Note that the ordinary derivative $\xi^a_{,i} \equiv \frac{\partial \xi^a}{\partial x^i}$, but

$$\xi^a_{;i} \equiv (\nabla_i X)^a = \xi^a_{,i} + \Gamma^a_{ib}\xi^b \tag{3.9}$$

is the covariant derivative of the components of the vector field. Similarly, from Eq. 3.6

$$\nabla_i \omega = \omega_{a;i}\theta^a, \qquad \omega_{a;i} \equiv (\nabla_i \omega)_a = \partial_i \omega_a - \Gamma^b_{ia}\omega_b.$$

Derivative of the components of any tensor field can be defined in the same way. For example, covariant derivative of the components of (1,1)-type tensor $K = K^a_b \theta^b \otimes e_a$ is

$$\nabla_i K = K^a_{b;i}\theta^b \otimes e_a, \qquad K^a_{b;i} = \partial_i K^a_b - \Gamma^c_{ib}K^a_c + \Gamma^a_{ic}K^c_b. \tag{3.10}$$

The following is a comparison between covariant derivative and Lie derivative. For a (1, 1)-type tensor field K in the natural frame

$$(\nabla_i K)^j_k = K^j_{k;i} = \partial_i K^j_k - \Gamma^l_{ik}K^j_l + \Gamma^j_{il}K^l_k,$$

$$(L_X K)^j_k = \xi^i \partial_i K^j_k + \partial_k \xi^l K^j_l - \partial_l \xi^j K^l_k. \tag{3.11}$$

From these two equations, we see that covariant derivative of tensor field K at point p depends on connection coefficients at p only and covariant derivatives along different directions at the same point p can be linearly added together. However, Lie derivative of tensor field K along vector field X at p depends not only on the tangent vector X_p at p but also on tangent field $X(p) \in \mathcal{X}(M)$ at the neighborhood of p, both the vector field itself and the derivative of the vector field appear as in Eq. 3.11 . For Lie derivative, there is no similar linear addition as in Eq. 3.8 for covariant derivative. Only the following relation is true:

$$L_{aX+bY}K = aL_XK + bL_YK, \qquad a, b \in \mathbf{R}.$$

To carry out covariant derivative for tensor fields on the manifold, an additional structure on the manifold has to be introduced. The connection coefficients $\{\Gamma^a_{ib}(x)\}$ has to be given first of all, which are continuous functions on M and have three indices: i is the index of the local natural bases on base manifold which labels directions on the base manifold; a, b are indices of tangent bundle frames. Are these indices indeed the indices of a tensor? To answer this question, transformation properties of these indices are considered. Supposing that there is a linear transformation of the bases $\{e_a(x)\}$ of tangent vector field

$$e_a(x) \to e'_a(x) = L^b_a(x)e_b(x), \qquad \det(L^b_a(x)) \neq 0,$$

the components $\xi^a(x)$ of vector field $X = \xi^a(x)e_a(x)$ are contravariant

$$\xi^a(x) \to \xi'^a(x) = (L^{-1}(x))^a{}_b\xi^b(x). \tag{3.12}$$

As covariant derivative ∇_iX of tangent vector field X is still a tangent vector field

$$\nabla_iX = \xi^a_{;i}e_a,$$

the components $\xi^a_{;i}$ must be also contravariant:

$$\xi^a_{;i} \to \xi'^a_{;i} = (L^{-1})^a{}_b\xi^b_{;i}. \tag{3.13}$$

Because the transformation matrix $(L^a_b(x))$ depends on the coordinate, when $\xi^a(x)$ are transformed by Eq. 3.12, $\xi^a_{,i} \equiv \frac{\partial \xi^a}{\partial x^i}$, which is the ordinary derivative with respect to the coordinates, transforms as

$$\xi'^a_{,i} = (L^{-1})^a{}_b\xi^b_{,i} + (L^{-1})^a{}_{b,i}\xi^b. \tag{3.14}$$

Compare Eq. 3.12 with Eq. 3.13, we can see that $\xi^a_{;i}$ and ξ^a are transformed in the same way owing to the introduction of connection. Hence, vector field $\{\xi^a_{;i}\}$ is called the covariant derivative of vector field $\{\xi^a\}$ and ∇_i is called covariant derivative operator.

To make sure that Eq. 3.13 is correct, Eq. 3.9 and Eq. 3.14 are put into Eq. 3.12, we can get

$$\begin{aligned}(L^{-1})^a{}_b(\xi^b_{,i} + \Gamma^b_{ic}\xi^c) &= \xi'^a_{,i} + \Gamma'^a_{ic}\xi'^c \\ &= (L^{-1})^a{}_b\xi^b_{,i} + (L^{-1})^a{}_{b,i}\xi^b + \Gamma'^a_{ib}(L^{-1})^b{}_c\xi^c,\end{aligned}$$

hence,

$$\Gamma'^{a}_{id} = (L^{-1})^{a}{}_{b}\Gamma^{b}_{ic}L^{c}_{d} + (L^{-1})^{a}{}_{c}L^{c}_{d,i}, \tag{3.15}$$

which can be written in matrix form:

$$\Gamma' = L^{-1}(\Gamma + d)L, \tag{3.16}$$

where matrices

$$\Gamma = (\Gamma^{a}{}_{b}), \qquad L = (L^{a}_{b}), \qquad L^{-1} = ((L^{-1})^{a}{}_{b}).$$

Because of the extra secound term in Eq. 3.15, indices of the connection 1-form $\Gamma^{a}{}_{b}$ are not tensor indices. Therefore, the connection is not a tensor field and it is possible to choose a local frame such that the connection at point p is zero but its derivative is not. This local frame is called the normal frame field at point p. The normal frame field often helps us to simplify calculation with tensor fields.

The connection coefficients are a set of continuous functions on manifold M and are transformed as Eq. 3.15 under the transformation of tangent frame. There is a great of arbitrariness in the determination of connection coefficients. For any vector bundle on M, connections which satisfying Eq. 3.16 always exist. There is a theorem.

Theorem 3.1 *On any vector bundle of manifold M, affine connection always exists.*

Proof: Let $\{U_\alpha\}$ be an open covering of M and $\{\rho_\alpha\}$ be a partition of unity subordinate to the covering. For each open set U_α there is a tangent frame field $\{e^{(\alpha)}_a\}$. At the overlapping region where $U_\alpha \cap U_\beta \neq 0$, the bases are transformed by an $n \times n$ non-singular matrix

$$e^{(\alpha)}_a = A^{b}_{(\alpha\beta)a} e^{(\beta)}_b,$$

or in matrix form

$$E_\alpha = A_{\alpha\beta} E_\beta,$$

where $A_{\alpha\beta} = (A^{b}_{(\alpha\beta)a})$ is an $n \times n$ non-singular matrix and $E_\alpha = (e^{(\alpha)}_a)$ is a column matrix. At the overlapping region where $U_\alpha \cap U_\beta \cap U_\gamma \neq 0$, $A_{\alpha\beta}$ satisfies the compatibility condition:

$$A_{\alpha\beta}A_{\beta\gamma} = A_{\alpha\gamma}.$$

At each open set U_α, there exists an $n \times n$ matrix function $\sigma_\alpha(x)$ which is of n^2 arbitrary smooth functions. Let

$$\omega_\beta \equiv \sum_\alpha \rho_\alpha (A_{\alpha\beta})^{-1}(\sigma_\alpha + d)A_{\alpha\beta},$$

it is easy to prove that

$$\begin{aligned}(A_{\beta\gamma})^{-1}\omega_\beta A_{\beta\gamma} &= \sum_\alpha \rho_\alpha (A_{\beta\gamma})^{-1}(A_{\alpha\beta})^{-1}(\sigma_\alpha + d)A_{\alpha\beta}A_{\beta\gamma} - (A_{\beta\gamma})^{-1}dA_{\beta\gamma} \\ &= \omega_\gamma - (A_{\beta\gamma})^{-1}dA_{\beta\gamma}\end{aligned}$$

satisfies the transformation rule Eq. 3.15 of connection 1-form at the overlapping region. This completes our proof. It is clear that the determination of connection 1-forms has a great deal of arbitrariness.

Linear differential operators which act on tensor fields on the manifold are mainly the exterior differentiation d, Lie derivative L_X and covariant differentiation ∇. The comparison of these three fundamental differential operators is given in table 3.1.

Table 3.1 The Comparison of Three Differential Operators On the Tensor Fields

exterior diff. d	Lie deriv. L_X	covariant diff. ∇
1)linear: $\Lambda^r \to \Lambda^{r+1}$ 2)skew diff.: $d(\alpha_p \wedge \beta_q) = d\alpha_p \wedge \beta_q$ $+(-1)^p \alpha_p \wedge d\beta_q$ 3)$df = f_{,i}dx^i$ 4)$d^2 = 0$	1)linear: $T^r_s \to T^r_s$ 2)even diff.: Leibniz rule OK 3)$L_X f = Xf = \xi^i f_{,i}$ 4)$L_X Y = [X, Y]$	1)linear :$T^r_s \to T^r_{s+1}$ 2)even diff.: Leibniz rule OK 3)$\nabla f = df$ 4)$\nabla X = (d\xi_i + \Gamma^i{}_j \xi^j)\partial_i$
no additional structure, act on diff. forms only	given vector field X act on all tensor fields	given conn. 1-form $\Gamma^i{}_j$ act on all tensor fields

The connection is closely related to the dynamics of a particle moving on a manifold M. When a particle is moving along a curve $x(t)$ on M, the velocity of the particle is determined by the tangent vector X of the curve:

$$X = \frac{d}{dt} = \frac{dx^i}{dt}\frac{\partial}{\partial x^i},$$

where the derivative $\frac{dx^i}{dt}$ along the coordinate axis is the component of the velocity vector. However, $\frac{d^2x^i}{dt^2}$ is not the component of the acceleration vector which depends on the forces, as it is not the component of any vector field and depends on the choice of the local coordinate systems so is neither a geometric object nor physical object.

How can the acceleration vector of the particle moving on manifold M be determined? Because the acceleration vector is the directional derivative of the velocity vector, the key is to get the directional derivative of the vector field X (for instance, the velocity of the particle) along direction Y (for instance, the direction of the force) so that the result is still a vector. To achieve this, our first candidate is the Lie derivative $L_Y X$. However, Lie derivative depends not only on the vector field at point p where the differentiation is taken, but also on the rate of variation of the vector field Y at the neighborhood of p. But the acceleration of a particle moving along a curve depends only on the properties of the points where the curve lies. In addition, Lie derivative of tangent vector field along the tangent direction of the corresponding integral curve is zero: $L_X X = [X, X] = 0$. Therefore, when we deal with the acceleration of a particle moving along the curve, the application of Lie derivative is improper and covariant derivative of the velocity must be applied, i.e., affine connections must be introduced on the manifold.

3.3 The curvature 2-form and the curvature tensor

To carry out covariant differentiation to tensor fields on the manifold, we must have the connection 1-forms on the tangent bundle of the manifold a priori:

$$\Gamma^a{}_b(x) = \Gamma^a_{ib}(x)dx^i.$$

With the help of connections, relations among tangent spaces at different points on the manifold can be established. However, from Eq. 3.15 we know that connection $\Gamma^a{}_b(x)$ itself is not a tensor field so is not a geometric object. How can we get the geometrically meaningful tensor field from the connection? The set $\{\Gamma^a{}_b(x)\}$ can be written as a $n \times n$ matrix $\Gamma(x)$ and Pfaff equation Eq. 3.16 in matrix form can be obtained. Through Frobenius theorem in §2.3 and applying the exterior differential operator d to all equations, geometric object satisfying the local integrable condition can be obtained. Eq. 3.16 can be written as

$$L\Gamma' = \Gamma L + dL. \tag{3.17}$$

Exterior differentiation of this equation gives

$$dL \wedge \Gamma' + Ld\Gamma' = d\Gamma L - \Gamma \wedge dL,$$

and use the equation 3.17 again, we have

$$L(\Gamma' \wedge \Gamma' + d\Gamma') = (d\Gamma + \Gamma \wedge \Gamma)L.$$

Let

$$\Omega \equiv d\Gamma + \Gamma \wedge \Gamma, \tag{3.18}$$

which is called the curvature matrix on the manifold M determined by connection Γ. Its components are:

$$\Omega^a{}_b = d\Gamma^a{}_b + \Gamma^a{}_c \wedge \Gamma^c{}_b, \tag{3.19}$$

and is called the curvature 2-form. These equations show that under a linear transformation of the local frame

$$\Omega' = L^{-1}\Omega L,$$

i.e., curvature 2-form is a tensor field and is of great significance on the manifold. The indices a, b of $\Omega^a{}_b$ are tensor indices. By use of $\{\ \Omega^a{}_b\}$, differential forms which are well defined globally can be constructed on M, of which the properties are closely related to the global topology of the manifold. One important subject of differential geometry is to study the relation between curvature form and the topology of the manifold.

Eq. 3.18 is called the structure equation for connection. Exterior differentiation of the structure equation gives

$$d\Omega = \Omega \wedge \Gamma - \Gamma \wedge \Omega, \tag{3.20}$$

which is the integrable condition of the structure equation and is called the Bianchi identity. Connection on the manifold must satisfy the integrable condition which is equivalent to the Jacobi identity of the commutator of covariant derivatives:

$$[[\nabla_X, \nabla_Y], \nabla_Z] + [[\nabla_Y, \nabla_Z], \nabla_X] + [[\nabla_Z, \nabla_X], \nabla_Y] = 0. \tag{3.21}$$

In the following we will give some tensor operators induced from connection. The curvature 2-form $\Omega^a{}_b$ has one covariant index and one contravariant index, for a given local coordinate system, the 2-forms can be written as:

$$\Omega^a{}_b = \frac{1}{2} R^a_{bjk} dx^j \wedge dx^k, \qquad R^a_{bjk} \in \mathcal{F}(M),$$

where the coefficients R^a_{bjk} are smooth functions on the manifold, called the coefficients of the curvature tensor. The indices of R^a_{bjk} are tensor indices. R^a_{bjk} are components of the (1, 3)-type curvature tensor R. From Eq. 3.19, R^a_{bjk} can be written in terms of the connection coefficients Γ^a_{ib} and their derivative with respect to the coordinate:

$$R^a_{bik} = \Gamma^a_{kb,i} - \Gamma^a_{ib,k} + \Gamma^a_{ic}\Gamma^c_{kb} - \Gamma^a_{kc}\Gamma^c_{ib}. \tag{3.22}$$

From now on, we encounter second order covariant derivatives of tensor fields. Two ordinary derivatives commute but covariant derivatives do not commute in general:

$$[\partial_i, \partial_k] = 0, \qquad [\nabla_i, \nabla_k] \neq 0.$$

Suppose that $X, Y \in \mathcal{X}(M)$ are two arbitrary tangent vector fields on M, the commutator $[\nabla_X, \nabla_Y]$ of covariant derivatives, which is also an operator on tangent vector fields. However, if $[X, Y] \neq 0$, the commutator is not a linear operator on tangent vector fields, i.e.,:

$$[\nabla_X, \nabla_Y] fZ \neq f[\nabla_X, \nabla_Y]Z.$$

Operator $[\nabla_X, \nabla_Y]$ does not preserve the linear structure of the space of vector fields so is not a tensor operator. To get a linear tensor operator, we try to cancel the nonlinear parts while preserving the skewsymmetry with respect to tangent fields X, Y. Considering a combination of the commutator of covariant derivatives $[\nabla_X, \nabla_Y]$ and covariant derivative with respect to the $[X, Y]$, we introduce an operator

$$R(X, Y) \equiv [\nabla_X, \nabla_Y] - \nabla_{[X,Y]}. \tag{3.23}$$

Operator $R(X,Y)$ keeps the linear structure of tensor fields. For any tangent field Z, $R(X,Y)$ satisfies:

$$R(X,Y)fZ = fR(X,Y)Z.$$

$R(X,Y)$ is a (1,1)-type tensor operator. This operator is skew bilinear with respect to the tangent fields X, Y:

$$R(X,Y) = -R(Y,X),$$

$$R(fX+gY,Z) = fR(X,Z) + gR(Y,Z),\ \forall f,g \in \mathcal{F}(M),$$

here R is a (1,3)-type tensor field. When given the local coordinate system and the local frame fields, the components of R can be obtained by its action on the bases. Therefore, we have

$$R(\partial_i,\partial_k)e_b = R^a_{bik}e_a.$$

It is easy to prove that these coefficients R^a_{bik} are the same as those in Eq. 3.22 when written in terms of connection coefficients. Hence, the curvature tensor R which is obtained by the curvature operator $R(X,Y)$ is just the curvature tensor obtained by the curvature 2-form $\Omega^a{}_b$. The existence of the curvature operator is a result of non-commutativity covariant derivatives.

3.4 Torsion tensor

After the connection on the manifold is defined, parallel transport of tensor fields along any curve can be analyzed. In a local coordinate system, tangent vector X of curve C can be written as

$$X = \frac{d}{dt} = \frac{dx^i}{dt}\frac{\partial}{\partial x^i}.$$

If vector field $Y = \eta^i \frac{\partial}{\partial x^i}$ satisfies:

$$\nabla_X Y = \frac{dx^i}{dt}\left(\frac{\partial \eta^k}{\partial x^i} + \Gamma^k_{ij}\eta^j\right)\frac{\partial}{\partial x^k} = 0,$$

i.e., the components of Y satisfy

$$\frac{d\eta^k}{dt} + \Gamma^k_{ij}\eta^j\frac{dx^i}{dt} = 0, \tag{3.24}$$

we say that the vector field Y is parallel transport along curve C. By parallel transport along curve C, isomorphism among tangent spaces on each point of C can be established.

If a vector field Y which is a section of the tangent bundle of manifold M satisfies:

$$\nabla Y = 0, \tag{3.25}$$

then it is called a parallel section of the tangent bundle. The existence condition of a parallel section is also the integrable condition of Eq. 3.25. We can rewrite this equation in differential forms as:

$$d\eta^k + \Gamma^k_j \eta^j = 0.$$

According to Frobenius theorem, the integrable condition of the above Pfaff system of equations require that the ideal spanned by $\omega^{(k)}$

$$I = \{\omega^{(k)} = d\eta^k + \Gamma^k_j \eta^j\} \subset \Lambda^*(M)$$

should be a closed ideal. Exterior differentiation of $\omega^{(k)}$ gives

$$\begin{aligned} d\omega^{(k)} &= d\Gamma^k_j \eta^j - \Gamma^k_j \wedge d\eta^j \\ &= d\Gamma^k_j \eta^j - \Gamma^K_l \wedge (\omega^{(l)} - \Gamma^l_j \eta^j) \\ &= \Omega^k_j \wedge \eta^j - \Gamma^k_l \wedge \omega^{(l)}. \end{aligned}$$

So we can conclude that the condition that an ideal is a closed one is the same as that the curvature equals zero. There exists parallel sections on the tangent bundle and the parallel transport of the corresponding vector field has no relation with the path chosen only if the curvature equals zero. Otherwise, we can only discuss the parallel transport of a vector field along a given curve.

When the tangent vector X of a curve is parallel transported along the curve itself, the curve is called the geodesic curve. For each point on the geodesic curve, there is:

$$\nabla_X X = 0.$$

Taking $\eta^k = \frac{dx^k}{dt}$ in the Eq. 3.24, the equations for the geodesic curve are

$$\frac{d^2x^i}{dt^2} + \Gamma^i_{jk}\frac{dx^j}{dt}\frac{dx^k}{dt} = 0. \tag{3.26}$$

This is a system of second order ordinary differential equations as is well known. Through any point $p \in M$, there is only one geodesic line along a certain tangent direction. Notice that this equation for geodesics is insensitive to the skewsymmetric part of connection coefficients $\{\Gamma^i_{jk}\}$. Two connections defined on a manifold with identical symmetric parts give identical geodesic curves.

Given two connections on the manifold, there are two different covariant derivatives: ∇ and ∇'. By the difference of these two connections, a mapping can be defined

$$\begin{aligned} B : \mathcal{X} \otimes \mathcal{X} &\to \mathcal{X} \\ B(X,Y) &\equiv \nabla'_X Y - \nabla_X Y. \end{aligned}$$

From the definition of connections, mapping B satisfies

$$B(fX,Y) = fB(X,Y).$$

It is easy to prove that

$$B(X,fY) = (Xf)Y + f\nabla'_X Y - (Xf)Y - f\nabla_X Y = fB(X,Y).$$

That is to say, B is bilinear over $\mathcal{F}(M)$ and is a (1,2)-type tensor field independent of the choice of coordinate systems.

In local natural bases, we have

$$B(X,Y) = \xi^k \eta^l (\Gamma'^i_{kl} - \Gamma^i_{kl})\partial_i.$$

If connections giving identical geodesic curves are put in one equivalence class, the difference of the connections within one equivalence class manifests the skewsymmetric part of tensor B. Suppose the skewsymmetric part of B is A, then

$$\begin{aligned} A(X,Y) &= \tfrac{1}{2}\{B(X,Y) - B(Y,X)\} \\ &= \tfrac{1}{2}\{[\nabla'_X(Y) - \nabla'_Y(X)] - [\nabla_X(Y) - \nabla_Y(X)]\} \\ &= \tfrac{1}{2}\{C'(X,Y) - C(X,Y)\}, \end{aligned} \tag{3.27}$$

where

$$C(X,Y) \equiv \nabla_X(Y) - \nabla_Y(X),$$

is skewsymmetric with respect to X, Y, but is not bilinear over $\mathcal{F}(M)$. The following equation is true only for constants a, b:

$$C(aX, bY) = abC(X,Y), \qquad a, b \text{ are constants.}$$

Given two vector fields X, Y, a new vector field $L_X Y$ can be obtained, which is not bilinear over $\mathcal{F}(M)$ either. The new vector field satisfies:

$$L_{aX} bY = abL_X Y, \qquad a, b \text{ are constants.}$$

Combining $C(X,Y)$ and $L_X Y$, we obtain a bilinear mapping which is skew bilinear with respect to X, Y on $\mathcal{F}(M)$:

$$\begin{aligned} T : \mathcal{X} \otimes \mathcal{X} &\to \mathcal{X} \\ T(X,Y) &\equiv \nabla_X Y - \nabla_Y X - [X,Y]. \end{aligned} \tag{3.28}$$

satisfying:

$$T(X,Y) = -T(Y,X),$$

$$T(fX, gY) = fgT(X,Y), \qquad f, g \in \mathcal{F}(M).$$

T is a (1,2)-type tensor field, and is called torsion tensor. Eq. 3.27 can be written as

$$A(X,Y) = \frac{1}{2}\{T'(X,Y) - T(X,Y)\},$$

i.e., the skewsymmetric part of tensor B of the difference of connections is the difference of the torsion tensor.

The connection with vanishing torsion is called symmetric connection. In an equivalence class of connections giving identical geodesic curves, there is only one symmetric connection and torsion is used to distinguish different connections in a class, and the symmetric part of the tensor B of the difference of connections in a class is zero, so B(X,Y)=A(X,Y). So arbitrary connection can be written as the sum of symmetric connection and the half torsion tensor. Suppose that ∇'_X is the covariant derivative of symmetric connection, the covariant derivative ∇_X for arbitrary connection in one equivalence class is

$$\begin{aligned}\nabla_X Y &= \nabla'_X Y - B(X,Y) = \nabla'_X Y - A(X,Y) \\ &= \nabla'_X Y + \frac{1}{2}T(X,Y).\end{aligned}$$

In the neighborhood of point p on manifold M, the bases of the moving frame of tangent field are $\{e_a(x), a = 1, 2, \cdots, n\}$, which are the linear combination of the bases of the natural frame. The corresponding bases of cotangent field are $\{\theta^a(x), a = 1, 2, \cdots, n\}$, and:

$$< \theta^a, e_b >= \delta^a_b.$$

The bases $\{e_a(x)\}$ do not commute:

$$[e_a(x), e_b(x)] = f^c_{ab}(x)e_c(x),$$

where $f^c_{ab}(x) \in \mathcal{F}(M)$ are called structure functions of the moving frame. For the bases of cotangent field, by Eq. 2.34, we have

$$d\theta^a(e_b, e_c) = -\theta^a([e_b, e_c]) = -f^a_{bc},$$

i.e.,

$$d\theta^a = -\frac{1}{2}f^a_{bc}(x)\theta^b \wedge \theta^a. \tag{3.29}$$

When one choice the bases $\{e_a, a = 1, 2, \cdots, n\}$, Eq. 3.28 becomes

$$T(e_a, e_b) = \nabla_{e_a} e_b - \nabla_{e_b} e_a - [e_a, e_b] = T^c_{ab} e_c,$$

where $\{T^c_{ab}\}$ are coefficients of the torsion tensor in moving frame. From the definition of connections, we have

$$\nabla_{e_a} e_b = e^i_a \nabla_i e_b = e^i_a \Gamma^c_{ib} e_c.$$

Hence, we have

$$T^c_{ab} = e^i_a \Gamma^c_{ib} - e^i_b \Gamma^c_{ia} - f^c_{ab}. \tag{3.30}$$

As structure functions are skewsymmetric with respect to the lower indices, the components T^c_{ab} of the torsion tensor are also skewsymmetric with respect to the lower indices. Hence, a torsion 2-form τ^c can be constructed from T^c_{ab} as

$$\tau^c = \frac{1}{2} T^c_{ab} \theta^a \wedge \theta^b. \tag{3.31}$$

Put Eq. 3.30 into Eq. 3.31, we have

$$\tau^c = d\theta^c + \Gamma^c{}_b \wedge \theta^b. \tag{3.32}$$

Eq. 3.32 and Eq. 3.18 are called Cartan's structure equations for connections on the manifold. The forms of the two equations are independent of the choice of frames. Exterior derivatives of these two structure equations yield integrable conditions of structure equations, which are called Bianchi identities. The Bianchi identities are Eq. 3.20 and

$$d\tau^a = \Omega^a{}_b \wedge \theta^b - \Gamma^a{}_b \wedge \tau^b, \tag{3.33}$$

which are integrable conditions of connections on the manifold.

The skewsymmetric part of the commutator of two covariant derivatives of arbitrary tensor fields is completely determined by the curvature tensor and the torsion tensor. For example, for a function $f \in \mathcal{F}(M)$,

$$f_{;ij} - f_{;ji} = -T^k_{ij} f_{;k}.$$

For a vector field $X = \xi^i \partial_i \in \mathcal{X}(M)$,

$$\xi^i_{;kl} - \xi^i_{;lk} = R^i_{jkl} \xi^j - T^j_{kl} \xi^i_{;j}. \tag{3.34}$$

For (1, 2)-type tensor field K,

$$K^{ij}_{k;pq} - K^{ij}_{k;qp} = K^{lj}_k R^i_{lpq} + K^{il}_k R^j_{lpq} - K^{ij}_l R^l_{kpq} - K^{ij}_{k;l} T^l_{pq}.$$

If we use ∇_i for the covariant derivative components of tensor fields, the above three equations can be written in one formula

$$[\nabla_p, \nabla_q] = R^\times_{\times pq} - T^j_{pq} \nabla_j,$$

where the first term in the right hand side contributes one term for each tensor index and has no contribution for scalar field. For symmetric connection with no torsion, the Bianchi identities Eq. 3.33 and Eq. 3.20 can be written by components of the curvature tensor under the natural frame as follows:

$$R^k_{lij} + R^k_{ijl} + R^k_{jli} = 0,$$
$$R^k_{lij;m} + R^k_{ljm;i} + R^k_{lmk;j} = 0.$$

3.5 Covariant exterior differential

Curvature 2-form $\Omega^a{}_b$ and torsion 2-form τ^a are differential forms with tensor indices. For differential forms with tensor indices, covariant exterior differential operator D and the related Lie derivative L_X can be defined. In the following, covariant exterior differential D, linear operator D^2 and Lie derivative L_X are studied.

Covariant exterior differential D is a generalization of exterior derivative d. The operator d acts on differential forms, while D acts on tensor-valued differential forms.

$$\begin{aligned} D: \ & T^r_s \otimes \Lambda^q && \to T^r_s \otimes \Lambda^{q+1} \\ & K\omega && \to D(K\omega) = (DK)\wedge\omega + K d\omega. \end{aligned}$$

Covariant exterior differential D can be viewed as generalization of covariant differential ∇ defined in Eq. 3.2, 3.3 and 3.4, and it satisfies:

1).

$$D\omega = d\omega, \qquad \omega \in \Lambda^k(M).$$

2).

$$D(X\omega) = DX\wedge\omega + Xd\omega, \qquad X \in \mathcal{X}(M),$$

where DX is as defined in Eq. 3.3 and Eq. 3.4:

$$DX = D(\xi^a e_a) = d\xi^a e_a + \xi^a \Gamma^b_a e_b = (d\xi^a + \xi^b \Gamma^a{}_b) e_a.$$

The connection 1-forms $\{\Gamma^a{}_b\}$ can be written in matrix form. The components $\{\xi^a(x)\}$ of vector $X = \xi^a(x)e_a$ can be written as a column vector $\xi(x)$, then the above equation can be written as

$$D\xi = d\xi + \Gamma\xi,$$

or

$$D\xi^a \equiv (D\xi)^a = d\xi^a + \Gamma^a{}_b\xi^b.$$

Under a change of bases of the tangent space of manifold M, we have

$$X = \xi^a e_a = \xi'^a e'_a,$$

where

$$e'_a = L^b_a e_b, \qquad \xi'^a = (L^{-1})^a{}_b \xi^b. \tag{3.35}$$

In matrix form the transformation of the components of vector field is

$$\xi' = L^{-1}\xi,$$

and

$$\begin{aligned} D\xi' &= d\xi' + \Gamma'\xi' \\ &= d(L^{-1}\xi) + (L^{-1}\Gamma L + L^{-1}dL)L^{-1}\xi \\ &= L^{-1}D\xi, \end{aligned} \tag{3.36}$$

where we have used the transformation formulas Eq 3.16 of connection matrix. This equation shows that $D\xi$ and ξ obey the same transformation rule. Covariant exterior differential is a mapping from tensor-valued differential forms to differential forms of one more order over the same order tensor.

For a general (r,s) type tensor valued k-forms Σ, its components can be written in the local frame as

$$\Sigma^{a_1\cdots a_r}_{b_1\cdots b_r} = \Sigma^{a_1\cdots a_r}_{b_1\cdots b_r i_1\cdots i_k} dx^{i_1} \wedge \cdots \wedge dx^{i_k}.$$

When the moving frame $\{e_a\}$ and $\{\theta^a\}$ are transformed according to Eq. 3.35, then Σ transforms as:

$$\Sigma'^{a_1\cdots a_r}_{b_1\cdots b_r} = (L^{-1})^{a_1}_{a'_1}\cdots(L^{-1})^{a_r}_{a'_r}\Sigma^{a'_1\cdots a'_r}_{b'_1\cdots b'_r} L^{b'_1}_{b_1}\cdots L^{b'_r}_{b_r}.$$

In the following, for simplicity, we assume that Σ is a $(1,1)$-type tensor valued k-form. Therefore Σ transforms as:

$$\Sigma' = L^{-1}\Sigma L,$$

here Σ can be written as a tensor product of a (1,1)-type tensor and a k-form ω

$$\Sigma = K\omega.$$

Covariant exterior differential of Σ is

$$\begin{aligned} D\Sigma &= D(\omega K) \\ &= d\omega K + (-1)^k\omega\wedge DK \\ &= d\Sigma + (-1)^k\omega\wedge(\Gamma K - K\Gamma) \\ &= d\Sigma + \Gamma\wedge\Sigma - (-1)^k\Sigma\wedge\Gamma \\ &\overset{def}{=} d\Sigma + [\Gamma,\Sigma], \end{aligned}$$

where the bracket $[\Gamma,\Sigma]$ means that differential forms take the exterior product which are abstracted from the bracket with the remainder matrix parts taking the commutator. It can be proved that thus defined covariant exterior differential $D\Sigma$ transforms in the same way as Σ under a change of frames

$$(D\Sigma)' = L^{-1}(D\Sigma)L.$$

The covariant exterior differential for any tensor valued differential form can be defined similarly

$$D = d + [\Gamma,\ \ ], \tag{3.37}$$

where the bracket means that Γ is applied to the left hand side of the given differential form for each contravariant index, and Γ is applied to the right hand side of the given differential form with the additional coefficient $(-1)^{k+1}$ for each covariant index (If the given differential forms has no tensor index, there is no contribution from Γ.).

Covariant exterior differential ensures that tensor-valued k-form is mapped to $(k+1)$-form over the same order tensor. Covariant exterior differential defined by Eq. 3.37 preserves transformation properties of tensors.

Many systems of equations can be expressed more simply in terms of covariant exterior differential. For example, the two Cartan's structure equations and the corresponding two Bianchi identities can be written and remembered as:

	Formule by using D	Tensor expression
Structure equations	$\Omega = D\Gamma$(Formally)*	${\Omega^a}_b = d{\Gamma^a}_b + {\Gamma^a}_c \wedge {\Gamma^c}_b$
	$\tau = D\theta$	$\tau^a = d\theta^a + {\Gamma^a}_c \wedge \theta^c$
Bianchi Identities	$D\Omega = 0$	$d{\Omega^a}_b = \Omega^a_c \wedge {\Gamma^c}_b - {\Gamma^a}_c \wedge \Omega^c_b$
	$D\tau = \Omega \wedge \theta$	$d\tau^a = \Omega^a_c \wedge \theta^c - {\Gamma^a}_c \wedge \tau^c$

*As connection Γ itself is not a tensor valued differential form, it does not make much sense to talk about covariant exterior differential of Γ, here is formally only.

For exterior differential d, we have $d^2 = 0$. However, for covariant exterior differential, it is easy to check:

$$\begin{aligned} D^2(\omega X) &= D(d\omega X + (-1)^n \omega \wedge DX) \\ &= (-1)^{k+1} d\omega \wedge DX + (-1)^k d\omega \wedge DX + \omega \wedge D^2 X \\ &= \omega \wedge D^2 X. \end{aligned} \tag{3.38}$$

When $k = 0$,

$$D^2(fX) = f(D^2 X), \tag{3.39}$$

which shows that D^2 is a linear operator. Now we calculate D^2X from the definition of covariant exterior differential Eq. 3.37.

$$\begin{aligned} D^2 X &= \xi^a (D^2 e_a) = \xi^a D(\Gamma^b_a e_b) \\ &= \xi^a (d\Gamma^b_a + {\Gamma^b}_c \wedge \Gamma^c_a) e_b \\ &= \xi^a \Omega^b_a e_b. \end{aligned}$$

In matrix form, it reads

$$D^2\xi = \Omega\xi, \quad \text{or} \quad D^2\xi^a \equiv (D^2\xi)^a = {\Omega^a}_b \xi^b. \tag{3.40}$$

Operator D^2 is equivalent to a linear operator (It is not a differential operator.). When D^2 is applied to tensor-valued differential forms, D^2 can be written as

$$D^2 = [\Omega, \ ], \tag{3.41}$$

where the bracket means that Ω is applied to the left hand side of the given differential form for each contravariant index and Ω is applied to the right hand side of the given differential form with an additional coefficient -1 for each covariant index. For instance, for the (1,1)-type tensor-valued differential forms Σ,

$$D^2\Sigma = [\Omega, \Sigma]. \tag{3.42}$$

Finally, we introduce the Lie derivative which can be applied to tensor-valued differential forms. It is defined as

$$L_X : T^r_s \otimes \Lambda^k \to T^r_s \otimes \Lambda^k.$$

Lie derivative of k-form $\omega \in \Lambda^k$ is

$$L_X \omega = (d \cdot i_X + i_X \cdot d)\omega, \quad \omega \in \Lambda^k(M).$$

For tensor valued differential forms, Cartan formula can be generalized to

$$L_X = d \cdot i_X + i_X \cdot d - [L_\xi, \ ], \tag{3.43}$$

where L_ξ is a matrix with elements $\{\xi^k_{,i}\}$ (here ξ^k are components of vector field $X = \xi^k \partial_k$). The bracket in Eq. 3.43 means that $L_\xi = (\xi^k_{,i})$ is applied to the left hand side of the given differential form for each contravariant index and L_ξ is applied to the right hand side of the given differential form with an additional coefficient -1 for each covariant index. For example, for differential form Σ on (1,1)-type tensor,

$$L_X \Sigma = d \cdot i_X \Sigma + i_X \cdot d\Sigma - L_\xi \Sigma + \Sigma L_\xi. \tag{3.44}$$

In ordinary tensor analysis, tensors are often given in components. The components of tensors are 0-forms with tensor indices. Covariant derivative of the components is given in Eq. 3.10, which is equivalent to Eq. 3.37. Lie derivative of the components is given in Eq. 3.11, which is equivalent to Eq. 3.44. In this case, the first term on the right hand side of Eq. 3.44 is zero as i_X applied to 0-form gives zero.

3.6 Holonomy group of connections

Tangent space $T_p(M)$ at point p on an n-dimensional manifold M is an n-dimensional vector space. There are n linearly independent vectors $\{e_a(p)\}$ which are called the bases of the vector space $T_p(M)$. Any tangent vector X_p is a linear combination of the bases:

$$X_p = \sum \xi^a(p) e_a(p),$$

where the set of bases $\{e_a(p), a = 1, 2, \cdots, n\}$ is called the frame of the tangent field at point p.

When connection is given on the manifold, parallel transport of a vector along a curve on the manifold is defined and so is the parallel transport of the frame along a curve on the manifold. Now choose a closed curve through point p. Parallel transport of the given frame along the closed curve which starting from p and returning to p will lead to a new frame at point p. This new frame is a linear transformation of the old frame so that the parallel transport of a frame along a closed curve is an automorphism of tangent space $T_p(M)$. For different closed curves through point p, we have different automorphism of tangent space $T_p(M)$. The set of the automorphism of $T_p(M)$ constitutes a group which is called the holonomy group of connections with reference point p. The holonomy group associated with affine connections is a subgroup of $GL(n, \mathbf{R})$.

For a connected manifold, the holonomy group is independent of the choice of the starting point p and is determined by the topological properties of the manifold. If the holonomy group of a manifold can be reduced to the subgroup G of $GL(n, \mathbf{R})$, we say that the manifold is of G structure. In the follows, we give some examples to highlight the idea of the holonomy group.

For an orientable manifold, parallel transport of the frame along a closed curve preserves the orientation of the frame and holonomy group is a connected subgroup of $SL(n, \mathbf{R})$.

The phase space in classical mechanics is an even dimensional orientable manifold with symplectic structure. The holonomy group of the manifold can be reduced to $SP(n, \mathbf{R})$ and the manifold is called symplectic manifold.

An n-dimensional complex manifold is a real $2n$-dimensional orientable analytic manifold. The holonomy group can be reduced to $SL(n, C)$. The manifold is of complex structure.

The manifold whose holonomy group is the identity element is called a parallelizablel manifold. For example, there exists n left (right) invariant vector fields for the simple Lie group manifold so it is a parallelizable manifold. The n-dimensional torus T^n is also a parallelizable manifold.

There are n linear independent tangent fields everywhere on a parallelizable manifold M. This is a very strong condition. Many manifolds have no so good property. Of n-dimensional sphere S^n, only S^1, S^3 and S^7 are parallelizable manifolds, which are related to the existence of the complex number, the quaternion and the octonion (Ref. Appendix H). The manifold of the unit complex is S^1 ($U(1)$ group manifold). The manifold of the unit quaternion is S^3 ($SU(2)$ group manifold), and the manifold of the unit octonion is S^7 ($SP(2)/SP(1)$ homogeneous space).

3.7 Berry phase, holonomy in physical system

The time dependent Schrödinger equation in quantum mechanics is:

$$i\hbar\frac{\partial}{\partial t}|\psi(t)>= H|\psi(t)> . \tag{3.45}$$

For a system described by two sets of variables

$$H = \frac{P^2}{2M} + \frac{p^2}{2m} + V(R, r) \equiv \frac{P^2}{2M} + h(R),$$

in which the capitalized coordinates and momentum R, P are the so called "slow moving" variables, and r, p are called the "fast moving" variables. In the adiabatic approximation (Born-Oppenheimer approximation), one starts by ignoring the kinetic energy (the motion of the slow coordinates) and take R approximately as a constant, then solves the stationary Schrödinger equation for the fast Hamiltonian $h(R)$ depending on the slow coordinate R

$$h(R)|n, R> = E_n|n, R> .$$

Notice here that the fast moving makes the differences between excitation eigenvalues sufficiently large so that although R changes slowly with time, $R = R(t)$, it is not quick enough to induce transition from one eigenstate to another.

The adiabatic approximation implies that the time evolution of a non-degenerate eigenstate can be written as:

$$|\psi(t)> = e^{-\frac{i}{\hbar}\int_0^t d\tau E_n(R(\tau))} e^{i\gamma_n(t)}|n, R(t)>, \tag{3.46}$$

i.e., the final eigenstate $|\psi(t)>$ is related with the static $|n, R(t)>$ by two phase factors, $\frac{i}{\hbar}\int_0^t d\tau E_n(R(\tau))$ comes from dynamics, the additional phase $\gamma_n(t)$ shows up if the time evolution is periodic so that the parameter $R(t)$ returns to its original value at a later time T, $R(T) = R(0)$, i.e., a non-trivial phase may arise from a closed path in parameter space. Insert the adiabatic approximation ansatz Eq. 3.46 into Eq. 3.45 yields

$$\begin{aligned}
\dot{\gamma}_n(t) &= i < n, R(t)|\frac{d}{dt}|n, R(t)>, \\
\gamma_n(t) &= \int_0^T dt\ \dot{\gamma}_n(t) \\
&= i\int_0^T dt\ \dot{R}(t) < n, R(t)|\nabla_R|n, R(t)> \\
&= \oint d\vec{R}\cdot\vec{A}(R), \\
\text{where}\quad \vec{A}(R) &= i < n, R|\nabla_R|n, R> .
\end{aligned}$$

In general the eigenstate $|n, R>$ has an arbitrary phase, if

$$|n, R> \longrightarrow e^{i\alpha(R)}|n, R>,$$

then $A(R)$ is changed correspondingly as

$$\vec{A}(R) \longrightarrow \vec{A}(R) - \nabla\alpha,$$

which is quite similar to the transformation of the connection 1-form. Thus the gauge invariant Berry phase can be obtained as:

$$\gamma_n = \oint d\vec{R}\cdot\vec{A}(R) = \int dS\cdot\nabla\times\vec{A},$$

or in terms of differential forms, let $A = \vec{A} \cdot d\vec{R}$ be a 1-form in parameter space,

$$\gamma_n = \int_{\partial S} A = \int_S dA = \int_S B.$$

Here $B = dA$ is a "curvature" 2-form in parameter space. This "geometric phase" is independent of the choice of phase of the wave function and can be measured experimentally. It is just the $U(1)$-holonomy we have studied in this chapter.

Next we consider the case where the state is degenerate, so the system no longer necessarily returns to the original eigenstate, in general it ends up in one of the degenerate states so the accumulated "phase" for an n-fold degenerate eigenvalue is a $U(n)$ matrix:

$$|\psi_\alpha(t) >= e^{-\frac{i}{\hbar}\int_0^T dt\ E(t)} U_{\alpha\beta}|\psi_\beta(0)>,$$

where the unitary matrix $U_{\alpha\beta}$ can be written as a path-ordered exponential integral:

$$\begin{aligned} U &= P\exp\int_0^T dt\ \vec{A}\cdot\dot{\vec{x}} \\ &= 1 + \int_0^T dt\ \vec{A}\cdot\dot{\vec{x}} + \int_0^T dt \int_0^T dt' \vec{A}(t)\cdot\dot{\vec{x}}(t)\vec{A}(t')\cdot\dot{\vec{x}}(t') + \cdots \ , \end{aligned}$$

which is called the Wilson line integral in gauge theory. The path-order P is so defined that all the non-Abelian matrices $\vec{A}(x)$, in the expansion of the exponential, are ordered as points in the path are, with the later times on the right. Or, in this case, the connection 1-form $A = \vec{A} \cdot d\vec{x}$ is a matrix valued 1-form in the degenerate case, $A = (A_{ab})$, where

$$A_{ab} = i < n, a, R|d_R|n, b, R >$$

is a 1-form in the parameter space. The curvature 2–form B is a matrix valued 2–form,

$$B = dA + A \wedge A,$$

it is gauge covariant and corresponds to the Yang-Mills gauge field strength in non-Abelian gauge theory. Matrix

$$U = (U_{\alpha\beta})$$

is the $U(n)$-holonomy over the parameter space.

Berry phase arises from non-trivial topological properties of the parameter space, forms A and B are the corresponding characteristic forms of the associated bundle on the parameter space.

Notations and Formulea

	Natural frame	Moving frame
Bases of vector	$\{\partial_i\}_1^n$ $[\partial_i, \partial_j] = 0$	$\{e_a = e_a^i(x)\partial_i\}_{a=1}^n$ $[e_a, e_b] = f_{ab}^c(x)e_c$
Bases of covector	$\{dx^i\}_1^n$ $d(dx^i) = 0$	$\{\theta^a = \theta_i^a(x)dx^i\}_{a=1}^n$ $d\theta^a = -\frac{1}{2}f_{bc}^a(x)\theta^b \wedge \theta^c$
duality	$< dx^i, \partial_j >= \delta_j^i$	$< \theta^a, e_b >= \delta_b^a$

- The covariant differential $\nabla : T_s^r \to T_{s+1}^r$

$$\begin{aligned} \nabla e_a &= \Gamma_a^b e_b, \\ \nabla \theta^a &= -\Gamma^a{}_b \theta^b. \end{aligned}$$

The connection 1-forms:

$$\Gamma^a{}_b =< \theta, \nabla e_b >= \Gamma_{ib}^a dx^i \quad \in \Lambda^1(M).$$

The connection coefficients:

$$\Gamma_{ib}^a(x) =< \Gamma_b^a, \partial_i >=< \theta^a, \nabla_i e_b > \ \in \ \mathcal{F}(M).$$

- The covariant derivation orientating along X: $\nabla_X : T_s^r \to T_s^r$

$$\nabla_X K =< \nabla K, X >, \quad \nabla_i K =< \nabla K, X >,$$

$$\nabla_{fX+gY} K = f\nabla_X K + g\nabla_Y K, \quad \nabla_i K =< \nabla K, \partial_i >,$$
$$\nabla_X(K + K') = \nabla_X K + \nabla_X K', \quad \nabla_X(fK) = f\nabla_X K + X(f)\nabla_X K,$$
$$\nabla_X(\alpha \wedge \beta) = \nabla_X \alpha \wedge \beta + \alpha \wedge \nabla_X \beta,$$
$$\nabla_X(i_Y \alpha) = i_{\nabla_X Y}\alpha + i_Y \nabla_X \alpha,$$
$$K_{b;i}^a \equiv (\nabla_i K)_b^a = \partial_i K_b^a - \Gamma_{ib}^c K_c^a + \Gamma_{ic}^a K_b^c, \text{ for } K = K_b^a \theta^b \otimes e_a.$$

- Curvature operator $\quad R(X,Y) = [\nabla_X, \nabla_Y] - \nabla_{[X,Y]}$.

  Torsion operator $\quad T(X,Y) = \nabla_X Y - \nabla_Y X - [X,Y]$.

  Curvature tensor coefficients:

$$R_{bik}^a =< \theta^a, R(\partial_i, \partial_k)e_b >= \Gamma_{kb,i}^a - \Gamma_{ib,k}^a + \Gamma_{ic}^a \Gamma_{kb}^c - \Gamma_{kc}^a \Gamma_{ib}^c.$$

  Torsion tensor coefficients:

$$T_{ab}^c =< \theta^c, T(\partial_a, \partial_b) >= e_a^i \Gamma_{ib}^c - e_b^i \Gamma_{ia}^c - f_{ab}^c.$$

Curvature 2-forms

$$\Omega^a{}_b = d\Gamma^a{}_b + \Gamma^a{}_c \wedge \Gamma^c{}_b = \frac{1}{2} R^a_{bik} dx^i \wedge dx^k.$$

Torsion 2-forms

$$\mathcal{T}^a = d\theta^a + \Gamma^a{}_c \wedge \theta^c = \frac{1}{2} T^a_{bc} \theta^b \wedge \theta^c.$$

- The covariant exterior differential act on the tensor-value differential form

$$\begin{aligned} D &= d + [\Gamma,\ ], \\ D^2 &= [\Omega,\ ]. \end{aligned}$$

- The Lie derivation

$$L_X = d \cdot i_X + i_X \cdot d - [L_\xi,\ ].$$

Exercises

1. In a natural bases, a vector X can be written as $X = \xi^i \partial_i$. Are the quantities $\frac{\partial \xi^i}{\partial x^j}$ components of a tensor? Why?

2. Please derive the integrable conditions of the two Cartan structure equations of connections.

3. For $SU(2)$ Yang-Mills field, both gauge potential A and gauge field strength F take values on the Lie algebra of Lie group $SU(2)$

$$A(x) = A^a_\mu(x) \frac{\sigma^a}{2i} dx^\mu, \quad F(x) = \frac{1}{2} F^a_{\mu\nu}(x) \frac{\sigma^a}{2i} dx^\mu \wedge dx^\nu = dA + A \wedge A,$$

where σ^a are the usual three Pauli matrices. Please prove that the covariant exterior differentiation of F satisfies the following relation

$$DF \equiv dF = [A, F] = 0.$$

Chapter 4

Riemannian Manifold

4.1 Metric tensor field, Hodge star and codifferentiation

In previous chapters, there has been no mention of the distance between two points on a manifold, i.e., metric has not been introduced. In this chapter, the metric structure of a manifold will be defined by a nonsingular symmetric order-2 covariant tensor field, so that the manifold becomes a Riemannian manifold.

Let $G \in \mathcal{T}_2(M)$ be a symmetric order-2 covariant tensor field, i.e., G is a bilinear mapping

$$\begin{aligned} G : \mathcal{X} \otimes \mathcal{X} &\to \mathcal{F}(M) \\ X\ ,\ Y &\to G(X,Y) \in \mathcal{F}(M) \end{aligned}$$

where $G(X,Y) = G(Y,X)$ and $X, Y \in \mathcal{X}(M)$. At the neighborhood of a point p there is a local coordinate system. Given the natural bases, we have

$$X = \xi^i(x)\partial_i, \qquad Y = \eta^i(x)\partial_i,$$

$$G(X,Y) = g_{ij}(x)\xi^i(x)\eta^j(x),$$

where

$$g_{ij}(x) = G(\partial_i, \partial_j) = g_{ji}(x).$$

If we write vector X by a column matrix $\widehat{X} = (\xi^1, \cdots, \xi^n)^t$. $\widehat{G} = (g_{ij})$ is a symmetric $n \times n$ matrix. Then,

$$G(X,Y) = \widehat{X}^t\widehat{G}\widehat{Y}.$$

Due to the symmetry of G, the following bilinear form

$$G(X,X) = \widehat{X}^t\widehat{G}\widehat{X} = g_{ij}\xi^i\xi^j$$

can be transformed through a coordinate transformation

$$\widehat{X} = \widehat{L}\widehat{X}', \qquad \det\ \widehat{L} \neq 0,$$

then

$$G(X,X) = \widehat{X}'^t\widehat{L}^t\widehat{G}\widehat{L}\widehat{X}'.$$

The matrix $\widehat{G}' = \widehat{L}^t\widehat{G}\widehat{L}$ can be diagonalized through this coordinate transformation, so that

$$G(X,X) = \sum_{i=1}^{n} a_i(\xi'^i)^2$$

is a real quadratic form on real manifold. Therefore, the positive factor $\sqrt{|a_i|}$ can be put into ξ'^i ($\xi''^i = \sqrt{|a_i|}\xi'^i$) to transform $G(X,X)$ into a standard diagonal formation. This is a well known theorem in linear algebra.

Theorem 4.1 *A real symmetric matrix $\widehat{G}$ is similar to a diagonal matrix in which the elements in the diagonal are $+1, -1$ or 0.*

Namely,

$$\widehat{G} = \widehat{L}^t\{+1, +1, \cdots, +1, -1, -1, \cdots, -1, 0, 0, \cdots, 0\}\widehat{L},$$

where $\det\widehat{L} \neq 0$, the number of $+1$ elements is k and the number of -1 elements is l. The rank of matrix $\widehat{G}$ is $r = k + l$. $s = k - l$ is called the signature of matrix $\widehat{G}$. Rank r and signature s of the $\widehat{G}$ are invariant under coordinate transformations.

Suppose that there exists a symmetric non-singular order-2 covariant tensor field, i.e., matrix $\widehat{G} = (g_{ij})$ satisfies

$$g_{ij} = g_{ji}, \quad \det(g_{ij}) \neq 0.$$

Under the transformations of the coordinates, rank r and signature s of $\widehat{G}$ are invariant. If both the rank and the signature is n, the matrix (g_{ij}) is positive definite.

Definition 4.1 *If there is a symmetric order-2 covariant tensor field G which is differentiable and positive definite everywhere on a manifold M, M is called a Riemannian manifold and G is called the basic tensor field or metric tensor field of M.*

In the above definition, if positive definite is replaced by non-singular(signature of the metric tensor field matrix (g_{ij}) is different from n), the corresponding manifold M is called a generalized Riemannian manifold. In physics, the ordinary four dimensional space-time (four dimensional Minkowski space) is a generalized Riemannian manifold. In this section, all formulas are applicable to generalized Riemannian manifold.

The metric structure on the manifold can be defined by the above non-singular symmetric order-2 covariant tensor field $g_{ij}(x)$, define:

$$ds^2 \equiv g_{ij}dx^i dx^j.$$

The arc length of a curve $x(t)$ on manifold M is:

$$\Delta s = \int (g_{ij} \frac{dx^i}{dt} \frac{dx^j}{dt})^{\frac{1}{2}} dt,$$

which is invariant under coordinate transformations so is a geometric quantity. For instance, for a curve $x(t)$ on the manifold as shown in figure below

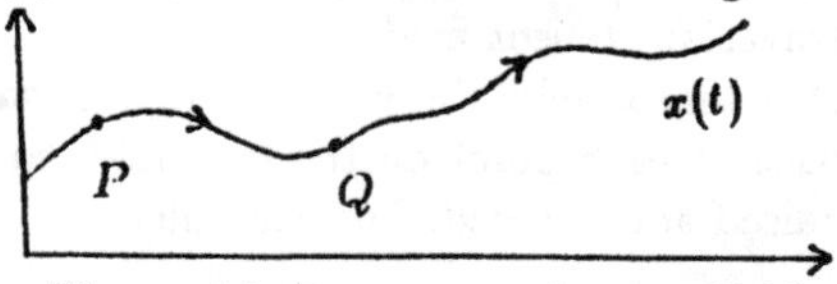

Figure 4.1 A curve on the manifold

The arc length between points P and Q on the curve is

$$l = \int_P^Q \sqrt{g_{ij}(x) \frac{dx^i}{dt} \frac{dx^j}{dt}} dt \equiv \int_P^Q L(t) dt.$$

In classical mechanics, variational principle is often used. For two given points P and Q on the manifold, which one of all these curves joining them gives an extremum? The answer is that it is the curve satisfying Euler-Lagrange equation:

$$\frac{d}{dt} \frac{\partial L}{\partial \dot{x}^i} - \frac{\partial L}{\partial x^i} = 0.$$

Put $L = \sqrt{g_{ij}(x) \frac{dx^i}{dt} \frac{dx^j}{dt}}$ into this equation, it can be proved that for two given point P and Q, the curve with the shortest arc length is just the geodesic curve defined by the Riemannian connection. Namely, the above equation is just the geodesic equation Eq. 3.26 in which the arc length is a parameter.

From this example, we see the close relations between the global and the local properties. In differential geometry, from properties at the neighborhood of a point, global properties for the manifold can be obtained. On the other hand, using variational principle, local properties of the manifold can be derived from global properties.

Through the metric tensor field, scalar product for any two tangent vectors in tangent space $T_p(M)$ at each point p on the manifold can be defined, i.e.,

$$X_p \cdot Y_p = G(X, Y)_p = g_{ij}(p) \xi^i(p) \eta^j(p),$$

where $X = \xi^i \partial_i, \quad Y = \eta^j \partial_j \in \mathcal{X}(M)$. When the metric tensor field is positive definite, we can define the length of the tangent vector as

$$|X_p| = \sqrt{g_{ij}(p) \xi^i(p) \xi^j(p)},$$

and the angle between tangent vectors X_p and Y_p is

$$\cos\theta = \frac{X_p \cdot Y_p}{|X_p| \, |Y_p|}.$$

As $\det(g_{ij}) \neq 0$, the inverse matrix (g^{ij}) of matrix (g_{ij}) is:

$$g^{ij} = \frac{A^{ji}}{\det(g_{ij})},$$

where A^{ji} is the algebraic cofactor of matrix element g_{ji}. $g^{ij}(x)$ are components of a symmetric order-2 contravariant tensor field.

As a result of the metric tensor field, there is a one to one mapping between tangent space and cotangent space at each point on the manifold, so indices of components of tensor fields can be raised and lowered. For example,

$$g_{ij}\xi^j = \xi_i, \qquad \xi^i = g^{ij}\xi_j.$$

Therefore, covariant vector and contravariant vector can be viewed as different manifestations of the same vector.

With metric tensor, inner product between differential forms of the same order can be defined. To do this, we first define the Hodge $\star$ operator which is a linear operator on differential forms. On an n-dimensional manifold, the Hodge $\star$ is a mapping from r-form to $(n-r)$-form, for

$$\alpha_r = \frac{1}{r!} f_{i_1\cdots i_r} dx^{i_1} \wedge \cdots \wedge dx^{i_r},$$

$$\star\alpha = \frac{1}{r!(n-r)!} f_{i_1\cdots i_r} \epsilon^{i_1\cdots i_r}{}_{k_1\cdots k_{n-r}} dx^{k_1} \wedge \cdots \wedge dx^{k_{n-r}},$$

where $\epsilon^{i_1\cdots i_r}{}_{k_1\cdots k_{n-r}}$ is the generalized Levi-Civita symbol, its properties and applications are given in appendix G. We define

$$\epsilon_{i_1\cdots i_n} = \sqrt{|g|}\delta^{1\cdots n}_{i_1\cdots i_n}, \qquad g = \det(g_{ij}),$$

where the factor $\sqrt{|g|}$ is introduced to make sure indices of ϵ are tensor indices. For the n-dimensional Euclidean space,

$$g_{ij} = \delta_{ij}, \qquad g = 1, \qquad \epsilon_{i_1\cdots i_n} = \delta^{1\cdots n}_{i_1\cdots i_n},$$

it is just the ordinary Levi-Civita symbol.

Metric tensors g_{ij} or g^{ij} can be used to raise and lower indices of the generalized Levi-Civita symbol. For instance,

$$\begin{aligned}
\epsilon^{i_1}{}_{i_2\cdots i_n} &= g^{i_1 k}\epsilon_{k i_2\cdots i_n} \\
\epsilon^{i_1\cdots i_n} &= g^{i_1 k_1}\cdots g^{i_n k_n}\epsilon_{k_1\cdots k_n} \\
&= sgn(g)\,|g|^{-\frac{1}{2}}\,\delta^{i_1\cdots i_n}_{1\cdots n},
\end{aligned}$$

where $g = det(g_{ij})$, and $\mathrm{sgn}(g) = g/|g| = \pm 1$. So the generalized Levi-Civita symbol is an order-n tensor and satisfies:

$$\epsilon_{i_1 \cdots i_n} \epsilon^{k_1 \cdots k_n} = sgn(g) \delta^{k_1 \cdots k_n}_{i_1 \cdots i_n}.$$

Now we can construct the Hodge $\star$ operator which is a linear operator which preserves the linear structure of tensors. The operator is completely determined by its action on the bases of differential forms. For instance,

$$\star dx^{i_1} \wedge \cdots \wedge dx^{i_r} = \frac{1}{(n-r)!} \epsilon^{i_1 \cdots i_r}{}_{k_1 \cdots k_{n-r}} dx^{k_1} \wedge \cdots \wedge dx^{k_{n-r}}.$$

Example 1. Two-dimensional Euclidean space, $\epsilon_{12} = -\epsilon_{21} = 1$,

$$\star dx^1 = dx^2, \qquad \star dx^2 = -dx^1,$$

$$\star 1 = dx^1 \wedge dx^2, \qquad \star dx^1 \wedge dx^2 = 1.$$

Example 2. Two-dimensional Minkowski space, $(g_{ij}) = \begin{pmatrix} 1 & 0 \\ 0 & -1 \end{pmatrix}$,

$$g = \det(g_{ij}) = -1, \qquad \mathrm{sgn}(g) = -1,$$

$$\epsilon_{01} = -\epsilon_{10} = 1 = \epsilon^{10} = -\epsilon^{01},$$

$$\star dt = g^{0i} \epsilon_{ij} dx^j = dx, \quad \star dx = g^{1i} \epsilon_{ij} dx^j = dt,$$

$$\star 1 = dt \wedge dx.$$

For a general n-dimensional manifold, due to the factor g in the definition of Hodge $\star$, the Hodge $\star$ operation is a tensor operation. And

$$\star 1 = \frac{1}{n!} \epsilon_{i_1 \cdots i_n} dx^{i_1} \wedge \cdots \wedge dx^{i_n} = |g|^{\frac{1}{2}} dx^1 \wedge \cdots \wedge dx^n \equiv \omega, \tag{4.1}$$

is the preferred n-form on the manifold. Under transformations of the coordinate system, we have

$$\begin{aligned} dx'^1 \wedge \cdots \wedge dx'^n &= \tfrac{\partial(x'^1, \cdots, x'^n)}{\partial(x^1, \cdots, x^n)} dx^1 \wedge \cdots \wedge dx^n \\ &= J^{-1} dx^1 \wedge \cdots \wedge dx^n, \end{aligned}$$

where J is the Jacobi matrix for coordinate transformation under which

$$g'_{ij} = \frac{\partial x^k}{\partial x'^i} \frac{\partial x^l}{\partial x'^j} g_{kl},$$

hence,

$$g' = J^2 g,$$

consequently the n-form ω introduced in Eq. 4.1 is invariant under coordinate transformations, i.e., it is a geometric object on the manifold and that is why it is called the preferred n-form on the manifold.

For two arbitrary r-forms α_r, β_r on manifold M,

$$\alpha_r = \frac{1}{r!} f_{i_1 \cdots i_r} dx^{i_1} \wedge \cdots \wedge dx^{i_r}, \quad \beta_r = \frac{1}{r!} h_{i_1 \cdots i_r} dx^{i_1} \wedge \cdots \wedge dx^{i_r},$$

it is easy to prove that

$$\begin{aligned} \alpha_r \wedge \star\beta_r &= \tfrac{1}{(r!)^2(n-r)!} f_{i_1 \cdots i_r} h_{j_1 \cdots j_r} \epsilon^{j_1 \cdots j_r}{}_{k_1 \cdots k_{n-r}} \\ &\qquad dx^{i_1} \wedge \cdots \wedge dx^{i_r} \wedge dx^{k_1} \wedge \cdots \wedge dx^{k_{n-r}} \\ &= \tfrac{1}{(r!)^2(n-r)!} f_{i_1 \cdots i_r} h^{j_1 \cdots j_r} \delta^{i_1 \cdots i_r k_1 \cdots k_{n-r}}_{j_1 \cdots j_r k_1 \cdots k_{n-r}} \star 1 \\ &= \tfrac{1}{r!} f_{i_1 \cdots i_r} h^{i_1 \cdots i_r} \star 1 \\ &= \beta_r \wedge \star\alpha_r, \end{aligned}$$

where we used

$$\begin{aligned} dx^{i_1} \wedge \cdots \wedge dx^{i_r} \wedge dx^{k_1} \wedge \cdots \wedge dx^{k_{n-r}} &= \delta^{i_1 \cdots i_r k_1 \cdots k_{n-r}}_{1 \cdots\cdots\cdots n} dx^1 \wedge \cdots \wedge dx^n \\ &= \mathrm{sgn}(g)\, |g|^{\frac{1}{2}} \epsilon^{i_1 \cdots i_r k_1 \cdots k_{n-r}} |g|^{-\frac{1}{2}} \star 1. \end{aligned}$$

Hence, when the manifold is compact, the inner product of two differential forms of the same order can be defined:

$$(\alpha, \beta) \equiv \frac{1}{v} \int_M \alpha \wedge \star\beta = \frac{1}{v} \int_M \beta \wedge \star\alpha, \quad v = \int_M \star 1, \tag{4.2}$$

where α and β are differential forms of the same type. The inner product thus defined satisfies:

$$(\alpha, \beta) = (\beta, \alpha) = \mathrm{sgn}(g)(\star\alpha, \star\beta).$$

If the metric tensor field is positive definite, then

$$(\alpha, \alpha) \geq 0, \qquad \text{equal iff } \alpha = 0.$$

From the above discussions, we see that when the metric tensor field is given on a manifold, inner product can be defined not only between two tensor fields, but also between two differential forms of the same type using the Hodge $\star$ operation.

The consecutive two Hodge $\star$ operations is still a linear operation on differential forms

$$\begin{aligned} \star\star\,\alpha_r &= \frac{1}{(r!)^2(n-r)!} f_{i_1 \cdots i_r} \epsilon^{i_1 \cdots i_r}{}_{j_1 \cdots j_{n-r}} \cdot \epsilon^{j_1 \cdots j_{n-r}}{}_{k_1 \cdots k_r} \cdot dx^{k_1} \wedge \cdots \wedge dx^{k_r} \\ &= \frac{\mathrm{sgn}(g)}{(r!)^2(n-r)!} (-1)^{r(n-r)} f_{i_1 \cdots i_r} \delta^{i_1 \cdots i_r j_1 \cdots j_{n-r}}_{k_1 \cdots k_r j_1 \cdots j_{n-r}} \cdot dx^{k_1} \wedge \cdots \wedge dx^{k_r} \\ &= \mathrm{sgn}(g)(-1)^{r(n-r)} \alpha_r \\ &= \mathrm{sgn}(g) \begin{cases} \alpha_r, & \text{when } n \text{ is odd,} \\ (-1)^r \alpha_r, & \text{when } n \text{ is even.} \end{cases} \end{aligned}$$

Combining the Hodge $\star$ operator with exterior differential operator d, the codifferential operator δ acting on r-form can be defined:

$$\begin{aligned}\delta &= \text{sgn}(g)(-1)^{nr+n-1} \star d\star \\ &= \text{sgn}(g)\begin{cases} -\star d\star, & \text{when } n \text{ is even,} \\ (-1)^r \star d\star, & \text{when } n \text{ is odd,}\end{cases}\end{aligned}$$

$$\delta\alpha_r = -\frac{sgn(g)}{r!(r-1)!}\delta^{j_1\cdots\cdots j_r}_{kk_1\cdots k_{r-1}} g^{kj} f_{j_1\cdots j_r;j} \cdot dx^{k_1} \wedge \cdots \wedge dx^{k_{r-1}},$$

where the subscript semicolon(i.e., ;) means the covariant derivative with the Riemannian connection on components of tensors(please refer to next section). The operator δ is a (-1)-degree skew differential operator which takes a r-form to $(r-1)$-form. It is a differential operator in the orthogonal complement subspace of the given differential form. The factor $\text{sgn}(g)(-1)^{nr+n-1}$ in the definition of δ is chosen such that the scalar product defined in Eq. 4.2 satisfies

$$(\alpha_r, d\beta_{r-1}) = (\delta\alpha_r, \beta_{r-1}),$$

i.e., δ is the adjoint operator of the exterior differential d with respect to the inner product of differential forms. It is easy to check that operator δ satisfies:

$$\delta^2 = 0.$$

Parallel to the case of exterior differential, differential form α is called co-closed if α satisfies $\delta\alpha = 0$. In addition, α is called a co-exact form if the r-form α can be written as codifferentiation of a $(r+1)$-form β, i.e., $\alpha = \delta\beta$. A co-exact form must be a co-closed form, but the inverse is not true in general.

In the following, we will take differential forms in three dimensional Euclidean space as examples to demonstrate the application of codifferential operator δ. In addition, a comparison with formulas of vector analysis is given.

Example 3. In E^3 space, for 0-form $f(x) \in \mathcal{F}(M)$

$$\delta f(x) = 0;$$

For 1-form $A(x) = A_i(x)dx^i \in \Lambda^1(M)$

$$\begin{aligned}\delta A &= -\star d\star(A_i dx^i) \\ &= -\star d(\frac{1}{2}\epsilon_{ijk}A_i dx^j \wedge dx^k) \\ &= -\star(\nabla\cdot\mathbf{A})dx^1\wedge dx^2\wedge dx^3 \\ &= (\nabla\cdot\mathbf{A});\end{aligned}$$

For 2-form $B(x) = \frac{1}{2}\epsilon_{ijk}B_i dx^j \wedge dx^k \in \Lambda^2(M)$

$$\begin{aligned}\delta B &= \star d \star (\frac{1}{2}\epsilon_{ijk}B_i dx^j \wedge dx^k) \\ &= \star d(B_i dx^i) \\ &= \star[\frac{1}{2}\epsilon_{ijk}(\nabla \times \mathbf{B})_i dx^j \wedge dx^k] \\ &= (\nabla \times \mathbf{B})_i dx^i;\end{aligned}$$

For 3-form $C = \varphi(x)dx \wedge dy \wedge dz$

$$\begin{aligned}\delta C &= -\star d \star (\varphi(x)dx \wedge dy \wedge dz) \\ &= -\star((\nabla\varphi)_i dx^i) \\ &= -\frac{1}{2}\epsilon_{ijk}(\nabla\varphi)^i dx^j \wedge dx^k.\end{aligned}$$

The Laplace operator is defined as

$$\Delta = (d + \delta)^2 = d\delta + \delta d.$$

The Laplace operator is a mapping from r-form to r-form, i.e., a second order differential operator in the differential forms of the same order. The Laplace operator Δ is also called the harmonic operator. If a r-form α_r satisfies:

$$\Delta\alpha_r = 0,$$

then α_r is called a harmonic r-form. 0-harmonic forms are just the usual harmonic functions.

Example 4. In E^3 space, for zero form $f(x)$

$$\Delta f = -(\frac{\partial^2 f}{\partial x^2} + \frac{\partial^2 f}{\partial y^2} + \frac{\partial^2 f}{\partial z^2});$$

For 1-form $A = A_i dx^i$,

$$\Delta A = -\frac{\partial^2 A_i}{\partial x^k \partial x^k} dx^i.$$

The Laplace operator which acts on differential forms on the manifold has the following properties:

1). It commutes with Hodge $\star$ operator

$$\star\Delta = \Delta\star.$$

2). It is a positive and self adjoint operator

$$(\Delta\alpha, \beta) = (\alpha, \Delta\beta),$$

where α and β are differential forms of the same type. And

$$(\alpha, \Delta\alpha) \geq 0, \quad \Delta\alpha = 0 \Longleftrightarrow d\alpha = 0, \quad \delta\alpha = 0.$$

3). Harmonic forms on compact, connected and orientable Riemannian manifold are constant forms.

The actions of operators d, δ, Δ on differential forms in E^3 are listed in Table (4.1) for reference.

Table 4.1 The actions of the operators d, δ, Δ on differential forms in E^3

form	$\Lambda^r(M)$	d	δ	Δ
0	$f(x)$	$(\nabla f)_i dx^i$	0	$-\frac{\partial^2 f}{\partial x^k \partial x^k}$
1	$A_i dx^i$	$(\nabla \times \mathbf{A})_i \star dx^i$	$-(\nabla \cdot \mathbf{A})$	$-\frac{\partial^2 A_i}{\partial x^k \partial x^k} dx^i$
2	$\frac{1}{2}\epsilon_{ijk} B_i dx^j \wedge dx^k$	$-(\nabla \cdot \mathbf{B}) \star 1$	$(\nabla \times \mathbf{B})_i dx^i$	$-\frac{1}{2}\epsilon_{ijk} \frac{\partial^2 B_i}{\partial x^l \partial x^l} dx^j \wedge dx^k$
3	$\varphi(x) \star 1$	0	$-(\nabla\varphi)_i \star dx^i$	$-\frac{\partial^2 \varphi}{\partial x^k \partial x^k} \star 1$

From table (4.1) we see that for 1-form $A \in \Lambda^1(M)$ the exterior differential d is equivalent to the curl of vector $\mathbf{A}$, the codifferential operator is equivalent to the divergence of vector $\mathbf{A}$. In three-dimensional space, the Hodge $\star$ operator maps 1-form and 2-form to each other such that the actions of d and δ are just inverse. Acting on 2-form $\mathbf{B}$ the exterior differential operator d is equivalent to the divergence of vector $\mathbf{B}$, the codifferential operator δ is equivalent to the curl of vector $\mathbf{B}$.

Differential form is a very effective tool in dealing with many physics problems. To show this, we look at electromagnetism field in E^4 in this last part of this section.

Example 5. The $U(1)$ gauge field in four-dimensional Euclidean space.

The vector potential $\mathbf{A}(x)$ and the scalar potential $\varphi(x)$ can be combined with the corresponding cotangent frame to obtain a gauge 1-form:

$$A = A_i(x)dx^i + \varphi(x)dt = A_\mu dx^\mu,$$

here and after summation over repeated indices convention is adopted. Let $\mu, \nu = 1, 2, 3, 4, \quad i, j = 1, 2, 3$ and $x^4 = t, A_4(x) = \varphi(x)$.

The gauge field 2-form is

$$F = dA = A_{\mu,\nu} dx^\nu \wedge dx^\mu = \frac{1}{2} F_{\mu\nu} dx^\mu \wedge dx^\nu,$$

where

$$F_{\mu\nu} = \partial_\mu A_\nu - \partial_\nu A_\mu.$$

Let the electric field and the magnetic field be

$$\mathbf{E} = \nabla\varphi - \frac{\partial \mathbf{A}}{\partial t}, \quad \mathbf{B} = \nabla \times \mathbf{A}.$$

Then gauge field 2-form is:

$$F = E_i dx^i \wedge dt + \frac{1}{2}\epsilon_{ijk} B_i dx^j \wedge dx^k.$$

As $F = dA$ is an exact form, the Bianchi identity must be satisfied:

$$dF = 0. \tag{4.3}$$

Its tensor formulation is

$$\nabla \times \mathbf{E} + \frac{\partial \mathbf{B}}{\partial t} = 0, \quad \nabla \cdot \mathbf{B} = 0.$$

Taking the Hodge $\star$ for F, one obtains

$$\star F = \frac{1}{2}\epsilon_{ijk} E_i dx^j \wedge dx^k + B_i dx^i \wedge dt.$$

Comparing with the F, we see that the Hodge $\star$ operation is equivalent to the dual transformation between electric and magnetic fields $\mathbf{E} \Leftrightarrow \mathbf{B}$.

The field equation satisfied by the gauge field is:

$$\delta F = J = J_\mu dx^\mu. \tag{4.4}$$

In tensor formulation

$$-\nabla \times \mathbf{B} + \frac{\partial \mathbf{E}}{\partial t} = \mathbf{J}, \quad \nabla \cdot \mathbf{E} = J_4 \equiv \rho.$$

The codifferentiation of Eq. 4.4 gives

$$\delta J = 0,$$

which is just the equation of current conservation

$$\partial^\mu J_\mu = 0,$$

more explicitly

$$\nabla \cdot \mathbf{J} + \frac{\partial \rho}{\partial t} = 0.$$

Gauge transformation of the gauge potential does not change field strength, i.e.,

$$A \to A' = A + df.$$

in tensor formulation

$$\mathbf{A} \to \mathbf{A}' = \mathbf{A} + \nabla f, \quad \varphi \to \varphi' = \varphi + \frac{\partial f}{\partial t}.$$

Choosing a gauge is equivalent to imposing constraints on gauge potential A. For instance, the covariant gauge is

$$\delta A = 0, \Rightarrow \partial^\mu A_\mu = 0, \qquad \nabla \cdot \mathbf{A} + \frac{\partial \varphi}{\partial t} = 0.$$

For the solution of the field equation without source, put $J = 0$ into Eq. 4.4, then

$$\delta F = 0. \tag{4.5}$$

together with the Bianchi identity $dF = 0$, we see that:

$$\Delta F = 0,$$

i.e., the field strength without source is a harmonic form.

If the field satisfies the self-dual condition

$$F = \pm \star F,$$

the field equation from the Bianchi identity $dF = 0$ gives $\delta F = 0$. Namely the solution of the field equation Eq. 4.5 without source (The corresponding differential equations of the gauge potential is the second order.) can be obtained from solutions of the self-dual equation (The corresponding differential equation of gauge potential is the first order.).

The above results are listed in table (4.2) where we see the advantages of differential forms more clearly. In dealing with physics problems, differential forms are more convenient than tensors. In addition, the former is an intrisic approach and has clear geometric and physics interpretations.

Table 4.2. Two formulations of the $U(1)$ gauge field in E^4

	differential forms	general tensors
gauge transformation	$A' = A + df$	$\mathbf{A}' = \mathbf{A} + \nabla f$, $\varphi' = \varphi + \frac{\partial f}{\partial t}$
gauge field strength	$F = dA$	$E = \nabla \varphi - \frac{\partial \mathbf{A}}{\partial t}$, $\mathbf{B} = \nabla \times \mathbf{A}$
Bianchi identity	$dF = 0$	$\nabla \times \mathbf{E} + \frac{\partial \mathbf{B}}{\partial t} = 0$, $\nabla \cdot \mathbf{B} = 0$
covariant gauge condition	$\delta A = 0$	$\nabla \cdot \mathbf{A} + \frac{\partial \varphi}{\partial t} = 0$
field equation	$\delta F = J$	$-\nabla \times \mathbf{B} + \frac{\partial \mathbf{E}}{\partial t} = \mathbf{J}$, $\nabla \cdot \mathbf{E} = \rho$
continuous equation	$\delta J = 0$	$\nabla \cdot \mathbf{J} + \frac{\partial \rho}{\partial t} = 0$

Example 6. The electrodynamics in four dimensional Minkowski space.

In four-dimensional Minkowski space the metric tensor for inertial frame is

$$g_{\mu\nu} = diag(+++-).$$

Similar analysis in example 6 can be applied to this case but covariant indices and contravariant indices are different this time. The gauge field strength is:

$$F = dA.$$

Maxwell equations are

$$dF = 0, \quad \delta F = J. \tag{4.4}$$

The tensor formulation of these two equations are just the Maxwell equations which we are familiar with. $dF = 0$ gives two equations without source. If F is expressed by the potential A, $dF = 0$ is the identity which A must satisfy, and is called the Bianchi identity which is determined by the local expression $F = dA$.

$\delta F = J$ gives two equations with matter. In physics, equation of motion can be obtained by using the variation of the action I with respect to field variable A, which depends on the choice of the metric of the space. Action I can be written as the integration of Lagrangian density over space time:

$$I = \int d^4x \mathcal{L},$$

$$\begin{aligned} \mathcal{L} &= \frac{1}{2}(\mathbf{E}^2 - \mathbf{B}^2) = -\frac{1}{2}|F|^2 \\ &= -\frac{1}{4} g^{\mu\nu} g^{\rho\sigma} F_{\mu\rho} F_{\nu\sigma}, \end{aligned}$$

which is Lorentz invariant with respect to metric tensor. For different inertial frames (i.e., taking the Lorentz transformation), $|\mathbf{E}|$ and $|\mathbf{B}|$ are changed but $\mathbf{E}^2 - \mathbf{B}^2$ is invariant. Both electric field $\mathbf{E}$ and magnetic field $\mathbf{B}$ depend on the choice of coordinates and are quantities without substantial physics meaning. However, the 2-form F is a physically significant covariant quantity.

In the following, we discuss the pure gauge field without matter ($J = 0$). The equation of motion can be obtained through the variation of the action. Due to the Bianchi identity, any variation $\triangleright F$ of field F is the exterior differential of a 1-form α

$$\triangleright F = d\,\alpha.$$

Action I of the pure Yang-Mills field is

$$I = -\frac{1}{2}\int_M F \wedge \star F = -\frac{1}{2}(F, F).$$

The variation of I is

$$\triangleright I = -(\triangleright F, F) = -(d\,\alpha, F) = -(\alpha, \delta F).$$

Because the variation must be zero with respect to any α, the field strength must satisfy:

$$\delta F = 0,$$

which is just the sourceless equation of motion. Substitute $F = dA$ into this equation, we have

$$\delta F = \delta dA = 0.$$

The gauge potential A which satisfies this equation is not unique. In Lorentz gauge:

$$\delta A = 0,$$

A satisfies:

$$(d\delta + \delta d)A = \triangle A = 0.$$

If the Minkowski inertial frame is chosen, the above equation can be written as:

$$\triangle A = (\partial_x^2 + \partial_y^2 + \partial_z^2 - \partial_t^2)A = 0.$$

That is to say, A satisfies the ordinary wave equation and is a harmonic 1-form.

The Lorentz gauge $\delta A = 0$ does not completely determine the gauge potential. If φ is a harmonic function, i.e., $\triangle\varphi = 0$, we have

$$\delta(A + d\varphi) = 0.$$

In other words, there exists a gauge transformation which does not destroy the Lorentz gauge

$$A \to A + d\varphi.$$

4.2 Riemannian connection

On an n-dimensional smooth manifold M, when the local frame in the neighborhood of point p and the metric tensor field G are given, the square of the infinitesimal arc length ds on the manifold M can be written as

$$ds^2 = g_{ij}(x)dx^i dx^j.$$

ds is equivalent to the length of the infinitesimal tangent vector. If one requires that the length does not change, i.e., ds^2 is invariant when parallel transported along a curve, covariant derivative of the metric tensor field must be zero

$$\nabla G = 0.$$

In the local frame, this can be written as

$$\nabla_k g_{ij} = g_{ij;k} = \partial_k g_{ij} - \Gamma^l_{ki} g_{lj} - \Gamma^l_{kj} g_{li} = 0, \tag{4.6}$$

or simply:

$$dg_{ij} = \Gamma_{ij} + \Gamma_{ji},$$

where

$$\Gamma_{ij} = g_{il}\Gamma^l_j.$$

The connection which satisfies Eq.4.6 is called metric compatible permissible connection of Riemannian manifold.

The compatible condition between connection and metric can be expressed in another way: for any two vector fields, scalar product can be defined through metric tensor,

$$(X,Y) \equiv G(X,Y) \in \mathcal{F}(M).$$

The compatible condition between connection and metric is

$$d(X,Y) = (DX,Y) + (X,DY).$$

Taking exterior derivative of the above equation (i.e., the integrable condition) again:

$$0 = (D^2X,Y) + (X,D^2Y), \quad \text{or} \quad (\Omega X,Y) + (X,\Omega Y) = 0,$$

which is true for any vector field. Namely, the curvature 2-form is skew self-adjoint

$$\Omega_{ij} = -\Omega_{ji}.$$

The connection which satisfies the above condition can be expressed as

$$\Gamma^m_{ik} = \left\{ \begin{matrix} m \\ ik \end{matrix} \right\} + K^m_{ik},$$

where $\left\{ \begin{matrix} m \\ ik \end{matrix} \right\}$ is called the Christoffel symbol:

$$\left\{ \begin{matrix} m \\ ik \end{matrix} \right\} = \frac{1}{2} g^{mj}(\partial_k g_{ij} + \partial_i g_{kj} - \partial_j g_{ik}).$$

and K^m_{ik} is called the contorsion tensor:

$$K^m_{ik} = \frac{1}{2} g^{mj}(T^l_{ik} g_{lj} + T^l_{jk} g_{li} + T^l_{ji} g_{lk}),$$

where T^l_{ik} are the skew symmetric parts of connection coefficients:

$$T^l_{ik} = \Gamma^l_{ik} - \Gamma^l_{ki},$$

i.e., components of torsion tensor in the natural frame. If

$$\nabla_k g_{ij} = 0, \quad T^l_{ik} = \Gamma^l_{ik} - \Gamma^l_{ki} = 0,$$

are satisfied simultaneously, connection Γ^m_{ik} is called Riemannian connection and can be written as

$$\Gamma^m_{ik} = \left\{ \begin{array}{c} m \\ ik \end{array} \right\} = \frac{1}{2} g^{mj} (\partial_k g_{ij} + \partial_i g_{kj} - \partial_j g_{ik}).$$

Riemannian connection is a symmetric connection which is determined completely by the metric tensor and its derivative.

From the analysis above, we see that on a (generalized) Riemannian manifold with metric tensor field, if connection satisfies the following two conditions

1). Compatible with metric: connection must satisfy $\nabla_k g_{ij} = 0$.

2). Torsion free: connection must ensure that torsion is zero.

Then connection is determined uniquely and is called Riemannian connection. There is a theorem:

Theorem 4.2 *On a (generalized) Riemannian manifold, there exists a unique connection which is compatible with metric and torsion free.*

It is due to this theorem that the holonomy group of Riemannian connection is also called the holonomy group of metric tensor field. For Riemannian connection, since parallel transport preserves the orthogonality of the frame field, the holonomy group of an n-dimensional manifold with positive definite metric belong to orthogonal group, i.e., $H \in O(n)$.

If the signature of metric tensor field is $n-2$, the generalized Riemannian manifold is called Lorentz manifold. For instance, the four dimensional Minkowski space is a Lorentz manifold. For an n-dimensional Lorentz manifold, the holonomy group is $H \subset O(n-1,1)$.

In Chapter 8 we will see that $O(n)$ group is the maximum compact subgroup of $GL(n,R)$ and they have the same homotopy type, namely, $GL(n,R)$ group manifold can be continuously deformed to $O(n)$ group manifold. Hence, for arbitrary n-dimensional differentiable manifold, Riemannian metric can be chosen such that the corresponding holonomy group is $O(n)$. And we have the following theorem:

Theorem 4.3 *Riemannian metric exists on any dimensional smooth manifold.*

However, on a general smooth manifold, Lorentz metric may not exist. Whether there exists a generalized Riemannian metric or not depends on the topological properties of the manifold.

In terms of the Riemannian connection one can defined the covariant derivation of a vector field X, which denote by ∇X. When choicing the dual basis $\{e_a\}_1^n$ and $\{\theta^a\}_1^n$, i.e., $<\theta^a, e_b>= \delta^a_b$, one can introduced the covariant divergence of X denoted by

$$DivX = \sum_a <\theta^a, \nabla_a X>,$$

which has the following properties:

$$Div(fX) = fDivX + Xf,$$

$$\int_M DivX * 1 = \int_M \sum_i \left(\frac{\partial \xi^i}{\partial x^i} + \xi^i \frac{\partial \log |g|^{\frac{1}{2}}}{\partial x^i}\right)|g|^{\frac{1}{2}} dx^1 \wedge \cdots \wedge dx^n = 0.$$

Now the differential operator d and codifferential operator δ of a r-form $\alpha \in \Lambda^r(M)$ can be given as

$$d\alpha = \theta^a \wedge \nabla_a \alpha \quad \in \Lambda^{r+1}(M),$$
$$\delta\alpha = -Sgn(g) g^{ab} i_b \nabla_a \alpha \quad \in \Lambda^{r-1}(M),$$

which is independent of the choice of the basis, and satisfy

$$(\delta\alpha, \beta) = (\alpha, d\beta),$$

and the Laplace operator $\Delta = d\delta + \delta d$ can be determined, for example:

$$\begin{aligned} \Delta f &= \delta df = \delta(f_{,i} dx_i) \\ &= -sgn(g) g^{jk} < \nabla_k (f_{,i} dx^i), \partial_j >= -sgn(g) g^{jk} (f_{,jk} - \Gamma^l_{kj} f_{,l}) \\ &= -sgn(g) \frac{1}{\sqrt{|g|}} \frac{\partial}{\partial x^j} (\sqrt{|g|} g^{jk} \frac{\partial f}{\partial x^k}), \end{aligned}$$

$$\Delta\alpha_r = -\frac{sgn(g)}{r!}[g^{jk} f_{l_1 \cdots l_r ;jk} + \frac{1}{r!}\delta^{jj_1\cdots j_r}_{ll_1\cdots l_r} g^{lk}(f_{j_1\cdots j_r;jk} - f_{j_1\cdots j_r;kj})] dx^{l_1} \wedge \cdots \wedge dx^{l_r}.$$

4.3 Riemannian curvature

Riemannian curvature tensor can be obtained by putting Riemannian connection into Eq.3.23. It is completely determined by metric tensor and its second order derivatives . In terms of $g_{ij}(x)$, indices of the components of Riemannian curvature tensor can be raised or lowered, such that the rank-4 covariant curvature tensor can be obtained

$$R_{lkij} = g_{lm} R^m{}_{kij}.$$

This Riemannian curvature tensor has the following symmetries.
1). The last two indices are skew symmetric:

$$R_{lkij} = -R_{lkji}, \tag{4.7}$$

which results directly from the definition of curvature tensor:

$$R^l{}_{kij} = -R^l{}_{kji}.$$

2). Summation of the cyclically permuted last three indices is zero:

$$R_{l[kij]} = 0,$$

from the integrable condition Eq.3.33, and torsion free condition, one has

$$R^l{}_{kij} dx^i \wedge dx^j \wedge dx^k = 0.$$

Hence,

$$R^l_{[kij]} = 0 \quad (R_{l[kij]} = 0). \tag{4.8}$$

This equation is also called the first Bianchi identity.

3).From metric preserving condition $\Omega_{ij} = -\Omega_{ji}$, one obtains

$$R_{lkij} = -R_{klij}.$$

4).From $R_{l[ikj]} = 0$ one has

$$R_{lkij} + R_{lijk} + R_{ljki} = 0, \quad R_{klij} + R_{kijl} + R_{kjli} = 0.$$

Subtracting the above two equations, one has

$$2R_{lkij} + R_{lijk} + R_{ljki} - R_{kijl} - R_{kjli} = 0,$$

$$2R_{ijlk} + R_{ilkj} + R_{ikjl} - R_{jlki} - R_{jkil} = 0.$$

Using further $R_{lkij} = -R_{lkji}$ and $R_{lkij} = -R_{klij}$, we have:

$$R_{lkij} = R_{ijlk}. \tag{4.9}$$

Ricci tensor can be defined by contracting two indices of curvature tensor:

$$R^l_{klj} \equiv \Re_{kj} = < \theta^l, R(\partial_i, \partial_j) e_k > .$$

From Eq.4.9, $R_{lkij} = R_{ijlk}$, Ricci tensor is a symmetric tensor:

$$\Re_{kj} = \Re_{jk}.$$

For the natural frame, from Eq.3.22, we have

$$\Re_{kj} = \Gamma^l_{jk,l} - \Gamma^l_{lk,j} + \Gamma^l_{lm}\Gamma^m_{jk} - \Gamma^l_{jm}\Gamma^m_{lk}.$$

The further contraction between metric tensor and Ricci tensor gives Ricci scalar curvature $\Re$

$$\Re = g^{jk}\Re_{jk}.$$

In an orthogonal moving frame, the curvature tensor components can be decomposed uniquely into parts which are irreducible representations of the rotation group of the moving frame, i.e., these parts are also geometric objects. Let

$$S_{ab} = \mathfrak{R}_{ab} - \frac{1}{n} g_{ab}\mathfrak{R},$$

$$S_{abcd} = \frac{1}{n-2}(g_{ac}S_{bd} + g_{bd}S_{ac} - g_{ad}S_{bc} - g_{bc}S_{ad}),$$

where S_{ab} are the traceless part of Ricci tensor, and are also the trace parts of S_{abcd}, i.e.,

$$g^{ac}S_{abcd} = S_{bd} = \mathfrak{R}_{bd} - \frac{1}{n} g_{bd}\mathfrak{R},$$

then Riemannian curvature tensor can be decomposed as

$$R_{abcd} = W_{abcd} + S_{abcd} + G_{abcd},$$

where

$$G_{abcd} = \frac{\mathfrak{R}}{n(n-1)}(g_{ac}g_{bd} - g_{ad}g_{bc}),$$

which have the same symmetries and the same double traces as R_{abcd},

$$g^{ac}g^{bd}G_{abcd} = \mathfrak{R}.$$

The Weyl conformal tensor

$$W_{abcd} = R_{abcd} - S_{abcd} - G_{abcd}$$

is the linear combination of the traceless parts of Riemannian curvature, and have the same symmetries as R_{abcd}. The Weyl tensor is completely traceless, i.e., contraction with respect to each pair of indices vanishes.

The Riemannian curvature tensor has a very nice symmetry. It is easy to check that the number of independent components of such kind of curvature tensor is

$$N = \frac{1}{12}n^2(n^2-1).$$

The curvature tensor of one-dimensional manifold must be zero, and that of two-dimensional manifold R_{ijkl} has only one independent component:

$$R_{abcd} = G_{abcd} = \frac{1}{2}\mathfrak{R}(g_{ac}g_{bd} - g_{ad}g_{bc}).$$

The number of independent components of curvature tensor of three-dimensional manifold is six, this is the same as Ricci tensor $\mathfrak{R}_{\mu\nu}$. So in this case, $\mathfrak{R}_{\mu\nu}$ can only be expressed by $R_{abcd} = S_{abcd} + G_{abcd}$, here the Weyl tensor equals to zero. The non-zero Weyl conformal tensor exists only when $n \geq 4$.

4.4 Bianchi identity and Einstein field equation of gravity

Riemannian curvature tensor components R_{lkij} satisfy the Bianchi identity (the second Bianchi identity):

$$R_{lkij;m} + R_{lkmi;j} + R_{lkjm;i} = 0,$$

which is the result in terms of the integrability condition of the second Cartan's structure equation (see Eq.3.20)

$$d\Omega^l{}_k = \Omega^l{}_h \wedge \Gamma^h{}_k - \Gamma^l{}_h \wedge \Omega^h{}_k.$$

This equation can be written in terms of curvature tensor components as

$$\frac{\partial R^l_{kij}}{\partial x^m} dx^m \wedge dx^i \wedge dx^j = (R^l_{hmi}\Gamma^h_{jk} - \Gamma^l_{mh}R^h_{kij})dx^m \wedge dx^i \wedge dx^j.$$

Remembering the rules of covariant derivative of tensor components, one has

$$R^l_{kij;m} = \frac{\partial R^l_{kij}}{\partial x^m} + \Gamma^l_{mh}R^h_{kij} - \Gamma^h_{mk}R^l_{hij} - \Gamma^h_{mi}R^l_{khj} - \Gamma^h_{mj}R^l_{kih},$$

combined with torsion free condition, one has

$$\begin{aligned} R^l_{kij;m} dx^m \wedge dx^i \wedge dx^j &= -(\Gamma^h_{mi}R^l_{khj} + \Gamma^h_{mj}R^l_{kih})dx^m \wedge dx^i \wedge dx^j \\ &= 0. \end{aligned}$$

Therefore,the Bianchi identity can expressed as:

$$R^l_{k[ij;m]} \equiv R^l_{kij;m} + R^l_{kjm;i} + R^l_{kmi;j} = 0. \qquad (4.10)$$

As covariant derivative of metric tensor is zero, from the contraction of the Bianchi identity, we have

$$\Re_{kj;m} - \Re_{km;j} + \Re^l_{kjm;l} = 0.$$

Further contraction gives the contracted Bianchi identity:

$$\nabla^k(\Re_{jk} - \frac{1}{2}g_{jk}\Re) = 0.$$

Now we give a simple introduction to the Einstein gravitation theory. In this theory, gravity is related to the properties of space-time, the metric tensor field $g_{\mu\nu}$ is looked upon as the "gauge potential" of gravitational field, and the world line of a free point particle in a gravitational field is timelike.

As a result of invariance of all physical systems under parallel transports, energy and momentum is conserved, which can be expressed as:

$$\partial_\mu T^{\mu\nu} = 0.$$

When considering curvature of space-time caused by gravity, the ordinary derivative should be changed to covariant derivative, i.e.,

$$\nabla^\mu T_{\mu\nu} = 0.$$

Comparing with contracted Bianchi identity, Einstein proposed the famous Einstein Equation:

$$\Re_{\mu\nu} - \frac{1}{2} g_{\mu\nu} \Re = -8\pi G T_{\mu\nu}, \tag{4.11}$$

where G is the Newton gravitational constant. We can see that this is a second order differential equation of metric tensor field $g_{\mu\nu}$ just like the usual field equation in physics. But this time it is a non-linear equation.

Having compared with Newton's gravity theory, Einstein proposed some new phenomena of gravitational field, such as the gravitational red-shift of spectra, the perihelion procession of Mercury, and the bending of light caused by the Sun. They are proved soon by experiments. It led to the recognition and popularity of general relativity and Riemannian geometry, and the subsequent intensive studies about them.

Noticing that Riemannian connection is compatible with metric $\nabla^\mu g_{\mu\nu} = 0$, another term which is proportional to $g_{\mu\nu}$ can be added to the gravitation equation, so Einstein generalized the equation to

$$\Re_{\mu\nu} - \frac{1}{2} g_{\mu\nu} \Re = -8\pi G T_{\mu\nu} - \lambda g_{\mu\nu},$$

where λ is a constant. As $\Re_{\mu\nu}$ involves second order derivative of $g_{\mu\nu}$, the dimension of λ must be $[L]^{-2}$. After comparing with experiments, the new term is found to be very small, only in cosmology could it have significant effect, and can usually be neglected in other cases. So this term is called the cosmological term and λ is called the cosmological constant.

At the place where there is no matter field, the right hand side of the equation 4.10 is zero. Contracting with $g_{\mu\nu}$, we can prove $\Re = 0$. So

$$\Re_{\mu\nu} = 0.$$

This is called Einstein's vacuum field equation. When $n < 4$, using $\Re_{\mu\nu} = 0$, it can be proved that the Weyl tensor is also zero, i.e., all components of the curvature tensor R_{ijkl} are zero. Therefore, there is no gravitational field and space-time is flat. There exists non-trivial vacuum gravitational field only when $n > 3$.

If the Ricci tensor is a multiple of the metric: $\Re_{\mu\nu} = f g_{\mu\nu}$, then this Riemannian metric be called the Einstein's metric. In this case, if the dimension of the manifold $n \geq 3$, it can be proved that f is a constant.

4.5 Isometry, conformal transformation and constant curvature space

Euclidean space $\mathbf{E}^n$ has very high symmetry and the metric relation is not affected by the translation or rotation of the space. A general Riemannian manifold does not have such high symmetry as Euclidean space. Metric tensor field on the manifold is a order-2 covariant tensor field. Under transformation of coordinate systems, for a given point, the new metric is

$$g'_{ij}(x') = \frac{\partial x^k}{\partial x'^i}\frac{\partial x^l}{\partial x'^j}g_{kl}(x).$$

If the transformed metric as a function of x' and the original metric as a function of x are expressed by the same function, i.e.,

$$g'_{ij}(x) = g_{ij}(x),$$

the metric tensor is called formal invariant. In this case,

$$g_{ij}(x') = \frac{\partial x^k}{\partial x'^i}\frac{\partial x^l}{\partial x'^j}g_{kl}(x).$$

and the coordinate transformation is called isometric transformation. The isometry group consists of all isometric transformations of the Riemannian manifold.

A special type of coordinate transformation, which leaves the metric invariant up to a scale change

$$g'_{ij}(x) = f(x)g_{ij}(x)$$

is called conformal transformation, which preserves the angle between two vector fields. The comformal group is the subgroup of generalized coordinate transformations, and has the isometry group as its subgroup, i.e.,

Group of coordinate transformation $\supset$ conformal group $\supset$isometry group.

Weyl tensor components W_{abcd} are invariant under conformal transformations so it is called a conformal tensor. A space which dimension is greater than three is conformally flat if and only if its Weyl tensor vanishes.

Now we study the isometry group and its generators: the Killing vectors.

On a manifold with some structures, such as Riemannian manifold, there may exist a group of transformation preserving these structures, which can be called the group of motion. For Riemannian manifold case, the generator of group of motion preserving the metric is called Killing vector, i.e.,

Definition 4.2 *For a Riemannian manifold with a given metric tensor field g_{ij}, there exist a vector field $X = \xi^i(x)\partial_i$ satisfying*

$$L_X g_{ij} = \nabla_i \xi_j + \nabla_j \xi_i = 0, \textit{or} \quad L_X G = 0 \tag{4.12}$$

The vector field X be called the Killing vector with respect to the metric tensor.

For the given metric tensor field G, the set of all Killing vectors spans a linear space. In addition, the set of all Killing vectors constitutes a Lie algebra. If X and Y are two Killing vectors, then $[X, Y]$ is also a Killing vector because

$$L_{[X,Y]}G = L_X L_Y G - L_Y L_X G = 0.$$

It can be proved that on an n-dimensional manifold, the maximum number of independent Killing vectors is $\frac{1}{2}n(n+1)$, in the following discussion, it will be showed clearly. From the commutator Eq. 3.34 of two covariant derivatives of a vector field

$$\xi_{i;kl} - \xi_{i;lk} = -R^j_{ikl}\xi_j,$$

and the first Bianchi identity Eq. 4.8 of the curvature tensor

$$R^j_{ikl} + R^j_{kli} + R^j_{lik} = 0,$$

it is easy to prove that the second order covariant derivative of the Killing vector obeys

$$\xi_{l;ki} = -R^j_{ikl}\xi_j. \tag{4.13}$$

The higher order derivative of ξ_i can be obtained from Eq. 4.13. All higher order derivatives of the Killing vector ξ_i at point p can be expressed as the linear combination of $\xi_i(p)$ and $\xi_{i;k}(p)$. Hence, the functions of $\xi_i(x)$ at point p are determined by $\xi_i(p)$ and $\xi_{i;k}(p)$. Therefore, the number of independent Killing vectors which satisfy the Killing equation (Eq. 4.12) is $\frac{1}{2}n(n+1)$ at most.

In the following, we deal with the anisotropic problem at a point on the manifold. From above, we know that the Riemannian curvature tensor has much higher symmetries (Eq. 4.7 to Eq. 4.9). For four arbitrary vector fields on the manifold:

$$X = \xi^i\partial_i, \quad Y = \eta^i\partial_i, \quad Z = \zeta^i\partial_i, \quad T = \tau^i\partial_i$$
$$R(X,Y,Z,T) \equiv (R(Z,T)Y, X) \; = R_{ijkl}\xi^i\eta^j\zeta^k\tau^l$$

is of much higher symmetries, too. In addition, the fourth order contravariant tensor $\widehat{G}$ can be defined by the Riemannian metric tensor G

$$\widehat{G}(X,Y,Z,T) = G(X,Z)G(Y,T) - G(Y,Z)G(X,T).$$

Its symmetry is same as that of $R(X,Y,Z,T)$. That is

1). For the last pair and for the first pair variables are skew symmetric.

2). Between the last pair and the first pair variables there exist permutation symmetry.

3). The sum of the cyclic permutation For the last three variables is zero. And when $X = Z, \quad Y = T$,

$$\widehat{G}(X,Y,X,Y) = G(X,X)G(Y,Y) - [G(X,Y)]^2, \tag{4.14}$$

is the power of the area of the parallelogram spanned by X, Y. When X and Y are independent, Eq. 4.14 is not zero.

$$K(S_p) = \frac{R(X,Y,X,Y)}{\widehat{G}(X,Y,X,Y)} = \frac{R_{ijkl}\xi^i\eta^j\zeta^k\tau^l}{|X|^2\,|Y|^2 - (X\cdot Y)^2},$$

which is called the section curvature at the point p on Riemannian manifold, where S_p represents the plane spanned by $X, Y \in T_p(M)$. The section curvature is also called the Riemannian curvature and only depends on the plane S_p but not on the selection of X, Y. The plane S_p is tangent to the manifold M at the point p. The set of all geodesic lines which is tangent to S_p at the point p forms a two dimensional surface. The section curvature along the S_p direction at the point p on the manifold M is just the Gauss curvature of this two dimensional surface (refer section 4.7). The anisotropic properties of the Riemannian manifold at the point p can be described by the Riemannian curvature along the various directions at the point. The curvature tensor at the point p on the manifold is uniquely determined by the section curvature of all two dimensional tangent subspaces at the point p.

When $K(S_p)$ is independent of the directions of S_p, we call that the manifold M is isotropic at the point p. In this case, the section curvature is written as $K(p)$ which is the scalar function on M. It can be proved that

$$R_{ijkl}(p) = K(p)[g_{ik}(p)g_{jl}(p) - g_{il}(p)g_{jk}(p)], \tag{4.15}$$

$$\Omega_{ij}(p) = \frac{1}{2}R_{ijkl}(p)\theta^k \wedge \theta^l = K(p)\theta_i \wedge \theta_j,$$

where $\theta_k = g_{ki}\theta^i$.

If the section curvature K at all point on the manifold is a constant, the manifold M is called the constant curvature space. By use of the Bianchi identity which the Riemannian curvature tensor satisfies, we can prove the following theorem:

Theorem 4.4 *Let M be a connected Riemannian manifold with dimension $d \geq 3$. If M is isotropic at all point, i.e., the section curvature $K(S_p) = K(p)$ is a scalar function at all point on M, then manifold M is a space with constant curvature.*

Proof: When M is isotropic at all point, i.e., the Eq. 4.15 is satisfied:

$$R_{ijkl}(p) = K(p)[g_{ik}(p)g_{jl}(p) - g_{il}(p)g_{jk}(p)],$$

put this into the Bianchi identity Eq. 4.10, we have

$$R_{ij[kl;m]} = \nabla_m K[g_{ik}g_{jl} - g_{il}g_{jk}] + \nabla_k K[g_{il}g_{jm} - g_{im}g_{jl}]$$

$$+\nabla_l K[g_{im}g_{jk} - g_{ik}g_{jm}] = 0,$$

where we have used the fact that the covariant derivative of the metric tensor is zero. The contraction of indices i, k gives

$$(n-2)(\nabla_m K g_{jl} - \nabla_l K g_{jm}) = 0.$$

The contraction of indices j, l gives

$$(n-2)(n-1)\nabla_j K = 0.$$

Therefore, when $n \geq 3$, $\nabla_j K = \frac{\partial K}{\partial x^j} = 0$, and K is a constant. This completes the proof.

The constant curvature space, which the manifold M is isotropic at all point, is called the homogeneous space which has maximum symmetries with respect to the isometric transformation. For arbitrary point p on the manifold, there is a infinitesimal isometric transformation which takes the point p to any other point in its neighborhood and there are $\frac{1}{2}n(n+1)$ Killing vectors. In contrary, on an n-dimensional manifold, if the metric can be defined such that there are $\frac{1}{2}n(n+1)$ Killing vectors, then the manifold M has maximum symmetric, it must be a homogeneous Riemannian space which is isotropic at all points and a constant curvature space. The metric of the maximum symmetric space is uniquely determined by the Riemannian constant curvature and the signature of the metric tensor.

The sphere, plane and pseudosphere are two dimensional constant curvature Riemannian spaces. For an n-dimensional metric which the signature is zero, the constant curvature metric can be written as

$$ds^2 = |d\mathbf{x}|^2 + \frac{K(\mathbf{x}\cdot d\mathbf{x})^2}{1 - K|\mathbf{x}|^2}, \qquad \mathbf{x} \in R^n.$$

4.6 Orthogonal frame field and spin connection

For local coordinate transformations on an n-dimensional manifold M, the induced transformation matrix of the natural bases at a point is an element of the $GL(n, R)$ group. The $GL(n, R)$ group has only tensor representations and no spin representation. When analyzing spin field on Riemannian manifold, one must use the moving frame (called vierbein if n = 4). On Riemannian manifold, the orthogonal moving frame field $\{e_a(x)\}$ which plays an important role can be chosen. An orthogonal frame changes to another orthogonal frame under local orthogonal rotation. The transformation group of tangent bundle can be reduced to $O(n)$. Orthogonal transformation group $O(n)$ has spin representations. For field with half integer spin, we must use spin representations. Spin connection is just the Riemannian connection on the bundle of orthogonal frames.

On Riemannian manifold, tangent frame fields $e_a(x)$ which are orthogonal to each other everywhere can be introduced:

$$e_a(x) = e^i_a(x)\frac{\partial}{\partial x^i}, \quad a = 1, \cdots, n,$$

satisfying:

$$g_{ij}e^i_a e^j_b = \eta_{ab}, \qquad g^{ij} = \eta^{ab}e^i_a e^j_b,$$

where $\eta_{ab} = \eta^{ab} = \pm\delta^a_b$ is the flat metric, which for Euclidean space is +1 for all indices and for Minkowski space is -1 for the last index and +1 for all other indices.

The dual bases of the orthogonal frame fields $\{e_a(x)\}$ are cotangent frame 1-forms $\{\theta^a(x)\}$

$$\theta^a(x) = \theta^a_i(x)dx^i, \quad a = 1, \cdots, n,$$

satisfying:

$$< \theta^a, e_b > = \sum_i \theta^a_i e^i_b = \delta^a_b,$$

$$g_{ij} = \eta_{ab}\theta^a_i\theta^b_j, \qquad \eta^{ab} = g^{ij}\theta^a_i\theta^b_j.$$

In this case,

$$ds^2 = g_{ij}dx^i dx^j = \eta_{ab}\theta^a_i\theta^b_j dx^i dx^j = \eta_{ab}\theta^a\theta^b.$$

There are two kinds of tensor indices: One is the tensor indices $i, j, \cdots$, of the local transformation group of coordinates on the manifold which can be raised and lowered by metric tensor field g_{ij} and g^{ij}; The other is the tensor indices $a, b, \cdots$, of the orthogonal frame of tangent space at each point which can be raised and lowered by the flat metric tensor η_{ab} and η^{ab}. In terms of these two kinds of metric tensor fields, the one to one correspondence between tangent frame fields and cotangent frame fields can be established

$$e^i_a(x) = \eta_{ab}g^{ij}(x)\theta^b_j(x).$$

Frame field automatically manifests two kinds of local transformation groups:

1). General local transformation of coordinates on the manifold:

$$\{x^i\} \to \{x'^i(x)\}.$$

In gravitation theory , this is equivalent to transformations between arbitrary non-inertial coordinate systems. Under the above coordinate transformations, tangent frame fields $e^i_a(x)$ transform according as components of tangent vector fields and cotangent frame fields $\theta^a_i(x)$ transform as components of cotangent vector fields:

$$e^i_a(x') = \frac{\partial x'^i}{\partial x^j}e^j_a(x), \quad \theta^a_i(x') = \frac{\partial x^j}{\partial x'^i}\theta^a_j(x).$$

To analyse invariance of physics laws under arbitrary coordinate transformations, i.e., principle of general covariance, ordinary derivative ∂_i of tensor with respect to components of the natural bases in the equation of motion must be replaced by covariant derivative which involves connection. For instance, for tangent vector field $X = \xi^i \partial_i$, covariant derivative of ξ^i is

$$\nabla_k \xi^i \equiv \xi^i_{;k} = \xi^i_{,k} + \Gamma^i_{kj}\xi^j,$$

where covariant derivative $\nabla_k \xi^i$ which involves connection is a component of a (1,1)-type tensor.

e_a and θ^b given above show that by using frame fields $\{e^i_a(x), \theta^b_j(x)\}$, an arbitrary coordinate system can be changed to local orthogonal moving frames. In general relativity, the natural bases of non-inertial coordinate system can be transformed to natural bases of local inertial coordinate system. In addition, there exist local (space-time dependent) orthogonal rotations between local inertial coordinate systems, i.e., the local Lorentz transformations. Therefore, there is another kind of transformation.

2). Rotations between local orthogonal frames of tangent space in the neighborhood of each point on the manifold M:

$$e^i_a(x) \to e'^i_a(x) = L^b_a(x) e^i_b(x), \quad \theta^a_i \to \theta'^a_i(x) = L^{-1}(x)^a{}_b \theta^b_i(x).$$

Transformations between different inertial systems are Lorentz transformations. The transformation is called global Lorentz transformation if the transformation matrix (L^a_b) is a space-time independent constant matrix . Invariance of physics laws with respect to global Lorentz transformation is just the principle of the invariance of the speed of light in special relativity. As pointed out by Einstein, local Lorentz transformations which are space-time dependent must also be considered. As in the case of conventional gauge field theory, when invariance of physics laws under local Lorentz transformations is required, ordinary derivative in the equations of motion must be replaced by covariant derivative which involves connection. Covariant derivative corresponding to local Lorentz transformations is written as D_i and the corresponding connection is called spin connection:

$$D_i e_a = \omega^b_{ia} e_b,$$

where $\omega^a_{ib}(x)$ are called coefficients of spin connection, and i is a vector index tangent to the manifold, a, b are indices of orthogonal frame.

$$\omega^a{}_b(x) \equiv \omega^a_{ib} dx^i,$$

is called spin connection 1-form, which can be written in a matrix form:

$$\omega \equiv (\omega^a{}_b),$$

where indices a, b are indices of the space of adjoint representation of rotation group $O(n)$ of the orthogonal frame. Spin connection 1-form ω takes value in the space of the adjoint representation of rotation group $O(n)$. Covariant derivative

$$D_i = \partial_i + \omega_i,$$

which contains spin connection can be applied to quantities which have tensor indices of the orthogonal rotation group. Covariant derivative with spin connection is a linear differential operator on the space of the adjoint representation of $O(n)$ group and has the following properties:

1). Under general transformation of local coordinate system: $x^i \to x'^i(x)$,

$$D_i \to \frac{\partial x^k}{\partial x'^i} D_k.$$

2). Under rotation of local orthogonal frame: $e^i_a(x) \to L^b_a(x) e^i_b(x)$,

$$D_i \to L^{-1} D_i L,$$

such that when $\psi(x) \to L^{-1}\psi(x)$, the covariant derivative of $\psi(x)$ also transform homogeneously: $D_i\psi \to L^{-1} D_i \psi$.

For an arbitrary tangent vector field X, it can be written in the natural bases of coordinate system or the orthogonal frame fields of tangent bundle,

$$X = \xi^i \partial_i = \xi^a e_a,$$

where $\{\xi^i\}$ and $\{\xi^a\}$ are components of the same tangent vector in different coordinate systems, which are related by components of frame fields

$$\xi^i = \xi^a e^i_a, \qquad \xi^a = \xi^i \theta^a_i.$$

Covariant derivative of $\{\xi^a\}$ is

$$D_i \xi^a = (D_i X)^a = \xi^a_{,i} + \omega^a_{ib} \xi^b.$$

Through spin connection, coupling between spin field ψ and gravity (tensor field or frame field) can be studied, which is why frame fields must be introduced. Let Σ_{ab} be the generators of spin representation of Lorentz group. We have commutation relations

$$[\Sigma_{ab}, \Sigma_{cd}] = \eta_{bc}\Sigma_{ad} + \eta_{ad}\Sigma_{bc} - \eta_{ac}\Sigma_{bd} - \eta_{bd}\Sigma_{ac}.$$

Covariant derivative of spin field ψ can be written as:

$$D_i \psi = \partial_i \psi + \frac{1}{2}\omega^{ab}_i \Sigma_{ab} \psi.$$

In the following, parallel transport of tangent vector is taken as an example to show the geometric meaning of spin connection and we compare it with non-Abelian gauge field to highlight the similarity between non-Abelian gauge symmetry and local Lorentz invariance in general relativity.

If a tangent vector $X = \xi^a e_a$ is parallel moved along a curve $C(t)$ on the base manifold, components of tangent vector ξ^a satisfy:

$$D_Y \xi^a = 0, \qquad Y = \frac{d}{dt}.$$

If $C(t)$ is a closed curve and X starts from point p on the curve, when X returns to p, in general the new tangent vector X'_p is different from X_p and there is a rotation with respect to X_p. For example, the manifold is a two dimensional sphere S^2 in E^3. As shown in Figure 4.2.

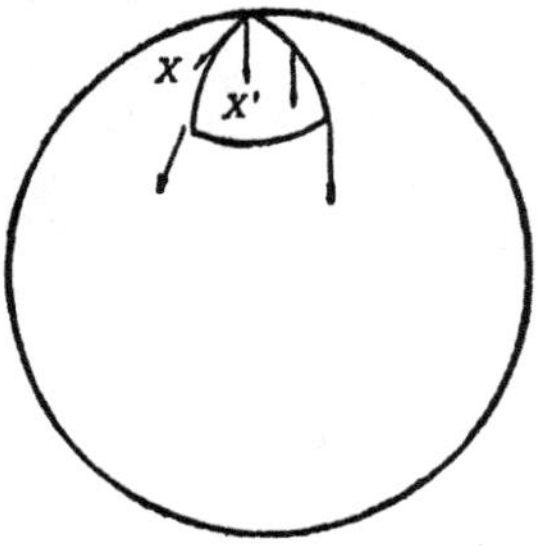

Figure 4.2 X is parallel transported along the edges of a spherical triangular through the north pole.

If tangent vector $X = \xi^a e_a$ is parallel transported along an infinitesimal closed curve, when X returns to the starting point, the infinitesimal rotation of the vector is

$$\delta\xi^a = \int dx^i \wedge dx^k R^a_{bik}\xi^b,$$

where curvature coefficients R^a_{bik} can be expressed by spin connection coefficients as in Eq. 3.22

$$R^a_{bik} = \partial_i\omega^a_{kb} - \partial_k\omega^a_{ib} + \omega^a_{ic}\omega^c_{kb} - \omega^a_{kc}\omega^c_{ib}. \tag{4.16}$$

Connection and curvature can be expressed by operators in adjoint representation space of $O(n)$ group. In matrix form,

$$\omega_k \equiv (\omega^a_{kb}), \qquad \Omega_{ik} \equiv (R^a_{bik}),$$

where a, b are indices of matrix elements and k, i are indices of the matrix. Then Eq. 4.16 can be written as

$$\Omega_{ik} = \partial_i\omega_k - \partial_k\omega_i + [\omega_i, \omega_k].$$

This is quite similar to the expression of the field strength of Yang-Mills field(refer to Exercise 3 in Chapter 3):

$$F_{\mu\nu} = \partial_\mu A_\nu - \partial_\nu A_\mu + [A_\mu, A_\nu].$$

In terms of differential forms, the above two equations can be written as

$$\Omega = d\omega + \omega \wedge \omega, \quad F = dA + A \wedge A.$$

From the above comparison, we see that to ensure invariance under local Lorentz transformation in general relativity, spin connection ω must be introduced. Spin connection is equivalent to gauge potential of $O(n)$ gauge field, and the corresponding curvature form Ω is equivalent to the field strength of $O(n)$ gauge field.

Now we look at the principle of general covariance in general relativity. The laws of physics can be written in arbitrary non-inertial system, i.e., the laws of physics can be written in arbitrary local coordinates. If there is a smooth infinitesimal transformation of local coordinate on the manifold,

$$x^i \to x'^i(x) = x^i + \xi^i(x),$$

the transformation of frame field is

$$\delta\theta_i^a \equiv \theta_i'^a - \theta_i^a = \xi^k \partial_k \theta_i^a + \xi^k_{,i}\theta_k^a,$$

which is equivalent to Lie derivative of indices of the natural bases of the frame field (The indices of the moving frame field is invariant.). When considering the approximate flat space which is far from matter, we can choose an orthogonal frame system in the infinitesimal neighborhood of a point such that components of the frame field θ^a are along directions of the natural bases ∂_k, i.e., $\theta_k^a = \delta_k^a$. In which case,

$$\delta\theta_i^a = \partial_i \xi^a.$$

This is equivalent to the change of gauge potential as a result of the local nature of the gauge parameter ξ^a. That is to say, in linear approximation, due to the invariance of the transformation of local translation (local diffeomorphism transformation), the required gauge potential is equivalent to θ_i^a and the corresponding gauge field strength is

$$T_{ik}^a = D_i\theta_k^a - D_k\theta_i^a,$$

which is just the torsion of spin connection.

The torsion of connection has no contribution to determining geodesic curves on the manifold. Therefore, in general relativity, connection is chosen such that torsion is zero. This is the Riemannian connection. If one requires that spin connection is Riemannian connection, the following conditions are needed

1). Metric compatible; i.e., it preserves the orthogonal relations of frame fields:

$$\omega_{ab} \equiv \eta_{ac}\omega^c{}_b = -\omega_{ba}. \tag{4.17}$$

2). Torsion free; i.e.,

$$D_i\theta_k^a - D_k\theta_i^a = 0,$$

or in differential forms

$$d\theta^a + \omega^a{}_b \wedge \theta^b = 0.$$

Comparing with Eq. 3.29, structure functions of frame fields are:

$$f_{bc}^a(x) = \omega_{c,b}^a - \omega_{b,c}^a,$$

where

$$\omega^a_{c,b} \equiv e^i_b \omega^a_{ic}.$$

From Eq. 4.17 and sum over cyclical permutation of $f^a_{bc}(x)$, one has

$$\omega^a_{b,c} = \frac{1}{2}(f^c_{ab} - f^a_{bc} - f^b_{ca}).$$

It can be proved that coefffecients of spin connection satisfying conditions of Riemannian connection is

$$\omega^{ab}_i = \frac{1}{2}\theta^{ak}(\partial_i\theta^b_k - \partial_k\theta^b_i) - \frac{1}{2}\theta^{bk}(\partial_i\theta^a_k - \partial_k\theta^a_i) - \frac{1}{2}\theta^{aj}\theta^{bl}(\partial_j e_{cl} - \partial_l e_{cj})\theta^c_i,$$

i.e, spin connection which is metric compatible and is torsion free can be determined completely by frame fields and their derivatives. This is similar to the case of Christoffel connection which is determined by metric tensor field and its derivatives.

There are two kinds of Riemannian connections on Riemannian manifold. One is Christoffel connection. Another is spin connection. Now we analyse relations between them. From connection 1-form, curvature 2-form can be obtained, of which the indices are tensor indices. The relation between the two kinds of curvature 2-forms is

$$\Omega^a{}_b(x) = d\omega^a{}_b + \omega^a{}_c \wedge \omega^c{}_b = \theta^a_i(d\Gamma^i{}_j + \Gamma^i{}_k \wedge \Gamma^k{}_j)e^j_b = \theta^a_i \Omega^i{}_j e^j_b.$$

Indices of spin connection 1-form $\omega^a{}_b$ are not tensor indices. The relation between spin connection 1-form and Christoffel connection 1-form is nonlinear. As

$$\omega^a{}_b \wedge \theta^b = \omega^a_{jb}\theta^b_i dx^j \wedge dx^i = -d(\theta^a_i dx^i) = -(\nabla_j \theta^a_i)dx^j \wedge dx^i, \tag{4.18}$$

where in the last step the symmetry condition of the connection is used. From metric compatible condition, one has

$$0 = \nabla_i(g^{jk}\theta^a_j\theta^b_k) = g^{jk}\theta^a_j\nabla_i\theta^b_k + (\nabla_i\theta^a_j)g^{jk}\theta^b_k,$$

i.e.,

$$\theta^{ak}\nabla_i\theta^b_k = -\theta^{bk}\nabla_i\theta^a_k. \tag{4.19}$$

Combining Eq. 4.18 and Eq. 4.19, one gets

$$\omega^a_{jb} = -e^k_b(\nabla_j\theta^a_k) = \theta^a_k\nabla_j e^k_b = \theta^a_k\partial_j e^k_b + \theta^a_k\Gamma^k_{jl}e^l_b \equiv \theta^a_k e^k_{b;j},$$

or

$$\omega^a{}_b = \theta^a_k(\delta^k_l d + \Gamma^k_l)e^l_b \equiv \theta^a_k(\nabla e_b)^k.$$

In the same way, Christoffel connection can be expressed by spin connection:

$$\Gamma^k_l = e^k_a\omega^a{}_b\theta^b_l + e^k_a d\theta^a_l. \tag{4.20}$$

For components of tensor field, there are relations between covariant derivative ∇_i related to Christoffel-Riemannian connection Γ, and covariant derivative D_i related to spin connection ω:

$$D_i = \partial_i + \omega^a_{ib} = \theta^a_k(\delta^k_j \partial_i + \Gamma^k_{ij})e^j_b.$$

$$\nabla_i = \partial_i + \Gamma^j_{ik} = e^j_a(\delta^a_b \partial_i + \omega^a_{ib})\theta^b_k.$$

When there are tensor indices, index of connection contract with corresponding index of the tensor. For instance, for tangent vector field $X = \xi^i \partial_i = \xi^a e_a$,

$$D_i \xi^a = (\delta^a_b \partial_i + \omega^a_{ib})\xi^b = \theta^a_k(\delta^k_j \partial_i + \Gamma^k_{ij})e^j_b \theta^b_l \xi^l = \theta^a_k \nabla_i \xi^k.$$

From the above calculations, we can see that covariant derivative of components of tensor field in the orthogonal moving frame is equivalent to extracting frame field and replacing spin connection by Christoffel connection. For simplicity, covariant derivative of both kinds of Riemannian connections are written as D_i and we take the convention that when it is applied to frame field, the frame field can be taken as a constant:

$$D_i \theta^a_j = 0,$$

and

$$\begin{cases} D_i \xi^j \equiv \nabla_i \xi^j = \partial_i \xi^j + \Gamma^j_{ik}\xi^k, \\ D_i \xi^a = \partial_i \xi^a + \omega^a_{ib}\xi^b. \end{cases}$$

As frame field θ^a_j has two tensor indices, $D_i \theta^a_j = 0$ can be written as

$$D_i \theta^a_j = \partial_i \theta^a_j - \Gamma^k_{ij}\theta^a_k + \omega^a_{ib}\theta^b_j = 0, \tag{4.21}$$

which is just Eq. 4.20. From this equation, relations between the two kinds of Riemannian connection can be determined. Eq. 4.21 has n^3 independent equations while the number of independent components of spin connection coefficients and Christoffel connection coefficients is n^3 also. Therefore, relation between them can be completely determined. A consequence of the above equation and $g_{ij} = \eta_{ab}\theta^a_i \theta^b_j$ is the fact that $D_k g_{ij} = 0$ as is indeed the case as Eq. 4.6.

The spin connection, which obey $D_i \theta^a_j = 0$, be called the minimal spin connection. Some time it is useful to consider other nonminimal spin connection. The difference between minimal and nonminimal spin connection can be measured by the torsion $\tau^a_{ik} = D_i \theta^a_k - D_k \theta^a_i$ of spin connection. Note that the torsion above defined depends only on the spin connection, not on the Christoffel symbol, which cancel out by virtue of symmetry of Γ^j_{ik} in the lower indices.

4.7 Surfaces and curves in 3-dimensional Euclidean space

Differential geometry, on the one hand, applies calculus (analytic method) to geometry, and on the other hand, provides the geometric interpretations of fundamental calculus. In physics, solutions of various differential equations usually have many free parameters such that the set of all solutions for a given problem constitutes a solution manifold (moduli space) with definite dimension. The solution manifold is usually a Riemannian manifold and the geometric interpretation can be given by studying the manifold. Riemannian metrics can be constructed by embedding the manifold in a Euclidean space and considering the induced metric. There is a theorem in standard differential geometry course (For instance, M. W. Hirsch: Differential Topology, GTM 33, Springer-Verlag, NY 1976.):

Theorem 4.5 Whitney: *Any n-dimensional manifold can be embedded as a closed submanifold of* E^{2n+1} *(and can be immersed in* E^{2n}*).*

Therefore, an arbitrary Riemannian manifold can be embedded in a higher dimensional Euclidean space as a submanifold. Euclidean space has trivial connection and zero curvature, its metric is also trivial. Since Riemannian connection, curvature and metric can be induced from the trivial ones of Euclidean space so we will pay more attention to submanifolds of Euclidean space. In this section, we will study submanifolds of Euclidean space with the help of moving frame. We first take a look at general features of this problem, then we will focus on curve and surfaces in three-dimensional Euclidean space E^3, where the embedding as well as the intrinsic characteristics will be studied.

For an n-dimensional Euclidean space E^n, there is an orthogonal moving frame

$$(x; e_1(x), \cdots, e_n(x)).$$

For Euclidean space, the frame of tangent space is the same as the frame of cotangent space and both have trivial parallel transport. Both the infinitesimal displacement dx and the infinitesimal transformation de_a of the corresponding frame at point x can be expanded in the above moving frame:

$$dx = \sum_{\alpha=1}^{n} e_\alpha \theta_\alpha, \tag{4.22}$$

$$de_\alpha = \sum_{\beta=1}^{n} e_\beta \omega_{\beta\alpha}, \tag{4.23}$$

where coefficients θ_α and $\omega_{\beta\alpha}$ are 1-forms, which are the frame 1-forms and the connection 1-forms respectively. These 1-forms satisfy the zero curvature and torsion-free conditions for trivial connections of Euclidean space:

$$d\theta_\alpha + \sum_{\beta=1}^{n} \omega_{\alpha\beta} \wedge \theta_\beta = 0, \quad d\omega_{\alpha\beta} + \sum_{\beta=1}^{n} \omega_{\alpha\gamma} \wedge \omega_{\gamma\beta} = 0.$$

The orthogonal frame completely describes the rigid motion. These two equations are called the structure equations of group manifold of rigid motion in $\mathbf{E}^n$ or structure equations of the orthogonal moving frame. From orthogonality of the frame, we have

$$(de_\alpha, e_\beta) + (e_\alpha, de_\beta) = 0, \quad \Longrightarrow \quad \omega_{\alpha\beta} + \omega_{\beta\alpha} = 0. \tag{4.24}$$

Now we look at the m-dimensional oriented smooth submanifold M embedded in $\mathbf{E}^m$. If the origin of the frame is on M, we have a moving frame with m parameters. At each point x on M, we can choose an orthogonal frame

$$(x; e_1, e_2, \cdots, e_m, e_{m+1}, \cdots, e_n),$$

such that $e_i (i \leq m)$ are tangent vector at point x on M. In addition, the orientation of $(e_1, e_2, \cdots, e_m)$ is consistent with that of M and $e_\alpha (m < \alpha \leq n)$ is the normal vector at point x on M. This local orthogonal frame field is called the Darboux frame of submanifold M.

As e_i are tangent vectors at point x on M, Eq. 4.22 can be simplified as

$$dx = \sum_{i=1}^{m} e_i \theta_i,$$

and

$$I = (dx, dx) = ds^2 = \sum_{i=1}^{m} (\theta_i)^2$$

is called the first fundamental form of manifold M. I is the Riemannian metric of M induced from E^n. The second equation of Eq. 4.23 for the Darboux frame can be written as

$$de_i = \sum_{j=1}^{m} e_j \omega_{ji} + \sum_{a=m+1}^{n} e_a \omega_{ai}, \tag{4.25}$$

$$de_b = \sum_{j=1}^{m} e_j \omega_{jb} + \sum_{a=m+1}^{n} e_a \omega_{ab}.$$

Eq. 4.24 can be written as:

$$d\theta_i + \sum_{j=1}^{m} \omega_{ij} \wedge \theta_j = 0, \quad \sum_{j=1}^{m} \omega_{aj} \wedge \theta_j = 0, \tag{4.26}$$

$$d\omega_{ij} + \sum_{k=1}^{m} \omega_{ik} \wedge \omega_{kj} + \sum_{a=m+1}^{n} \omega_{ia} \wedge \omega_{aj} = 0,$$

$$d\omega_{ib} + \sum_{k=1}^{m} \omega_{ik} \wedge \omega_{kb} + \sum_{a=m+1}^{n} \omega_{ia} \wedge \omega_{ab} = 0,$$

$$d\omega_{ab} + \sum_{k=1}^{m} \omega_{ak} \wedge \omega_{kb} + \sum_{c=m+1}^{n} \omega_{ac} \wedge \omega_{cb} = 0. \tag{4.27}$$

From Eq. 4.24 and Eq. 4.26, ω_{ij} must be a metric compatible torsion free symmetric connection, i.e., it must be a Riemannian connection on M. In addition, from Eq. 4.25, if we omit the second term on the right hand side and define:

$$De_i = \sum_{j=1}^{m} e_j\omega_{ji},$$

where De_i is the covariant derivative of e_i, which is the orthogonal projection of the infinitesimal transformation de_i of the frame to tangent plane T_xM on submanifold M, so ω_{ji} is the connection. By using the parallel transport of vectors in E^n (trivial connection) and projecting orthogonally to the tangent plane of submanifold M, one obtains the parallel transport of tangent vectors on M such that Riemannian connection on M can be induced.

For simplicity, supposing that M is an orientable hypersurface in E^n, i.e., $n = m+1$, then there is only one normal vector e_n for the Darboux frame of M. In this case, Eqs.4.25-4.27 can be simplified as

$$de_i = \sum_{j=1}^{m} e_j\omega_{ji} + e_n\omega_{ni},$$

$$de_n = \sum_{j=1}^{m} e_j\omega_{jn}, \tag{4.28}$$

$$d\theta_i + \sum_{j=1}^{m} \omega_{ij}\wedge\theta_j = 0, \quad \sum_{j=1}^{m} \omega_{nj}\wedge\theta_j = 0,$$

$$d\omega_{ij} + \sum_{k=1}^{m} \omega_{ik}\wedge\omega_{kj} + \omega_{in}\wedge\omega_{nj} = 0, \tag{4.29}$$

$$d\omega_{in} + \sum_{k=1}^{m} \omega_{ik}\wedge\omega_{kn} = 0. \tag{4.30}$$

As $\{\theta^i\}$ are linearly independent, from $\sum_{j=1}^{m}\omega_{nj}\wedge\theta_j = 0$ and the Cartan's Lemma (refer to § 1.5 Theorem 1.3), we have

$$\omega_{ni} = \sum_{j=1}^{m} h_{ij}\theta_j, \qquad h_{ij} = h_{ji}. \tag{4.31}$$

The second fundamental form of the hypersurface can be introduced

$$II = -(dx, de_n) = \sum_{i=1}^{m}\omega_{ni}\theta_i = \sum_{i,j=1}^{m} h_{ij}\theta_i\theta_j.$$

Riemannian curvature form of the hypersurface M can be obtained subsequently from Eqs.4.29 and 4.31

$$\Omega_{ij} \;=\; d\omega_{ij} + \sum_{k=1}^{m}\omega_{ik}\wedge\omega_{kj}$$

$$\overset{4.29}{=} -\omega_{in} \wedge \omega_{nj}$$
$$\overset{4.31}{=} \frac{1}{2} \sum_{k,l=1}^{m} R_{ijkl} \theta_k \wedge \theta_l, \tag{4.32}$$

where

$$R_{ijkl} = h_{ik} h_{jl} - h_{il} h_{jk}. \tag{4.33}$$

Eqs. 4.29 4.32 and 4.33 are called Gauss equation. Eq. 4.30 is equivalent to the Codazzi equation in the theory of surfaces.

As living beings in a three-dimensional space, we have clear geometric pictures of three dimensional Euclidean space. In the following, we look at curves and surfaces in three-dimensional Euclidean space by the method of moving frame. The geometric pictures will be very clear there.

For a given curve in three dimensional Euclidean space, the frame can be chosen such that the curve can be expressed in a simple way. Let the origin of the frame be on the curve and $\mathbf{e}_1$ along the direction of the tangent (The vectors in three-dimensional space will be expressed by bold faced letters). Therefore,

$$d\mathbf{x} = \mathbf{e}_1 \theta_1,$$

$$I = ds^2 = (d\mathbf{x}, d\mathbf{x}) = (\theta_1)^2,$$

and $\theta_1 = \pm ds$, where the sign determines the direction of the curve. For an orientable curve, the sign can be chosen such that

$$\theta_1 = ds \neq 0,$$

i.e., $\frac{d\mathbf{x}}{ds} = \mathbf{e}_1$ is the unit tangent vector

$$\left| \frac{d\mathbf{x}}{ds} \right| = 1.$$

The plane, which is perpendicular to the tangent vector at a point on the curve, is called the normal plane of this curve. As $|\mathbf{e}_1|$=1, one get

$$(\mathbf{e}_1, \frac{d\mathbf{e}_1}{ds}) = 0,$$

i.e., at the point where $\frac{d\mathbf{e}_1}{ds} \neq 0$, $\frac{d\mathbf{e}_1}{ds}$ must be on the normal plane. This $\frac{d\mathbf{e}_1}{ds}$ is called the curvature vector of the curve, its direction is called the main normal direction

$$\frac{d\mathbf{e}_1}{ds} = \kappa \mathbf{e}_2,$$

where $\mathbf{e}_2$ is the unit main normal vector, and $\kappa = \kappa(s)$ is called the curvature of the curve, which is the rate of change of the unit tangent vector with respect to the length of the arc s. The faster the direction of the tangent vector changes, the larger the curvature κ is. At the point with $\kappa \neq 0$, the radius of curvature is defined as:

$$\rho \equiv 1/\kappa.$$

The curve, of which κ equals to zero identically, is called a straight line. For straight line, the main normal direction can not be determined unambiguously. In the following, let $\kappa \neq 0$ everywhere, i.e., the main normal direction $\mathbf{e}_2$ is definite. Using the right hand rule to choose the moving frame, let

$$\mathbf{e}_3 = \mathbf{e}_1 \times \mathbf{e}_2$$

be the unit auxiliary normal vector, such that the moving frame with a single parameter along the curve is obtained. From Eq. 4.23, where

$$\omega_{\alpha\beta} = P_{\alpha\beta} ds,$$

$$(P_{\alpha\beta}) = \begin{pmatrix} 0 & \kappa & 0 \\ -\kappa & 0 & \tau \\ 0 & -\tau & 0 \end{pmatrix},$$

i.e.,

$$\frac{d\mathbf{e}_2}{ds} = -\kappa \mathbf{e}_1 + \tau \mathbf{e}_3,$$

where $\tau = \tau(s)$ is called the torsion of the curve. The plane, which is determined by the tangent vector and the main normal direction of the curve, is called the osculate plane of the curve. Torsion τ represents the rate of rotation of the osculate plane. The curve is a plane curve if τ is zero everywhere.

For a closed curve c, the total torsion $\tau(c)$ is defined as

$$\tau(c) = \frac{1}{2\pi} \oint \tau ds,$$

which is conformal invariant. $\tau(c) = 0$ on the sphere. And we have a theorem:

Theorem 4.6 *If the total torsion for any closed curve on a surface is zero, this surface must be a sphere.*

Now, we consider surface $X(u, v)$ in three dimensional Euclidean space. On the tangent plane at each point, we choose an orthogonal frame $\mathbf{e}_1$ and $\mathbf{e}_2$ (determined up to a rotation), such that

$$d\mathbf{x} = \mathbf{e}_1 \theta_1 + \mathbf{e}_2 \theta_2, \tag{4.34}$$

$\theta_3 = 0$. From $d\theta_3 = 0$(i.e., Eq. 4.26),

$$\omega_{31} \wedge \theta_1 + \omega_{32} \wedge \theta_2 = 0,$$

and Cartan's Lemma(Eq. 4.31), we have

$$\omega_{31} = a\theta_1 + b\theta_2, \quad \omega_{32} = b\theta_1 + c\theta_2. \tag{4.35}$$

The two fundamental forms of the surface are

$$\begin{aligned} I &= ds^2 = (d\mathbf{x}, d\mathbf{x}) = (\theta_1)^2 + (\theta_2)^2 \\ &= Edu^2 + 2Fdudv + Gdv^2 \\ &= \sum_{i,j=1}^{2} g_{ij} du^i du^j, \\ II &= -(d\mathbf{x}, d\mathbf{e}_3) = -\theta_1\omega_{13} - \theta_2\omega_{23} \\ &= a(\theta_1)^2 + 2b\theta_1\theta_2 + c(\theta_2)^2 = Ldu^2 + 2Mdudv + Ndv^2 \\ &= \sum_{i,j=1}^{2} h_{ij} du^i du^j, \end{aligned}$$

where u^1 and u^2 are the local coordinates of the surface. In the first fundamental form of the surface,

$$g_{ij}(u) = \sum_{k=1}^{3} \frac{\partial x^k}{\partial u^i} \frac{\partial x^k}{\partial u^j}$$

is the metric of the surface which is induced by the Euclidean metric in E^3 and x^1, x^2, x^3 are the Euclidean coordinates of E^3. If the surface is a two dimensional Riemannian manifold, the induced metric in general is not Euclidean. The first fundamental form determines the intrinsic properties of the surface, and the area of a given region D on the surface can be computed as:

$$S_D = \int_D \sqrt{g}\, dudv,$$

where $g = |\det(g_{ij}(u))|$. For instance, for Cartesian coordinates x, y, z in E^3, the area of region D, which is determined by the surface $z = f(x, y)$, is

$$S_D = \int_D \sqrt{1 + f_x^2 + f_y^2}\, dxdy.$$

The second fundamental form, which depends on the choice of coordinates, determines the embedding properties of the surface. We mainly deal with the intrinsic properties of the surface which is independent of the choice of frames. In the following, the algebraic method is used. We will focus on its geometric interpretations.

For given coordinate $\{u^i\}$ on the surface, the fundamental forms can be written as

$$\hat{G} = \begin{pmatrix} g_{11} & g_{12} \\ g_{21} & g_{22} \end{pmatrix}, \qquad \hat{H} = \begin{pmatrix} h_{11} & h_{12} \\ h_{21} & h_{22} \end{pmatrix},$$

where $\hat{G}$ is a positive definite symmetric matrix, $\hat{H}$ is a symmetric matrix:

$$g_{12} = g_{21}, \qquad h_{21} = h_{12}.$$

The equation

$$\det(\hat{H} - \kappa \hat{G}) = 0$$

has two eigenvalues κ_1 and κ_2, which are called eigenvalues of the pair of quadratic forms. When the two eigenvalues are not equal $\kappa_1 \neq \kappa_2$, the corresponding two eigenvectors are orthogonal. Hence, a real symmetric quadratic form can be transformed to diagonal form by a rotation of the given orthogonal coordinate system. In the following, we further highlight its geometric significance.

A displacement on the surface along the tangent direction $d\mathbf{x}$ will lead to a corresponding change $d\mathbf{e}_3$ of the normal vector $\mathbf{e}_3$ which is perpendicular to $d\mathbf{e}_3$. Define a linear transformation of the tangent plane, which is a mapping from $d\mathbf{x}$ to $d\mathbf{e}_3$, called the Weingarten transformation:

$$d\mathbf{e}_3 = -W(d\mathbf{x}).$$

From Eq. 4.28,

$$d\mathbf{e}_3 = \mathbf{e}_1\omega_{13} + \mathbf{e}_2\omega_{23}.$$

Making use of Eq. 4.34 and Eq. 4.35, we obtain

$$W(\mathbf{e}_1) = a\mathbf{e}_1 + b\mathbf{e}_2, \quad W(\mathbf{e}_2) = b\mathbf{e}_1 + c\mathbf{e}_2.$$

The transformation matrix $\begin{pmatrix} a & b \\ b & c \end{pmatrix}$ is a symmetric matrix. Therefore the Weingarten transformation is a self-conjugate transformation, which has two real eigenvalues κ_1 and κ_2

$$W(\mathbf{u}) = \kappa_1\mathbf{u}, \qquad W(\mathbf{v}) = \kappa_2\mathbf{v},$$

and the corresponding eigenvectors $\mathbf{u}, \mathbf{v}$ are called the main directions at this point. When two eigenvalues are not equal, the two vectors are orthogonal to each other

$$(\mathbf{u}, \mathbf{v}) = 0, \text{ if } \kappa_1 \neq \kappa_2.$$

The eigenvalues κ_1 and κ_2 of the Weingarten transformation are called the main curvature. The product of this two eigenvalues (the determinant of the transformation matrix) is called the total curvature K :

$$K = \kappa_1\kappa_2 = ac - b^2 = \frac{LN - M^2}{EG - F^2} = \frac{R_{1212}}{g}$$

which is also called the Gauss curvature, which is equal to the ratio of the determinant of the second fundamental form over that of the first fundamental form. Half of the trace of the Weingarten transformation matrix is called the mean curvature H at this point:

$$H = \frac{1}{2}(\kappa_1 + \kappa_2) = \frac{1}{2}(a + c).$$

The surface with $H = 0$ is called the minimal surface. Both the Gauss curvature K and the mean curvature H are geometric quantities which are independent of the choice of frames. When the orientation of the surface is changed, K is invariant and H changes its sign.

In the following, we look at the canonical connection and curvature of the surface. The surface is embedded in the three dimensional Euclidean space $\mathbf{E}^3$. For given moving frame $\{\mathbf{e}_i\}$, the projection $D\mathbf{e}_i$ of $d\mathbf{e}_i$ on the tangent plane is called the covariant derivative of the frame:

$$D\mathbf{e}_1 = \omega_{12}\mathbf{e}_2, \quad D\mathbf{e}_2 = -\omega_{12}\mathbf{e}_1,$$

where ω_{12} is the Riemannian connection of the surface. The corresponding curvature 2-form is

$$d\omega_{12} = -\omega_{13} \wedge \omega_{32} = K\theta_1 \wedge \theta_2. \tag{4.36}$$

For surfaces in three-dimensional Euclidean space, there is an important theorem:

Theorem 4.7 Gauss-Remarkable Theorem *The Gauss curvature K depends on the first fundamental form only.*

Proof: First, we prove that ω_{12} depends only on the first fundamental form I. As

$$d\theta_1 = -\omega_{12} \wedge \theta_2, \quad d\theta_2 = -\theta_1 \wedge \omega_{12}, \tag{4.37}$$

if there exists ω'_{12} satisfying:

$$d\theta_1 = -\omega'_{12} \wedge \theta_2, \quad d\theta_2 = -\theta_1 \wedge \omega'_{12},$$

then

$$(\omega'_{12} - \omega_{12}) \wedge \theta_i = 0 \quad (i = 1, 2).$$

As θ_1 and θ_2 are linearly independent, hence, $\omega'_{12} = \omega_{12}$ which depends only on the Riemannian metric of the surface. From Eq. 4.36, as the volume element $\theta_1 \wedge \theta_2$ also depends on the Riemannian metric of the surface only, so the Gauss curvature K depends only on the first fundamental form through $\{g_{ij}\}$ of the surface.

The Gauss-Remarkable theorem shows that even though the definition of the Gauss curvature K depends on the embedding space $\mathbf{E}^3$(ambient space), K depends only on the intrinsic properties of the surface. In 1827, Gauss guessed that there exists intrinsic geometry of the surface which is independent of the properties of the ambient space. At that time, one had no tools such as the concept of differentiable manifold. It is only after Riemann proposed the Riemannian geometry in 1852 that one can vigorously prove that the Gauss curvature K depends only on the intrinsic geometry of the surface and is not related to the methods of embedding. It is because of this, that the Gauss curvature and its higher dimensional counterparts play an important role in general relativity, as general relativity deals with the intrinsic properties of the four dimensional manifold.

For a homeomorphic mapping of the surface M, the mapping is called the isometric mapping if we require that the arc length of the image of arbitrary smooth curve C on M is equal to the arc length of curve C. For example, the deformation of a piece of paper without stretching or compressing is an isometric mapping. The invariant properties of surface M under isometric transformations are called intrinsic properties of the surface, which can be expressed by the first fundamental quantity $\{g_{ij}\}$ and its derivative. For instance, the arc length on the surface, the angle between two tangent vectors and the area of the surface etc. are all intrinsic properties of the surface. The Gauss-Remarkable theorem shows that the Gauss curvature is an intrinsic property of the surface. This is one of the important contributions of Gauss.

In the following, we look at the curve C on surface M in three-dimensional Euclidean space. The unit tangent vector is

$$\mathbf{T} = \frac{d\mathbf{x}}{ds}.$$

In this case, the normal vector $\mathbf{e}_3$ at each point on surface M is given. Let $\mathbf{N}$ be the unit main normal vector of C:

$$\frac{d\mathbf{T}}{ds} = \kappa\mathbf{N} = \mathbf{k_g} + \kappa_n\mathbf{e}_3,$$

where $\mathbf{k_g}$ is the projection of $\kappa\mathbf{N}$ on the tangent plane, and is called the geodesic curvature vector, and $\kappa_n\mathbf{e}_3$ is called the normal curvature vector. The normal curvature

$$\begin{aligned}\kappa_n &= (\frac{d\mathbf{e}_1}{ds}, \mathbf{e}_3) = -(\frac{d\mathbf{x}}{ds}, \frac{d\mathbf{e}_3}{ds}) = \frac{(d\mathbf{x}, W(d\mathbf{x}))}{ds^2} \\ &= \frac{II \text{ (second fundamental form)}}{I \text{ (first fundamental form)}}.\end{aligned}$$

When taking the special Darboux frame on surface M, such that $\mathbf{e}_1, \mathbf{e}_2$ are the characteristic directions (main directions) of the Weingarten transformation at point x on M, we have

$$\omega_{13} = \kappa_1\theta_1, \qquad \omega_{23} = \kappa_2\theta_2.$$

The second fundamental form becomes

$$II = \kappa_1\theta_1^2 + \kappa_2\theta_2^2.$$

and the normal curvature is

$$\kappa_n = \frac{II}{I} = \kappa_1 \cos^2\theta + \kappa_2 \sin^2\theta,$$

which is called the Euler formula, where θ is the angle between tangent vector $d\mathbf{x}$ and the main direction $\mathbf{e}_1$, i.e.,

$$\frac{d\mathbf{x}}{ds} = \cos\theta \mathbf{e}_1 + \sin\theta \mathbf{e}_2.$$

The normal curvature κ_n depends only on the direction of the tangent vector of curve C and is independent of the length of the tangent vector. If two curves are tangent at point x, then the normal curvatures of two curves at point x are equal. The plane curve, which is the intersection between surface M and the plane, which is determined by the normal vector of M and the tangent vector $d\mathbf{x}$ of curve C, is called the normal intersection curve of surface M along the direction of $d\mathbf{x}$. The normal curvature κ_n is just the curvature of the normal intersection curve at this point.

The geodesic curvature vector $\mathbf{k_g}$ of curve C is just the covariant derivative of the tangent vector of C. When $\mathbf{k_g} = 0$, i.e., the tangent vector of the curve is parallel transported along the curve, the curve is the geodesic curve. The main normal vector is parallel to the normal vector of the surface. This condition can be written as

$$(dT, \mathbf{e}_3 \times \mathbf{T}) \equiv (d\mathbf{T}, \mathbf{e}_3, \mathbf{T}) = 0.$$

Geodesic curve gives the extremum of the arc length between two points on the surface. For any direction $d\mathbf{x}$ at point p on surface M, there is a geodesic curve. Torsion τ_g of the geodesic curve at this point is called the geodesic torsion at p on surface M along $\mathbf{T}$:

$$\begin{aligned}\tau_g &= (\frac{d\mathbf{N}}{ds}, \mathbf{T} \times \mathbf{N}) = (\frac{d\mathbf{N}}{ds}, \mathbf{T}, \mathbf{N}) \\ &= (\frac{d\mathbf{e}_3}{ds}, \mathbf{T}, \mathbf{e}_3),\end{aligned}$$

where in the last step we used the fact that the main normal vector $\mathbf{N}$ on the geodesic curve is parallel to the normal vector $\mathbf{e}_3$ of the surface.

On surface M, there are two main directions through each point p. $\mathbf{e}_1(\mathbf{e}_2)$ is the direction along which the normal curvature of the surface takes the maximum (minimum) value. The integral curve of the main vector field is called the main curvature curve, of which the tangent direction is the characteristic direction of the Weingarten transformation, i.e., variations of the normal vector is along the tangent direction:

$$d\mathbf{e}_3 = -W(d\mathbf{x}) = -\lambda(s)d\mathbf{x}.$$

Hence, geodesic torsion along the tangent direction of the main curvature curve is zero

$$(d\mathbf{x}, \mathbf{e}_3, d\mathbf{e}_3) = 0.$$

This is the necessary and sufficient condition for a main curvature curve. For an arbitrary surface in $\mathbf{E}^3$, all such main curvature curves can be obtained in this way, which the direction at each point on the surface coincides with one of the main directions. This family of curves are also called the curvature curves on the surface, which constitute a congruence of curves on the surface. Two curvature curves are perpendicular at generic point p on surface M, such that the set of curvature curves form a family of orthogonal curves on M. But all curves are main curvature curves on surface S^2.

When κ_1 and κ_2 have different signs at point p, i.e.,

$$K = ac - b^2 \leq 0,$$

the second fundamental form at p can be decomposed as a product of two real monomials. At point p, there is a direction $\mathbf{u}$ such that

$$(\mathbf{u}, W(\mathbf{u})) = 0,$$

in which case $\mathbf{u}$ is called the asymptotic direction of point p.

If the tangent vector at each point on the curve is the asymptotic direction, the curve is called an asymptotic curve. It is easy to prove that the osculate plane of the asymptotic curve coincides with the tangent plane of the surface. However, the osculate plane of the geodesic curve intersects with the normal of the surface. The properties of the asymptotic curve and the geodesic curve are just opposite.

In the following table, the basic properties and the condition of existence of three kinds of curves on surfaces in $\mathbf{E}^3$ are listed for comparison. $\mathbf{T}$ is the unit tangent vector of the curve, $\mathbf{e}_3$ is the unit normal vector of the surface.

Curve	Basic Properties	Condition of Existence
Geodesic Curve	$(\mathbf{T}, \mathbf{e}_3, d\mathbf{T}) = 0$; The main normal vector of the curve is parallel to the normal vector of the surface	For a generic point there is a geodesic curve in any given direction.
Main Curvature curve	$(\mathbf{T}, \mathbf{e}_3, d\mathbf{e}_3) = 0$; The tangent vector of the curve is parallel to the main direction of the surface	It is the integral curve of the vector field of the main direction
Asymptotic Curve	$(\mathbf{T}, d\mathbf{e}_3) = 0$; The osculate tangent plane is parallel to the tangent plane of the surface	Only for the case of negative curvature. It is the integral curve on asymptotic direction

4.8 The computation of Riemannian curvature tensor

In this section, we briefly look at the computations of Riemannian curvature tensors. In the sections 4.2 and 4.3, we see that for given metric tensor field $g_{ij}(x)$

on the manifold, Riemannian connection (Christoffel symbol) can be obtained and so is Riemannian curvature tensor. However, this kind of calculation could be very tedious. A more efficient method is Cartan's method of moving frame in term of Cartan's two structure equations. Especially when the manifold under study has certain symmetries, both the curvature tensor and the frame fields can be obtained. Broadly speaking this method consists of the following three steps:

1. Choose the moving frame fields θ^i according to symmtries of the system. The best is an orthonormal frame.

2. Calculate the connection 1-forms $\Gamma^a{}_b = \theta^c \Gamma^a_{cb}$. Riemannian connection can be determined uniquely by the following two conditions:

(i) metric compatible: $dg_{ab} = \Gamma_{ab} + \Gamma_{ba}$,

(ii) torsion free: $d\theta^a + \Gamma^a{}_b \wedge \theta^b = 0$.

3. Calculate the curvature 2-forms

$$\Omega^a{}_b = d\Gamma^a{}_b + \Gamma^a{}_c \wedge \Gamma^c{}_b = \frac{1}{2} R^a{}_{bcd} \theta^c \wedge \theta^d$$

So the components $R^a{}_{bcd}$ with respect to a moving frame θ^a can be obtained.

Now we give some examples.

Example 1. Sphere S^2, which is the surface defined by $r =$ constant in three-dimensional Euclidean space $\mathbf{E}^3$. We use the polar coordinates

$$ds^2 = r^2 d\theta^2 + r^2(\sin\theta d\varphi)^2.$$

1°. choose a orthogonal moving frame field,

$$\sigma^1 = r d\theta, \qquad \sigma^2 = r \sin\theta d\varphi.$$

2°. use Cartan's first structure equation and the condition that the metric is torsion free,

$$\begin{cases} d\sigma^a = -\omega^a{}_b \wedge \sigma^b, \\ \omega_{ab} + \omega_{ba} = 0, \end{cases} \tag{4.38}$$

connection 1-forms can be solved. Comparing Eq. 4.38 with exterior derivative of σ_1, σ_2, we have he connection 1-form:

$$\omega^1{}_2 = -\cos\theta d\varphi.$$

3°. use Cartan's second structure equation, we have the curvature 2-form

$$\begin{aligned} \Omega^1{}_2 &= \frac{1}{2} R^1{}_{2ab} \sigma^a \wedge \sigma^b = d\omega^1{}_2 \\ &= \frac{1}{r^2} \sigma^1 \wedge \sigma^2. \end{aligned}$$

The Ricci curvature tensor is

$$\mathfrak{R}_{ab} = \frac{1}{r^2} \begin{pmatrix} 1 & 0 \\ 0 & 1 \end{pmatrix},$$

and the Ricci scalar curvature is

$$\Re = R^a_{bac}\eta^{bc} = \mathrm{Tr}(\Re_{ab}) = \frac{2}{r^2}.$$

The Gauss curvature is

$$K = R_{1212} = \frac{1}{r^2}.$$

We see S^2 is a surface with positive constant curvature. Comparing this example with the magnetic monopole in §1.7 , we see that the gauge field F is equivalent to the curvature 2-form and the gauge potential A is equivalent to the connection 1-form (up to a gauge transformation).

Example 2. Pseudosphere in three-dimensional Euclidean space: The surface of the revolving tractrix.

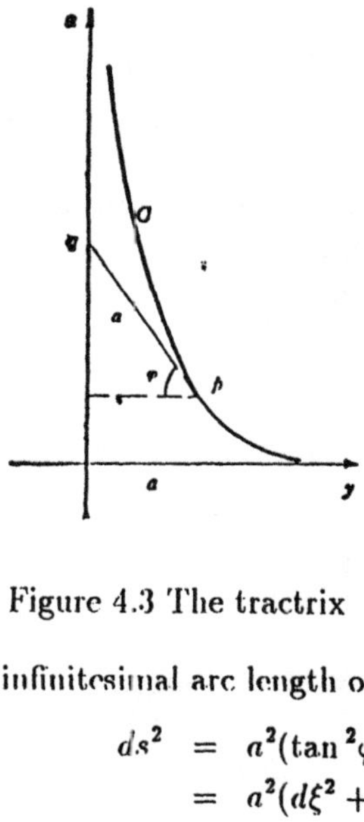

Figure 4.3 The tractrix

As shown in Figure 4.3, on the y-z plane, the tangent line at point p on curve C intersects the z-axis at point q. The length of the segment $\overline{pq}$ is a constant a.

Curve C is called the tractrix, its curvature is negative and constant. The surface of tractrix revolving around the z-axis can be expressed by the three-dimensional Euclidean coordinates as:

$$x = a\cos\varphi\cos v,$$
$$y = a\cos\varphi\sin v,$$
$$z = a\ln\tan(\tfrac{\varphi}{2}+\tfrac{\pi}{4}) - a\sin\varphi.$$

The infinitesimal arc length on the surface is

$$\begin{aligned} ds^2 &= a^2(\tan^2\varphi d\varphi^2 + \cos^2\varphi dv^2) \\ &= a^2(d\xi^2 + d\eta^2)/\eta^2, \qquad \xi = av, \quad \eta = a/\cos\varphi. \end{aligned}$$

1). The frame 1-forms are taken as

$$\theta^1 = \frac{a}{\eta}d\xi, \qquad \theta^2 = \frac{a}{\eta}d\eta,$$

$$d\theta^1 = -\frac{a}{\eta^2}d\eta \wedge d\xi = \frac{1}{\eta}d\xi \wedge \theta^2.$$

2). Thus the connection 1-form is:

$$\omega^1{}_2 = -\frac{1}{\eta}d\xi.$$

3). The curvature two form is obtained as:

$$\Omega^1_2 = d\omega^1_2 = \frac{1}{\eta^2} d\eta \wedge d\xi = -\frac{1}{a^2}\theta^1 \wedge \theta^2.$$

4).The Gauss curvature is

$$K = -\frac{1}{a^2}.$$

A surface with negative constant curvature like this is called pseudosphere.

Example 3. Sphere S^3, the radius of which is a constant a in four-dimensional Euclidean space. From Appendix K, we know that $S^3 \sim SU(2)$, we use the left invariant 1-forms $\{\theta_i\}$ on the group manifold $SU(2)$ as the moving frame:

$$\theta = \mathbf{x}dt - td\mathbf{x} + \mathbf{x} \times d\mathbf{x}.$$

The components are

$$\begin{aligned} \theta_x &= xdt - tdx + ydz - zdy, \\ \theta_y &= ydt - tdy + zdx - xdz, \\ \theta_z &= zdt - tdz + xdy - ydx, \end{aligned}$$

which satisfy $d\theta_x = -\theta_y \wedge \theta_z$, and the cyclical permutation.

1). Choose an orthogonal frame $a\theta_x, a\theta_y, a\theta_z$,

$$ds^2 = a^2(\theta_x^2 + \theta_y^2 + \theta_z^2).$$

2). The connection 1-forms are

$$\omega^1_2 = -\theta_z, \quad \omega^2_3 = -\theta_x, \quad \omega^3_1 = -\theta_y.$$

Because of the orthogonal frame is just the left invariant forms on the group manifold, so that the structure functions become constants. In this case, the connection 1-forms are proportional to the frame 1-forms so that they are also left invariant forms.

3). The curvature form is obtained as:

$$\Omega^2{}_3 = \frac{1}{a^2}\theta_y \wedge \theta_z, \quad \text{and the cyclical permutation.}$$

Example 4. The Robinson-Trautman manifold[14] ,

$$ds^2 = \frac{r^2}{x^3}(dx^2 + dy^2) + 2dudr + \frac{3}{2}xdu^2.$$

1). Select an orthogonal frame field $\{\theta^a\}$, such that

$$ds^2 = \eta_{ab}\theta^a\theta^b, \quad \eta_{ab} = (+,+,+,-).$$

Let $\alpha \equiv \sqrt{x}/r, (x > 0)$, it is easy to prove that

$$\begin{cases} \theta^1 = & \frac{1}{\alpha x} dx, \\ \theta^2 = & \frac{1}{\alpha x} dy, \\ \theta^3 = & (\frac{3x}{2})^{\frac{1}{2}} du + (\frac{2}{3x})^{\frac{1}{2}} dr, \\ \theta^4 = & (\frac{2}{3x})^{\frac{1}{2}} dr. \end{cases}$$

2). From Eq. 4.38, there are six independent connection 1-forms. Let

$$\omega_{12} = f_a \theta^a, \quad \omega_{13} = g_a \theta^a, \quad \cdots,$$

put these into Eq. 4.38 and comparing with the exterior derivative of Eq. 4.8, we have

$$\omega_{12} = \frac{3}{2}\alpha\theta^2, \qquad \omega_{13} = -\frac{\alpha}{2}(\theta^3 - \theta^4), \qquad \omega_{14} = \sqrt{\frac{3}{2}}\alpha\theta^1 + \frac{\alpha}{2}\theta^3 - \frac{\alpha}{2}\theta^4,$$

$$\omega_{23} = 0, \qquad \omega_{24} = \sqrt{\frac{3}{2}}\alpha\theta^2, \qquad \omega_{34} = \frac{\alpha}{2}\theta^1.$$

3). The curvature forms can be obtained from Cartan's second structure equation:

$$\Omega_2^1 = \frac{3}{2}d(\alpha\theta^2) - \frac{3}{2}\alpha^2\theta^1 \wedge \theta^2 - \frac{\alpha^2}{2}\sqrt{\frac{3}{2}}(\theta^3 \wedge \theta^2 - \theta^4 \wedge \theta^2),$$

and other similar equations.

Example 5. The Schwarzshild solution[15].

$$ds^2 = -(1 - \frac{2M}{r})dt^2 + (1 - \frac{2M}{r})^{-1}dr^2 + r^2(d\theta^2 + \sin^2\theta d\varphi^2),$$

$$(0 \le \theta < \pi, \ \ 0 \le \varphi < 2\pi).$$

1). Choose an orthogonal frame $\eta_{ab} = (-, +, +, +)$

$$\theta^0 = (1 - \frac{2M}{r})^{\frac{1}{2}} dt, \qquad \theta^1 = (1 - \frac{2M}{r})^{-\frac{1}{2}} dr,$$

$$\theta^2 = r d\theta, \qquad \theta^3 = r \sin\theta d\varphi.$$

2). The connection 1-forms are

$$\omega_1^0 = \frac{M}{r^2} dt, \qquad \omega_2^0 = 0, \quad \omega_3^0 = 0,$$

$$\omega_3^2 = -\cos\theta d\varphi, \qquad \omega_1^3 = (1 - \frac{2M}{r})^{\frac{1}{2}} \sin\theta d\varphi,$$

$$\omega_2^1 = -(1 - \frac{2M}{r})^{-\frac{1}{2}} d\theta.$$

3). The curvature 2-forms can be obtained:

$$\Omega^0_1 = \frac{2M}{r^3}\theta^0 \wedge \theta^1, \quad \Omega^0_2 = -\frac{M}{r^3}\theta^0 \wedge \theta^2,$$

$$\Omega^0_3 = -\frac{M}{r^3}\theta^0 \wedge \theta^3, \quad \Omega^2_3 = \frac{2M}{r^3}\theta^2 \wedge \theta^3,$$

$$\Omega^3_1 = -\frac{M}{r^3}\theta^3 \wedge \theta^1, \quad \Omega^1_2 = -\frac{M}{r^3}\theta^1 \wedge \theta^2.$$

Example 6. The self-dual manifold and the vacuum solution of Einstein field equation.

For curvature 2-forms, a dual operator $*$ is introduced:

$$*\Omega_{ab} = \frac{1}{2}\epsilon_{ab}{}^{cd}\Omega_{cd}.$$

The curvature 2-forms are self(anti-self)-dual, if

$$*\Omega_{ab} = \pm\Omega_{ab}.$$

From the torsion-free condition we have

$$*\Omega_{ab} \wedge \theta^b = 0.$$

which is

$$\frac{1}{2}\epsilon_{ab}{}^{cd}R_{cdmn}\theta^m \wedge \theta^n \wedge \theta^b = 0.$$

That is to say,

$$\begin{aligned} 0 &= \epsilon_{ab}{}^{cd}R_{cdmn}\epsilon^{mnbp} \\ &= -\delta^{mnbp}_{abcd}\Re^{cd}{}_{mn} \\ &= -4(\delta^p_a 2\Re - 4\Re^p_a) \\ &= 16(\mathcal{R}^p_a - \frac{1}{2}\delta^p_a\mathcal{R}). \end{aligned}$$

The left hand side is zero and this is called the vacuum Einstein field equation. Combining the self-dual condition and the Bianchi identity, we can obtain the vacuum solution of Einstein field equation. This is equivalent to the discussion of gauge field in four dimensional Euclidean space in §4.1.

Example 7, the Eguchi-Hanson solution[16]:

$$ds^2 = (1-(\frac{a}{r})^4)^{-1}dr^2 + r^2[\theta_x^2 + \theta_y^2 + (1-(\frac{a}{r})^4)\theta_z^2].$$

1). Frame fields are taken as

$$\{\theta^a\} = \{(1-(\frac{a}{r})^4)^{-\frac{1}{2}}dr, r\theta_x, r\theta_y, r(1-(\frac{a}{r})^4)^{\frac{1}{2}}\theta_z\}.$$

2). Connection 1-forms are

$$\omega_1^0 = -\omega_3^2 = -(1-(\frac{a}{r})^4)^{\frac{1}{2}}\theta_x,$$

$$\omega_2^0 = -\omega_1^3 = -(1-(\frac{a}{r})^4)^{\frac{1}{2}}\theta_y,$$

$$\omega_3^0 = -\omega_2^1 = -(1-(\frac{a}{r})^4)\theta_z.$$

3). The self-dual curvature 2-forms are

$$\Omega_1^0 = -\Omega_3^2 = \frac{2a^4}{r^6}(-\theta^0\wedge\theta^1+\theta^2\wedge\theta^3),$$

$$\Omega_2^0 = -\Omega_1^3 = \frac{2a^4}{r^6}(-\theta^0\wedge\theta^2+\theta^3\wedge\theta^1),$$

$$\Omega_3^0 = -\Omega_2^1 = \frac{2a^4}{r^6}(-\theta^0\wedge\theta^3+\theta^1\wedge\theta^2).$$

4.9 Pseudosphere and Bäcklund transformation

Both the Yang-Mills theory and the theory of the gravity are nonlinear, where the exact integrable solutions play an important role. Exact integrable theories have many common characteristics. For example, they have soliton solutions with nontrivial topology. These soliton solutions can often be described by the pseudosphere[17–23] and the new soliton solutions can be obtained through the Bäcklund transformation. In this section, we analyse the pseudosphere ($K=-1$, with negative constant curvature) in $\mathbf{E}^3$ by the method of moving frame and the Bäcklund transformation[24]. A thorough understanding of this section is helpful to understanding the above exact integrable physical problems.

In the last section, we deal with the surface of the revolving tractrix in three dimensional Euclidean space. It is the pseudosphere with

$$K = -\frac{1}{a^2}.$$

For pseudosphere with $K=-1$, the frame can be taken to be main curvature curve, i.e.,

$$\theta_1 = f du, \quad \theta_2 = g dv,$$
$$\omega_{31} = \kappa_1 f du, \quad \omega_{32} = \kappa_2 g dv. \tag{4.39}$$

From Eq. 4.37, we have

$$d\theta_1 = -\omega_{12}\wedge\theta_2 = -f_{,v} du\wedge dv,$$

$$d\theta_2 = -\theta_1 \wedge \omega_{12} = -g_{,u} du \wedge dv.$$

Hence

$$\omega_{12} = \frac{f_{,v}}{g} du - \frac{g_{,u}}{f} dv. \tag{4.40}$$

From Codazzi equation Eq. 4.30, we have

$$d\omega_{13} = -\omega_{12} \wedge \omega_{23}.$$

Put the Eq. 4.39 and Eq. 4.40 into the above equation, we have:

$$(\kappa_1 - \kappa_2) f_{,v} + \kappa_{1,v} f = 0. \tag{4.41}$$

As $K = \kappa_1 \kappa_2 = -1$, let

$$\kappa_1 = \cot \psi, \qquad \kappa_2 = -\tan \psi,$$

and put them into Eq. 4.41, we have

$$(\ln f)_{,v} + (-\ln \sin \psi)_{,v} = 0, \qquad f = U(u) \sin \psi.$$

Similarly, let

$$g = V(v) \cos \psi.$$

By changing the parameters of the main curvature curve, $U(u)$ and $V(v)$ can be dropped. Finally, we get

$$\theta_1 = \sin \psi du, \qquad \theta_2 = \cos \psi dv, \tag{4.42}$$

$$\omega_{31} = \cos \psi du, \qquad \omega_{32} = -\sin \psi dv. \tag{4.43}$$

And the two fundamental forms of the pseudosphere are

$$I = \sin^2 \psi du^2 + \cos^2 \psi dv^2,$$

$$II = \cos \psi \sin \psi (du^2 - dv^2).$$

II can be decomposed into two real monomials, i.e., there are two asymptotic directions and the angle between them is 2ψ. Put Eq. 4.42 and Eq. 4.43 into Eq. 4.40, one has

$$\omega_{12} = \psi_{,v} du + \psi_{,u} dv. \tag{4.44}$$

From the Gauss equation (Eq.4.32), one obtains

$$d\omega_{12} = K \theta_1 \wedge \theta_2 = -\theta_1 \wedge \theta_2.$$

Combining with Eq. 4.42 and Eq. 4.44, we have

$$\psi_{,uu} - \psi_{,vv} = \sin \psi \cos \psi. \tag{4.45}$$

Let $\varphi = 2\psi$, then φ satisfies

$$\varphi_{,uu} - \varphi_{,vv} = \sin\varphi. \tag{4.46}$$

This is the well known Sine-Gordon equation. When a solution is given, from Eqs.4.42, 4.43 and 4.44, one can get $\theta_1, \theta_2, \omega_{31}, \omega_{32}, \omega_{12}$, which satisfy the structure equations. The end points of these frame fields constitute a pseudosphere with $K = -1$. Thus, solutions of Sine-Gordon equation have a 1-1 correspondence with the pseudosphere.

The transformation carried out to obtain another solution of an equation from a known solution, is called the Bäcklund transformation. Each solution of Sine-Gordon equation corresponds to a pseudosphere, which is called the soliton surface. Now we analyse the transformation from one soliton surface M to another soliton surface M'.

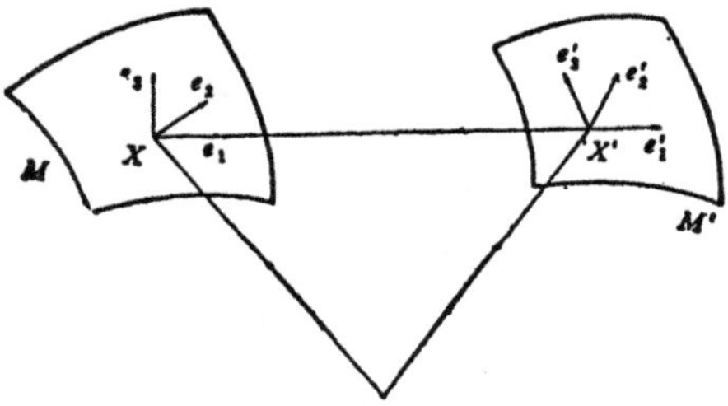

Figure 4.4. Bäcklund transformation between two pseudospheres

The set of all common tangent lines of two surfaces forms a congruence of lines. We choose

$$\mathbf{e}_1 = \mathbf{e}_1' \tag{4.47}$$

along the common tangent line, and

$$\mathbf{x}' = \mathbf{x} + r\mathbf{e}_1.$$

Suppose the angle between normal vectors $\mathbf{e}_3$ and $\mathbf{e}_3'$ of two surfaces is τ

$$(\mathbf{e}_3, \mathbf{e}_3') = \cos\tau. \tag{4.48}$$

If r and τ are constants, the corresponding congruence of lines is called the family of the pseudosphere.

Theorem 4.8 (Bäcklund) *The Gauss curvatures of two focal surfaces of the family of pseudosphere are equal,*

$$K = K' = -\frac{\sin\tau}{r^2} < 0.$$

Proof: Take a right hand frame, from Eq. 4.47 and Eq. 4.48 we have

$$\mathbf{e}_2' = \cos\tau\mathbf{e}_2 + \sin\tau\mathbf{e}_3, \qquad \mathbf{e}_3' = -\sin\tau\mathbf{e}_2 + \cos\tau\mathbf{e}_3.$$

Therefore, we get

$$d\mathbf{x}' = \mathbf{e}_1'\theta_1' + \mathbf{e}_2'\theta_2' = \mathbf{e}_1\theta_1' + (\cos\tau\mathbf{e}_2 + \sin\tau\mathbf{e}_3)\theta_2'.$$

In addition, one has

$$d\mathbf{x}' = d\mathbf{x} + rd\mathbf{e}_1 = \mathbf{e}_1\theta_1 + \mathbf{e}_2\theta_2 + r(\mathbf{e}_2\omega_{21} + \mathbf{e}_3\omega_{31}).$$

The comparison gives

$$\begin{cases} \theta_1' = \theta_1, \\ \cos\tau\theta_2' = \theta_2 + r\omega_{21}, \\ \sin\tau\theta_2' = r\omega_{31}. \end{cases} \tag{4.49}$$

Eliminating θ_2', one has

$$\sin\tau(\theta_2 + r\omega_{21}) = r\cos\tau\omega_{31}. \tag{4.50}$$

From Eq. 4.35 and Eq. 4.49, it is easy to prove that

$$\theta_1' \wedge \theta_2' = \frac{r}{\sin\tau} b\theta_1 \wedge \theta_2. \tag{4.51}$$

In addition, we have

$$\begin{aligned} \omega_{31}' &= (d\mathbf{e}_1', \mathbf{e}_3') = (d\mathbf{e}_1, -\sin\tau\mathbf{e}_2 + \cos\tau\mathbf{e}_3) \\ &= -\sin\tau\omega_{21} + \cos\tau\omega_{31} \\ &= \frac{1}{r}[-r\sin\tau\omega_{21} + r\cos\tau\omega_{31}] \\ &= \frac{\sin\tau}{r}\theta_2, \end{aligned}$$

$$\begin{aligned} \omega_{32}' &= (d\mathbf{e}_2', \mathbf{e}_3') = \cos^2\tau(d\mathbf{e}_2, \mathbf{e}_3) - \sin^2\tau(d\mathbf{e}_3, \mathbf{e}_2), \\ &= \omega_{32} = b\theta_1 + c\theta_2. \end{aligned}$$

Therefore,

$$\omega_{31}' \wedge \omega_{32}' = -\frac{\sin\tau}{r} b\theta_1 \wedge \theta_2.$$

If $b \neq 0$, and compare with Eq. 4.51, one has

$$\begin{aligned} d\omega_{12}' &= -\omega_{13}' \wedge \omega_{32}' = -\frac{\sin\tau}{r} b\theta_1 \wedge \theta_2 \\ &= -\frac{\sin^2\tau}{r^2}\theta_1' \wedge \theta_2'. \end{aligned}$$

That is, the Gauss curvature is

$$K = -\frac{\sin^2\tau}{r^2}.$$

As the two focal surfaces are on the same footing, one has

$$K = K' = -\frac{\sin^2\tau}{r^2}.$$

This completes our proof.

Starting from a pseudosphere (negative constant Gauss curvature, $K = -\frac{\sin^2\tau}{r^2}$), and choose a unit vector $\mathbf{e}_1$ which is not in the main direction, so that the tangent lines of M along the tangent vector form a family of pseudosphere, i.e., satisfying Eq. 4.50 which is completely integrable. From its solutions, one can get the family of pseudospheres and subsequently a soliton surface M' from soliton surface M, such that soliton surface M' has the same negative constant curvature as M.

The moving frame $\{\mathbf{e}_\alpha\}$, which θ_α, $\omega_{\alpha\beta}$ satisfy Eq. 4.50, is called the common tangent frame. For simplicity, let the soliton surface be the pseudosphere with $K = -1$, i.e., $r = \sin\tau$. Eq.4.50 becomes

$$\omega_{21} = \cot\tau\omega_{31} - \csc\tau\theta_2. \tag{4.52}$$

Let $\mathbf{a}_1, \mathbf{a}_2, \mathbf{e}_3$ be the moving frame of the main curvature on the soliton surface M, the corresponding components are denoted by $\sigma_\alpha, \lambda_{\alpha\beta}$

$$\sigma_1 = \sin\psi du, \quad \sigma_2 = \cos\psi dv,$$

$$\lambda_{31} = \cos\psi du, \quad \lambda_{32} = -\sin\psi dv. \tag{4.53}$$

On the direction of the common tangent, one has

$$\mathbf{e}_1 = \cos\alpha\mathbf{a}_1 + \sin\alpha\mathbf{a}_2,$$

$$\mathbf{e}_2 = -\sin\alpha\mathbf{a}_1 + \cos\alpha\mathbf{a}_2,$$

$$d\mathbf{e}_1 = \cos\alpha d\mathbf{a}_1 + \sin\alpha d\mathbf{a}_2 + \mathbf{e}_2 d\alpha,$$

hence,

$$\begin{cases} \omega_{21} = (\mathbf{e}_1, d\mathbf{e}_1) = d\alpha + \lambda_{21}, \\ \omega_{31} = (\mathbf{e}_3, d\mathbf{e}_1) = \cos\alpha\lambda_{31} + \sin\alpha\lambda_{32}, \\ \theta_2 = (\sigma_1\mathbf{a}_1 + \sigma_2\mathbf{a}_2, \mathbf{e}_2) = -\sin\alpha\sigma_1 + \cos\alpha\sigma_2. \end{cases}$$

Put these three equations into Eq.4.52 and use Eq.4.53, one obtains

$$\sin\tau(\alpha_{,u} - \psi_{,v}) = \cos\tau\cos\alpha\cos\psi + \sin\alpha\sin\psi,$$

$$\sin\tau(\alpha_{,v} - \psi_{,u}) = -\cos\tau\sin\alpha\sin\psi - \cos\alpha\cos\psi. \tag{4.54}$$

If we take the asymptotic curves as parameters, i.e., suppose

$$u = x - y, \qquad v = x + y,$$

we have

$$\frac{\partial(\alpha-\psi)}{\partial x} = \gamma\cos(\alpha+\psi),$$
$$\frac{\partial(\alpha+\psi)}{\partial y} = \frac{1}{\gamma}\cos(\alpha-\psi), \tag{4.55}$$

where

$$\gamma = \cot\tau - \csc\tau = -\frac{1}{\cot\tau+\csc\tau}.$$

If ψ is a solution of sine-Gordon equation, then α gives a new solution of the equation. The angle between two asymptotic directions of the pseudosphere M' corresponding to the new solution is 2α. Eq.4.55 (or Eq. 4.54) is called the equation of Bäcklund transformation, which is a first order differential equation. But, Sine-Gordon equation Eq. 4.46(or Eq. 4.45) is a second order differential equation. If we know one special solution of Sine-Gordon equation, other new solutions can be obtained by the Bäcklund transformation. This method has been used to find new solutions of the chiral field[19],the axial symmetric gravitation field[20], the self-dual Yang-Mills field[21] etc..

Notations and Formulae

$$\epsilon_{i_1\cdots i_n} = |g|^{\frac{1}{2}}\,\delta^{1\cdots n}_{i_1\cdots i_n}.$$
$$\epsilon^{i_1\cdots i_n} = sgn(g)\,|g|^{-\frac{1}{2}}\,\delta^{i_1\cdots i_n}_{1\cdots n}.$$
$$\epsilon_{i_1\cdots i_n}\epsilon^{k_1\cdots k_n} = sgn(g)\delta^{k_1\cdots k_n}_{i_1\cdots i_n}.$$

$$\star dx^{i_1}\wedge\cdots\wedge dx^{i_r} = \tfrac{1}{(n-r)!}\epsilon^{i_1\cdots i_r}{}_{k_1\cdots k_{n-r}}dx^{k_1}\wedge\cdots\wedge dx^{k_{n-r}}.$$
$$\star 1 = |g|^{\frac{1}{2}}\,dx^1\wedge\cdots\wedge dx^n,\ g = det(g_{ij}).$$
$$\star\star = \mathrm{sgn}(g)(-1)^{r(n-r)}.$$
$$\delta = \mathrm{sgn}(g)(-1)^{nr+n-1}\star d\star.$$
$$\Delta = (d+\delta)^2 = d\delta + \delta d.$$
$$(\alpha,\beta) = \tfrac{1}{v}\textstyle\int_M \alpha\wedge\star\beta = \tfrac{1}{v}\int_M \beta\wedge\star\alpha, \quad v = \int_M \star 1.$$
$$(\alpha_r, d\beta_{r-1}) = (\delta\alpha_r, \beta_{r-1}).$$
$$\alpha_r = \tfrac{1}{r!}f_{i_1\cdots i_r}dx^{r_1}\wedge\cdots\wedge dx^{r_{n-r}}.$$
$$\star\alpha_r = \tfrac{1}{r!(n-r)!}f_{i_1\cdots i_r}\epsilon^{i_1\cdots i_r}{}_{k_1\cdots k_{n-r}}dx^{k_1}\wedge\cdots\wedge dx^{k_{n-r}}.$$

- Riemannian connection satisfies:

(1) Metric preserve: $\nabla_k\, g_{ij} = 0$,
or $\quad X(G(Y,Z)) = G(\nabla_X Y, Z) + G(Y, \nabla_X Z).$

(2) Torsion free

$$\Gamma^m_{ik} = \frac{1}{2} g^{mj}(\partial_k g_{ij} + \partial_i g_{kj} - \partial_j g_{ik}),$$

or $\qquad \nabla_X Y - \nabla_Y X = [X, Y].$

- Riemannian curvature tensor has following symmetries:

(1) The last two indices are skewsymmetric: $R_{lkij} = -R_{lkji}$.

(2) The first two indices are skewsymmetric: $R_{lkij} = -R_{klij}$.

(3) $R_{lkij} = R_{ijlk}$.

(4) $R_{l[kij]} = R_{lkij} + R_{lijk} + R_{ljki} = 0$.

(5) $R_{lk[ij;m]} = R_{lkij;m} + R_{lkjm;i} + R_{lkmi;j} = 0$.

- Ricci tensor: $\qquad \mathfrak{R}_{kj} = R^l_{klj} = \mathfrak{R}_{jk}$.
Ricci scalar curvature: $\qquad \mathfrak{R} = g^{jk}\mathfrak{R}_{jk}$.

- The decomposition w.r.t. irr. rep. of $O(n)$

$$R_{abcd} = W_{abcd} + S_{abcd} + G_{abcd},$$

where

$$\begin{aligned} G_{abcd} &= \tfrac{\mathfrak{R}}{n(n-1)}(g_{ac}g_{bd} - g_{ad}g_{bc}), \\ S_{abcd} &= \tfrac{1}{n-2}(g_{ac}S_{bd} + g_{bd}S_{ac} - g_{ad}S_{bc} - g_{bc}S_{ad}), \\ S_{ab} &= \mathfrak{R}_{ab} - \tfrac{1}{n}g_{ab}\mathfrak{R}. \end{aligned}$$

The Weyl tensor W_{abcd} is the traceless parts of the Riemannian tensor.

- Riemannian section curvature at point p is

$$K(S_p) = \frac{R(X,Y,X,Y)}{|X|^2|Y|^2 - (X,Y)^2},$$

where S_p represent the plane spanned by X and Y.

- The first fundamental form : $I = (dx, dx) = ds^2 = \sum_i (\theta^i)^2 = \sum g_{ij} dx^i dx^j$.

The second fundamental form : $II = -(dx, de_n) = \sum_{ij} h_{ij} \theta^i \theta^j$.

Exercises

1. Using the exterior differential operator d and Hodge $*$ operation, prove the following identities in three-dimensional Euclidean space.

$$\begin{aligned} grad\ fg &= (grad\ f)g + f(grad\ g), \\ curl(f\mathbf{V}) &= (grad\ f)\mathbf{V} + f(curl\ \mathbf{V}), \\ div(f\mathbf{V}) &= (grad\ f) \bullet \mathbf{V} + f\ div\ \mathbf{V}, \\ div(\mathbf{V} \times \mathbf{W}) &= \mathbf{W} \bullet curl\mathbf{V} - \mathbf{V}\ curl\mathbf{W}. \end{aligned}$$

2. Find the tensor formulation of the $U(1)$ gauge field equation in four-dimensional Euclidean space

$$\delta F = J_\mu dx^\mu \equiv J,$$

and prove the current conservation equation $\delta J = 0$.

3. Prove

$$\begin{aligned} & * * \alpha_r = sgn(g)(-1)^{r(n-r)} \alpha_r, \\ & \alpha_r \wedge * \beta_r = \beta_r \wedge * \alpha_r, \\ & (\alpha_r,\ d\beta_{r-1}) = (\delta\alpha_r, \beta_{r-1}). \end{aligned}$$

4. Prove the following identities for Riemannian manifold:

$$\begin{aligned} & g^{\alpha\beta}_{,\sigma} = -g^{\alpha\mu} g^{\beta\nu} g_{\mu\nu,\sigma}, \\ & g_{,\nu} = g g^{\lambda\mu} g_{\lambda\mu,\nu} \quad (g = det(g_{\mu\nu})), \\ & \Gamma^\mu_{\nu\mu} = \tfrac{1}{2}(ln\ g)_{,\nu}, \\ & g^{\mu\nu}\Gamma^\lambda_{\mu\nu} = -\tfrac{1}{\sqrt{g}} \partial_\rho(\sqrt{g} g^{\rho\lambda}). \end{aligned}$$

5. Prove Riemann curvature satisfies the Bianchi identity,

$$R^l_{k[ij,m]} = 0,$$

and derive its reduced form

$$\nabla_\mu(\mathcal{R}^k_j - \frac{1}{2}\delta^k_j \mathcal{R}) = 0.$$

6. Prove the Riemann curvature is completely determined by the Ricci tensor in two and three dimensions.

7. Is spin connection a Riemannian connection? Find the relation between spin connection and Christopher connection on Riemannian manifold.

8. Using the natural metric, derive the Riemannian connection and curvature for S^2.
9. Using the $SU(2)$ left invariant vector fields as frame fields on S^3, derive the corresponding connection and curvature.
10. Prove the metric compatible vector field $X = \xi_i \partial_i$ ($L_X g_{ij} = 0$), the so called Killing vector, satisfies Killing equation

$$\xi_{i;j} + \xi_{j;i} = 0.$$

Chapter 5

Symplectic Manifold and Contact Manifold

In analytical mechanics, the Lagrangian $L = L(q, \dot{q})$ of a mechanic system is a function defined on the tangent bundle of the configuration space, and the Hamiltonian $H = H(q,p)$ of the corresponding system is a function defined on the cotangent bundle of the configuration space. The cotangent bundle of the configuration space is called the phase space, which is a smooth symplectic manifold. The Hamiltonian formulation of classical mechanics (analytical dynamics) is described by the canonical equations of motion in phase space, i.e., by symplectic geometry. Contact geometry is the odd-dimensional twin of symplectic geometry. The state space (extended phase space) of the system is described by variables $\{q^i, p^i, t\}$, the Hamiltonian $H(q,p,t)$ is a function defined on the state space. The state space is an odd-dimensional contact manifold. Canonical transformations, which preserve the canonical formulation of the equations of motion, are described by contact geometry.

5.1 Symplectic manifold

A symplectic structure on a smooth manifold M is defined by a differential 2-form ω satisfying:

(i) ω is non-degenerate. In local coordinate, $\omega = \frac{1}{2}\omega_{ij}dx^i \wedge dx^j$, nondegeneracy is written as

$$det(\omega_{ij}) \neq 0.$$

(ii) ω is closed, i.e.,

$$d\omega = 0.$$

A manifold endowed with a symplectic structure (M, ω) is called a symplectic manifold.

Manifold M is called almost symplectic if ω satisfies (i) but not necessarily (ii).

Manifold M is called presymplectic if ω satisfies (ii) but not necessarily (i).

The tangent space $T_x(M)$ at each point of an almost symplectic manifold is a symplectic vector space, i.e., a linear space endowed with a skew-scalar product:

$$\omega(X,Y) = -\omega(Y,X) = <\omega(x); X, Y>, \qquad \forall X, Y \in T_x(M)$$

The nondegeneracy condition

$$\omega(X,Y) = 0 \text{ for all Y} \quad \Longrightarrow \quad X = 0$$

implies that the space must be even dimensional $d = 2n$, and

$$\omega^n \equiv \wedge_n \omega \neq 0.$$

This shows that there exists a 2n-form which vanishes nowhere on the 2n-dimensional almost symplectic manifold, which is called the volume element. So almost symplectic manifold is an even dimensional orientable manifold.

Darboux proved that any two symplectic spaces of the same dimension are symplectically isomorphic, i.e., there exists a linear isomorphism between them preserving the skew-scalar product. The closedness of symplectic structure implies that this structure is integrable on the symplectic manifold, the skew-scalar products in the tangent spaces of neighboring points are defined such that the local geometry of symplectic manifold is universal. The diffeomorphism which preserves the symplectic structure is called symplectomorphism. There is the following theorem:(the proof see ref.[5] Arnol'd.V.I. Mathematical method of classical mechanics. 1978. P230)

Theorem 5.1 (Darboux) *Symplectic manifolds of the same dimension are locally symplectomorphic.*

In terms of this theorem under a suitable choice of local coordinates in the neighborhood of any point $x \in M$, the symplectic structure has the form:

$$\omega = \sum_{i=1}^{n} dp_i \wedge dq^i,$$

and the local coordinates $\{q^i, p_i\}$ are called the Darboux coordinates.

Note that on a Riemannian manifold, the Riemann metric can not be transformed to a flat diagonal metric through a coordinate transformation if the Riemann tensor R_{ijkl} is non-vanishing. The metric g_{ij} could be transformed to the standard diagonal form $g_{ij} = \delta_{ij}$ at a point but not in a neighborhood of the point. To determine when a flat coordinate system can exists one has to study the Riemann tensor field R_{ijkl} derived from g_{ij}. There is a locally flat coordinate system if and only if $R_{ijkl} = 0$. But symplectic structures are locally integrable, and are locally symplectomorphic.

We take the following three types of Symplectic manifolds as examples: the cotangent bundle; the coadjoint orbits of Lie group; and the Kähler manifolds.

Example 1. The cotangent bundle.

Let N be an arbitrary n-dimensional manifold, called the configuration manifold, $T(N) = \cup_{x\in N} T_x(N)$ be its tangent bundle, $T^*(N) = \cup_{x\in N} T_x^*(N)$ be its cotangent bundle. A point of $T_x^*(N)$ is defined by a linear function on $T_x(N)$. The symplectic structure of the cotangent bundle $T^*(N) = M$ can be determined by the smooth structure of manifold N as follows. Locally there are coordinates $\{q^i, p_i\}$ on the 2n-dimensional manifold $M = T^*(N)$, and the canonical 1-form

$$\theta = \sum_i p_i dq^i \quad \in \Lambda^1(M)$$

is independent of the choice of the local coordinates on N. The differential 2-form

$$\omega = d\theta = \sum_i dp_i \wedge dq^i \quad \in \Lambda^2(M)$$

gives a symplectic structure on $T^*(N)$. The canonical 1-form θ on a cotangent bundle can be defined globally, and is called the symplectic potential. In classical mechanics, the base manifold N is called the configuration space, the covector $\{p_i\}$ are called the generalized momenta, and the cotangent bundle $T^*(N)$ are called the phase space of the Hamiltonian system.

Example 2. The coadjoint orbits.

Let G be a given Lie group, $\mathbf{g}$ be its Lie algebra, $\mathbf{g}^*$ be the space dual to $\mathbf{g}$. If $\mathbf{g}$ is realized as the algebra of left invariant vector fields on G, then $\mathbf{g}^*$ is realized as the left invariant 1-forms on G. G acts on the space of left invariant 1-forms by right translations, from which one can obtain a coadjoint representation of G:

$$Ad^* : \ G \rightarrow GL(\mathbf{g}^*)$$

Its tangent map gives the corresponding coadjoint representation of $\mathbf{g}$:

$$\begin{aligned} ad^* : \ \mathbf{g} &\rightarrow End(\mathbf{g}^*) \\ X &\rightarrow ad_X^* \end{aligned}$$

Let $\sigma \in \mathbf{g}^*$, N_σ be the orbit through σ in $\mathbf{g}^*$ by G acting, and let $\xi_X(\sigma)$ denote the fundamental vector field on N_σ corresponding to the element $X \in \mathbf{g}$. A bilinear skew symmetric 2-form Ω_σ on N_σ can be defined by the following equation

$$\Omega_\sigma(\xi_X, \xi_Y) = <\sigma, \ [X, Y]> .$$

Let G_σ denote the stable subgroup of point σ, $\mathbf{g}_\sigma$ denote the corresponding Lie algebra. Then the tangent space $T_\sigma N$ is identified with $\mathbf{g}/\mathbf{g}_\sigma$. It is easy to prove that the skew product $\Omega_\sigma(\xi_X, \ \xi_Y)$ on $T_\sigma N$ is nondegenerate, so the 2-form Ω_σ on N_σ is nondegenerate. Furthermore, through the Jacobi identity of the Lie algebra one can prove that the 2-form Ω_σ on N_σ is closed. So the coadjoint orbit (N, Ω) is a symplectic manifold.

Example 3. Complex projective manifold and Kähler manifold.

A symplectic structure can be introduced as the imaginary part of an Hermitian metric. Please refer to the next chapter for details.

5.2 Special submanifolds of symplectic manifold

Let (M,ω) be a 2n-dimensional symplectic manifold. Let $V = T_x(M)$ be the tangent space at point $x \in M$, which is a symplectic vector space. For any two vectors $X, Y \in V$, there is the skew scalar product

$$< X,\ Y >= - < Y,\ X >= \omega(X,\ Y)|_x,$$

X and Y are called skew orthogonal if $< X, Y >= 0$.

For an arbitrary subspace $W \subset V$, there is the skew orthogonal complement $W^{\perp} \subset V$ such that

$$< X,\ Y >= 0, \quad \forall\, X \in W, \quad Y \in W^{\perp}.$$

I.e., $W^{\perp} = \{Y \in V \mid \omega(X,Y) = 0, \forall X \in W\}$. By virtue of the nondegeneracy of the skew-scalar product, we get

$$dim\ W\ +\ dim\ W^{\perp} = 2n.$$

But unlike Euclidean space, the skew orthogonal complement may be interect the original space, since any vector is self orthogonal.

The subspace of Euclidean space has only one invariant: its dimension. But for symplectic subspace, in addition to the dimension, the rank of the restriction of the symplectic structure to subspace is also an intrinsic quantity. Let ω_W denote the form induced by ω on W, its kernel is

$$ker\ \omega_W = \{X \in W | i_X\omega_W = 0\} = W \cap W^{\perp}.$$

Let $k = dim\ ker\ \omega_W$, $s = dim\ W$, the rank of ω_W is equal to the codimension of $ker\ \omega_W$ in W, i.e., let r=rank of ω_W, then

$$r + k = s.$$

For example, in a 2n-dimensional symplectic space V, with a suitable choice of Darboux coordinates $\{p_i, q^i\}$, the symplectic structure ω can be written as:

$$\omega = \sum_{i=1}^{n} dp_i \wedge dq^i,$$

in this case, a subspace W of rank $r = 2l$ and dimension $s = 2l + k$ is given by

$$q^{l+1} = \cdots = q^n = 0, \quad p_{l+k+1} = \cdots = p_n = 0,$$

the skew-orthogonal complement $W^{\perp}$ is given by

$$q^1 = \cdots = q^{l+k} = 0, \quad p_1 = \cdots = p_l = 0,$$

which is their intersection along the k-dimensional kernel of the symplectic form.

Subspaces, which lie within their skew-orthogonal complement, are called isotropic. Subspaces, which contain their skew-orthogonal complement, are called coisotropic. Subspaces, which are isotropic and coisotropic at the same time, are called Lagrangian. A subspace W, whose induced 2-form ω_W by ω is nondegenerate on W, i.e., $ker\ \omega_N$ is zero, is called symplectic subspace. These four types of subspaces of symplectic space and their characteristics are listed in the table below.

Subspace W	Properties	rank of ω_W	dim. of W
Isotropic	$W \subset W^\perp \quad \omega_w = 0$	$r = 0$	$s \leq n$
Coisotropic	$W \supset W^\perp \quad \omega_{w^\perp} = 0$	$r = 2s - 2n$	$s \geq n$
Lagrangian	$W = W^\perp \quad \omega_w = \omega_{W^\perp} = 0$	$r = 0$	$s = n$
Symplectic	$W \cap W^\perp \quad \omega_w$ nondegenerate	$r = 2l > 0$	$s = 2l < 2n$

Now let us consider the submanifold N of a symplectic manifold (M, ω), and denote the canonical injection by $i: \ N \to M$, the 2-form induced by ω on N is denoted by

$$\omega_N = i^*\omega.$$

Here ω_N is obviously closed.

A manifold on which a closed 2-form is defined is called a presymplectic manifold. The submanifold (N, ω_N) of symplectic manifold (M, ω) is a presymplectic manifold, the induced 2-form ω_N is closed but could be degenerate.

At any point $x \in N$, the tangent space $T_x(N)$ has a skew orthogonal complement $(T_x(N))^\perp$ in the symplectic space $T_x(M)$, and the kernel of the induced 2-form ω_N

$$ker_x\omega_N = T_x(N) \cap (T_x(N))^\perp, \quad \forall x \in N.$$

The dimension $k(x)$ of $ker_x\omega_N$ is nonnegative and bounded by the dimensions both $T_x(N)$ and $(T_x(N))^\perp$. The rank $r(x)$ of ω_N at point x is an even integer $2l(x)$, equal to the codimension of $ker_x\omega_N$ in the tangent space $T_x(N)$. $T(N)$ is a completely integrable vector subbundle of $T(M)$ if 2-form ω_N is of constant rank $2l$, in which case we say that N is an isotropic (respectively, coisotropic, Lagrangian, or symplectic) submanifold if $T_x(N)$ is an isotropic (respectively, coisotropic, Lagrangian, or symplectic) subspace of $(T_x(M), \omega)$ at each point $x \in N$. Alternatively, a submanifold N of $2n$-dimensional symplectic manifold (M, ω) is called isotropic, if ω vanishes on N, the dimension of an isotropic submanifold does not exceed n. The maximum isotropic submanifold having dimension n is called the Lagrangian manifold. If the induced 2-form ω_N by ω is nondegenerate on N, then N is a symplectic submanifold.

Next we consider the case where the closed 2-form ω_N is of constant rank $2l$. Then subbundle $Ker\ \omega_N$ is completely integrable. Denote the quotient manifold $N/Ker\ \omega_N$ by P and the canonical projection by π,

$$\pi: \ N \to P = N/Ker\ \omega_N,$$

then there exists a unique symplectic 2-form ω_P on P such that $\omega_N = \pi^*\omega_P$, and (P,ω_P) is the reduced symplectic manifold associated with the presymplectic manifold (N,ω_N).

5.3 Symplectic and Hamiltonian vector fields, Poisson bracket

Let (M,ω) be a symplectic manifold, the nondegenerate closed 2-form ω gives an isomorphism μ between sections of the tangent and cotangent bundles, which sends each vector field $X \in \mathcal{X}(M)$ into a 1-form $\alpha \in \Lambda^1(M)$ by

$$\begin{aligned} \mu: \mathcal{X}(M) &\to \Lambda^1(M) \\ X &\to \alpha = \mu(X) \equiv -i_X\omega. \end{aligned}$$

Taking a local coordinate system $\{x^i\}$, the symplectic structure can be written as

$$\omega = \frac{1}{2}\omega_{ij}(x)dx^i \wedge dx^j.$$

Let $X = \xi^i(x)\partial_i \in \mathcal{X}(M)$, then the corresponding 1-form is

$$\alpha = -i_X\omega = -\omega_{ij}\xi^i dx^j,$$

the nondegeneracy condition $\omega^n \neq 0$ implies

$$det(\omega_{ij}) \neq 0.$$

As 2-form ω is bilinear and nondegenerate, map μ is a linear isomorphism at any point $x \in M$, and the inverse map μ^{-1} exists uniquely. Let (λ^{ij}) be the transposed inverse matrix of (ω_{ij}), which is also skew : $\lambda^{ij}(x) = -\lambda^{ji}(x)$,

$$\lambda^{ij}\omega_{jk} = -\delta^i_k. \tag{5.1}$$

Then one can constructed a bi-vector $\bigwedge$:

$$\bigwedge = \lambda^{ij}(x)\frac{\partial}{\partial x^i} \otimes \frac{\partial}{\partial x^j}, \quad \lambda^{ij}(x) = -\lambda^{ji}(x).$$

From this bi-vector the inverse map μ^{-1} can be expressed as:

$$\begin{aligned} \mu^{-1}: \Lambda^1(M) &\to \mathcal{X}(M) \\ \alpha_i dx^i &\to <\alpha_i dx^i, \bigwedge>= \lambda^{ij}\alpha_i\partial_j. \end{aligned}$$

A vector field X which is an infinitesimal automorphism of the symplectic structure, is called a Symplectic vector field, i.e., it satisfies

$$L_X\omega = 0.$$

The set of all symplectic vector fields on (M,ω) is denoted by $Sym(M,\omega)$. As $[L_X, L_Y] = L_{[X,Y]}$, the $Sym(M,\omega)$ forms a subalgebra of the vector field Lie algebra $\mathcal{X}(M)$.

For symplectic structure ω, $d\omega = 0$, so

$$L_X\omega = di_X\omega = d(\omega_{ij}\xi^i dx^j).$$

A vector field $X = \xi^i\partial_i$ is a symplectic vector field if $i_X\omega = \omega_{ij}\xi^i dx^j$ is a closed form. And we have the following proposition.

Proposition: Let $X \in \mathcal{X}(M)$ be a vector field on the symplectic manifold (M,ω), the following conditions are equivalent:
(1) X is a symplectic vector field. $X \in Sym(M,\omega)$.
(2) Lie derivative of ω along X is zero. $L_X\omega = 0$.
(3) $i_X\omega$ is a closed form. $d(i_X\omega) = 0$.

Furthermore if $i_X\omega$ is an exact form,

$$i_X\omega = \omega_{ij}\xi^i dx^j = -df, \quad f \in \mathcal{F}(M),$$

then vector field $X_f = \mu^{-1}(df) = \lambda^{ij}f_{,i}\partial_j$ is called the Hamiltonian vector field associated with the Hamiltonian function $f \in \mathcal{F}(M)$.

We know that an exact form is also closed, but the inverse is not true in general. Correspondingly, a Hamiltonian vector field is symplectic, but the inverse is not true in general. Let $Ham(M,\omega)$ denote the set of all Hamiltonian vector fields, then

$$Ham(M,\omega) \subset Sym(M,\omega) \subset \mathcal{X}(M).$$

It is easy to prove that if $X, Y \in Sym(M,\omega)$, then $[X,Y] \in Ham(M,\omega)$.

Proof: Let $X, Y \in Sym(M,\omega)$, i.e., $\mu(X) = -i_X\omega$ and $\mu(Y) = -i_Y\omega$ are closed forms, then from $[L_X, i_Y] = i_{[X,Y]}$,

$$\begin{aligned}
\mu([X,Y]) &= -i_{[X,Y]}\omega = -[L_X, i_Y]\omega \\
&= -L_X i_Y\omega = -d\, i_X i_Y\omega,
\end{aligned}$$

which is an exact form. So $[X,Y] \in Ham(M,\omega)$. Not only $Sym(M,\omega)$ forms a subalgebra of vector field Lie algebra $\mathcal{X}(M)$, but also $Hom(M,\omega)$ forms an ideal of Lie algebra $Sym(M,\omega)$.

Next we look at an important notion on the space of differentiable functions on a symplectic manifold, the Poisson bracket. A symplectic structure (M,ω) allows one to introduce a Lie algebra structure on the space $\mathcal{F}(M)$, i.e., we can define a bilinear map from $\mathcal{F}(M) \times \mathcal{F}(M)$ into $\mathcal{F}(M)$ denoted by $\{\, ,\, \}$. For any two functions $f, g \in \mathcal{F}(M)$,

$$\begin{aligned}
\{f,g\} &= \omega(X_f, X_g) \\
&= < i_{X_f}\omega, X_g > = - < i_{X_g}\omega, X_f > \\
&= < dg, X_f > = X_f g = -X_g f \\
&= \lambda^{ij}\partial_i f \partial_j g = < df \otimes dg, \Lambda >,
\end{aligned}$$

which is called the Poisson bracket between f and g.

In the local Darboux coordinate on the symplectic manifold M, the symplectic form

$$\begin{aligned} \omega &= \sum_{i=1}^{n} dp_i \wedge dq^i, \\ \text{Hamiltonian vector}\quad X_f &= \sum_{i=1}^{n}\left(\frac{\partial f}{\partial p_i}\frac{\partial}{\partial q^i} - \frac{\partial f}{\partial q^i}\frac{\partial}{\partial p_i}\right), \\ \text{Poisson bracket}\quad \{f,g\} &= \sum_{i=1}^{n}\left(\frac{\partial f}{\partial p_i}\frac{\partial g}{\partial q^i} - \frac{\partial f}{\partial q^i}\frac{\partial g}{\partial p_i}\right). \end{aligned}$$

Classical Hamiltonian mechanics is described by the phase space in terms of dynamical variables q^i and p_i. Phase space is a symplectic manifold (M,ω). Each state of the mechanical system is given by a point on the symplectic manifold. Physical quantities are functions on M. The evolution of the state is described by the Hamiltonian vector field, with the Hamiltonian $H(p,q)$ of the system as the generating function. The evolution of physical quantity $f(p,q)$ is determined by $H(p,q)$,

$$\frac{d}{dt}f(p,q) = \{H,f\} = X_H f.$$

For example, the integral curve $\{q^i(t), p_i(t)\}$ of the Hamiltonian vector field X_H satisfies

$$\frac{d}{dt}q^i = X_H q^i = \{H,q^i\} = \frac{\partial H}{\partial p_i}, \quad \frac{d}{dt}p_i = X_H p_i = \{H,p_i\} = -\frac{\partial H}{\partial q^i}.$$

These are just the Hamiltonian equations in classical mechanics. From the definition of Hamiltonian vector field, we have

$$L_{X_H}\omega = 0.$$

This equation shows that the volume element is invariant along the Hamiltonian vector field, which is the well known Liouville theorem, which simply states that the volume of the phase space does not change with time.

5.4 Poission manifold and symplectic leaves

In this section we consider Poission structure on the manifold M from the dual view point, and start from the bivector $\bigwedge$, which can be used to repressed the endomorphism of the differential forms $\Lambda^*(M)$. $\bigwedge$ is a mapping $\Lambda^*(M) \to \Lambda^*(M)$, and reduced the order of $\Lambda^*(M)$ by two:

$$\begin{aligned} \bigwedge:\ \Lambda^k(M) &\to \Lambda^{k-2}(M) \\ \alpha_k &\to \bigwedge(\alpha_k) = <\alpha_k, \bigwedge>. \end{aligned}$$

For example, the eqn. 5.1 can be rewritten as:

$$\begin{aligned}\bigwedge(\omega) &= <\omega, \bigwedge> &= <\tfrac{1}{2}\omega_{ij}dx^i\wedge dx^j, \lambda^{kl}\partial_k\otimes\partial_l> \\ &= \tfrac{1}{2}\omega_{ij}\lambda^{kl}\delta^i_k\delta^j_l &= <1>,\end{aligned}$$

here $<1>$ denotes the trace of unit matrix and equals the dimension of the manifold.

Analogously we can introduce k-vector field, which is the k-order completely skew symmetric tensor product of vector fields. by using them we can repressed the endomorphism of differential form, but the k-vector field reduced the order of $\Lambda^*(M)$ by k units.

As is wellknown, one can defined the Lie derivative L_X for vector field Y as Lie bracket:

$$L_XY = [X, Y] \in \mathcal{X}(M), \quad for X, Y \in \mathcal{X}(M).$$

Now we can generalized the Lie bracket to multiptic vector field, the result is also an endomorphism of $\Lambda^*(M)$. Let A, B be two k-vector fields, then the Schouten bracket [A, B] is a $(k+l-1)$-vector field. In particular, for any vector field $X \in \mathcal{X}$, the Schouten bracket

$$[X, A] = L_XA.$$

If A, B are two bi-vector fields on the manifold M, its Schouten bracket [A,B] is a 3-vector field on the manifold, which can be defined as: (for $\sigma_i \in \Lambda^1(M)$)

$$\begin{aligned}[A, B]: \quad \Lambda^3(M) &\to \mathcal{F}(M) \\ \sigma_1\wedge\sigma_2\wedge\sigma_3 &\to [A, B](\sigma_1, \sigma_2, \sigma_3) \equiv <\sigma_1\wedge\sigma_2\wedge\sigma_3, [A, B]> \quad \in \mathcal{F}(M)\end{aligned}$$

$$[A, B](\sigma_1, \sigma_2, \sigma_3) \equiv <\sigma_1, A(L_{B(\sigma_2)}\sigma_3)> + <\sigma_1, B(L_{A(\sigma_2)}\sigma_3)> + c.p.,$$

here c. p. denote the cyclic permutation. In particular for $\sigma_i = df_i$, $A = B = \bigwedge$,

$$\bigwedge(df_1, df_2) = \{f_1, f_2\}.$$

$$\begin{aligned}[\bigwedge, \bigwedge](df_1, df_2, df_3) &= 2 < df_1, \bigwedge(L_{\bigwedge(df_2)}df_3) > + c.p. \\ &= 2 < df_1, \bigwedge(d(\bigwedge(df_2)f_3)) > + c.p. \\ &= 2\{f_1, \{f_2, f_3\}\} + c.p.,\end{aligned}$$

so

$$[\bigwedge, \bigwedge] = 0 \implies \{\ ,\ \} \text{ satisfy the Jacobi identity.}$$

These properties mean that bracket { , } such defined by the bivector $\bigwedge$ is just the Poisson bracket introduced in the last section, which satisfies the following conditions:

(1) Bilinear, $\{af + bg, h\} = a\{f, h\} + b\{g, h\}, \forall a, b \in \mathbf{R}, f, g, h \in \mathcal{F}(M)$.

(2) Skew symmetry, $\{f, g\} = -\{g, f\}$.

(3) Jacobi identity, $\{f, \{g, h\}\} + \{g, \{h, f\}\} + \{h, \{f, g\}\} = 0$.

(4) Leibniz rule, $\{f, gh\} = g\{f, h\} + \{f, g\}h$.

The condition (1)-(3) mean that { , } is a Lie algebra on $\mathcal{F}(M)$. Furthermore the Leibniz rule mean that for $\forall f, g \in \mathcal{F}(M)$, the mapping $f \to \{f, g\}$ is a derivation of

$\mathcal{F}(M)$, hence there exist a vector field X_f on M such that $X_f(g) = \{f, g\}$. In a local coordinate, the bivector $\bigwedge$ can be written as

$$\bigwedge = \lambda^{ij}\,\partial_i \otimes \partial_j, \qquad \lambda^{ij} = -\lambda^{ji},$$

then for any $f \in \mathcal{F}(M)$, the vector field X_f can be obtained as

$$X_f = \bigwedge(df) = \lambda^{ij}\, f_{,i}\, \partial_j,$$

and the Poisson bracket between two functions $f, g \in \mathcal{F}(M)$ can be expressed as

$$\{f, g\} = X_f(g) = \bigwedge(df, dg) = \lambda^{ij}(x) f_{,i}\, g_{,j}.$$

In particular, $\{x^i, x^j\} = \lambda^{ij}$ and consequently $\{f, g\} = \{x^i, x^j\} f_{,i}\, g_{,j}$, i.e., on a coordinate chart U the knowledge of the Poisson bracket of coordinate functions permits to determine the Poisson bracket of any pair of function. Thus $\mathcal{F}(M)$ endowed with Poisson bracket as composition law forms a Lie algebra $(\mathcal{F}(M), \{\ ,\ \})$, there exist homomorphisms from $(\mathcal{F}(M), \{\ ,\ \})$ into the Lie algebra $\mathcal{X}(M)$

$$f \to X_f = \lambda^{ij}\,\partial_i f\,\partial_j, \qquad [X_f, X_g] = X_{\{f,g\}}.$$

This algebra be called the Piosson-Lie algebra.

Proof: Let $f, g, h \in \{\mathcal{F}\}(\mathcal{M})$,

$$\begin{aligned} [X_f, X_g]h &= X_f X_g h - X_g X_g h \\ &= \{f, \{g, h\}\} - \{g, \{f, h\}\} \\ &= \{\{f, g\}, h\} = X_{\{f,g\}} h, \end{aligned}$$

This equation is true for any funtion $h \in \{\mathcal{F}\}(\mathcal{M})$, so one get $[X_f, X_g] = X_{\{f,g\}}$.

A Piosson structure on a smooth manifold M is defined by a differential skew symmetric bi-vector $\bigwedge$ satisfying the Schoulen bracket $[\bigwedge, \bigwedge] = 0$.

A manifold endowed with a Piosson structure $(M, \bigwedge)$ is called Piosson manifold. The rank of a Piosson structure at a point $x \in M$ is defined as the rank of the linear mapping

$$\bigwedge :\ \Lambda^1(M) \to \mathcal{X}(M),$$

i.e., the rank of matrix $(\lambda_{ij}(x))$.

A symplectic manifold is a Piosson manifold , but the inverse may be not true. A Piosson manifold appears as a natural generalization of symplectic manifold. Any manifold admits the trivial Piosson structure: $\{f, g\} = 0$, for any $f, g \in \mathcal{F}(M)$, i.e. the Piosson bivector $\bigwedge = 0$. At the opposite extreme, if there are an everywhere non-degenerate bivector field $\bigwedge(x)$, such that the mapping $\bigwedge(x)$ is an isomorphism, then there exists an inverse mapping, which can be expressed by a everywhere non-degenerate 2-form ω on M. Furthermore it is easy to show that the Schouten bracket

$[\Lambda, \Lambda] = 0$ is equivalent to the ω being closed, then in this case, the manifold is symplectic. Thus a Poisson structure is symplectic if and only if its rank is everywhere equal to the dimension of M. So in some sense the Poisson structure is the dual of that of a presymplectic structure.

Let (M_1, Λ_1) and (M_2, Λ_2) be two Piosson manifold, a differential mapping $\varphi : M_1 \to M_2$ be called the Poisson morphism if it satisfy

$$\varphi^*\{f, g\} = \{\varphi^* f, \varphi^* g\}, \quad \forall\, f, g \in \mathcal{F}(M_2).$$

Now on a Piosson manifold (M, Λ), the Poisson infinitesimal automorphism can be denoted by a vector field X, whose flow φ is a Poisson morphism, and the following properties are equivalent:

(1) Vector field X is a infinitesimal automorphism:

$$\varphi^*\{f, g\} = \{\varphi^* f, \varphi^* g\}, \quad \forall\, f, g \in \mathcal{F}(M_2).$$

(2) The Lie derivative of the Poisson bivector Λ with respect to X is zero:

$$L_X \Lambda = 0.$$

(3) The vector field X is a derivation of the Poisson-Lie algebra $(\mathcal{F}(M), \{\ ,\ \})$:

$$X\{f, g\} = \{Xf, g\} + \{f, Xg\}.$$

For any $f \in \mathcal{F}(M)$, the vector field $X_f = \Lambda(df)$ be called the Hamiltonian vector field related to the function f, which is a Poisson infinitesimal automorphism . The integral curve starting from a point $p \in M$ of a Hamiltonian vector field on M is denoted by $C_p(f)$. The set of $C_p(f)$ thorough p for all function on M form a symplectic Poisson submanifold of M, called the symplectic leaves. So one get that every Poisson manifold is a disjoint union of immersed symplectic submanifolds, each leaf has an unique symplectic structure, called the symplectic leaves of the Poisson structure, and the canonical injection of this leaf into (M, Λ) is a Poisson morphism. Note that the symplectic leaves of a Poisson manifold may not have the same dimension. For example, in section 5.1 example 2, $\mathbf{g}$ is a Lie algebra with Lie bracket $[\ ,\ ]$, $\mathbf{g}^*$ is a vector space, the orbit of the coadjoint of G is a symplectic manifold, which precisely the symplectic leaf of $\mathbf{g}^*$, thus there exist a nondegenerate Poisson bracket $\{,\}$ on it, it is called the Lie-Poisson structure.

5.5 Homogeneous symplectic manifold and the reduced phase space

A diffeomorphism

$$f: M_1 \to M_2$$

from a symplectic manifold (M_1, ω_1) to another Symplectic manifold (M_2, ω_2) of the same dimension, preserving the symplectic structure

$$f^*\omega_2 = \omega_1,$$

is called symplectomorphism.

The set of symplectomorphisms of a symplectic manifold (M, ω) onto itself forms a group, which is a subgroup of the diffeomorphism group. The primary object of symplectic geometry is to study the invariances of a symplectic manifold under symplectomorphisms.

Let us look at the action of a Lie group G on a Symplectic manifold (M, ω), just as in §2.6, a left action of G on M can be expressed as

$$\begin{aligned} \Phi : G \times M &\rightarrow M \\ g\ ,\ x &\rightarrow \Phi(g, x) \equiv \Phi_g(x) \equiv gx. \end{aligned}$$

Thus we can obtain the fundamental vector field X_M associated with element $X \in \mathbf{g}$:

$$X_M f(x) = \frac{d}{dt} f(e^{-tX}x)|_{t=0}, \quad \forall f \in \mathcal{F}(M), X \in \mathbf{g}.$$

Φ is called symplectic if G acts on the Symplectic manifold (M, ω) by symplectomorphism, i.e.,

$$\begin{aligned} \Phi_g^*\omega = \omega \quad &, \quad \forall g \in G. \\ \text{or} \quad L_{X_M}\omega = 0 \quad &, \quad \forall X \in \mathbf{g}. \end{aligned}$$

So for the symplectomorphism, the fundamental vector field X_M for $X \in \mathbf{g}$ is a symplectic vector field, and the Poisson structure is also invariant:

$$\begin{aligned} \Phi_g^*\{f, g\} &= \{\Phi_g^* f, \Phi_g^* h\}, \ \forall f, h \in \mathcal{F}(M). \\ X_M\{f, h\} &= \{X_M f, h\} + \{f, X_M h\}, \ \ \forall X \in \mathbf{g}. \end{aligned}$$

Now suppose that Lie group G acts transitively on the manifold in such a way that all the transformations of this group are symplectomorphisms, then manifold (M, ω) is said to be a homogenous symplectic manifold. To every element X of the Lie algebra $\mathbf{g}$ of G, there corresponds a symplectic fundamental vector field X_M on M.

If all of the fundamental vector fields X_M are Hamiltonain vector fields, then the symplectic action of group G is called Hamiltonian, i.e., for every element $X \in \mathbf{g}$, the fundamental vector field X_M satisfies

$$i_{X_M}\omega = -dH_X, \ \forall\ X \in \mathbf{g},$$

here $H_X(x) \in \mathcal{F}(M)$ is the generating function of the vector field X_M, called the Hamiltonian for X and can be chosen only up to a constant.

In the following we assume that all fundamental vector fields have single valued Hamiltonians, choose such Hamiltonians to form a bases, we get a linear map $\mathcal{H}$

$$\begin{aligned} \mathcal{H}: \mathbf{g} &\rightarrow \mathcal{F}(M) \\ X &\rightarrow H_X, \end{aligned}$$

called the Hamiltonian map. Poisson bracket $\{H_X, H_Y\}$ may differ from $H_{[X,Y]}$ by a constant on each connected component of M

$$\{H_X, H_Y\} = H_{[X,Y]} + C(X,Y), \ \forall X, Y \in \mathbf{g}.$$

Where constant $C(X,Y)$ satisfies

$$\begin{gathered} C(X,Y) = -C(Y,X) \qquad , \\ C([X,Y],Z) + C([Y,Z],X) + C([Z,X],Y) = 0. \end{gathered}$$

A bilinear skew-symmetric function on a Lie algebra satisfies these equations are called a 2-cocycle of the Lie algebra(see App.R.)

If the bases of Hamiltonian $\{H_X\}$ can be chosen such that $C(X,Y) = 0$ for all $X, Y \in \mathbf{g}$, i.e.,

$$\{H_X, H_Y\} = H_{[X,Y]}, \quad \forall X, Y \in \mathbf{g},$$

then the Hamiltonian action of G on (M, ω) is called a Poisson or a strict Hamiltonian action, and the manifold (M, ω) is called a strictly homogenous symplectic manifold.

A Poisson action also defines a moment map $J: M \rightarrow \mathbf{g}^*$, with $\mathbf{g}^*$ the space dual to $\mathbf{g}$, moment map J is dual to the Hamiltonian map $\mathcal{H}: \mathbf{g} \rightarrow \mathcal{F}(M)$:

$$\begin{aligned} J: M &\rightarrow \mathbf{g}^* \\ x &\rightarrow J(x), \end{aligned}$$

satisfying

$$< J(x), \ X >= H_X(x), \quad \forall X \in \mathbf{g}.$$

Under the moment map, a Poisson action of G on M can induce a coadjoint action Ad^* of G on $\mathbf{g}^*$, the following diagram commutes

$$\begin{array}{ccc} M & \xrightarrow{g} & M \\ J\downarrow & & J\downarrow \\ \mathbf{g}^* & \xrightarrow{Ad_g^*} & \mathbf{g}^* \end{array}$$

At point $\sigma \in \mathbf{g}^*$, the inverse image under moment map $J^{-1}(\sigma)$ is denoted by

$$M_\sigma = J^{-1}(\sigma).$$

In many cases the set M_σ is a manifold (if σ is a regular point). Denote the stable subgroup of point σ under the coadjoint action Ad^* of G on $\mathbf{g}^*$ by G_σ. The set of orbits of the action of G_σ on M_σ form a quotient space $N = M_\sigma/G_\sigma$, called the reduced phase space of the system with symmetry. The reduced phase space has the structure of smooth manifold if G_σ is compact and the action of G_σ on M_σ has no fixed points. Manifold N is a homogenous manifold under the G action. The tangent space $T_\sigma(N)$ is isomorphic to $\mathbf{g}/\mathbf{g}_\sigma$. Let $X \in \mathbf{g}/\mathbf{g}_\sigma$ be a generator of the transitive action, ξ_X denote the fundamental vector field on N associated with X, the tangent space $T_\sigma(N)$ is spanned by $\{\xi_X\}_{X\in\mathbf{g}/\mathbf{g}_\sigma}$. As the action of G on (M,ω) is Poisson, the Hamiltonian $\{H_X\}$ is invariant with respect to the action of G_σ, so $\{H_X\}$ is also well defined on N, and we can define a differential 2-form ω_N on N as follows

$$\omega_N(\xi_X, \xi_Y) = H_{[X,Y]}$$

it is easy to prove that this 2-form ω_N on N is nondegenerate and closed, so the reduced phase space (N, ω_N) is an even-dimensional symplectic manifold.

5.6 Contact manifold

Contact structure is an analog of symplectic structure for odd-dimensional manifold.

Let θ be a differential 1-form on a smooth manifold M of dimension $m = 2n+1$ satisfying

$$\theta \wedge (d\theta)^n \neq 0, \qquad \forall x \in M. \tag{5.2}$$

then θ (up to an invertible function) is called the contact structure on M, or simply the Pfaffian.

A manifold M endowed with the contact structure is called the contact manifold (M, θ). The no-where vanishing $(2n+1)$-form $\theta \wedge (d\theta)^n$ is a volume element of the manifold, so contact manifold (M, θ) is an odd-dimensional orientable manifold.

In classical mechanics, symplectic manifold appears as the cotangent bundle on configuration manifold, and contact manifold appears as manifold of contact elements of configuration manifold. First we introduce the concept of contact elements.

In three dimensional Euclidean space E^3, a contact element is a pair (x, σ) of a point $x \in E^3$ and a plane σ passing through x, and the transformation of space E^3 which maps one contact element to another contact element is called contact transformation.

A contact element on a smooth $(n+1)$-dimensional manifold N is a pair (x, σ) of a point $x \in N$ and an n-dimensional hyperplane $\sigma \subset T_x(N)$ which is an element of a projective space of dimension n, and each point of this projective space RP(n) is a line passing through the origin x of $T_x(N)$. A family of hyperplanes is defined locally by a Pfaffian θ which vanishes nowhere and

$$< \theta, X > |_x = 0, \qquad X \in \sigma \subset T_x(N),$$

which is the equation defining the hyperplane σ at point x.

The set of all contact elements of an $n+1$-dimensional manifold N forms a $(2n+1)$-dimensional smooth manifold M, which has a smooth Pfaffian θ satisfying

$$\theta \wedge (d\theta)^n \neq 0.$$

This Pfaffian defines a smooth field of hyperplanes satisfying the nondegeneracy condition. So M is a contact manifold.

A contact form is an algebraic linear form on the tangent space of the point of contact, which is still a contact form when multiplied by an invertible function. Let M' denote the one-dimensional linear boundle over M, its fibre on $x \in M$ is the non-zero linear function of the tangent space $T_x(M)$.The total boundle space M' is a manifold of even dimension m+1=2n+2. Now let π denote the projection from M' to M, then on M' a differential 1-form $\theta' = \pi^*\theta$ can be defined. In general the Pfaffian θ' on M' defines a Symplectic structure $d\theta'$, and the Symplectic manifold $(M', d\theta')$ is called the symplectification of the contact manifold (M, θ).

A diffeomorphism of a contact manifold to itself is called a contactomorphism if it preserves the contact structure, i.e., keeps the form invariant up to a multiplication factor.

Just like in symplectic manifold, there is the Darboux theorem for contact manifold, contact manifolds of the same dimension are locally contactomorphic, i.e., there is local coordinate $(z, q^1, \cdots, q^n, p_1, \cdots, p_n)$ such that the contact structure can be written as

$$\theta = dz - \sum_{i=1}^{n} p_i dq^i.$$

A vector field $X \in \mathcal{X}(M)$ on a contact manifold (M, θ) is called a contact vector field (an infinitesimal contact transformation) if it is the generator of a one-parameter group of contactomorphism, i.e.,

$$L_X\theta = f\theta, \quad f \in \mathcal{F}(M).$$

Now denote the set of all contact vector field by $Cont(M, \theta)$.

A vector field $X \in \mathcal{X}(M)$ on a contact manifold (M, θ) is called an invariant vector field (an infinitesimal automorphism of contact structure) if it keeps the contact structure invariant, i.e.,

$$L_X\theta = 0.$$

And denote the set of all invariant vector field by $Cont_0(M, \theta)$.

As $L_{[X,Y]} = [L_X, L_Y]$, both $Cont(M, \theta)$ and $Cont_0(M, \theta)$ are Lie algebras.

On a contact manifold, there exists a unique vector field $Y \in \mathcal{X}(M)$ which is nowhere vanishing and satisfies

$$i_Y\theta = 1, \quad i_Y d\theta = 0.$$

This vector is called the Reeb vector field of θ. It is easy to prove that $Y \in Cont_0(M,\theta)$. In local Darboux coordinates, the Reeb vector is

$$Y = \frac{\partial}{\partial z}$$

If $X \in Cont(M,\theta)$, then for $f \in \mathcal{F}(M)$ with $L_X\theta = f\theta$ is given by

$$f = Y < \theta, X > .$$

For contact manifold (M,θ), its tangent bundle $T(M)$ can be decomposed as

$$T(M) = ker\ \theta \oplus ker\ d\theta,$$

where $ker\ \theta$ is of rank $2n$, vector field X is called horizontal if $X \in ker\ \theta$, i.e., $< \theta, X >= 0$. $ker\ d\theta$ is of rank 1, and is generated by the Reeb vector field Y. Each vector $X \in \mathcal{X}(M)$ may be decomposed as

$$X = (i_X\theta)Y + (X - (i_X\theta)Y),$$

where $(i_X\theta)Y$ is vertical, and $X - (i_X\theta)Y$ is horizontal.

A differential p-form is called invariant if $L_Y\alpha_p = 0$.

A differential p-form (for $p > 0$) is called dynamic if $L_Y\alpha_p = 0$, $i_Y\alpha_p = 0$. And is sometimes also called absolute integral invariant of Y.

As $[L_Y, d] = 0$, the set of all invariant forms forms a ring with differential operator d. It is clear that the set of all dynamic forms with differential operator d also forms a ring.

The dynamics of an n freedoms system in classical mechanics can be described by the Lagrangian on the tangent bundle of the configuration space:

$$L = L(q, \dot{q}, t).$$

On another hand, the dynamics of the system can be described by the Hamiltonian formulation. The Hamiltonian function is

$$H = H(q, p, t).$$

which is a function on the cotangent bundle of the configuration space. More precisely, H is the function on state space (with one additional variable t). On the state space, the action 1-form can be introduced

$$\theta = \sum_{i=1}^{n} p_i dq^i - H dt. \tag{5.3}$$

It is easy to prove that

$$\theta \wedge (d\theta)^n = (\sum_{i=1}^{n} p_i \frac{\partial H}{\partial p_i} - H) \wedge_n (dp_i \wedge dq^i) \wedge dt.$$

Let $\dot{q}^i = \frac{\partial H}{\partial p_i}$, the requirement that the state space be a contact manifold leads to

$$L = \sum_{i=1}^{n} p_i \dot{q}^i - H \neq 0.$$

In which case the action 1-form defined in Eq. 5.3 is called the contact structure of the state space.

Transformations which keep the contact structure invariant (up to an invertible function) are called contact transformations. In classical dynamics, canonical transformations are just the contact transformations.

From Eq. 5.2, it is easy to see that $\theta + d\phi$ also satisfies this equation, i.e., equivalent classes of contact structure can differ by an exact form. We take the canonical transformation of the state space in classical mechanics as an example. Let

$$Q^i = Q^i(q, p, t), \quad P_i = P_i(q, p, t),$$

and the corresponding Hamiltonian function is

$$\mathcal{H} = \mathcal{H}(Q, P, t).$$

The canonical transformation which preserves the contact structure of the state space satisfies

$$\sum_{i=1}^{n} p_i dq^i - H dt - \sum_{i=1}^{n} P_i dQ^i + \mathcal{H} dt = d\phi, \tag{5.4}$$

where function ϕ is a smooth function on the $(2n+1)-$ dimensional manifold and can be taken as a function of the $2n+1$ independent variables $\{q^i, Q^i, t\}$, called the generating function of the canonical transformation:

$$\phi = \phi(q, Q, t).$$

Namely, the generating function depends on the old configuration space coordinate variables $\{q^i\}$, the new configuration space coordinate variables $\{Q^i\}$ and time t. If $\{q^i, Q^i, t\}$ are changed independently, then from Eq. 5.4

$$p_i = \frac{\partial \phi}{\partial q^i}, \quad P_i = \frac{\partial \phi}{\partial Q^i}, \quad \mathcal{H} = H + \frac{\partial \phi}{\partial t}.$$

In this way, canonical transformation of the state space is obtained, which keeps the canonical formulation of the equation of motion invariant.

Notations and Formulae

<table>
<tr><td>Presymplectic manifold (M,ω)</td><td></td><td>Poission manifold $(M,\bigwedge)$</td></tr>
<tr><td>$\omega=\frac{1}{2}\omega_{ij}(x)dx^i\wedge dx^j$
Two-form $\omega:\ \mathcal{X}(M)\to\Lambda^1(M)$</td><td></td><td>$\bigwedge=\lambda^{ij}(x)\partial_i\otimes\partial_j,\quad \lambda^{ij}=-\lambda^{ji}$
Bivector $\bigwedge:\Lambda^1(M)\to\mathcal{X}(M)$</td></tr>
<tr><td>$d\,\omega=0$</td><td></td><td>$[\bigwedge,\bigwedge]=0$</td></tr>
<tr><td>ω nondegeneracy
$\det(\omega_{ij})\neq 0$</td><td>$\Longleftrightarrow$
$\Downarrow$
Symp.M.</td><td>$\bigwedge$ nondegeneracy
$\det(\lambda^{ij})\neq 0$</td></tr>
</table>

- **Poisson bracket**, which satisfies the following conditions:
  (1) Bilinear, $\{af+bg,h\}=a\{f,h\}+b\{g,h\},\forall a,b\in\mathbf{R},f,g,h\in\mathcal{F}(M)$.
  (2) Skew symmetry, $\{f,g\}=-\{g,f\}$.
  (3) Jacobi identity, $\{f,\{g,h\}\}+\{g,\{h,f\}\}+\{h,\{f,g\}\}=0$.
  (4) Leibniz rule, $\{f,gh\}=g\{f,h\}+\{f,g\}h$.
  Let
$$X_f=\lambda^{ij}(x)\,\partial_i f\,\partial_j,\qquad \lambda^{ij}(x)=\{x^i,x^j\}$$
$$\begin{aligned}\{f,g\} &= \bigwedge(df,dg)=\lambda^{ij}(x)f_{,i}\,g_{,j}=X_f g=-X_g f\\ &= \omega(X_f,X_g)=<i_{X_f}\omega,X_g>=-<i_{X_g}\omega,X_f>=<dg,X_f>\end{aligned}$$

- On symplectic manifold, in terms of local Darboux coordinate:
  the symplectic structure, $\omega=dp_i\wedge dq^i$.
  Hamiltonian vector field, $X_f=\frac{\partial f}{\partial p_i}\frac{\partial}{\partial q^i}-\frac{\partial f}{\partial q^i}\frac{\partial}{\partial p_i}$.
  Poisson bracket, $\{f,g\}=(\frac{\partial f}{\partial p_i}\frac{\partial g}{\partial q^i}-\frac{\partial f}{\partial q^i}\frac{\partial g}{\partial p_i})$.

- The set of symplectic v.f.: $Sym(M,\omega)=\{X\in\mathcal{X}(M)|L_X\omega=0\}$.
  The set of Hamiltonian v.f.: $Ham(M,\omega)=\{X\in\mathcal{X}(M)|i_X\omega=-df\}$.

- On contact manifold,
  The contact form $\theta=dz-p_idq^i$.
  Reeb vector $Y=\frac{\partial}{\partial z}$.
  The set of generators of infinitesimal contact transformation:
$$Cont(M,\theta)=\{X\in\mathcal{X}(M)|L_X\theta=f\theta,\quad f=Y<\theta,X>\in\mathcal{F}(M)\}.$$
  The set of generators of infinitesimal automorphism of contact structure:
$$Cont_0(M,\theta)=\{X\in\mathcal{X}(M)|L_X\theta=0\}.$$

Exercises

1. Please show that the almost symplectic manifold is an even dimensional orientable manifold.

2. A symplectic manifold is a Poisson manifold, is the inverse true ?

3. The submanifold of a symplectic manifold is a presymplectic manifold, is which a Poisson manifold ?

4. Let (M,ω) be a symplectic manifold, in the local Darboux coordinate, $\omega = \sum_{i=1}^{n} dp_i \wedge dq^i$. Prove that the Hamiltonian vector field X_f with respect to function f can be expressed as:

$$X_f = \sum_{i=1}^{n}\left(\frac{\partial f}{\partial p_i}\frac{\partial}{\partial q_i} - \frac{\partial f}{\partial q_i}\frac{\partial}{\partial p_i}\right),$$

meanwhile give the explicit expression for the Poisson bracket of any two function $f, g \in \mathcal{F}(M)$.

5. Let (M,ω) be a contact manifold, please show that the set of contact vector fields $Cont(M,\theta)$ is a Lie algebra.

6. Prove Reeb vector field $Y \in Cont_0(M,\theta)$ and if $X \in Cont(M,\theta)$, then

$$L_X\theta = Y < \theta, X > \theta.$$

Chapter 6

Complex Manifolds

6.1 Complex structure of manifolds, almost complex manifolds

Formally, the definition of complex manifolds is similar to that of real manifolds. An n-dimensional complex manifold M is locally homeomorphic to an open set of an n-dimensional complex linear space $\mathbf{C}^n$ with holomorphic local coordinate. i.e., this manifold can be parameterized in terms of complex coordinates $\{z^\alpha, \overline{z}^\alpha\}_1^n$, under a holomorphic transformation,

$$z'^\alpha = z'^\alpha(z^1, \cdots, z^n), \quad \overline{z'}^\alpha = \overline{z'}^\alpha(\overline{z}^1, \cdots, \overline{z}^n), \quad (\alpha = 1, 2, \cdots, n).$$

Holomorphy is a much stronger requirement than real analyticity, so complex manifold has more interesting and richer structures than real manifold. First we consider the complex structure of complex manifold.

An n-dimensional complex manifold is equivalent to a $2n$-dimensional real manifold under certain conditions. We can use $2n$ real coordinates to represent the n complex coordinates:

$$z^\alpha = x^\alpha + iy^\alpha, \quad \bar{z}^\alpha = x^\alpha - iy^\alpha,$$

$$dz^\alpha = dx^\alpha + idy^\alpha, \quad d\bar{z}^\alpha = dx^\alpha - idy^\alpha,$$

$$\frac{\partial}{\partial x^\alpha} = \frac{\partial}{\partial z^\alpha} + \frac{\partial}{\partial \bar{z}^\alpha}, \quad \frac{\partial}{\partial y^\alpha} = i\frac{\partial}{\partial z^\alpha} - i\frac{\partial}{\partial \bar{z}^\alpha},$$

$$\frac{\partial}{\partial z^\alpha} \equiv \partial_\alpha = \frac{1}{2}(\frac{\partial}{\partial x^\alpha} - i\frac{\partial}{\partial y^\alpha}), \quad \frac{\partial}{\partial \bar{z}^\alpha} \equiv \bar{\partial}_\alpha = \frac{1}{2}(\frac{\partial}{\partial x^\alpha} + i\frac{\partial}{\partial y^\alpha}).$$

Given the local coordinates, exterior differentiation of an arbitrary complex function f is

$$df = \partial_\alpha f dz^\alpha + \bar{\partial}_\alpha f d\bar{z}^\alpha \equiv \partial f + \bar{\partial} f.$$

Function f is called holomorphic if and only if

$$\bar{\partial} f = 0, \tag{6.1}$$

This definition is independent of the choice of holomorphic coordinates. As any complex function f can be represented by two real valued functions u, v,

$$f = u + iv,$$

the requirement of holomorphy Eq. 6.1 yields the so called Cauchy-Riemann equations:

$$\frac{\partial u}{\partial x^\alpha} = \frac{\partial v}{\partial y^\alpha}, \quad \frac{\partial u}{\partial y^\alpha} = -\frac{\partial v}{\partial x^\alpha}.$$

For two local coordinate systems $\{z^\alpha; \alpha = 1, 2, \cdots, n\}, \{z'^\alpha; \alpha = 1, 2, \cdots, n\}$ in the neighborhood of a point on manifold M, local coordinate transformation can be defined:

$$z^\alpha \to z'^\alpha = f^\alpha(z), \quad \alpha = 1, 2, \cdots, n.$$

If we require that the transformation is holomorphic, then each $f^\alpha = u^\alpha + iv^\alpha$ satisfies Cauchy-Riemann equations:

$$\frac{\partial u^\alpha}{\partial x^\beta} = \frac{\partial v^\alpha}{\partial y^\beta}, \quad \frac{\partial u^\alpha}{\partial y^\beta} = -\frac{\partial v^\alpha}{\partial x^\beta}, \quad \alpha, \beta = 1, 2, \cdots, n.$$

Let

$$y^\alpha \equiv x^{n+\alpha}, \quad v^\alpha = u^{n+\alpha},$$

and define the $2n \times 2n$ matrix J:

$$J = \begin{pmatrix} 0 & -I_n \\ I_n & 0 \end{pmatrix}, \quad (J^2 = -I_{2n}) \tag{6.2}$$

where I_n is the $n \times n$ unit matrix. Cauchy-Riemann equations can be represented as one formula

$$J^j_k \frac{\partial u^k}{\partial x^l} = \frac{\partial u^j}{\partial x^k} J^k_l; \quad j, k, l = 1, 2, \cdots, 2n.$$

under a local coordinate transformation, matrix elements J^k_l satisfy

$$J^j_l = \frac{\partial u^j}{\partial x^k} J^k_n \frac{\partial x^n}{\partial u^l}.$$

Here J is a (1,1)-type tensor and is called the complex structure of the complex manifold,which is an automorphism transformation operator on the tangent bundle of the manifold. Now we consider the action of J on the tangent bundle. From Eq. 6.2, we have

$$J\frac{\partial}{\partial x^\alpha} = \frac{\partial}{\partial y^\alpha}, \quad J\frac{\partial}{\partial y^\alpha} = -\frac{\partial}{\partial x^\alpha}.$$

Hence,

$$\begin{aligned} J\tfrac{\partial}{\partial z^\alpha} &= \tfrac{1}{2}\left(\tfrac{\partial}{\partial y^\alpha} + i\tfrac{\partial}{\partial x^\alpha}\right) = i\tfrac{\partial}{\partial z^\alpha}, \\ J\tfrac{\partial}{\partial \bar{z}^\alpha} &= \tfrac{1}{2}\left(\tfrac{\partial}{\partial y^\alpha} - i\tfrac{\partial}{\partial x^\alpha}\right) = -i\tfrac{\partial}{\partial \bar{z}^\alpha}. \end{aligned} \tag{6.3}$$

In other words, $\frac{\partial}{\partial z^\alpha}$ and $\frac{\partial}{\partial \bar{z}^\alpha}$ are eigenvectors of J corresponding to eigenvalues $\pm i$. Eigenvector fields ($\frac{\partial}{\partial z^\alpha}$ as base vectors) with eigenvalue i are called [1,0]-type vector fields. And eigenvector fields ($\{\frac{\partial}{\partial \bar{z}^\alpha}\}$ as base vectors) with eigenvalue $-i$ are called [0,1]-type vector fields. This classification of vector fields is independent of the choice of holomorphic coordinates. The type of vector field is unchanged by local holomorphic transformations, which keeps the complex structure invariant.

The complex structure matrix J, defined on the tangent space at each point on the manifold, induces an adjoint matrix defined on the cotangent space. The adjoint matrix is still denoted by J:

$$<\omega, JX> = <J\omega, X>. \tag{6.4}$$

In cotangent space, J is the transpose matrix of Eq. 6.2:

$$Jdx^\alpha = -dy^\alpha, \quad Jdy^\alpha = dx^\alpha.$$

Hence,

$$Jdz^\alpha = -dy^\alpha + idx^\alpha = idz^\alpha, \quad Jd\bar{z}^\alpha = -dy^\alpha - idx^\alpha = -id\bar{z}^\alpha,$$

so dz^α and $d\bar{z}^\alpha$ are 1-forms with eigenvalue i and $-i$ of J respectively, which are called $[1,0]$ type 1-forms and $[0,1]$ type 1-forms respectively. The classification of cotangent vector fields is also independent of the choice of local coordinates. From 6.4 it is clear that the $[1,0]$-type 1-form annihilates $[0,1]$-type vector and vice versa.

Now we discuss some more general cases: consider the almost complex structure.

Definition 6.1 *An almost complex structure on a real differentiable manifold is a smooth (1,1)-type tensor field $J \in End(TM)$ satisfying $J^2 = -I$.*

A real manifold with almost complex structure is called an almost complex manifold.

The smooth (1,1)-type tensor field J is an automorphism operator on the set of tangent fields $\mathrm{X}(M)$. That is, for any tangent field $X \in \mathrm{X}(M)$:

$$JX \in \mathcal{X}(M).$$

At an arbitrary point p on M, J is a linear mapping from tangent space $T_p(M)$ to itself. At the neighborhood U of point p there is a local coordinate system $(x^i; i = 1, \cdots, m)$. In terms of the natural bases of tangent fields, the action of the almost complex structure J on the bases can be represented as

$$J\frac{\partial}{\partial x^i} = J_i^k(x)\frac{\partial}{\partial x^k}, \quad J_i^k(x) \in \mathcal{F}(M),$$

where functions $\{J_i^k(x)\}$ satisfy

$$J_i^k(x)J_k^j(x) = -\delta_i^j.$$

which is just $J^2 = -I$ in component form. The eigenvalues of the matrix $\{J_i^k(x)\}$ are $\pm i$ and always in pairs. Therefore, the tangent space $T_p(M)$ must be even-dimensional. We can set $m = 2n$, i.e., the almost complex manifold M is an even-dimensional real manifold.

As eigenvalues of J are the pair of pure imaginary number $\pm i$, the analysis could be simplified if we complexity the tangent and the cotangent spaces. The cotangent space $T_p^*(M)$ is a set of linear functions on $T_p(M)$:

$$T_p^*(M) = \mathrm{Hom}(T_p(M), R)$$

The complexified cotangent space is a set of complex linear functions on $T_p(M)$:

$$T_p^*(M) \times C = \mathrm{Hom}(T_p(M), C),$$

which is an m-dimensional complex vector space. $T_p^*(M) \times C$ can adopt the bases of the original real m-dimensional real vector space $T_p^*(M)$ but with complex coefficients. Any element $\theta \in T_p^*(M) \times C$ in complexified vector space can be represented as

$$\theta = \sigma + i\omega, \quad \sigma, \omega \in T_p^*(M).$$

The linear mapping J on $T_p(M)$ induces a linear mapping on cotangent space $T_p^*(M)$ which is also called J :

$$< \sigma, JX > = < J\sigma, X > .$$

so the almost complex operator, which acts on tangent space, can be naturally extended to the complexified cotangent space. We define

$$J\theta \equiv \sigma + iJ\omega.$$

The eigenvalues of J are $\pm i$. Cotangent vectors with eigenvalues i and $-i$ of J are called [1,0]-type cotangent vector and [0,1]-type cotangent vector respectively. The set of [1,0]-type cotangent vectors is denoted as $T_{10}^*(M)$ and the set of [0,1]-type cotangent vectors is denoted as $T_{01}^*(M)$. It is easy to prove (subscript p is omitted.):

1).

$$\begin{aligned} T^*(M) \times C &= T_{10}^*(M) \oplus T_{01}^*(M) \\ &\equiv T_c^*(M) \oplus \bar{T}_c^*(M). \end{aligned} \tag{6.5}$$

2) Complex conjugate on the complexified vector space establishes an isomorphism between $T_{10}^*(M)$ and $T_{01}^*(M)$ (please refer to Exercise 1).

From these two points, we can see that both $T_{10}^*(M)$ and $T_{01}^*(M)$ are n-dimensional complex vector spaces, called the n-dimensional complex cotangent spaces $T_c^*(M)$ and $\bar{T}_c^*(M)$ respectively. If $\omega \in T^*(M)$ is a real cotangent field, under transformation J, $J\omega = \sigma$ also a real cotangent field, then $J\sigma = -\omega$, and

$$\theta = (J + i)\omega = \sigma + i\omega \quad \in T_{10}^*(M), \qquad \bar{\theta} = (J - i)\omega = \sigma - i\omega \quad \in T_{01}^*(M).$$

If we take n linearly independent cotangent vectors $\{\theta^\alpha, \alpha = 1, \cdots, n\}$ as bases of $T^*_{10}(M)$, then $\{\theta^\alpha, \bar{\theta}^\alpha\}$ constitutes a set bases of $T^*(M) \times C$. In this bases, components of almost complex structure J are

$$J^\alpha_\beta = i\delta^\alpha_\beta, \quad J^{\bar{\alpha}}_{\bar{\beta}} = -i\delta^{\bar{\alpha}}_{\bar{\beta}}, \quad J^{\bar{\alpha}}_\alpha = 0 \ \ (\alpha, \bar{\alpha} = 1, \cdots, n). \tag{6.6}$$

This set of bases $\{\theta^\alpha, \bar{\theta}^\alpha\}$ is called the canonical bases. it is easy to see that the canonical form of J are invariant under the holomorphic transformation. Thus the tensor field J can be defined globally on all of M.

θ^α is a complex linear function on the tangent space which can be decomposed into the real part and the imaginary part:

$$\theta^\alpha = \sigma^\alpha + i\omega^\alpha, \quad \alpha = 1, \cdots, n,$$

where both σ^α and ω^α are real cotangent vectors which have not been complexified. Thus:

$$\sigma^\alpha = \frac{1}{2}(\theta^\alpha + \bar{\theta}^\alpha), \quad \omega^\alpha = -\frac{i}{2}(\theta^\alpha - \bar{\theta}^\alpha). \tag{6.7}$$

This shows that the $2n = m$ real cotangent vectors $\{\sigma^\alpha, \omega^\alpha\}$ can be a set of bases of the complexified cotangent space, which are linearly independent and can also be a set of bases of the original real cotangent space. In terms of the bases $\{\sigma^\alpha, \omega^\alpha\}$, an $m = 2n$ dimensional volume element of the cotangent space on the manifold can be given

$$\tau_{2n} = \bigwedge_{1 \le \alpha \le n} (\sigma^\alpha \wedge \omega^\alpha).$$

When taking another set of bases $\{\theta'^\alpha, \bar{\theta}'^\alpha\}$, where θ'^α are $[1,0]$-type cotangent vectors and are the linear combinations of $\{\theta^\alpha\}$, we have the corresponding transformation matrix with Jacobian $\Im \neq 0$:

$$\theta'^1 \wedge \cdots \wedge \theta'^n = \Im \theta^1 \wedge \cdots \wedge \theta^n.$$

Hence,

$$\bigwedge_{1 \le \alpha \le n} (\theta'^\alpha \wedge \bar{\theta}'^\alpha) = \Im\bar{\Im} \bigwedge_{1 \le \alpha \le n} (\theta^\alpha \wedge \bar{\theta}^\alpha),$$

$$\Im\bar{\Im} = |\Im|^2 > 0.$$

In terms of the bases $\{\sigma'^\alpha, \omega'^\alpha\}$ of the real cotangent space which are obtained from $\{\theta'^\alpha, \bar{\theta}'^\alpha\}$, the m-dimensional volume element is

$$\tau'_{2n} = \bigwedge_{1 \le \alpha \le n} (\sigma'^\alpha \wedge \omega'^\alpha) = |\Im|^2 \, \tau_{2n}.$$

The two m-dimensional volume elements differ only by a positive valued factor. This shows that the two sets of bases give identical orientation of the cotangent space of the manifold. That is, the volume elements τ_{2n}, constructed from any two sets of bases according to procedures above in terms of the almost complex structure J, have the same orientation. Thus a theorem is proved:

Theorem 6.1 *Each and every almost complex manifold is an orientable even-dimensional real manifold.*

From the analysis of complex manifold so far, we know that on a complex manifold there must exist an almost complex structure. A complex manifold is an almost complex manifold but an almost complex manifold may not be a complex manifold. Let M have an almost complex structure J, can one choose coordinate to put J in the canonical form (6.6) not just at the point p but in a neighborhood containing p? The coordinates with this property be called local holomorphic coordinates. If local holomorphic coordinates exist, then the almost complex structure J is said to be integrable. In this section, we should considered the integrable condition for the almost structure J in terms of differential forms. Various equivalent descriptions of integrable condition for the almost complex structure from different view points are given in the next section.

The discussions of tensor space at each point on the manifold can be generalized to the tensor bundle and tensor field on the manifold. On the $m = 2n$-dimensional almost complex manifold, if $\{\theta^\alpha\}$ is a set of n linearly independent $[1,0]$-type 1-forms, then $\{\bar{\theta}^\alpha\}$ is a set of n linearly independent $[0,1]$-type 1-forms. In term of $\{\theta^\alpha\}$, $\{\bar{\theta}^\alpha\}$ and smooth complex functions on M, the $[p,q]$-type complex differential forms $\alpha_{p,q} \in \Lambda^{p,q}(M)$ can be obtained:

$$\alpha_{p,q} = \frac{1}{p!q!} f_{\alpha_1 \cdots \alpha_p \beta_1 \cdots \beta_q} \theta^{\alpha_1} \wedge \cdots \wedge \theta^{\alpha_p} \wedge \bar{\theta}^{\beta_1} \wedge \cdots \wedge \bar{\theta}^{\beta_q}.$$

The set of $[p,q]$-type complex differential forms $\Lambda^{p,q}$ is a module over the ring of complex functions which has the following properties

1). If $\alpha \in \Lambda^{p,q}$, then $\bar{\alpha} \in \Lambda^{q,p}$.

2). If $\alpha \in \Lambda^{p,q}$, $\beta \in \Lambda^{r,s}$, then $\alpha \wedge \beta \in \Lambda^{p+r,q+s}$.

3). $\Lambda^{p,q} = 0$ if p or $q > n$.

Exterior differentiation of $[1,0]$-type differential forms $\{\theta^\alpha\}$, which yields 2-forms, can be written as

$$d\theta^\alpha = f^\alpha_{\beta\gamma} \theta^\beta \wedge \theta^\gamma + g^\alpha_{\beta\gamma} \theta^\beta \wedge \bar{\theta}^\gamma + h^\alpha_{\beta\gamma} \bar{\theta}^\beta \wedge \bar{\theta}^\gamma, \tag{6.8}$$

$$f^\alpha_{\beta\gamma},\quad g^\alpha_{\beta\gamma},\quad h^\alpha_{\beta\gamma} \in \mathcal{F}(M).$$

From Frobenius theorem in §2.3, we see that the almost complex structure is integrable if

$$d\theta^\alpha \subset \Lambda^{2,0} \oplus \Lambda^{1,1}, \tag{6.9}$$

which means that all the coefficients $h^\alpha_{\beta\gamma} = 0$ in Eq. 6.8. The integrable condition can also be written as:

$$d\theta^\alpha = 0 \bmod\{\theta^\alpha\},$$

i.e., the ideal consisting of $\{\theta^\alpha\}$ is a closed ideal.

The almost complex structure J is called integrable if it arises from atlas with holomorphic transition mappings on manifold M. i.e., there exists a set of local complex coordinates $\{z^\alpha\}$ such that $[1,0]$-type differential forms are the complex linear combinations of $\{dz^\alpha\}$ if the almost complex structure is integrable. In the natural frame of this local coordinates, the components of the almost complex structure take the canonical form of Eq. 6.6.

A manifold often needs more than one open set $\{U_\alpha\}$ to cover it. There is one J_α corresponding to each open set U_α. In the overlapping region $U_\alpha \cap U_\beta$ of two open sets, if both J_α and J_β are integrable, one can take the local complex coordinates $\{z^\alpha\}$ and $\{w^\alpha\}$ such that J_α and J_β are equal and take the canonical form. In this case, $\{dz^\alpha\}$ is a linear combination of $\{dw^\alpha\}$. This shows that $\{w^\alpha\}$ are holomorphic functions of $\{z^\alpha\}$. Therefore, there exists complex analytic compatible coordinate atlases. Hence, the manifold is a complex manifold. There is a theorem:

Theorem 6.2 *Manifold M is a complex manifold iff there is an integrable almost complex structure on M.*

A complex manifold is an almost complex manifold and an almost complex manifold is an orientable even-dimensional real manifold. However, the reverse of the two statements are not true in general. For instance, S^{2n} is an orientable even-dimensional real manifolds, but only S^2 which is holomorphic to $CP(1)$ is a complex manifold. In 1969, Adler [25] proved that S^6 is an almost complex manifold which is not integrable. All other $S^{2n} (n \neq 1, 3)$ have no almost complex structure.

On a compact closed manifold, there exist three important structure: metric structure, symplectic structure and complex structure, but only two of them are independent, and the third can be determined by the two others. If a metric tensor field is defined on a symplectic manifold, then this manifold is also a complex manifold. Following in the sections 6.3 and 6.4 we should discussed this problem in detail.

It can be proved that complex structures exist for two-dimensional real compact smooth orientable surfaces. On two-dimensional manifold, there is a order-2 covariant skewsymmetric tensor ϵ_{ij} . And there is a metric tensor field $\{g_{ij}\}$ on arbitrary smooth manifold. Combining these two tensors, we have

$$J^i_j = g^{ik} \epsilon_{kj},$$

which satisfies $J^2 = -I$. Therefore, in two dimension, a metric determines a complex structure. Conversely, a complex structure determines a conformal class of metric. Hence, the two-dimensional real compact smooth orientable manifold is a complex manifold which is called the Riemann surface. The neighborhood at each point on the Riemann surface is holomorphic to an open set of the one-dimensional complex manifold $\mathbf{C}^1$. In addition, the local coordinate transformation functions are holomorphic. This is the major topic of complex analysis.

There are manifolds that are equivalent as real manifolds but not as complex manifolds, i.e., it is possible to put inequivalent complex structure on the same real manifold. Take the Riemann surface as an example. From Appendix I, one learns the Levi-Civita symbol ϵ_{ik} is a tensor density. Under a conformal transformation

$$g_{ij} \to e^{2\varphi} g_{ij}, \quad \epsilon_{ik} \to e^{2\varphi} \epsilon_{ik},$$

and the complex structure $J^i_j = g^{ik}\epsilon_{kj}$ is conformal invariant. In other words, the complex structure of the Riemann surface is determined by the conformal invariant classes of the metric. There is a one to one correspondence between the complex structures of the Riemann surface and the conformal invariant classes of the metric. When the metric on the Riemann surface under goes a non-conformal transformation, the relevant complex structures are different in general. For example, the two-dimensional torus T_z, which can be viewed as the quotient space of the complex plane over the lattice $e_1 = (1,0)$ and $e_2 = (x,y)$, is a one-dimensional complex manifold. It can be proved that one can obtain another complex manifold $T_{z'}$ when changing the lattice e_2. But, when

$$z' = \frac{az+b}{cz+d}, \quad a,b,c,d \text{ are integers and } \quad ad-bc=1,$$

T_z and $T_{z'}$ are conformally equivalent, i.e., they have the same complex structure. However, if one changes the lattice continuously, the resulting tori have different complex structures even though they have the same topology.

Homeomorphisms may change the differential structure of the manifold. The diffeomorphic mapping which preserves the differential structure may change the complex structure. there are three kinds of special topological mappings :

1). Under homeomorphic mapping of the manifold M, the invariant properties of the manifold are called the topological properties of the manifold. The homeomorphic mapping keeps the topological structure of the manifold invariant.

2). Under diffeomorphic mapping of M, the invariant properties of the manifold are called the diffeomorphical properties of the manifold. The diffeomorphic mapping keeps the differential structure of the manifold invariant.

3). Under holomorphic mapping of the complex manifold M, the invariant properties of the complex manifold are called the holomorphic properties. The holomorphic mapping keeps the complex structure of the manifold invariant.

Holomorphic mapping is diffeomorphic and diffeomorphic mapping must be homeomorphic. Neither reverse of these two statements is true in general. The holomorphic mapping is the most smooth diffeomorphism, which keeps the complex structure invariant.

6.2 Integrable condition of almost complex structure

In the last section, we use the exterior differential operator to study the integrable condition of almost complex structure. In this section, we look at various equivalent definitions of integrable conditions of almost complex structure. There are essentially three kinds of them, and the following nine conditions are equivalence.

I. In terms of differential forms.

On the $m = 2n$ dimensional almost complex manifold, by using the almost complex structure on complex differential forms, one has a set of bases $\{\theta^\alpha, \bar{\theta}^\alpha; \alpha = 1, \cdots, n\}$ of complex differential 1-forms. To decide the integrability of the almost complex structure, there is a theorem

Theorem 6.3 *For an almost complex manifold, the following conditions are equivalent: almost complex structure is integrable if and only if:*

1).
$$d\theta^\alpha \in \Lambda^{2,0} \oplus \Lambda^{1,1}, \quad \forall \theta^\alpha \in \Lambda^{1,0}, \tag{6.9}$$

2).
$$d\bar{\theta}^\alpha \in \Lambda^{1,1} \oplus \Lambda^{0,2}, \quad \forall \bar{\theta}^\alpha \in \Lambda^{0,1},$$

3).
$$d\alpha_{p,q} \in \Lambda^{p+1,q} \oplus \Lambda^{p,q+1}, \quad \forall \alpha_{p,q} \in \Lambda^{p,q}, \quad p, q = 0, 1, \cdots, n\,,$$

4).
$$d = \partial + \bar{\partial}, \tag{6.10}$$

5).
$$\bar{\partial}^2 = 0,$$

6).
$$\partial^2 = 0.$$

II. In terms of tangent vector fields.

For almost complex manifold with almost complex structure J, the tangent space at each point of the manifold can be complexified. Following the action of J, the complexified space is the direct sum of two subspaces with eigenvalues $\pm i$ of J respectively. Correspondingly, we have the complexified $[1,0]$-type and $[0,1]$-type tangent vector fields, which are in one to one correspondence under complex conjugation. For almost complex manifold M with given almost complex structure, complex vector field on the manifold can be uniquely decomposed as the direct sum of the $[1,0]$-type and $[0,1]$-type tangent vector fields. Any $[1,0]$-type tangent vector field Z can be written as

$$Z = X - iJX,$$

where $X \in \mathcal{X}(M)$ is a real tangent vector field on M. Obviously

$$JZ = iZ,$$

that is, Z is the eigen-tangent vector field with eigenvalue i of the almost complex structure J. For the dual complex cotangent field, one has

$$< \bar{\theta}^\alpha, Z > = 0, \quad \forall \bar{\theta}^\alpha \in \Lambda^{0,1}. \tag{6.11}$$

Above three equations can be used as an alternative definition of $[1,0]$-type tangent field. For $[0,1]$-type tangent field, there is an analogous definition (with $i \to -i$, and $\bar{\theta}^\alpha \to \theta^\alpha$).

On the integrability of almost complex structure, there is a theorem which is equivalent to **Theorem 6.3**.

Theorem 6.4 *For an almost complex manifold, almost complex structure is integrable if and only if (These conditions are also equivalent to each other and equivalent to the conditions in* **Theorem 6.3***):*

7). The $[1,0]$-type tensor field is closed under Lie bracket operation, i.e., if X and Y are complex vector fields of $[1,0]$ type, then so is $[X,Y]$.

8). The $[0,1]$-type tensor field is closed under Lie bracket operation, i.e., if X and Y are complex vector fields of $[0,1]$ type, then so is $[X,Y]$.

Now we show that condition 7 is equivalent to condition 2. Let Z and W be the $[1,0]$-type tangent fields. Condition 7 shows that $[Z,W]$ is also a $[1,0]$-type tangent field. All of them satisfy Eq. 6.11. In addition, from Eq. 2.34, we have

$$d\bar{\theta}^\alpha(Z,W) = Z\bar{\theta}^\alpha(W) - W\bar{\theta}^\alpha(Z) - \bar{\theta}^\alpha([Z,W]) = 0.$$

Hence, $d\bar{\theta}^\alpha$ does not contain any $[0,2]$-type differential forms, which is just condition 2. On the other hand, from condition 2, if both Z and W are $[1,0]$-type tangent fields, substituting into Eq. 2.34, we have

$$\bar{\theta}^\alpha([Z,W]) = 0.$$

This is true for any $\bar{\theta}^\alpha \in \Lambda^{0,1}$. Hence, $[Z,W]$ is a $[1,0]$-type tangent field.

III. In term of the properties of almost complex structure itself.

From the $(1,0)$-type tensor J and its derivatives, we can construct a $(1,2)$-type tensor N, called Nijenhuis torsion tensor[26],

$$N(X,Y) = [JX,JY] - [X,Y] - J[X,JY] - J[JX,Y], \text{ for } X,Y \in \mathcal{X}(M).$$

Tensor N is also called the torsion of J, which is the obstruction to integrability of J.

Theorem 6.5 *Almost complex structure is integrable if and only if it is torsion free. i.e.,*

9). The Nijenhuis torsion of J is vanshing, i.e., $N(X,Y) = 0$ (for $X,Y \in \mathcal{X}(M)$).

Now we prove this condition,(the condition 9), is equivalent to condition 7. Let both Z and W be $[1,0]$-type tangent vector fields. From Eq. 6.10, they can be written as

$$Z = X - iJX, \quad W = Y - IJY, \quad X,Y \in \mathcal{X}(M).$$

It is easy to prove

$$[Z,W]+iJ[Z,W]=-N(X,Y)-iJN(X,Y).$$

Condition 7 requires that the left hand side of this equation is zero. For arbitrary tangent field, this is true if and only if $N(X,Y)$ is zero. Which is just the theorem 6.5.

This nine conditions above are mutually equivalent. For an almost complex manifold M, the almost complex structure is integrable if one of the above conditions is satisfied, and thus this manifold is a complex manifold (Theorem 6.2).

On a complex manifold the exterior differential operator, covariant derivative operator and Lie derivative operator all have good properties. For example, there exists torsion free almost complex affine connection satisfying: (see Sectionn 6.5)

(1) J is paralled transported by the covariant derivative:

$$\nabla_X(JY)=J\nabla_X Y,\qquad X,Y\in\mathcal{X}(M).$$

So one obtain

$$R(X,Y)\cdot J=J\cdot R(X,Y),\qquad R(JX,JY)=R(X,Y),$$

i.e., R(X,Y) is linear with respect to the complex structure$J$.

(2) Torsion free:

$$T(X,Y)=\nabla_X Y-\nabla_Y X-[X,Y]=\frac{1}{4}N(X,Y)=0.$$

On a complex manifold the infinitesimal automorphism Lie algebra of the almost complex structure is stable with respect to J, i.e.,

$$J[X,Y]=[JX,Y]=[X,JY],$$

then if let $JX=iX$, the Lie algebra $\mathcal{X}(M)$ becomes a complex Lie algebra.

6.3 Hermitian manifold

For an almost complex manifold M, a complex order-2 covariant skew linear tensor field can be defined:

$$H:\mathcal{X}(M)\times\mathcal{X}(M)\to C$$

which has the following properties

1).

$$H(aX+bY,cZ)=acH(X,Z)+bcH(Y,Z),$$
$$\forall X,Y,Z\in\mathcal{X}(M),\quad a,b,c\in R,$$

i.e., function $H(X,Y)$ is real bilinear with respect to its arguments.

2).
$$H(X,Y) = \overline{H(Y,X)}, \quad \forall X,Y \in \mathcal{X}(M),$$

3).
$$H(JX,Y) = iH(Y,X).$$

I.e., it is complex linear with respect to the first argument and complex anti-linear with respect to the second. The tensor field H which satisfies the above three conditions is called Hermitian metric of M. If further

4).
$$H(X,X) > 0,$$

Then this Hermitian metric H is called the positive definite Hermitian metric.

From 2 and 3, it is easy to prove that the Hermitian metric tensor field is invariant under the action of the almost complex structure J:

$$H(JX,JY) = H(X,Y), \quad \forall X,Y \in \mathcal{X}(M).$$

Definition 6.2 *Almost complex manifold with Hermitian metric is called almost Hermitian manifold.*

Hermitian metric tensor field can be explicitly decomposed into the real part and the imaginary part:

$$H(X,Y) = G(X,Y) - iK(X,Y),$$

where

$$G(X,Y) = G(Y,X) = \frac{1}{2}(H(X,Y) + H(Y,X)),$$

$$K(X,Y) = -K(Y,X) = \frac{i}{2}(H(X,Y) - H(Y,X)).$$

Both G and K are real tensor fields on the manifold. G is a order-2 covariant symmetric tensor field and K is a order-2 covariant skewsymmetric tensor field, i.e., an almost symplectic structure. Both of them are invariant under the action of the almost complex structure J:

$$G(JX,JY) = G(X,Y), \quad K(JX,JY) = K(X,Y),$$

and as a result of $H(JX,Y) = iH(X,Y)$,

$$G(JX,Y) = K(X,Y), \quad K(JX,Y) = -G(X,Y).$$

On a symplectic manifold (M,K), if $G(X,Y) = K(X,JY)$ defines a Riemannian metric on M, then the almost complex structure J is called compatible with K. Equivalently, for an almost complex manifold, if a Riemannian metric tensor G compatible with the almost complex structure is given, an Hermitian metric tensor field H

can be constructed. The triple (M, J, G) or (M, J, K) is called a Hermitian structure on manifold M, and its volume form can be given by

$$\tau = K^n/n!.$$

In addition, from Theorem 2 in §4.2, Riemannian metric tensor field g can be defined on any differentiable manifold. For almost complex manifold M, from Riemann metric g and the almost complex structure J, a metric tensor G, which is invariant under the action of J, can be defined as:

$$G(X,Y) = g(X,Y) + g(JX, JY), \quad \forall X,Y \in \mathcal{X}(M).$$

It is easy to prove that thus defined metric tensor field G is invariant under the action of J:

$$G(JX, JY) = G(X,Y).$$

From the analysis above, we have

Theorem 6.6 *There exists an Hermitian metric on any almost complex manifold.*

From the last section we see that for almost complex manifold, the tangent field and the cotangent field can be complexified and can be written as the direct sum of [1, 0]-type and [0, 1]-type tensors according to the eigenvalues $\pm i$ of J. In the following, Hermitian metric tensor is represented by the bases $\{\theta^\alpha\}$ and $\{\bar{\theta}^\alpha\}$ of [1, 0]-type and [0, 1]-type tensors of the complexified cotangent fields respectively.

From Eq. 6.7, the bases of the complexified cotangent fields can be taken as $\{\sigma^\alpha, \omega^\alpha; \alpha = 1, \cdots, n\}$, which are also bases of the original real cotangent field. We have

$$J\sigma^\alpha = -\omega^\alpha, \quad J\omega^\alpha = \sigma^\alpha, \quad \alpha = 1, \cdots, n. \tag{6.12}$$

Suppose that $\{e_\alpha, E_\alpha\}$ are the bases of $\mathcal{X}(M)$ of real tangent fields, which are dual to bases $\{\sigma^\alpha, \omega^\alpha\}$ of cotangent fields, as a result of duality and Eq. 6.12, it is easy to obtain

$$Je_\alpha = E_\alpha, \quad JE_\alpha = -e_\alpha.$$

Tangent vector fields $X, Y \in \mathcal{X}(M)$ can be written as

$$X = \xi^\alpha e_\alpha + \xi^{n+\alpha} E_\alpha, \quad Y = \eta^\alpha e_\alpha + \eta^{n+\alpha} E_\alpha.$$

From the dual relation between $\{e_\alpha, E_\alpha\}$ and $\{\sigma^\alpha, \omega^\alpha\}$, we have

$$< \theta^\alpha, X > = \xi^\alpha + i\xi^{n+\alpha}, \quad < \bar{\theta}^\alpha, Y > = \eta^\alpha - i\eta^{n+\alpha}.$$

So

$$\begin{aligned} H(X,Y) &= (\xi^\alpha + i\xi^{n+\alpha})(\eta^\beta - i\eta^{n+\beta}) H(e_\alpha, e_\beta) \\ &= H(e_\alpha, e_\beta) < \theta^\alpha, X > < \bar{\theta}^\beta, Y > . \end{aligned}$$

Let

$$H_{\alpha\bar{\beta}} \equiv H(e_\alpha, e_\beta),$$

the metric tensor field can be written as

$$H = H_{\alpha\bar{\beta}}\theta^\alpha \otimes \bar{\theta}^\beta.$$

The symmetric and the skewsymmetric parts of H are

$$G = H_{\alpha\bar{\beta}}\theta^\alpha\bar{\theta}^\beta \equiv \frac{1}{2}H_{\alpha\bar{\beta}}(\theta^\alpha \otimes \bar{\theta}^\beta + \bar{\theta}^\beta \otimes \theta^\alpha). \tag{6.13}$$

$$K = \frac{i}{2}H_{\alpha\bar{\beta}}\theta^\alpha \wedge \bar{\theta}^\beta \equiv \frac{i}{2}H_{\alpha\bar{\beta}}(\theta^\alpha \otimes \bar{\theta}^\beta - \bar{\theta}^\beta \otimes \theta^\alpha).$$

Both G and K are real two-order tensor fields, so the components $\theta^\alpha \otimes \theta^\beta$, and $\bar{\theta}^\alpha \otimes \bar{\theta}^\beta$ do not appear. K is a real 2-form, called the fundamental 2-form of Hermitian manifold.

The torsion free almost complex structure is just the complex structure. In the following, we will focus on complex manifold which has complex structure.

Definition 6.3 *Complex manifold with positive definite Hermitian metric is called Hermitian manifold.*

On a Hermitian manifold, it is usually convenient to study the skewsymmetric form K but the complex structure J.

In the neighborhood of each point on a complex manifold, there is a local complex coordinate $\{U, z^\alpha; \alpha = 1, \cdots, n\}$. The bases of cotangent vector fields are $\{dz^\alpha, d\bar{z}^\alpha\}$ which are related to the $[1,0]$-type and the $[0,1]$-type cotangent vector fields respectively. Eq. 6.13 can be written as

$$G = H_{\alpha\bar{\beta}}dz^\alpha d\bar{z}^\beta,$$

and the dual bases are $\{\frac{\partial}{\partial z^\alpha}, \frac{\partial}{\partial \bar{z}^\alpha}\}$ which are related to the $[1,0]$-type and the $[0,1]$-type tangent vector fields respectively. From Eq. 6.13 it is easy to prove

$$G(\frac{\partial}{\partial z^\alpha}, \frac{\partial}{\partial \bar{z}^\beta}) = G(\frac{\partial}{\partial \bar{z}^\beta}, \frac{\partial}{\partial z^\alpha}) = \frac{1}{2}H_{\alpha\bar{\beta}},$$

$$G(\frac{\partial}{\partial z^\alpha}, \frac{\partial}{\partial z^\beta}) = G(\frac{\partial}{\partial \bar{z}^\alpha}, \frac{\partial}{\partial \bar{z}^\beta}) = 0.$$

Let

$$h_{\alpha\bar{\beta}} \equiv G(\frac{\partial}{\partial z^\alpha}, \frac{\partial}{\partial \bar{z}^\beta}) = h_{\bar{\beta}\alpha} \equiv G(\frac{\partial}{\partial \bar{z}^\beta}, \frac{\partial}{\partial z^\alpha}) = \frac{1}{2}H_{\alpha\bar{\beta}},$$

then

$$dS^2 = 2h_{\alpha\bar{\beta}}dz^\alpha d\bar{z}^\beta,$$

$$K = ih_{\alpha\bar{\beta}}dz^\alpha \wedge d\bar{z}^\beta,$$

where we use dS^2 instead of G in order to show explicitly that the metric is a positive definite Hermitian metric. The coefficients $\{h_{\alpha\bar{\beta}}\}$ are elements of an $n\times n$ non-singular positive definite matrix

$$h \equiv \det(h_{\alpha\bar{\beta}}) > 0.$$

We use $h^{\bar{\alpha}\beta}$ as element of the inverse matrix of the $(h_{\alpha\bar{\beta}})$, i.e.,

$$h^{\bar{\alpha}\beta}h_{\beta\bar{\gamma}} = \delta^{\bar{\alpha}}_{\bar{\gamma}}, \quad h_{\alpha\bar{\beta}}h^{\bar{\beta}\gamma} = \delta^\gamma_\alpha.$$

For any two tangent fields on manifold M,

$$X = \xi^\alpha\frac{\partial}{\partial z^\alpha} + \bar{\xi}^\alpha\frac{\partial}{\partial \bar{z}^\alpha}, \quad Y = \eta^\alpha\frac{\partial}{\partial z^\alpha} + \bar{\eta}^\alpha\frac{\partial}{\partial \bar{z}^\alpha},$$

inner product between them can be defined by the metric tensor defined above

$$(X,Y) = h_{\alpha\bar{\beta}}(\xi^\alpha\bar{\eta}^\beta + \eta^\alpha\bar{\xi}^\beta).$$

Inner product between differential forms of the same type can also be defined. To do this, for the Hermite manifold, the complex Hodge * is introduced:

$$\begin{aligned}&*(dz^{\alpha_1}\wedge\cdots\wedge dz^{\alpha_p}\wedge d\bar{z}^{\beta_1}\wedge\cdots\wedge d\bar{z}^{\beta_q})\\ &= \frac{1}{(n-p)!(n-q)!}i^n(-1)^{\frac{1}{2}n(n-1)+np}\epsilon^{\alpha_1\cdots\alpha_p}{}_{\bar{\gamma}_1\cdots\bar{\gamma}_{n-p}}\epsilon^{\bar{\beta}_1\cdots\bar{\beta}_q}{}_{\delta_1\cdots\delta_{n-q}}\cdot\\ &\quad dz^{\delta_1}\wedge\cdots\wedge dz^{\delta_{n-q}}\wedge d\bar{z}^{\gamma_1}\wedge\cdots\wedge d\bar{z}^{\gamma_{n-p}},\end{aligned} \tag{6.14}$$

where ϵ is the generalized Levi-Civita symbol. In order that the indices are tensor indices, we define

$$\epsilon_{12\cdots n} = |h|^{\frac{1}{2}}, \quad \overline{\epsilon_{i_1\cdots i_n}} = \epsilon_{\bar{i}_1\cdots\bar{i}_n}, \quad \epsilon^{\bar{i}_1}{}_{i_2\cdots i_n} = h^{\bar{i}_1\alpha}\epsilon_{\alpha i_2\cdots i_n}.$$

It is easy to prove

$$\begin{aligned}*1 &= i^n(-1)^{\frac{1}{2}n(n-1)}h dz^1\wedge\cdots\wedge dz^n\wedge d\bar{z}^1\wedge\cdots\wedge d\bar{z}^n\\ &= 2^n h dx^1\wedge dy^1\wedge\cdots\wedge dx^n\wedge dy^n = \frac{K^n}{n!},\end{aligned}$$

where $K = ih_{\alpha\bar{\beta}}dz^\alpha\wedge d\bar{z}^\beta$ is a real form. For arbitrary (p,q)-form $\alpha_{p,q}\in\Lambda^{p,q}(M)$

$$\alpha_{p,q} = \frac{1}{p!q!}f_{\alpha_1\cdots\alpha_p\bar{\beta}_1\cdots\bar{\beta}_q}dz^{\alpha_1}\wedge\cdots\wedge dz^{\alpha_p}\wedge d\bar{z}^{\beta_1}\wedge\cdots d\bar{z}^{\beta_q},$$

one can prove

$$\overline{*\alpha_{p,q}} = *\bar{\alpha}_{p,q},$$

i.e., the Hodge $*$ operator is a real operator. In addition, it can be proved that

$$**\alpha_{p,q} = (-1)^{p+q}\alpha_{p,q},$$

$$\begin{aligned} \alpha_{p,q} \wedge *\overline{\beta_{p,q}} &= \frac{1}{p!q!} f_{\alpha_1\cdots\alpha_p\bar{\beta}_1\cdots\bar{\beta}_q} g^{\alpha_1\cdots\alpha_p\bar{\beta}_1\cdots\bar{\beta}_q} * 1 \\ &= \bar{\beta}_{p,q} \wedge *\alpha_{p,q} = \overline{\beta_{p,q} \wedge *\bar{\alpha}_{p,q}}, \end{aligned}$$

where

$$\beta_{p,q} = \frac{1}{p!q!} g_{\alpha_1\cdots\alpha_p\bar{\beta}_1\cdots\bar{\beta}_q} dz^{\alpha_1} \wedge \cdots \wedge dz^{\alpha_p} \wedge d\bar{z}^{\beta_1} \wedge \cdots d\bar{z}^{\beta_q},$$

$$g^{\alpha_1\cdots\alpha_p\bar{\beta}_1\cdots\bar{\beta}_q} = h^{\bar{\gamma}_1\alpha_1} \cdots h^{\bar{\gamma}_p\alpha_p} h^{\bar{\beta}_1\delta_1} \cdots h^{\bar{\beta}_q\delta_q} \overline{g_{\gamma_1\cdots\gamma_p\bar{\delta}_1\cdots\bar{\delta}_q}}.$$

Hence, inner product between two differential forms of the same type can be defined as

$$\begin{aligned} (\alpha_{p,q}, \beta_{p,q}) &= \int \alpha_{p,q} \wedge *\overline{\beta_{p,q}} \\ &= \frac{1}{p!q!} \int f_{\alpha_1\cdots\alpha_p\bar{\beta}_1\cdots\bar{\beta}_q} g^{\alpha_1\cdots\alpha_p\bar{\beta}_1\cdots\bar{\beta}_q} * 1. \end{aligned}$$

So defined inner product between two differential forms of the same type satisfy:
1). Skew linear.

$$(\alpha, \beta) = \overline{(\beta, \alpha)},$$

$$(a\alpha + b\beta, \gamma) = a(\alpha, \gamma) + b(\beta, \gamma).$$

2). Positive definite.

$$(\alpha, \alpha) \geq 0, \qquad \text{equal if and only if } \alpha = 0.$$

Hence the norm of differential forms can be defined

$$|\alpha| \equiv \sqrt{(\alpha, \alpha)}.$$

For complex differential forms, exterior differential operator can also be introduced:

$$\begin{aligned} d &= \sum_{i=1}^{n} (dx^i \frac{\partial}{\partial x^i} + dy^i \frac{\partial}{\partial y^i}) = \sum_{\alpha=1}^{n} (dz^\alpha \frac{\partial}{\partial z^\alpha} + d\bar{z}^\alpha \frac{\partial}{\partial \bar{z}^\alpha}) \\ &= \partial + \bar{\partial}, \end{aligned}$$

where

$$\partial : \Lambda^{q,r} \to \Lambda^{q+1,r},$$
$$\bar{\partial} : \Lambda^{q,r} \to \Lambda^{q,r+1}.$$

If a $(q,0)$-forms $\omega \in \Lambda^{q,0}$ satisfies

$$\bar{\partial}\omega = 0,$$

then ω is called holomorphic $(q,0)$-form which can be written as

$$\omega = \frac{1}{q!} f_{k_1\cdots k_q} dz^{k_1} \wedge \cdots \wedge dz^{k_q},$$

where f satisfies $\bar{\partial} f = 0$, i.e., f is a holomorphic function. A differential form is holomorphic if and only if its coefficients are holomorphic functions. It is easy to prove

$$\partial \cdot \partial = 0, \quad \bar{\partial} \cdot \bar{\partial} = 0, \quad \partial \cdot \bar{\partial} = -\bar{\partial} \cdot \partial.$$

The operators ∂ and $\bar{\partial}$ can be defined just in terms of the complex structure of the manifold M. $\partial\alpha$ and $\bar{\partial}\alpha$ transform homogeneously under holomorphic change of coordinates.

If α is a $[p,q]$-form with $p > 0$ and $\partial\alpha = 0$, then locally $\alpha = \partial\beta$ for some $[p-1,q]$-form β. It is an analog of Poincare lemma. But for complex manifold, the further statement is that if α is a $[p-1,q]$-form with $p > 0$, $q > 0$ and $\partial\alpha = \bar{\partial}\alpha = 0$, then locally $\alpha = \partial\bar{\partial}\beta$ for some $[p-1, q-1]$-form β.

The co-differential operator can be introduced through the Hermitian metric as:

$$\delta = - * d * = - * (\partial + \bar{\partial}) * = \theta + \bar{\theta},$$

where

$$\theta = - * \bar{\partial} * : \Lambda^{p,q} \longrightarrow \Lambda^{p-1,q},$$
$$\bar{\theta} = - * \partial * : \Lambda^{p,q} \longrightarrow \Lambda^{p,q-1}.$$

For M a compact manifold, it can be proved that

$$(\partial\alpha, \beta) = (\alpha, \theta\beta), \quad \alpha \in \Lambda^{p,q}, \quad \beta \in \Lambda^{p+1,q},$$
$$(\bar{\partial}\alpha, \beta) = (\alpha, \bar{\theta}\beta), \quad \alpha \in \Lambda^{p,q}, \quad \beta \in \Lambda^{p,q+1},$$
$$(d\alpha, \beta) = (\alpha, \delta\beta), \quad \alpha \in \Lambda^{r}, \quad \beta \in \Lambda^{r+1}.$$

The Laplace operator is defined as

$$\Delta = (d + \delta)^2 = d\delta + \delta d.$$

In addition, other Laplace operators can also be defined:

$$\Diamond = 2(\partial\theta + \theta\partial),$$
$$\bar{\Diamond} = 2(\bar{\partial}\bar{\theta} + \bar{\theta}\bar{\partial}).$$

These three Laplace operators are not equal in general. But all of them are self adjoint-operators. For instance

$$(\Diamond\alpha, \beta) = (\alpha, \Diamond\beta).$$

$\Diamond\alpha = 0$ if and only if $\partial\alpha = \theta\alpha = 0$.

6.4 Kähler manifold

For Hermitian metric on the an n-dimensional manifold M

$$dS^2 = 2h_{\alpha\bar{\beta}}dz^\alpha d\bar{z}^\beta,$$

a $[1,1]$-type differential form can be defined

$$K = ih_{\alpha\bar{\beta}}dz^\alpha \wedge d\bar{z}^\beta, \tag{6.15}$$

which is called the Kähler form. It is easy to prove that K is a real form:

$$K = \bar{K}.$$

If the Kähler form of the manifold M is a closed form

$$dK = 0, \tag{6.16}$$

M is called Kähler manifold. By the way, if $\partial K = \bar{\partial} K = 0$, then K can be written locally as

$$K = -\frac{i}{2}\partial\bar{\partial}F(z,\bar{z})$$

where F is a scalar function called a Kähler potential.

It can be proved that Kähler form of the Kähler manifold is an harmonic form, i.e., it satisfies

$$dK = \delta K = 0.$$

The integrabiliy of J implies that

$$d = \partial + \bar{\partial}, \qquad \delta = \theta + \bar{\theta}$$

$$d^2 = 0 \implies \partial^2 = 0, \ \bar{\partial}^2 = 0, \ \partial\bar{\partial} + \bar{\partial}\partial = 0$$

$$\delta^2 = 0 \implies \theta^2 = 0, \ \bar{\theta}^2 = 0, \ \theta\bar{\theta} + \bar{\theta}\theta = 0.$$

In this case the Laplace operator:

$$\Delta = d\delta + \delta d = \frac{1}{2}(\Diamond + \bar{\Diamond}) + \partial\bar{\theta} + \bar{\theta}\partial + \theta\bar{\partial} + \bar{\partial}\theta.$$

Operator $(\partial\bar{\theta}+\bar{\theta}\partial)$ acts from $\Lambda^{p,q}$ to $\Lambda^{p+1,q-1}$, and $(\theta\bar{\partial}+\bar{\partial}\theta)$ acts from $\Lambda^{p,q}$ to $\Lambda^{p-1,q+1}$. Note that the space $\Lambda^{p,q}$ are invariant under Δ. Hence this operator $(\partial\bar{\theta} + \bar{\theta}\partial)$ and $(\theta\bar{\partial} + \bar{\partial}\theta)$ equal zero. Moreover, the two operator $\Diamond$ and $\bar{\Diamond}$ agree, hence on Kähler manifold the three of Laplace operators, defined in the last section, are equal:

$$\Delta = \Diamond = \bar{\Diamond}.$$

The compact Kähler manifold has very strong topological restriction. The possible exterior product of Kähler form

$$K \wedge K\ ,\ K \wedge K \wedge K\ ,\ \cdots\ ,\ \underbrace{K \wedge \cdots \wedge K}_{n}$$

are non-zero harmonic forms and can not be represented as exact forms globally on the manifold. Because from Eq. 6.15, we have the exterior product of n Kähler forms K:

$$\bigwedge_n K = i^n n! h \bigwedge_\alpha (dz^\alpha \wedge d\bar{z}^\alpha) = n! * 1.$$

and $h = \det(h_{\alpha\bar{\beta}}) > 0$ for compact manifold, so we have

$$V = \int_M \bigwedge_n K > 0.$$

If the exterior product of l Kähler forms is an exact form

$$\bigwedge_l K \equiv K \wedge K \wedge \cdots \wedge K = d\alpha,$$

then

$$V = \int_M d\alpha \wedge \bigwedge_{n-l} K = \int_M d(\alpha \wedge K \wedge \cdots \wedge K) = 0.$$

This is clearly a self contradiction. Therefore, none of the exterior products of Kähler forms is an exact form. Hence, all even order Betti numbers(see the next chapter) of the Kähler manifold are non-zero. The above discussion shows that the existence of Kähler metric on a complex manifold is determined by the topological properties of the manifold. For an arbitrary complex manifold, there is an Hermitian metric. The condition for the manifold to be an Hermitian manifold is the same as that for the manifold to have a complex structure. For Kähler manifold, one needs a stronger condition. In the following, we give an example in which the manifold has complex structure(therefore an Hermitian metric) but no Kähler metric. The manifold is Hopf manifold:

In $\mathbf{C}^n - \{0\}$, the discrete group generated by the equivalent relation $z^{i'} = 2z^i (1 \leq i \leq n)$ is written as $\mathcal{R}$, the corresponding quotient manifold

$$C^n - \{0\}/\mathcal{R} \sim S^{2n-1} \times S^1$$

is called the Hopf manifold. This is a complex manifold but it is not a Kähler manifold.

In general, the product of any two odd-dimensional spheres $S^p \times S^q$ often has complex structure but it is not a Kähler manifold and admits no Kähler metric.

On Kähler manifold, for holomorphic $[r, q]$-form ω,

$$\bar{\partial}\omega = 0.$$

It can be proved that

$$\bar{\theta}\omega = 0,$$

hence

$$\bar{\Diamond}\omega = 0.$$

From Eq. 6.3, we have

$$\Delta\omega = 0,$$

i.e., ω is a harmonic form. On Kähler manifold, holomorphic forms respect to the Kähler metric are automatically harmonic forms and vice versa, i.e., from $\Delta\omega = 0$, we have $\bar{\partial}\omega = 0$, such that every harmonic r-form is a holomorphic r-form.

For Kähler manifold, from condition Eq. 6.16, we have

$$\frac{\partial h_{\alpha\bar{\beta}}}{\partial z^{\gamma}} = \frac{\partial h_{\gamma\bar{\beta}}}{\partial z^{\alpha}}, \quad \frac{\partial h_{\alpha\bar{\beta}}}{\partial \bar{z}^{\gamma}} = \frac{\partial h_{\alpha\bar{\gamma}}}{\partial \bar{z}^{\beta}},$$

i.e., $h_{\alpha\bar{\beta}}$ can be written locally as

$$h_{\alpha\bar{\beta}} = \frac{\partial^2 F(z,\bar{z})}{\partial z^{\alpha}\partial \bar{z}^{\beta}}, \tag{6.17}$$

where $F(z,\bar{z})$ is a real function which is called the Kähler potential. The Kähler potential completely determines Kähler manifold and Kähler manifold is often defined through this equation. Kähler potential $F(z,\bar{z})$ can only be defined locally, although the metric tensor $h_{\alpha\bar{\beta}}$ may be defined on the Kähler manifold globally, and Kähler potential is not unique. Changing the Kähler potential as

$$F(z,\bar{z}) \rightarrow F(z,\bar{z}) + G_1(z) + G_2(\bar{z}),$$

we still have the same metric tensor $h_{\alpha\bar{\beta}}$ according to Eq. 6.17, where $G_1(G_2)$ is any holomorphic (anti-holomorphic) function. Such a change is called a transformation of Kähler potential.

Now we look at some examples, in which the manifolds have Kähler metric. The simplest noncompact Kähler manifold is $\mathbf{C}^n$, the standard Kähler form on $\mathbf{C}^n$ is given by

$$K = \frac{i}{2}\sum_{\alpha=1}^{n} dz^{\alpha}\wedge d\bar{z}^{\alpha} = \frac{i}{2}\partial\bar{\partial}(\sum_{\alpha=1}^{n}|z^{\alpha}|^2)$$

Furthermore, It can be proved that the simplely connected, compact, homogeneous Kähler manifold M can be represented as the orbit space of a compact simple Lie group G on the adjoint representation space, namely,

$$M = G/H,$$

here H is a subgroup of G. In general, in the adjoint representation of G, the orbit space G/H may have more than one G-invariant complex structures. So there may be several inequivalent complex manifolds G/H corresponding to one real manifold G/H. The complex structure is unique only if the subgroup H has one $U(1)$ factor. For instance,

$$CP(n) = SU(n+1)/SU(n) \times U(1).$$

$CP(n)$ is a complex n-dimensional projective space which is the set of straight lines through the origin in $\mathbf{C}^{n+1}$. There exists the Fubini-Study[27,28] metric

$$dS^2 = (\bar{z}^i z^i)^{-1}(dz^\alpha - z^\alpha \frac{\bar{z}^b dz^b}{z^c \bar{z}^c})(d\bar{z}^\alpha - \bar{z}^\alpha \frac{z^d d\bar{z}^d}{z^e \bar{z}^e}).$$

This metric is invariant under scale transformation of z^α and is defined on $CP(n)$. At the neighborhood of a point with $z^{n+1} \neq 0$, Kähler form can be written as

$$K = \frac{i}{2}\partial\bar{\partial}\ln(1 + \sum_{\alpha=1}^{n} z^\alpha \bar{z}^\alpha),$$

i.e., $CP(n)$ manifold is a Kähler manifold.

When $n = 1$, for the $CP(1) \simeq S^2$ manifold, we have projective coordinates(radius $r = \frac{1}{2}$)

$$dS^2 = \frac{dz \cdot d\bar{z}}{(1 + z \cdot \bar{z})^2} = \frac{dx^2 + dy^2}{(1 + x^2 + y^2)^2},$$

$$K = \frac{i}{2}\frac{dz \wedge d\bar{z}}{(1 + z \cdot \bar{z})^2} = \frac{dx \wedge dy}{(1 + x^2 + y^2)^2}.$$

This Kähler form is a harmonic form and can be written as

$$K = \frac{i}{2}\partial\bar{\partial}\ln(1 + z\bar{z}).$$

In section 6.6 example 3, we should derived this result in terms of the generalized coherent states, which is will known for physicists.

In $CP(n)$, the orbit determined by a set of homogeneous polynomials is called the algebraic variety. The algebraic variety is usually a Kähler manifold and is the research objects of algebraic geometry.

6.5 Connections on complex manifold

Pick a local coordinate $\{z^\alpha\}$ in the neighborhood of point p on an n-dimensional manifold M. To simplify, Greek letters $\alpha, \beta, \cdots$ take values from 1 to n and Latin letters $j, k, \cdots$ take the values $1, 2, \cdots, n, \bar{1}, \bar{2}, \cdots, \bar{n}$. We use covariant derivative operator ∇ to define the affine connection

$$\nabla_{\partial_j}\partial_k = \Gamma^l_{jk}\partial_l.$$

Originally affine connection is defined on real vector fields. Now it is generalized to complex vector fields. Let we denote

$$\bar{\Gamma}^l_{jk} = \Gamma^{\bar{l}}_{\bar{j}\bar{k}}. \tag{6.18}$$

When defining connection on a manifold with almost complex structure J, we require that it preserves the almost complex structure, that is,

$$\nabla_X(JY) = J\nabla_X Y, \quad \forall X, Y \in \mathcal{X}(M).$$

Connections satisfying this requirement are called almost complex affine connections and are defined on complex linear frame bundle. Because

$$J\partial_\alpha = i\partial_\alpha, \quad J\partial_{\bar{\alpha}} = -i\partial_{\bar{\alpha}},$$

for almost complex connection which preserves the almost complex structure , from $\nabla_X JY = J\nabla_X Y$ we have

$$\Gamma^\alpha_{j\bar{\beta}} = \Gamma^{\bar{\alpha}}_{j\beta} = 0. \tag{6.19}$$

For complex manifold, the almost complex structure is torsion free and admits a torsion free almost complex affine connection. That is,

$$\Gamma^i_{jk} = \Gamma^i_{kj}.$$

For Hermitian manifold, connection is required to preserve Hermitian metric H. In other words, for two vector fields which are parallel transported along an arbitrary curve on the manifold

$$X = \xi^\alpha \partial_\alpha, \quad Y = \eta^{\bar{\alpha}} \partial_{\bar{\alpha}},$$

the scalar product

$$H(X, Y) = h_{\alpha\bar{\beta}} \xi^\alpha \eta^{\bar{\beta}}$$

is a constant. From Eq. 6.19, if X and Y are parallel transported, we have

$$\begin{aligned}\nabla X &= (d\xi^\alpha + \Gamma^\alpha_\beta \xi^\beta)\partial_\alpha = 0,\\ \nabla Y &= (d\eta^{\bar{\beta}} + \Gamma^{\bar{\beta}}_{\bar{\gamma}} \eta^{\bar{\gamma}})\partial_{\bar{\beta}} = 0.\end{aligned}$$

In order that the scalar product is also invariant under the above conditions, the connection must satisfy:

$$dh_{\alpha\bar{\beta}} - h_{\gamma\bar{\beta}}\Gamma^\gamma_\alpha - h_{\alpha\bar{\gamma}}\Gamma^{\bar{\gamma}}_{\bar{\beta}} = 0.$$

The matrix form of this equation is

$$dH = \Gamma^t \cdot H + H \cdot \bar{\Gamma}. \tag{6.20}$$

The exterior derivative of this equation gives

$$\Omega^t \cdot H + H \cdot \bar{\Omega} = 0,$$

i.e., for the Hermitian metric compatible connection, $\Omega^t \cdot H$ is Hermitian-skewsymmetry.

Connection 1-form $\Gamma^{\alpha}_{\bar{\beta}} = \Gamma^{\alpha}_{\gamma\beta} dz^{\gamma}$, is called [1,0]-type connection if

$$\Gamma^{\alpha}_{\bar{\gamma}\beta} = 0. \tag{6.21}$$

It can be proved that this definition is independent of coordinate system. Under a holomorphic transformation, the [1,0]-type connection is still [1,0]-type connection. For [1,0]-type Γ, $\bar{\Gamma}$ is [0,1]-type, so from Eq. 6.20, we see that

$$\partial H = \Gamma^t H,$$

and [1,0]-type connection Γ can be defined on Hermitian manifold through this equation. In other words,

$$\Gamma = (H)^{-1} \cdot \partial H, \quad \Gamma^{\gamma}_{\alpha\beta} = h^{\bar{\delta}\gamma} \partial_{\alpha} h_{\beta\bar{\delta}}, \tag{6.22}$$

and the corresponding curvature 2-form matrix is

$$\begin{aligned} \Omega &= d((H)^{-1}\partial H) + (H)^{-1}\partial H \wedge (H)^{-1}\partial H \\ &= -(H)^{-1}\bar{\partial} H \wedge (H)^{-1}\partial H + (H)^{-1}\bar{\partial}\partial H. \end{aligned} \tag{6.23}$$

The matrix elements of this curvature 2-form matrix are [1,1]-type differential forms. It can be further proved that connection on tangent bundle is of [1,0]-type if and only if torsion matrix elements are [2,0]-type differential forms.

In other words, there exists a unique [1,0]-type permissible connection on the holomorphic tangent bundle of Hermitian manifold. The corresponding torsion is [2,0]-type and curvature is [1,1]-type. This connection is called Hermitian connection. In terms of this form of the affine connection, the standard complex structure J in Eq.(6.6) is a covariant constant since J is covariant under the subgroup $U(n)$ of $O(2n)$. The holonomy group of the Hermitian connection is

$$U(n) \subset GL(n, C) \cap O(2n).$$

From Hermitian metric, a 4-order covariant curvature tensor in terms of the (1,3)-type curvature tensor can be obtained

$$R_{ijkl} = h_{im} R^m_{\ jkl}.$$

As a result of $\Gamma^{\alpha}_{j\bar{\beta}} = \Gamma^{\bar{\alpha}}_{j\beta} = 0$, we have

$$\begin{aligned} R^{\alpha}_{\bar{\beta}jk} &= R^{\bar{\alpha}}_{\beta jk} = 0, \\ R_{\bar{\alpha}\bar{\beta}jk} &= R_{\alpha\beta jk} = 0. \end{aligned}$$

From Eq. 6.23, for Hermitian connection, the curvature is (1,1)-type and

$$\begin{aligned} R^j_{k\alpha\beta} &= R^j_{k\bar{\alpha}\bar{\beta}} = 0, \\ R_{jk\alpha\beta} &= R_{jk\bar{\alpha}\bar{\beta}} = 0. \end{aligned}$$

So for Hermitian connection, the possible non-zero components of the curvature tensor are

$$R^{\alpha}_{\beta\gamma\bar{\delta}},\quad R^{\alpha}_{\beta\bar{\gamma}\delta},\quad R^{\bar{\alpha}}_{\bar{\beta}\gamma\bar{\delta}},\quad R^{\bar{\alpha}}_{\bar{\beta}\bar{\gamma}\delta}.$$

In covariant form, they are

$$R_{\bar{\alpha}\beta\gamma\bar{\delta}},\quad R_{\bar{\alpha}\beta\bar{\gamma}\delta},\quad R_{\alpha\bar{\beta}\gamma\bar{\delta}}, R_{\alpha\bar{\beta}\bar{\gamma}\delta}.$$

On Hermitian manifold, there exist the complex structure J and the Hermitian metric H which is invariant under the action of J. In general, the torsion tensor of the connection which preserves the Hermitian metric is non-zero and torsion free connection usually does not preserve the Hermitian metric. On Hermite manifold, there exist two different torsion free connections: The connection which parallel transports the almost complex structure is called the almost complex connection and the corresponding holonomy group is $GL(n, C)$. The torsion free connection which parallel transports the Riemannian metric tensor is called Riemannian connection and the corresponding holonomy group is $O(2n)$. These two connections are not compatible in general, as Riemannian structure is not compatible with complex structure in general, and for the [1,0]-type Hermitian connection, which has a [2,0]-type torsion, whose holonomy group is U(n).

If the Kähler form is a closed form on the Hermitian manifold, which is integrable, and the manifold becomes a Kähler manifold, For Kähler manifold, similar to Eq. 6.22, connection satisfies

$$\frac{\partial h_{\alpha\bar{\beta}}}{\partial z^{\gamma}} = h_{\delta\bar{\beta}}\Gamma^{\delta}_{\gamma\alpha},\quad \frac{\partial h_{\alpha\bar{\beta}}}{\partial \bar{z}^{\gamma}} = h_{\alpha\bar{\delta}}\Gamma^{\bar{\delta}}_{\bar{\gamma}\bar{\beta}}.$$

and the corresponding torsion is zero. In other words, Hermitian connection on Kähler manifold is torsion free. The torsion free of the Hermitian connection is the necessary and sufficient condition for the Hermitian manifold be a Kähler manifold.

Kähler metric is the best metric on complex manifold because the Riemannian structure and the complex structure are compatible. Riemannian connection on Kähler manifold preserves the complex structure. In the following, we analyse in detail the connection and curvature on Kähler manifold.

For Kähler manifold, the connection coefficients satisfy Eqs. 6.18,6.19, 6.21 and the torsion free condition:

$$\Gamma^{\alpha}_{\beta\gamma} = \Gamma^{\alpha}_{\gamma\beta} = h^{\alpha\bar{\delta}}\partial_{\beta}h_{\gamma\bar{\delta}},\quad \Gamma^{\bar{\alpha}}_{\bar{\beta}\bar{\gamma}} = \Gamma^{\bar{\alpha}}_{\bar{\gamma}\bar{\beta}} = \overline{\Gamma^{\alpha}_{\beta\gamma}}. \tag{6.24}$$

These are the only non-vanishing components of connection on a Kähler manifold.

Through the torsion free condition Eq. 6.24, one can prove

$$R^{\alpha}_{\beta\gamma\bar{\delta}} = -\frac{\partial\Gamma^{\alpha}_{\beta\gamma}}{\partial\bar{z}^{\delta}}. \tag{6.25}$$

For Kähler manifold, the curvature tensor has the following additional symmetry:

$$R^{\alpha}_{\beta\gamma\bar{\delta}} = R^{\alpha}_{\gamma\beta\bar{\delta}}.$$

The independent non-zero connection coefficients on Kähler manifold can be written as

$$\Gamma^{\alpha}_{\beta\gamma} = h^{\alpha\bar{\delta}} \frac{\partial h_{\beta\bar{\delta}}}{\partial z^{\gamma}}.$$

Substituting this equation into Eq. 6.25, one has

$$R_{\bar{\alpha}\beta\gamma\bar{\delta}} = -\frac{\partial^2 h_{\beta\bar{\alpha}}}{\partial z^{\gamma}\partial \bar{z}^{\delta}} + h^{\bar{\lambda}\rho} \frac{\partial h_{\beta\bar{\lambda}}}{\partial z^{\gamma}} \frac{\partial h_{\rho\bar{\alpha}}}{\partial \bar{z}^{\delta}}.$$

From Eq. 6.17, we have

$$R_{\bar{\alpha}\beta\gamma\bar{\delta}} = -\frac{\partial^4 F}{\partial \bar{z}^{\alpha}\partial z^{\beta}\partial z^{\gamma}\partial \bar{z}^{\delta}} + h^{\bar{\lambda}\rho} \frac{\partial^3 F}{\partial \bar{z}^{\lambda}\partial z^{\beta}\partial z^{\gamma}} \frac{\partial^3 F}{\partial z^{\rho}\partial \bar{z}^{\alpha}\partial \bar{z}^{\delta}}.$$

On Kähler manifold, the Ricci tensor $\Re_{jk} = R^l_{jlk}$ satisfies

$$\Re_{\alpha\beta} = \Re_{\bar{\alpha}\bar{\beta}} = 0, \tag{6.26}$$

$$\Re_{\alpha\bar{\beta}} = \overline{\Re_{\bar{\alpha}\beta}} = -\frac{\partial \Gamma^{\gamma}_{\alpha\gamma}}{\partial \bar{z}^{\beta}} = -\frac{\partial^2 \ln h}{\partial z^{\alpha}\partial \bar{z}^{\beta}}, \quad (h = \det(h_{\alpha\bar{\beta}})).$$

Ricci form can be introduced

$$\rho = -i\Re_{\alpha\bar{\beta}} dz^{\alpha} \wedge d\bar{z}^{\beta} = i\partial\bar{\partial} \ln h. \tag{6.27}$$

It is clear that Ricci form is a real closed form

$$d\rho = 0.$$

In general the Ricci form ρ is different from the Kähler form, which is also a closed form.

The holonomy group of the n-dimensional Riemannian manifold can be reduced to $SO(n)$. The reason is that Riemannian connection preserves the metric and does not change the scalar product of vectors and hence the length of vectors. That the connection preserves metric is manifested in the skewsymmetry of the Riemannian curvature:

$$R_{ijkl} = -R_{jikl}.$$

The Riemannian curvature 2-forms $\Omega_{ij} = \frac{1}{2} R_{ijkl} dx^k \wedge dx^l$ can be taken as the generators of the holonomy group.

For Kähler manifold, the Riemannian connection preserves the complex structure and so the $2n$-dimensional manifold is also an n-dimensional complex manifold. The holonomy group is

$$U(n) = U(1) \times SU(n).$$

On the other hand, if $h_{\alpha\bar{\beta}}$ is a Kähler metric, so is $h'_{\alpha\bar{\beta}} = h_{\alpha\bar{\beta}} + \partial_{\alpha}\bar{\partial}_{\beta} F$ for an arbitrary scalar function $F(z, \bar{z})$. Thus it is natural to ask whether or not the holonomy group

$U(n)$ can be reduced to its subgroup $SU(n)$ for some Kähler metric $h'_{\alpha\bar{\beta}}$ with suitable function $F(z, \bar{z})$, the answer to the question is that there is a topological obstruction to finding such a metric. Now we shall consider the reducible condition.

$U(n)$ group is not a semisimple Lie group, the generator of the $U(1)$ subgroup is the trace of the Riemannian curvature tensor, i.e., the Ricci form

$$-iR^{\alpha}_{\beta\alpha\bar{\gamma}}dz^{\beta} \wedge d\bar{z}^{\gamma}.$$

When the Ricci form vanishes identically, the holonomy group can be reduced from $U(n)$ to $SU(n)$ (see reference[7], S.Kobayashi and K.Nomizu, Volume 2, page 151.)

In physics, symmetry groups are often semisimple groups which has no $U(1)$ factor. But holonomy group of Kähler manifold is $U(n)$. If we require that the holonomy group is $SU(n)$, the metric on the manifold will be very complicated. On Kähler manifold, there are often topological obstructions to Ricci flat metric.

The trace of the curvature 2-form Ω of the Riemannian connection on the manifold is called the first Chern character c_1 of M:

$$c_1 = \frac{i}{2\pi}\text{Tr}\Omega,$$

which is a closed 2-form on M, and is determined by the topological property of M (see Chapter 10). When Kähler metric exists on the manifold, the Ricci form introduced in Eq. 6.27 is just the first Chern character c_1 (mod a constant factor). If $c_1 \neq 0$, (mod exact form), the Ricci form is not exact and there is no Ricci flat Kähler metric. In this case, the holonomy group must be $U(n)$. However, when $c_1 = 0$, it is not clear if Ricci flat metric and the corresponding holonomy group $SU(n)$ exist. In 1959, Calabi[29] conjectured that for Kähler manifold with $c_1 = 0$, there exists Ricci flat Kähler metric with holonomy group $SU(n)$. Calabi proved that Ricci flat metric is unique for given complex structure of the manifold. In 1977, Yau[30] proved the Calabi conjecture and showed that Ricci flat Kähler metric exists if the first Chern character $c_1 = 0$. Kähler manifold with $c_1 = 0$ is called Calabi-Yau manifold, on which there exists unique Ricci flat Kähler metric and the holonomy group is $SU(n)$. In other words, Calabi-Yau manifold is a complex manifold with holonomy group $SU(n)$. For an n-dimensional complex manifold, $c_1 = 0$ if and only if there exists a holomorphic n-form which is nonsingular and nonzero everywhere.

In summary, there are different affine connections on n-dimensional complex manifold:

1. Almost complex connection, which satisfies:
   (i) Torsion free: $\Gamma^i_{jk} = \Gamma^i_{kj}$,
   (ii) Complex structure preserving: $\Gamma^{\alpha}_{j\bar{\beta}} = \Gamma^{\bar{\alpha}}_{j\beta} = 0$,

its holonomy group is $GL(n, C)$.

2. Riemannian connection, which satisfies:
   (i) Torsion free: $\Gamma^i_{jk} = \Gamma^i_{kj}$,
   (iii) Riemannian metric preserving: $\nabla_i g_{kl} = 0$,

$$\text{i.e.,} \qquad dg_{kl} = g_{kj}\Gamma^j_l + g_{jl}\Gamma^j_k.$$

Consequently

$$\Gamma^i_{jk} = \frac{1}{2} g^{il}(\partial_j g_{kl} + \partial_k g_{jl} - \partial_l g_{jk}).$$

Its holonomy group is $SO(2n)$.

3. Hermitian connection, which satisfies:
   (ii) Complex structure preserving: $\Gamma^\alpha_{j\bar\beta} = \Gamma^{\bar\alpha}_{j\beta} = 0$,
   (iii') Hermitian metric preserving: $\nabla_j h_{\alpha\bar\beta} = 0$,

$$\text{i.e.,} \qquad dh_{\alpha\bar\beta} = h_{\gamma\bar\beta}\Gamma^\gamma_\alpha + h_{\alpha\bar\gamma}\Gamma^{\bar\gamma}_{\bar\beta},$$

   (iv) It can be $[1,0]$-type : $\Gamma^\alpha_{\bar\gamma\beta} = 0$.

This $[1,0]$-type connection on Hermitian manifold can be defined as:

$$\Gamma^\gamma_{\alpha\beta} = h^{\bar\delta\gamma}\partial_\alpha h_{\beta\bar\delta},$$

its holonomy group is $U(n)$.

We know that the transformation rule for the connection is:

$$\Gamma^\alpha_{\beta\gamma} = \frac{\partial z^\mu}{\partial z'^\beta}\frac{\partial z^\nu}{\partial z'^\gamma}\Gamma^\lambda_{\mu\nu} + \frac{\partial^2 z^\delta}{\partial z'^\beta \partial z'^\gamma}\frac{\partial z'^\alpha}{\partial z^\delta}, \qquad \Gamma^\alpha_{\bar\beta\gamma} = \frac{\partial \bar z^\mu}{\partial \bar z'^\beta}\frac{\partial z^\nu}{\partial z'^\gamma}\frac{\partial z'^\alpha}{\partial z^\lambda}\Gamma^\lambda_{\bar\mu\nu}.$$

The first equation shows that it is consistent to require torsion free:

$$\Gamma^\alpha_{\beta\gamma} = \Gamma^\alpha_{\gamma\beta}.$$

For Hermitian manifold, $(ii), (iii') and (iv)$ are compatible, we have a $[1,0]$-type connection, whose torsion is $[2,0]$-type, usually nonvanishing. But for Kähler manifold, the torsion of the Hermitian connection is zero, so conditions $(ii), (iii') and (iv)$ are consistent with (i)

The only nonvanishing components of connection on a Kähler manifold can be written as 6.26

$$\Gamma^\alpha_{\beta\gamma} = \Gamma^\alpha_{\gamma\beta} = h^{\alpha\bar\delta}\frac{\partial h_{\beta\bar\delta}}{\partial z^\gamma}$$

its holonomy group is $U(n)$.

For Calabi-Yau manifold, which is a Kähler manifold, with $\mathrm{Tr}\Omega = 0$ (mod exact form), on which there exists unique Ricci flat Kähler metric, the holonomy group is $SU(n)$.

6.6 Riemannian symmetric space, its Kähler structure and nonlinear realization

Symmetric space plays an important role in theoretical physics. In the following, we consider the algebraic structure of symmetric space. If for any point x on manifold M endowed with metric, there exists an isometry transformation

$$S_x : \quad M \to M$$

which makes x an isolated fixed point and the induced mapping of S_x on the tangent space of x reverses all tangent vectors at x, the manifold M is called Riemann symmetric space. The isometry transformation S_x is an involutive transformation such that S_x^2 is an identity transformation on the tangent space at point x. A symmetric Riemannian manifold is always homogeneous. The Lie group of all isometries of M will be denoted by G, and the isotropy group by H, so that we can identify M with the coset manifold G/H

$$M \simeq G/H$$

which has an involutive automorphism σ. A symmetric space consists of three elements (G, H, σ). For G a connected Lie group, with $\mathbf{g}$ its Lie algebra, the space $\mathbf{g}$ can be chosen to be a representation space of the adjoint representation. There exists a G-invariant quadratic form, the so called Killing-Cartan form

$$(X,Y) = Tr(Ad_X, Ad_Y), \qquad X, Y \in \mathbf{g} \tag{6.28}$$

where Ad_X is the adjoint representation matrix of the algebra element X. There exists an involutive automorphism operator σ which does not change the metric above:

$$(\sigma X, \sigma Y) = (X, Y),$$

where $\sigma^2 = 1$ but $\sigma \neq 1$. The eigenvalues of the automorphism σ are ± 1. The subspace with eigenvalue $+1$ is called η and the subspace with eigenvalue -1 is called κ. Obviously, the two spaces are orthogonal to each other with respect to the Killing-Cartan form. Hence $\mathbf{g}$ has the following decomposition(called Cartan decomposition):

$$\mathbf{g} = \eta + \kappa,$$
$$[\sigma, \eta] = 0, \qquad \{\sigma, \kappa\} = 0,$$

where [,] stands for commutator and { , } stands for anticommutator. Consequently there exist the following commutation relations between κ and η:

$$[\eta, \eta] \subset \eta, \quad [\eta, \kappa] \subset \kappa, \quad [\kappa, \kappa] \subset \eta.$$

Clearly η is a subalgebra, which is the Lie algebra of the closed subgroup H of Lie group G. The coset $G/H = K$ is a homogeneous manifold at which the tangent space of each point is isomorphic to κ. The second equation shows that H is a stable subgroup. Space κ is invariant under the operation of Ad_H and is a representation space of subalgebra η. All coset spaces must satisfy the first two equations, but the third equation shows the existence of the involutive automorphism σ, then $K = G/H$ is a symmetric space, and κ, which is a eigenspace of σ of eigenvalue -1, is a complementary space of η. The subspace κ can be identified with a space tangent to $K = G/H$ at point x_0. The map from $\mathbf{g}$ into group G maps κ into K. Let the rank of κ be m, i.e., the dimension of the maximal commutative subalgebra in space κ is m.

In general, there are three cases:

(1). $[\kappa, \kappa] = 0$, i.e., κ is an Abelian ideal in $\mathbf{g}$. Algebra $\mathbf{g}$ is not semisimple. Correspondingly, group G is not simple and there exists an invariant subgroup K. Group G is a semi-direct product of H and K,

$$G = H \otimes_s K.$$

Example 1. $G = \{q\}$ is the group of motion of the plane

$$q\mathbf{x} = r\mathbf{x} + \mathbf{a},$$

where $H = SO(2) = \{r\}$ is the group of rotation of the plane and $a \in \mathbf{R}^2$. Space $K = G/H$ is a two-dimensional Euclidean plane and a symmetric space with zero curvature. Group G acts as a group of affine transformations of K.

(2). When G is a semisimple noncompact Lie group.

For semisimple Lie algebra, the Killing-Cartan form is nondegenerate and it is possible to choose a positive definite quadratic form which is invariant with respect to the adjoint representation of the subgroup H. By using the group G of transformations, one can obtain such a form over the entire space $K = G/H$. The Killing-Cartan form Eq. 6.28 is such a form for semisimple noncompact algebra. Space K is of non-positive curvature.

Example 2. $G = SO(2,1)$, $H = SO(2)$,

$$K = SO(2,1)/SO(2),$$

is the Lobachevsky space which has negative constant curvature.

(3). When G is a semisimple compact Lie group, the Killing-Cartan form can be taken as Eq. 6.28 with a minus sign, so that the space G/H has a Riemannian structure, which has non-negative curvature. For example, $G = SO(3)$, $H = SO(2)$, then

$$K = SO(3)/SO(2) \sim S^2,$$

is a two dimensional sphere S^2 which has positive constant curvature.

Riemannian symmetric space is a Kähler manifold.

The coordinate systems on symmetric space K can be defined in terms of the generalized coherent states. Let $|\psi_0>$ be the most symmetric state, i.e., if G is compact, state $|\psi_0>$ is the lowest (highest) weight state, and if G is noncompact, $|\psi_0>$ is merely the lowest bound state. Then all states $|\psi>$ can be generated as

$$|\psi> = F(X_i^+)|\psi_0>$$

where $F(X_i^+)$ is a polynomial of $\{X_i^+\}^m$. Operators X_i^+ can be called the elementary exitation operators.

For example, the regular representation of involution automorphism operator can be expressed as:

$$\sigma = \begin{bmatrix} I_p & \\ & -I_q \end{bmatrix} = I_{pq}$$

The subspace $\mathbf{g} = \eta + \kappa$ can be expressed as:

$$R(\eta) = \begin{pmatrix} \mu & 0 \\ 0 & \nu \end{pmatrix}, \qquad R(\kappa) = \begin{pmatrix} 0 & \zeta \\ \pm\zeta^+ & 0 \end{pmatrix}.$$

We can obtain a unitary coset representation (called generalized coherent state) of $K = G/H$ through the exponential mapping for κ:

$$\begin{aligned} |\alpha> &= e^{\sum_{i=1}^{m}(\zeta^i X_i^+ - h.c.)}|\psi_0> = \exp\begin{pmatrix} & \zeta \\ \pm\zeta^+ & \end{pmatrix}|\psi_0> \\ &= \begin{pmatrix} (1\mp X^+X)^{\frac{1}{2}} & X \\ \mp X^+ & (1\mp X^+X)^{\frac{1}{2}} \end{pmatrix}|\psi_0>. \end{aligned}$$

For $g = \begin{pmatrix} A & B \\ C & D \end{pmatrix} \in G$ acting on $K = G/H$, $\begin{pmatrix} W & X \\ T & Y \end{pmatrix} \in K$,

$$\begin{pmatrix} A & B \\ C & D \end{pmatrix}\begin{pmatrix} W & X \\ T & Y \end{pmatrix} = \begin{pmatrix} W' & X' \\ T' & Y' \end{pmatrix}\begin{pmatrix} U & 0 \\ 0 & V \end{pmatrix}.$$

Since the coset representation is in terms of submatrix X, it is necessary to solve the following equation

$$X' = (AX + BY)V^{-1}.$$

The problem is complicated by the involvement of group operation V. Usually we choose coset coordinates

$$Z = XY^{-1} = X(1 \mp X^+X)^{\frac{1}{2}}.$$

There is a one to one correspondence between X and Z,

$$X = Z(1 \pm Z^+Z)^{-\frac{1}{2}}$$

Under the action of group G,

$$Z' = X'Y'^{-1} = (AZ + B)(CZ + D)^{-1}.$$

This transformation involves A, B, C, D only, with no complex conjugation, so G/H is a complex manifold and holomorphic coordinates are applicable.

$$\begin{aligned} |\alpha> &= K^{-\frac{1}{2}}(z, z^*)\exp\left(\sum_{i=1}^{m} z^i X_i^+\right)|\psi_0> \\ &= K^{-\frac{1}{2}}(z, z^*)|z>, \end{aligned}$$

where $|z>$ is the (unnormalized) coherent state, its normalization factor is the Bergmann kernel (Kähler potential)

$$K(z,z^*) = |<\psi_0|\alpha>|^{-2},$$

which determines the Kähler structure of G/H space.

Example 3. $S^2 = SU(2)/U(1)$

$su(2)$ algebra is $\{S_+, S_-, S_0\}$

$$[S_+, S_-] = 2S_0, \qquad [S_0, S_\pm] = \pm S_\pm.$$

For a $(2j+1)$-dimensional representation, we choose the common eigenstate of J^2 and J_0:

$$|\psi_0>=|j,-j>,$$

then

$$|j,m>=\frac{1}{(j+m)!}\begin{pmatrix} 2j \\ j+m \end{pmatrix}^{-\frac{1}{2}} J_+^{j+m}|j,-j>,$$

satisfies

$$J^2|j,m>=j(j+1)|j,m>,$$
$$J_0|j,m>=m|j,m>.$$

Coset $SU(2)/U(1)$ can be generated from J_+, J_-

$$\alpha = \exp(zJ_+ - z^*J_-) \quad \in \ SU(2)/U(1).$$

$$\begin{aligned} \alpha|j,-j> &= (1+zz^*)^{-j}e^{zj_+}|j,-j> \\ &= (1+zz^*)^{-j}|jz>. \end{aligned}$$

Bergmann kernel

$$F(z,z^*) = (1+zz^*)^{2j}$$

for $j=\frac{1}{2}$, the Kähler form of G/H at the neighborhood of $z \neq 0$ can be written as

$$K = \frac{i}{2}\partial\bar{\partial}\ln\,(1+zz^*) = \frac{i}{2}\frac{dz\wedge dz^*}{(1+zz^*)^2} = \frac{dx\wedge dy}{(1+x^2+y^2)^2},$$

and

$$dS^2 = \frac{dzdz^*}{(1+zz^*)^2} = \frac{dx^2+dy^2}{(1+x^2+y^2)^2}.$$

This is just the Fubini-stady[27,28] metric given in p.203.

Example 4. Lobachevsky plane $H^2 = SU(1,1)/U(1)$. The generators $\{K_i\}$ of $su(1,1)$ algebra satisfy:

$$[K_1, K_2] = -iK_3, \quad [K_2, K_3] = iK_1, \quad [K_3, K_1] = iK_2,$$

the infinite dimensional discrete representation of this algebra is generated from the eigenstate $|k,n>$ of K^2 and $K_3 \equiv K_0$,

$$K^2|k,n>= k(k-1)|k,n>, \qquad K_0|k,n>= (k+n)|k,n>,$$

$$|k,n>= \sqrt{\frac{\Gamma}{n!\Gamma(2k+n)}}(K^+)^n|k,0> .$$

$K_\pm$ generate the coset space $SU(1,1)/U(1)$,

$$\alpha = \exp(zK_+ - z^*K_-) \in SU(1,1)/U(1),$$

$$\alpha|k,0>= (1-zz^*)^k e^{zk_+}|k,0>,$$

so the Bergmann kernel is

$$F(z,z^*) = (1-zz^*)^{-2k},$$

and the Kähler form can be get as

$$\mathbf{K} = \frac{ik}{(1-zz^*)^2} dz \wedge dz^*.$$

Finally, we consider the nonlinear realization of Lie group of dimension n on a symmetric space of dimension d. A Riemannian symmetric space is a homogeneous space with an involutive automorphism. Any group element g of G can be written as

$$g = e^{-i\xi\cdot X} e^{-i\zeta\cdot Y} = \phi h,$$

where $\xi \cdot X = \sum_{i=1}^{d} \xi^i X_i$, $\zeta \cdot Y = \sum_{a=1}^{n-d} \zeta^a Y_a$, $\{Y_a\}$ is the Hermitian generator of H and $\{X_i\}$ is the Hermitian generator of G in G/H, and $\{\xi^i\}$, $\{\zeta^a\}$ are real parameters, then $h \in H,\ \phi \in G/H$.

When two group elements g_1 and g_2 are related by a left multiplication of $h \in H$, $g_1 = g_2 h$, g_1 and g_2 correspond to the same point in coset space G/H and $g_1 \sim g_2$. Therefore the coset space G/H can be parameterized by $\phi = e^{-i\xi X}$ and ξ is a coordinate parameter on the coset manifold.

In the following, we deal with the left multiplication of $g_0 \in G$ to the coset G/H.

$$g_0\, \phi = g_0\, e^{-i\xi X} = e^{-i\xi' X} e^{-i\zeta' Y}, \tag{6.29}$$

where $\xi' = \xi'(\xi, g_0)$ and $\zeta' = \zeta'(\xi, g_0)$ are functions determined by the structure of the group. When $g_0 = h' \in H$, the transformation is

$$\phi \to h'\phi = h'\phi h'^{-1} h' = \phi' h',$$

$$\phi' = h'\phi h'^{-1},$$

i.e.,

$$e^{-i\xi' X} = h' e^{-i\xi X} h'^{-1}.$$

Under the action of h' in the stable subgroup H, the coset is transformed according to the adjoint representation of H and the coset parameter ξ is transformed linearly

$$\xi \to \xi' = D(h')\xi \quad D(h') \in Ad_H.$$

However, when $g_0 \in G$ and $g_0 \notin H$, Eq. 6.29 can not be written in a compact form in general. Only when G/H is a symmetric space can the nonlinear transformations be written out explicitly.

Now we look at cases where G/H is a symmetric space. In group G, there exists an automorphism mapping:

$$Y \to Y, \quad X \to -X.$$

When $g_0 \in G$ and $g_0 \notin H$, the left action of g_0 is

$$\phi \to g_o \phi = \phi' h,$$

after the action of involutive automorphism mapping $g_0 \to g_0'$, $\phi \to \phi^{-1}$ and $h \to h$, we have

$$g_0' \phi^{-1} = \phi'^{-1} h,$$

so one can get

$$\phi'^2 = g_0 \phi^2 g_0'^{-1}, \quad e^{-2i\xi' X} = g_0 e^{-2i\xi X} g_0'^{-1}.$$

This is a nonlinear transformation. For instance, when $g_0 = e^{i\xi X}$, $g_0'^{-1} = g_0 = e^{i\xi X}$, then $\xi' = 0$. There exists a transformation from null to non-zero, which is nonlinear. This is one of the characteristics of symmetric spaces. We can get nonlinear realizations of group G by using symmetric space, which is very important to field theory models taking value on a symmetric space.

Now, we discuss the principal chiral field taking value on group manifold. Because there exist both a left and a right action of group G on the group manifold G itself, we treat the chiral field as a field taking the value on a symmetric manifold $G_L \times G_R/G_V$. For instance, we treat the $SU(3)$ chiral field as a field taking values on $SU(3)_L \times SU(3)_R/SU(3)_V$. For convenience, let

$$G = SU(3)_L \times SU(3)_R, \qquad H = SU(3)_V.$$

And element g of G can be written as

$$g = \Phi \overline{h} \quad (\overline{h} \in H, \Phi \in G/H).$$

There is an involutive automorphism σ,

$$\sigma \overline{h} \sigma = \overline{h}, \quad \sigma \Phi \sigma = \Phi^{-1}.$$

Taking

$$\sigma = \begin{pmatrix} 0 & I \\ I & 0 \end{pmatrix},$$

then

$$\overline{h} = \begin{pmatrix} h & 0 \\ 0 & h \end{pmatrix}, \quad \Phi = \begin{pmatrix} \phi & 0 \\ 0 & \phi^{-1} \end{pmatrix}.$$

Dynamical variable on symmetric space can be written as

$$N(x) = g(x)\sigma g(x)^{-1} = \Phi\sigma\Phi = \begin{pmatrix} 0 & \phi^2 \\ \phi^{-2} & 0 \end{pmatrix} \equiv \begin{pmatrix} 0 & U(x) \\ U^{-1}(x) & 0 \end{pmatrix},$$

where $N(x)$ satisfies

$$N^2(x) = I.$$

The current operator in terms of the field variable on the symmetric space is

$$N(x)dN(x) = \begin{pmatrix} U(x)dU^{-1}(x) & 0 \\ 0 & U^{-1}(x)dU(x) \end{pmatrix}.$$

Thus we can combine the left current and the right current together and take them as one field variable of the symmetric space.

6.7 Nonlinear σ-models, soliton solutions and their geometric meaning

In §4.7 we pointed out that for the chiral model, the self-dual Yang-Mills model and the axisymmetric gravitation model, there exist completely integrable exact solutions. These completely integrable physics systems can be expressed as field theories taking values on symmetric spaces[31]. In the following we take the completely integrable nonlinear σ-model as an example.

Nonlinear σ-model has many applications in theoretical physics. It is a model of which interaction is not introduced by adding an interaction term to the Lagrangian of free field, but instead the interaction is introduced through a geometric approach, i.e., by the connection and curvature which are determined via metric of the manifold. The characteristic of the interaction of this model are very similar to those of the nonabelian gauge field theory in many aspects. For instance, they are asymptotically free, have non trivial topological structure and have soliton solutions etc.. By comparing this section with Chapter 9, where the nonabelian Yang-Mills field is considered, we will have a better understanding of their characteristics.

Dynamics of a physics system is based upon the principle of the least action. The equation of motion can be obtained by the variation principle $\delta I = 0$. The action functional I of the σ-model is

$$I = \frac{1}{2}\int d^d x \; g_{ij} \; \partial_\mu u^i(x) \; \partial^\mu u^j(x), \quad i,j = 1,\cdots,n; \; \mu = 1,\cdots,d,$$

where $u^i (i = 1, \cdots, n)$ are the local coordinates of manifold M and $\{g_{ij}\}$ is its metric. The variation of the action I, i.e., $\delta I = 0$, gives the equation of motion

$$\partial_\mu \partial^\mu u^i + \Gamma^i_{jk}(u) \partial_\mu u^j(x) \partial^\mu u^k = 0, \tag{6.30}$$

where $\Gamma^i_{jk}(u)$ is the Riemann connection determined by the metric of the manifold:

$$\Gamma^i_{jk} = \frac{1}{2} g^{il} \left(\frac{\partial g_{lj}}{\partial u^k} + \frac{\partial g_{lk}}{\partial u^j} - \frac{\partial g_{jk}}{\partial u^l} \right).$$

Eq. 6.30 is a generalization of the usual geodesic equation. In the following, we mainly look at solutions of this equation which gives finite action. Suppose that when $|x| \to \infty$, $u(x)$ has finite limit u_0, we can add a point ∞ to space $\mathbf{R}^d$ such that the whole space can be taken as a d-dimensional sphere:

$$S^d = \mathbf{R}^d \bigcup \{\infty\}.$$

The field $u(x)$ is a mapping from S^d to manifold M

$$u : S^d \to M,$$

and can be labeled by the homotopy equivalence class of the homotopy group $\Pi_d(M)$. There may exist topologically nontrivial solutions if $\Pi_d(M) \neq 0$,(see Chapter 8).

When the manifold has certain symmetries, it could be more convenient to work in the moving frame. For σ-model taking value on the symmetric space $M = G/H$, the whole system has global symmetry group G and gauge group H. On the adjoint representation space of Lie algebra $\mathbf{g}$ of the group G, there exists an automorphism operator σ satisfying $\sigma^2 = 1$ and $\sigma \neq 1$, such that the Lie algebra $\mathbf{g}$ has a Cartan decomposition as

$$\mathbf{g} = \eta + \kappa$$

satisfying

$$[\sigma, \eta] = 0, \quad \{\sigma, \kappa\} = 0.$$

Suppose the coordinates of the base manifold of the field are $\{x^\mu, \mu = 1, 2\}$, field $u(x)$ taking values on the symmetric space G/H can be expressed as

$$u(x) = g(x)H,$$

where H appears as the gauge group and g as the map from the base space x to group G. Using the involutive operator σ of the symmetric space, we can represent the field $N(x)$ on G/H in terms of $g(x)$

$$N(x) = g(x)\sigma g(x)^{-1} \quad \in \quad G/H.$$

Obviously

$$N^2(x) = 1.$$

In this case the Lagrangian of the system can be written as

$$\mathcal{L} = \frac{1}{8} Tr(\partial_\mu N(x) \partial^\mu N(x)). \tag{6.31}$$

Taking the variation of the action under the constraint $N^2 = 1$, we get the equation of motion

$$[\partial_\mu \partial^\mu N(x), N(x)] = 0, \tag{6.32}$$

which can also be written as

$$\partial_\mu K^\mu(x) = 0,$$

where

$$K_\mu(x) = -\frac{1}{2} N(x) \partial_\mu N(x).$$

When taking the global transformation

$$g(x) \to g_0 g(x), \qquad N(x) \to g_0 N(x) g_0^{-1},$$

or the gauge transformation

$$g(x) \to g(x)h(x), \qquad N(x) \to N(x),$$

the Lagrangian is invariant. Using the involutive automorphism operator σ, we have the decomposition of the global G-invariant Maurer-Cartan form

$$a_\mu(x) = g^{-1}(x) \partial_\mu g(x) = h_\mu(x) + k_\mu(x),$$

where

$$h_\mu = \frac{1}{2}\{a_\mu(x), \sigma\}\sigma,$$

$$k_\mu = \frac{1}{2}[a_\mu(x), \sigma]\ \sigma = g^{-1}(x) K_\mu(x) g(x).$$

It is easy to prove the following relations

$$[\sigma, h_\mu(x)] = 0, \qquad \text{i.e., } h_\mu(x) \in \eta.$$

$$\{\sigma, k_\mu(x)\} = 0, \qquad \text{i.e., } k_\mu(x) \in \kappa.$$

The Lagrangian can be written as

$$\mathcal{L} = -\frac{1}{2}(k_\mu(x), k^\mu(x)),$$

where the (,) is a G-invariant metric on coset G/H, which is induced from the Killing-Cartan metric of the Lie algebra $\mathbf{g}$. Consequently, the equation of motion Eq. 6.32 can be rewritten as

$$\mathcal{D}_\mu k^\mu(x) \equiv \partial_\mu k^\mu(x) + [h_\mu(x), k^\mu(x)] = 0. \tag{6.33}$$

In addition, the Maurer-Cartan equation, satisfied by pure gauge $a_\mu(x)$, becomes the so called Codazzi equation

$$\partial_\mu h_\nu - \partial_\nu h_\mu + [h_\mu, h_\nu] + [k_\mu, k_\nu] = 0, \tag{6.34}$$

$$\mathcal{D}_\mu k_\nu - \mathcal{D}_\nu k_\mu = 0. \tag{6.35}$$

Let

$$^\star k_\mu = \varepsilon_{\mu\nu} k^\nu \quad (\varepsilon_{01} = -\varepsilon_{10} = 1),$$

the Codazzi equation Eq. 6.35 can be written as

$$\mathcal{D}_\mu \, {}^\star k^\mu = 0. \tag{6.36}$$

And Eq. 6.34 can be written as

$$\partial_\mu h_\nu - \partial_\nu h_\mu + [h_\mu, h_\nu] - [{}^\star k_\mu, {}^\star k_\nu] = 0. \tag{6.37}$$

The equation of motion Eq. 6.32 and the Codazzi equation Eq. 6.36 (correspondingly Eq. 6.34 and Eq. 6.37) are dual to each other under

$$k_\mu \leftrightarrow i \, {}^\star k_\mu.$$

If k_μ are self-dual

$$k_\mu = \pm i \, {}^\star k_\mu,$$

then Codazzi equation Eq. 6.36 and the equation of motion Eq. 6.33 are equivalent.

As mentioned above, both k_μ and $i \, {}^\star k_\mu$ satisfy the same linear equation. Through duality, we can introduce a continuous dual transformation, i.e., the real linear combination of k_μ and ${}^\star k_\mu$:

$$\tilde{k}_\mu(x, \gamma) = \cosh\varphi \; k_\mu(x) + \sinh\varphi \; {}^\star k_\mu(x),$$

where

$$\cosh\varphi = \tfrac{1}{2}(\gamma + \gamma^{-1}),$$
$$\sinh\varphi = \tfrac{1}{2}(\gamma - \gamma^{-1}),$$

and γ is a real parameter. When $\gamma = 1$, we have

$$\tilde{k}_\mu(x, 1) = k_\mu(x).$$

It is easy to prove that h_μ and $\tilde{k}_\mu(x, \gamma)$ satisfy the same equations as h_μ and k_μ

$$\mathcal{D}_\mu \tilde{k}^\mu(x, \gamma) = 0,$$
$$\partial_\mu h^\nu - \partial_\nu h^\mu + [h_\mu, h^\nu] + [\tilde{k}_\mu(x, \gamma), \tilde{k}_\nu(x, \gamma)] = 0.$$

Therefore $h_\mu(x) + \tilde{k}_\mu(x, \gamma)$ can be represented by a pure gauge,

$$h_\mu(x) + \tilde{k}_\mu(x, \gamma) = \psi(x, \gamma) \partial_\mu \psi^{-1}(x, \gamma), \tag{6.38}$$

it can be expressed as

$$\mathcal{D}_\mu\psi(x,\gamma) \equiv (\partial_\mu + h_\mu(x))\psi(x,\gamma) = -(\tilde{k})_\mu(x,\gamma)\psi(x,\gamma). \tag{6.39}$$

Take a gauge transformation of g by the multiplication of a local $v(x) \in H$ from right

$$g(x) \to g'(x) = g(x)v(x),$$

then the Maurer-Cartan potential a_μ and its projections $h_\mu(x)$ and $k_\mu(x)$ have the gauge transformation correspondingly

$$\begin{aligned} a_\mu(x) &\to a'_\mu(x) = v(x)^{-1}a_\mu(x)v(x) + v(x)^{-1}\partial_\mu v(x), \\ h_\mu(x) &\to h'_\mu(x) = v(x)^{-1}h_\mu(x)v(x) + v(x)^{-1}\partial_\mu v(x), \\ k_\mu(x) &\to k'_\mu(x) = v(x)^{-1}k_\mu(x)v(x), \end{aligned} \tag{6.40}$$

i.e., $a_\mu(x)$ and $h_\mu(x)$ transform as gauge potential while $k_\mu(x)$ is gauge covariant. In general, $v(x)$ takes values in the stable subgroup H. Now, we generalize $v(x)$ to taking values on the whole Lie group G. Let $v(x) = g(x)^{-1}$, then

$$\begin{aligned} a'_\mu(x) &= g(x)a_\mu(x)g(x)^{-1} + g(x)\partial_\mu g(x)^{-1}, \\ h'_\mu(x) &= g(x)h_\mu(x)g(x)^{-1} + g(x)\partial_\mu g(x)^{-1} \equiv H_\mu(x) = -K_\mu(x), \\ k'_\mu(x) &= g(x)k_\mu(x)g(x)^{-1} = K_\mu(x). \end{aligned}$$

Let

$$U(x,\gamma) = g(x)\psi(x,\gamma),$$

from Eq. 6.38 we have

$$\partial_\mu U(x,\gamma) = -(H_\mu(x) + \tilde{K}_\mu(x,\gamma))U(x,\gamma), \tag{6.41}$$

where

$$\begin{aligned} \tilde{K}_\mu(x,\gamma) &= g(x)\tilde{k}_\mu(x,\gamma)g^{-1}(x) \\ &= \cosh\varphi K_\mu(x) + \sinh\varphi\ {}^*K_\mu(x). \end{aligned}$$

When both $H_\mu(x)$ and $K_\mu(x)$ are in terms of $N(x)$, Eq. 6.41 is

$$\partial_\mu U(x,\gamma) = \frac{1}{2}(N\partial_\mu N(\cosh\varphi - 1) + \epsilon_{\mu\nu}N\partial^\nu N\sinh\varphi)U(x,\gamma). \tag{6.42}$$

From the integrable condition of the above linear equations, the nonlinear equation of motion for chiral field $N(x)$ can be obtained. Hence the linear equations above are called the auxiliary linear equations of the nonlinear σ-model. These auxiliary equations contain an arbitrary real parameter γ, this completely integrable field theory model can be solved by the inverse scattering method. Eq. 6.42 is usually called Lax pair and can be represented as

$$\partial_\mu U(x,\gamma) = L_\mu(x,\gamma)U(x,\gamma).$$

The Lax pair plays an important role in integrable field theory. In the last part of this section, we will take the $O(N)$ σ-model as an example to analyse its Poisson-Lie structure. Note that Eqs. (6.38), (6.41) are Lax pair representations in different frames. When $v(x) \in H$, the right action is equivalent to a gauge transformation. But when $v(x)$ takes values in the whole Lie group G, the right multiplication is equivalent to a change of the frame. One usually starts from the Lax pair Eq. 6.41, which is called the Lax pair in fixed frame. In the following we will see that one can also start from the Lax pair Eq. 6.38, which is called the Lax pair in moving frame. Sometimes starting from the covariant formulation Eq. 6.39 is more convenient, because it is easier to manifest the symmetry of the system and to take the dressing transformation to find new solutions in the gauge covariant formulation.

When taking the generalized gauge transformation Eq. 6.40 and letting $v(x)$ be the solution $\phi(x,\gamma)$ which satisfies Eq. 6.38, we get

$$\begin{aligned} \hat{H}_\mu <\gamma> &= \phi^{-1}(x,\gamma)h_\mu(x)\phi(x,\gamma) + \phi^{-1}(x,\gamma)\partial_\mu\phi(x,\gamma), \\ \hat{K}_\mu <\gamma> &= \phi^{-1}(x,\gamma)\tilde{k}_\mu(x,\gamma)\phi(x,\gamma). \end{aligned} \tag{6.43}$$

From Eq. 6.38 it is easy to prove

$$\hat{K}_\mu <\gamma> = -\hat{H}_\mu <\gamma> = -\frac{1}{2}\hat{N} <\gamma> \partial_\mu \hat{N} <\gamma>,$$

where

$$\hat{N} <\gamma> = \phi^{-1}(x,\gamma)\ \sigma\ \phi(x,\gamma).$$

From gauge covariance, we know that $\hat{K}^\mu <\gamma>$ satisfy

$$\partial_\mu \hat{K}^\mu <\gamma> + [\hat{H}_\mu <\gamma>, \hat{K}^\mu <\gamma>] = \partial_\mu \hat{K}^\mu <\gamma> = 0,$$

i.e., $\hat{N} <\gamma>$ is the new solution of the chiral field equation Eq. 6.32 and the corresponding conserved current is $\hat{K}^\mu <\gamma>$. The relation between the new solution $\hat{N} <\gamma>$ and the old solution $N(x)$ is

$$\hat{N} <\gamma> = U^{-1}(x,\gamma)N(x)U(x,\gamma),$$

where

$$U(x,\gamma) = g(x)\ \phi(x,\gamma) \qquad (U(x,1) = 1)$$

is called the dual transformation operator of solutions of the linear equation Eq. 6.41 or Eq. 6.42.

In the integrable field theory model, the Hamiltonian method is very useful. In terms of the Hamiltonian approach, one can study the Poisson brackets between L-matrices (the spatial components of the Lax Pair). For many integrable models, such as WZNW model and the Toda system, this approach usually leads to a Lie-Poisson algebra of the form

$$\{L(x,\lambda) \overset{\otimes}{,} L(y,\mu)\} = [r(\lambda,\mu), L(x,\lambda) \overset{\otimes}{,} 1 + 1 \overset{\otimes}{,} L(y,\mu)]\delta(x-y), \tag{6.44}$$

where the antisymmetric field-independent r-matrix $r \in g \otimes g$ satisfies the classical Yang-Baxter equation

$$[r_{12}(\lambda,\mu), r_{13}(\lambda,\nu)] + [r_{12}(\lambda,\mu), r_{23}(\mu,\nu)] + [r_{13}(\lambda,\nu), r_{23}(\mu,\nu)] = 0,$$

where

$$\begin{aligned} r_{12} &= \sum_{\alpha\beta} r^{\alpha\beta} T_\alpha \otimes T_\beta \otimes 1, \qquad T_\alpha \text{ and } T_\beta \in \mathbf{g}, \\ r_{13} &= \sum_{\alpha\beta} r^{\alpha\beta} T_\alpha \otimes 1 \otimes T_\beta, \\ r_{23} &= \sum_{\alpha\beta} r^{\alpha\beta} 1 \otimes T_\alpha \otimes T_\beta. \end{aligned}$$

But in many cases of the nonlinear σ-model, there appear the Schwinger terms (which contain the derivative of the delta function). this kind of Lie-Poisson structure has been developed by Maillet[32]. In his r,s-matrix formula, equation Eq. 6.44 is replaced by

$$\begin{aligned} \{L(x,\lambda) \overset{\otimes}{,} L(y,\mu)\} = &-[r(x,\lambda,\mu), L(x,\lambda)\otimes 1 + 1 \otimes L(x,\mu)]\delta(x-y) \\ &+[s(x,\lambda,\mu), L(x,\lambda)\otimes 1 - 1 \otimes L(x,\mu)]\delta(x-y) \\ &-([r(x,\lambda,\mu) - r(y,\lambda,\mu) + s(x,\lambda,\mu) + s(y,\lambda,\mu))\delta'(x-y), \end{aligned} \tag{6.45}$$

with an antisymmetric r-matrix and a symmetric s-matrix which in general depend explicitly on the fields of the theory. For example, the $O(N)$ σ-model, $CP(N)$ σ-model and the complex Sine-Gordon model etc. And that are called non-ultralocal models. By using the current algebra, Forger et. al.[33] find the general expressions for these r,s-matrices. Maillet starts from the Lax potential Eq. 6.42 and the Lagrangian Eq. 6.31, in which $N(x)$ satisfies constraint condition $N^2(x) = 1$ and the computation is exceedingly tedious as one has to use the Dirac bracket. Forger et al. use the local coordinate system on the manifold $M = G/H$, and only the canonical Poisson bracket is needed. In terms of the symmetries of the manifold and the moment map

$$\begin{aligned} M &\to \mathbf{g}^\star, \\ u &\to J(u), \end{aligned}$$

then taking the mapping $\mathbf{g}^\star \to \mathbf{g}$ by the Killing-Cartan metric on the Lie algebra, one first obtains the current algebra of Poisson brackets among the conserved currents J_μ, and then by the current algebra, one gets the r,s-matrices which contain the field variables.

We use the natural moving frame on the symmetric manifold $M = G/H$, and start directly from Eq. 6.38 instead of Eq. 6.41, so it is more straightforward to get the r,s-matrices explicitly. In the following we take the $O(N)$ σ-model as an example[34].

Let us introduce the notation

$$R_{jk}(\theta) = \exp(-\theta T_{jk}),$$

which is a finite rotation operation in the $e_j \wedge e_k$ plane, where T_{jk} are the skewhermitian generator of $SO(N)$ and satisfy

$$[T_{jk}, T_{lm}] = \delta_{ul} T_{jm} + \delta_{jm} T_{kl} - \delta_{jl} T_{km} - \delta_{km} T_{jl}.$$

Any $SO(N)$ finite rotation operation g can be expressed as

$$g = R(\Theta) D(\Phi),$$

$$R(\Theta) = R_1(\theta_1) R_2(\theta_2) \cdots R_{N-1}(\theta_{N-1}) \in \text{ coset space } O(N)/O(N-1),$$

where $R_k(\theta) \equiv R_{k,k+1}(\theta)$, θ_i have the boundary conditions

$$\begin{aligned} 0 \le \theta_1 < 2\pi, \\ 0 \le \theta_j < \pi \qquad (j \neq 1), \end{aligned}$$

and $D(\phi) \in O(N-1)$.

Fixing the gauge as (the Schwinger gauge),

$$D(\phi) = 1,$$

and we get the Maurer-Cartan form

$$g^{-1} dg = \sum_{i=1}^{N-1} d\theta_i T_i s_{i+1} s_{i+2} \cdots s_{N-1} + \sum_{i=1}^{N-1} d\theta_i \sum_{j=i+1}^{N-1} T_{ij} s_{i+1} \cdots s_{j-1} c_j,$$

where $T_i \equiv T_{iN}$, $c_i \equiv \cos\theta_i$ and $s_i \equiv \sin\theta_i$. In the fundamental representation, the matrix elements of T_{jk} are

$$(T_{jk})_{lm} = \delta_{jl}\delta_{km} - \delta_{kl}\delta_{jm} \equiv (E_{jk})_{lm} - (E_{kj})_{lm}.$$

If we choose the involutive operator

$$\sigma = \left(\begin{array}{c|c} I & 0 \\ \hline 0 & -1 \end{array} \right),$$

then

$$\{T_{ij},\ i < j < N\} \in \eta, \quad \{T_i,\ i = 1, \cdots, N-1\} \in \kappa.$$

So that we have

$$\begin{aligned} h_\mu &= \textstyle\sum_{i=1}^{N-2} \sum_{j=i+1}^{N-1} (\partial_\mu \theta_i) T_{ij} s_{i+1} \cdots s_{j-1} c_j, \\ k_\mu &= \textstyle\sum_{i=1}^{N-1} \partial_\mu \theta_i T_i s_{i+1} \cdots s_{N-1}. \end{aligned}$$

Then from the Lagrangian

$$\mathcal{L} = \frac{1}{2} \sum_{i=1}^{N-1} \partial^\mu \theta_i \partial_\mu \theta_i s_{i+1}^2 \cdots s_{N-1}^2,$$

we get the canonical momentum

$$\pi_i = \frac{d\theta_i}{dt} s_{i+1}^2 \cdots s_{N-1}^2,$$

here θ_i and π_j satisfy the canonical Poisson brackets:

$$\begin{gathered}\{\theta_i(x), \pi_j(y)\} = \delta_{ij}\delta(x-y),\\ \{\theta_i(x), \theta_j(y)\} = 0, \quad \{\pi_i(x), \pi_j(y)\} = 0.\end{gathered}$$

From Eq. 6.38, the Lax potential in this moving frame is

$$L(x, \lambda) = -h_1(x) - \cosh\varphi k_1(x) + \sinh\varphi k_0(x).$$

The Poisson bracket between the Lax potentials is

$$\begin{aligned}\{L(x,\lambda) \stackrel{\otimes}{,} L(y,\mu)\} &= \sinh a\varphi_1 \sinh\varphi_2 \{k_0(x) \stackrel{\otimes}{,} k_0(y)\} + \sinh\varphi_1 \cosh\varphi_2 \{k_0(x) \stackrel{\otimes}{,} k_1(y)\}\\ &+ \cosh\varphi_1 \sinh\varphi_2 \{k_1(x) \stackrel{\otimes}{,} k_0(y)\} - \sinh\varphi_1 \{k_0(x) \stackrel{\otimes}{,} h_1(y)\} - \sinh\varphi_2 \{h_1(x) \stackrel{\otimes}{,} k_0(y)\}\end{aligned} \tag{6.46}$$

After introducing the following notations

$$\begin{aligned}\Gamma^i(x) &\equiv \sum_{j=i+1}^{N-1} \frac{c_j}{s_j s_{j+1} \cdots s_{N-1}} T_{ij} \in \eta,\\ \Gamma_1(x) &\equiv \sum_{i=1}^{N-1} \Gamma^i(x) \otimes T_i, \qquad \Gamma_2(x) \equiv \sum_{i=1}^{N-1} T_i \otimes \Gamma^i(x),\\ P_\kappa(x) &\equiv \sum_{i=1}^{N-1} T_i \otimes T_i, \qquad P_\eta(x) \equiv \sum_{i=1}^{N-2} \sum_{j=i+1}^{N-1} T_{ij} \otimes T_{ij},\end{aligned}$$

and remembering the identity

$$(f(x) - f(y))\delta'(x-y) = -f'(x)\delta(x-y),$$

the Poisson bracket among k_0, k_1 and h_1 can be obtained as follows

$$\begin{aligned}\{k_0(x) \stackrel{\otimes}{,} k_0(y)\} &= ([\Gamma_2(x), 1 \otimes k_0(x)] - [\Gamma_1(x), k_0(x) \otimes 1])\delta(x-y),\\ \{k_0(x) \stackrel{\otimes}{,} k_1(y)\} &= ([\Gamma_2(x), 1 \otimes k_1(x)] + [P_k(x), 1 \otimes h_1(x)])\delta(x-y)\\ &\quad + P_k\delta'(x-y),\\ \{k_1(x) \stackrel{\otimes}{,} k_0(y)\} &= (-[\Gamma_1(x), k_1(x) \otimes 1] - [P_k(x), h_1(x) \otimes 1])\delta(x-y)\\ &\quad + P_k\delta'(x-y),\\ \{k_0(x) \stackrel{\otimes}{,} h_1(y)\} &= ([\Gamma_2(x), 1 \otimes h_1(x)] + [P_k(x), 1 \otimes k_1(x)])\delta(x-y)\\ &\quad + \Gamma_2(x)\delta'(x-y),\\ \{h_1(x) \stackrel{\otimes}{,} k_0(y)\} &= (-[\Gamma_1(x), h_1(x) \otimes 1] - [P_k(x), k_1(x) \otimes 1])\delta(x-y)\\ &\quad + \Gamma_1(x)\delta'(x-y),\\ \{h_1(x) \stackrel{\otimes}{,} k_1(y)\} &= \{k_1(x) \stackrel{\otimes}{,} h_1(y)\} = 0.\end{aligned}$$

Substituting the above results into Eq. 6.46 and comparing the Schwinger terms with those in Eq. 6.45, we immediately get the matrix $s(x, \lambda, \mu)$:

$$s(x, \lambda, \mu) = \frac{1}{2}\sinh(\varphi_1 + \varphi_2)P_k + \frac{1}{2}\sinh\varphi_1\Gamma_2(x) + \frac{1}{2}\sinh\varphi_2\Gamma_1(x).$$

Then propose the ansatz for $r(x, \lambda, \mu)$:

$$r(x, \lambda, \mu) = A\ P_k + B\ P_h + \frac{1}{2}\sinh\varphi_1\Gamma_2(x) - \frac{1}{2}\sinh\varphi_2\Gamma_1(x),$$

and use the following identities

$$\begin{cases} [P_k, k_\mu \otimes 1] + [P_h, 1 \otimes k_\mu] = 0, \\ [P_h, k_\mu \otimes 1] + [P_k, 1 \otimes k_\mu] = 0, \\ [P_k, h_\mu \otimes 1 + 1 \otimes h_\mu] = 0, \\ [P_h, h_\mu \otimes 1 + 1 \otimes h_\mu] = 0, \end{cases}$$

we arrive at the following solution:

$$A = -\frac{1}{2}\frac{\sinh^2\varphi_1 + \sinh^2\varphi_2}{\sinh(\varphi_1 - \varphi_2)}, \qquad B = -\frac{\sinh\varphi_1\sinh\varphi_2}{\sinh(\varphi_1 - \varphi_2)},$$

so $r(x, \lambda, \mu)$ matrix can be obtained as

$$\begin{aligned} r(x, \lambda, \mu) = & -\frac{\sinh\varphi_1\sinh\varphi_2}{\sinh(\varphi_1 - \varphi_2)}C - \frac{1}{2}\frac{(\sinh\varphi_1 - \sinh\varphi_2)^2}{\sinh(\varphi_1 - \varphi_2)}P_k \\ & + \frac{1}{2}\sinh\varphi_1\Gamma_2(x) - \frac{1}{2}\, sinh\varphi_2\Gamma_1(x), \end{aligned}$$

where $C = P_k + P_h$ is the Casimir tensor of the Lie algebra $so(N)$.

Let us take a look at how the r- and s- matrices change under gauge transformations. First we calculate the Poisson bracket between $k'_\mu(x), k'_\mu(y)$ after the gauge transformation Eq. 6.40. we get

$$\begin{aligned} &\{k'_0(x) \overset{\otimes}{,} k'_0(y)\} = v^{-1}(x) \otimes v_{-1}(y)(\{k_0(x) \overset{\otimes}{,} k_0(y)\} - [k_0(x) \otimes 1, \alpha] \\ &\quad -[1 \otimes k_0(y), \beta])v(x) \otimes v(y), \\ &\{k'_0(x) \overset{\otimes}{,} k'_1(y)\} = v^{-1}(x) \otimes v_{-1}(y)(\{k_0(x) \overset{\otimes}{,} k_1(y)\} \\ &\quad -[1 \otimes k_1(x), \beta])v(x) \otimes v(y), \\ &\{k_0(x) \overset{\otimes}{,} h_1(y)\} = v^{-1}(x) \otimes v_{-1}(y)(\{k_0(x) \otimes h_1(y)\} \\ &\quad -[1 \otimes h_1(y), \beta])v(x) \otimes v(y), \end{aligned}$$

where

$$\begin{aligned} &\alpha = v(x) \otimes 1\{v^{-1}(x) \otimes k_0(y)\} = \textstyle\sum_{i=1}^{N-1} + \frac{-v\partial_i v^{-1}}{2s_{i+1}\cdots s_{N-1}} \otimes T^i\delta(x-y), \\ &\beta = 1 \otimes v(y)\{k_0(x) \otimes v^{-1}(y)\} = \textstyle\sum_{i=1}^{N-1} T^i \otimes \frac{-v\partial_i v^{-1}}{2s_{i+1}\cdots s_{N-1}}\delta(x-y), \end{aligned}$$

so the r- and s- matrices are changed to

$$\begin{aligned}
&r(x,\lambda,\mu)\delta(x-y) \longrightarrow r^v(x,\lambda,\mu)\delta(x-y)\\
&= v^{-1}(x)\otimes v^{-1}(y)(r(x,\lambda,\mu)+\tfrac{1}{2}(\sinh\varphi_1\beta+\sinh\varphi_2\alpha))v(x)\otimes v(y),\\
&s(x,\lambda,\mu)\delta(x-y) \longrightarrow s^v(x,\lambda,\mu)\delta(x-y)\\
&= v^{-1}(x)\otimes v^{-1}(y)(s(x,\lambda,\mu)+\tfrac{1}{2}(\sinh\varphi_1\beta-\sinh\varphi_2\alpha))v(x)\otimes v(y).
\end{aligned}$$

Finally we have

$$\begin{aligned}
r^v(x,\lambda,\mu) &= -\frac{\sinh\varphi_1\sinh\varphi_2}{\sinh(\varphi_1-\varphi_2)}C-\frac{1}{2}\frac{(\sinh\varphi_1-\sinh\varphi_2)^2}{\sinh(\varphi_1-\varphi_2)}P_k\\
&+ \frac{1}{2}\sinh\varphi_1\Gamma_2^v(x)-\frac{1}{2}\sinh\varphi_2\Gamma_1^v(x),\\
s^v(x,\lambda,\mu) &= \frac{1}{2}\sinh(\varphi_1+\varphi_2)P_k+\frac{1}{2}\sinh\varphi_1\Gamma_2^v(x)+\frac{1}{2}\sinh\varphi_2\Gamma_1^v(x),
\end{aligned}$$

where

$$\begin{aligned}
\Gamma_1^v(x) &= \sum_{i=1}^{N-1}\Gamma_v^i(x)\otimes v^{-1}T_iv\\
\Gamma_2^v(x) &= \sum_{i=1}^{N-1}v^{-1}T_iv\otimes\Gamma_v^i(x)\\
\Gamma_v^i(x) &= v^{-1}(x)(\Gamma^i+\frac{1}{s_{i+1}s_{i+2}\cdots s_{N-1}}\partial_i)v(x). \qquad (6.47)
\end{aligned}$$

So we see that $\Gamma^i(x)$ is just a Riemannian connection on the target manifold, as the way it changes under a gauge transformation, it is the same as a connection. For instance, if one takes

$$v(x)=R_{N-2}(\mp\theta_{N-2})\cdots R_2(-\theta_2)R_1(-\theta_1),$$

from Eq. 6.47, we obtain

$$\Gamma_v^i(x)=\frac{\mp 1-\cos\theta_{N-1}}{\sin\theta_{N-1}}v(x)T^{i,N-1}v^{-1}(x),$$

which is exactly the Riemannian connection in geodesic gauge (Wu-Yang gauge). When one takes the generalized gauge transformation by replacing $v^{-1}(x)$ by $g(x)$, noting

$$\begin{cases}(1\otimes g^{-1}(y))\{k_0(x)\otimes g(y)\}=-(P_k+\Gamma_2(x))\delta(x-y),\\ (g^{-1}(x)\otimes 1)\{g(x)\otimes k_0(y)\}=(P_k+\Gamma_1(x))\delta(x-y),\end{cases}$$

and replacing $\sinh\varphi_1$, $\cosh\varphi_1$ and $\sinh\varphi_2$, $\cosh\varphi_2$ by λ and μ respectively, one gets

$$\begin{cases}r^g(x,\lambda,\mu)=-\frac{2\lambda\mu}{(1-\lambda\mu)(\lambda-\mu)}C-\frac{2(1+\lambda\mu)(\lambda-\mu)}{(1-\lambda\mu)(1-\lambda^2)(1-\mu^2)}J(x),\\ s^g(x,\lambda,\mu)=-\frac{2(\lambda+\mu)}{(1-\lambda^2)(1-\mu^2)}J(x),\end{cases}$$

where the field dependent term

$$J(x) = g(x) \otimes g(x) P_k g^{-1}(x) \otimes g^{-1}(x).$$

The r^g- and s^g- matrices are nothing but those given by M. Forger[33].

Notations and Formulae

	Type [1,0]	Type [0,1]
The bases of covector	$dz^\alpha = dx^\alpha + idy^\alpha$	$d\bar{z}^\alpha = dx^\alpha - idy^\alpha$
The bases of vector	$\partial_\alpha = \frac{1}{2}(\frac{\partial}{\partial_{x^\alpha}} - i\frac{\partial}{\partial_{y^\alpha}})$	$\partial_{\bar{\alpha}} = \frac{1}{2}(\frac{\partial}{\partial_{x^\alpha}} + i\frac{\partial}{\partial_{y^\alpha}})$

$$T^*(M) \otimes C = T^*_{10}(M) \oplus T^*_{01}(M) \equiv T^*_c(M) \oplus \bar{T}^*_c(M).$$

$$\theta_\alpha = \sigma^\alpha + i\omega^\alpha \quad \in \ T^*_{10}(M), \qquad \bar{\theta}_\alpha = \sigma^\alpha - i\omega^\alpha \quad \in \ T^*_{01}(M).$$

- The integrability condition of almost complex structure.

The following nine conditions are equivalent to each other.

I. The differential forms approach:

1).

$$d\theta^\alpha \in \Lambda^{2,0} \oplus \Lambda^{1,1}, \quad \forall \theta^\alpha \in \Lambda^{1,0}.$$

2).

$$d\bar{\theta}^\alpha \in \Lambda^{1,1} \oplus \Lambda^{0,2}, \quad \forall \bar{\theta}^\alpha \in \Lambda^{0,1},$$

3).

$$d\alpha_{p,q} \in \Lambda^{p+1,q} \oplus \Lambda^{p,q+1}, \quad \forall \alpha_{p,q} \in \Lambda^{p,q}, \quad p,q = 0,1,\cdots,n \ .$$

4).

$$d = \partial + \bar{\partial}, \qquad \partial = dz^\alpha \partial_\alpha, \qquad \bar{\partial} = d\bar{z}^\alpha \partial_{\bar{\alpha}}.$$

5).

$$\bar{\partial}^2 = 0,$$

6).

$$\partial^2 = 0.$$

II. The tangent vector fields approach:

7). The [1,0]-type tensor field is closed under Lie bracket operation.

8). The [0,1]-type tensor field is closed under Lie bracket operation.

III. The intrinsic approach:

9). The Nijenhuis torsion tensor N of the almost complex structure J is vanshing, i.e.,

$$N(X,Y) = [JX, JY] - [X,Y] - J[X, JY] - J[JX,Y] = 0, \quad \text{for } X,Y \in \mathcal{X}(M).$$

- On complex manifold, the affine connection satisfy:

  (1) J is paralled transported by the covariant derivative:

$$\nabla_X(JY) = J\nabla_X Y, \qquad X,Y \in \mathcal{X}(M).$$

$$\text{so} \quad R(X,Y)\cdot J = J\cdot R(X,Y), \qquad R(JX,JY) = R(X,Y),$$

  (2) Torsion free:

$$T(X,Y) = \nabla_X Y - \nabla_Y X - [X,Y] = \frac{1}{4}N(X,Y) = 0.$$

The infinitesimal automorphism Lie algebra is stable with respect to J, i.e.,

$$J[X,Y] = [JX,Y] = [X,JY],$$

- On Hermite manifold,
  let $\quad h_{\alpha\bar\beta} = h_{\bar\beta\alpha} = G(\partial_\alpha, \partial_{\bar\beta}).$

$$\begin{aligned} ds^2 &= 2h_{\alpha\bar\beta}dz^\alpha d\bar z^\beta. \\ K &= ih_{\alpha\bar\beta}dz^\alpha d\bar z^\beta = \bar K. \end{aligned}$$

$$*1 = \frac{K^n}{n!} = i^n(-1)^{\frac{1}{2}n(n-1)}h dz^1\wedge\cdots\wedge dz^n\wedge d\bar z^1\wedge\cdots\wedge d\bar z^n, \quad h = det\left(h_{\alpha\bar\beta}\right).$$

$$**\,\alpha_{p,q} = (-1)^{p+q}\alpha_{p,q}.$$

$$(\alpha_{p,q}, \beta_{p,q}) = \int \alpha_{p,q}\wedge *\overline{\beta_{p,q}} = \frac{1}{p!q!}\int f_{\alpha_1\cdots\alpha_p\bar\beta_1\cdots\bar\beta_q} g^{\alpha_1\cdots\alpha_p\bar\beta_1\cdots\bar\beta_q} * 1.$$

$$(\alpha,\beta) = \left(\bar\beta, \bar\alpha\right).$$

$$d = \partial + \bar\partial, \qquad \partial = dz^\alpha\partial_\alpha, \qquad \bar\partial = d\bar z^\alpha\partial_{\bar\alpha}.$$

$$\delta = -\star d\star = \theta + \bar\theta.$$

$$\theta = -\star\bar\partial\star: \quad \Lambda^{pq}\longrightarrow\Lambda^{p-1,q}.$$

$$\bar\theta = -\star\partial\star: \quad \Lambda^{pq}\longrightarrow\Lambda^{p,q-1}.$$

$$(d\alpha,\beta) = (\alpha,\delta\beta), \qquad (\partial\alpha,\beta) = (\alpha,\theta\beta), \qquad (\bar\partial\alpha,\beta) = (\alpha,\bar\theta\beta).$$

Here,

$$\delta = -\star d\star = \theta + \bar\theta, \qquad \theta = -\star\bar\partial\star, \qquad \bar\theta = -\star\partial\star.$$

- On Kähler manifold

$$dK = \delta K = 0, \qquad K = -\frac{i}{2}\partial\bar{\partial}F(z,\bar{z}), \qquad h_{\alpha\bar{\beta}} = \frac{\partial^2 F(z,\bar{z})}{\partial z^\alpha \partial \bar{z}^{beta}}.$$

$$\Delta \equiv d\delta + \delta d = \Diamond = \bar{\Diamond}, \qquad \Diamond \equiv 2(\partial\theta + \theta\partial), \qquad \bar{\Diamond} \equiv 2(\bar{\partial}\bar{\theta} + \bar{\theta}\bar{\partial}).$$

- On complex manifold there exist various affine connection satisfying different conditions:

	Torsion free	complex preserving	Metric preserving	Holonomy
almost complex	$\Gamma^i_{jk} = \Gamma^i_{kj}$	$\Gamma^\alpha_{j\bar{\beta}} = 0 = \Gamma^{\bar{\alpha}}_{j\beta}$	no	$GL(n,c)$
Riemann.	$\Gamma^i_{jk} = \Gamma^i_{kj}$	no	$\nabla_i g_{kl} = 0$	$SO(2n)$
Hermit.	$[2,0]-$ torsion	$\Gamma^\alpha_{j\bar{\beta}} = 0 = \Gamma^{\bar{\alpha}}_{j\beta}$ $[1,0]$ type$\Gamma^\alpha_{\bar{\gamma}\beta} = 0$	$\nabla_i h_{\alpha\bar{\beta}} = 0$ $\Gamma^\alpha_{\beta\gamma} = h^{\bar{\delta}\alpha}\partial_\beta h_{\gamma\bar{\delta}}$	$U(n)$
Kähler	$\Gamma^i_{jk} = \Gamma^i_{kj}$	$\Gamma^\alpha_{j\bar{\beta}} = 0 = \Gamma^{\bar{\alpha}}_{j\beta}$ $\Gamma^\alpha_{\bar{\gamma}\beta} = 0$	$\nabla_i h_{\alpha\bar{\beta}} = 0$ $\Gamma^\alpha_{\beta\gamma} = h^{\bar{\delta}\alpha}\partial_\beta h_{\gamma\bar{\delta}}$	$U(n)$

- Riemannian connection : $\Gamma^i_{jk} = \frac{1}{2}g^{jl}(\partial_j g_{kl} + \partial_k g_{jl} - \partial_l g_{jk})$.

- Hermitian $[1,0]$-connection: ([2.0]-torsion, [1.1]-curvature)

$$\Gamma^\alpha_{\beta\gamma} = h^{\bar{\delta}\alpha}\partial_\beta h_{\gamma\bar{\delta}}.$$

- On Kähler manifold

$$\Gamma^\alpha_{\beta\gamma} = \Gamma^\alpha_{\gamma\beta} = h^{\bar{\delta}\alpha}\partial_\beta h_{\gamma\bar{\delta}}, \qquad \Gamma^{\bar{\alpha}}_{\bar{\beta}\bar{\gamma}} = \Gamma^{\bar{\alpha}}_{\bar{\beta}\gamma}.$$

$$\begin{aligned} R_{\bar{\alpha}\beta\gamma\bar{\delta}} &= -\frac{\partial^2 h_{\beta\bar{\alpha}}}{\partial z^\gamma \partial \bar{z}^\delta} + h^{\bar{\lambda}\beta}\frac{\partial h_{\beta\bar{\lambda}}\partial h_{\rho\bar{\alpha}}}{\partial z^\gamma \partial \bar{z}^\delta} \\ &= -\frac{\partial^4 F}{\partial \bar{z}^\alpha \partial z^\beta \partial z^\gamma \partial \bar{z}^\delta} + h^{\bar{\lambda}\rho}\frac{\partial^3 F}{\partial \bar{z}^\lambda \partial z^\beta \partial z^\gamma}\frac{\partial^3 F}{\partial z^\rho \partial \bar{z}^\alpha \partial \bar{z}^\delta} \\ &= R_{\bar{\alpha}\gamma\beta\bar{\delta}}. \end{aligned}$$

$$\Re_{\alpha\bar\beta} = -\frac{\partial \Gamma^{\gamma}_{\alpha\gamma}}{\partial \bar z^{\beta}} = -\frac{\partial^2 \ln h}{\partial z^{\alpha}\partial \bar z^{\beta}}.$$
$$\rho = -i\Re_{abeta}dz^{\alpha}\wedge d\bar z^{\beta} = i\partial\bar\partial \ln h.$$

Exercises

1. For almost complex manifold M, if

$$\theta = \sigma + i\omega \in T^*_{10}(M), \quad \sigma,\omega \in T^*_p(M),$$

prove $\quad \bar\theta = \sigma - i\omega \in T^*_{01}(M)$.

2. Prove that almost complex 2-dimensional manifold is complex manifold.
3. Prove that $\nabla_{JX}J = 0$, if almost complex structure J is torsion free and $\nabla_X J = 0$.
4. Of n-dimensional spheres S^n, find those which are almost complex manifold, complex manifold, group manifold, and parallelizable manifold.
5. Prove the Kähler form is a real form.
6. Write K^n in terms of $dz^{\alpha_1}\wedge\cdots\wedge d\bar z^{\beta_n}$, and also in terms of real coordinate differential forms.
7. For any (p,q)-from $\alpha_{p,q}, \beta_{p,q}$, prove

$$\begin{aligned}&\alpha_{p,q}\wedge *\bar\beta_{p,q} = \beta_{p,q}\wedge *\bar\alpha_{p,q} = \bar\beta_{p,q}\wedge *\alpha_{p,q},\\ &\overline{*\alpha_{p,q}} = *\bar\alpha_{p,q},\\ &*\alpha_{p,q} = (-1)^{p+q}\alpha_{p,q}.\end{aligned}$$

8. Prove for Kähler manifold

$$\begin{aligned}\partial\alpha_{p,q} &= \tfrac{1}{p!q!}\nabla_{\alpha} f_{\alpha_1\cdots\alpha_p\beta_1\cdots\beta_q} dz^{\alpha}\wedge dz^{\alpha_1}\wedge\cdots\wedge d\bar z^{\beta_q} ,\\ \bar\partial\alpha_{p,q} &= \tfrac{1}{p!q!}\bar\nabla_{\alpha} f_{\alpha_1\cdots\alpha_p\beta_1\cdots\beta_q} d\bar z^{\alpha}\wedge dz^{\alpha_1}\wedge\cdots\wedge d\bar z^{\beta_q} ,\\ \theta\alpha_{p,q} &= -(-1)^{p}\tfrac{1}{p!q!}h^{\bar\beta\gamma}\nabla_{\gamma} f_{\alpha_1\cdots\alpha_p\beta\beta_1\cdots\beta_q} dz^{\alpha_1}\wedge\cdots\wedge dz^{\alpha_p}\wedge d\bar z^{\beta_2}\wedge d\bar z^{\beta_q} .\end{aligned}$$

9. Prove that for Kähler manifold,

$$\Gamma^{\alpha}_{\beta\gamma} = \Gamma^{\alpha}_{\gamma\beta}, \qquad R^{\alpha}_{\beta\gamma\bar\delta} = \bar\partial_{\gamma}\Gamma^{\alpha}_{\delta\beta}.$$

Chapter 7

Homology of Manifolds

From this chapter, we study the global topological properties of the manifold. The global topological analysis plays a very important role not only in the classical field theory but also in the modern quantum field theory. For a differentiable manifold M the properties of the neighborhood of a point is closely related to the global properties of M. To study the global properties of the manifold, two algebraic tools are often used: the homotopy group $\Pi_k(M)$ and the homology group $H_k(M)$. The k-homotopy group $\Pi_k(M)$ on M represents the topological obstruction of the continuous mapping from a k-dimensional sphere S^k to M. $\Pi_k(M) \neq 0$ indicates the existence of $(k+1)$-dimensional holes on M. The k-homology group $H_k(M)$ represents the number of homotopic class of independent and topologically nontrivial k-dimensional submanifolds on M. Both the homotopy groups and the homology groups are topological invariants, furthermore, they are both homotopically invariant. We will first study the homology groups and leave homotopy groups to the next Chapter.

In the first section, we will study the homotopic types of manifold. Then the homology group is introduced. In the theory of homology group, the de Rham cohomology theory plays an prominent role. It deals with the global topological properties of the manifold through the local analysis such as differential forms and differential operators on the manifold, so that the local properties and the global properties for the manifold are related. Both the characteristic classes and the Atiyah-Singer index theory, which will be discussed in this book, depend on de Rham theory. In physics, de Rham theory has a wide range of applications. It can be said that de Rham theory enriches topology, analysis and physics and lies at their crossing point. In this chapter, after a simple introduction of the homology group, we will focus our attention on de Rham cohomology theory and the related topics.

7.1 Homotopic mapping and manifolds with the same homotopy type

An important tool in studying the topological properties of manifold is mapping. All continuous mappings between manifolds can be classified according to their equivalence classes by using the concept of homotopy. Given two continuous mappings f_0 and f_1 between two manifolds M and N:

$$f_0, f_1 : \quad M \to N,$$

the two mappings f_0 and f_1 are called homotopic to each other if f_0 can be transformed to f_1 by continuous deformation. This definition can be restated by more rigorous mathematical terms as: If there exists a set of continuous mapping F,

$$\begin{array}{rcl} F : M \times [0,1] & \to & N \\ x \ , \ \lambda & \to & F(x,\lambda) \end{array}$$

when the parameter λ in $F(x,\lambda)$ continuously changes from 0 to 1 with

$$F(x,0) = f_0(x), \qquad F(x,1) = f_1(x),$$

such that mapping f_0 changes continuously to mapping f_1 , these two continuous mappings f_0 and f_1 are then called homotopic, which is written as

$$f_0 \sim f_1.$$

It is easy to prove that the homotopy relations are equivalent relations. Therefore, mappings from M to N can be classified according to their homotopy equivalence classes. The homotopy class of the mapping f can be denoted by $< f >$.

Using the homotopic mappings one can introduced the concepts such as the homotopy type and the retract of manifold.

Definition 7.1 *Two manifolds M and N are called homotopically equivalent or they have the same homotopy type if there exist continuous mappings f and g between M and N,*

$$\begin{array}{rcl} f : M & \to & N \\ g : N & \to & M \end{array}$$

such that $g \cdot f$ is homotopic to the identity mapping on M and $f \cdot g$ homotopic to the identity mapping on N:

$$g \cdot f \sim 1_M, \qquad f \cdot g \sim 1_N,$$

Definition 7.2 *$N \subset M$ is called the deformation retract of M, if there exists a continuous mapping $r : M \to N$ and a homotopy mapping F,*

$$\begin{aligned} &F : M \times [0,1] \to M, \\ &F(p,0) = p, \qquad \forall p \in M, \\ &F(p,1) = r(p), \quad \forall r(p) \in N, \\ &F(q,\lambda) = q, \quad \forall q \in N, \lambda \in [0,1). \end{aligned}$$

Namely, M can be continuously deformed to N during which N is unchanged. The deformation retract does not change the homotopy type of the manifold, therefore, M and N have the same homotopy type. When one applies homeomorphic transformations to manifolds M and N, the homotopy equivalence classes of the mappings between M and N are invariant and are topological invariants of the pair of manifolds. So the requirement of homotopy equivalence is weaker than that of homeomorphism (topological equivalence). For example, a circle and an annulus have the same homotopy type, but they are not homeomorphic. In the following, we give a few examples:

Example 1. The n-dimensional ball (disc) D^n in E^n. After the origin is removed, $D^n - \{0\}$ can be continuously deformed to the unit sphere S^{n-1}, i.e., there is a homotopy mapping F

$$\begin{aligned} F : (D^n - \{0\}) \times [0,1] &\to D^n - \{0\} \\ x \ , \ \lambda &\to F(x,\lambda) = (1-\lambda)x + \lambda\frac{x}{|x|}, \end{aligned}$$

hence $D^n - \{0\}$ and S^{n-1} have the same homotopy type

$$D^n - \{0\} \sim S^{n-1}.$$

Example 2. For n-dimensional space $\mathbf{R}^n$, the homotopy mapping F is introduced as

$$\begin{aligned} F : \mathbf{R}^n \times [0,1] &\to \mathbf{R}^n \\ x \ , \ \lambda &\to F(x,\lambda) = \lambda x \end{aligned}$$

such that the origin $\{0\}$ is the deformation retract of $\mathbf{R}^n$. Hence, $\mathbf{R}^n$ and the single point $\{0\}$ have the same homotopy type.

Definition 7.3 *A manifold which has the same homotopy type as single point is called retractable manifold.*

The retractable manifold is the simplest manifold, which is often called topologically trivial manifold. From example 2, we see that $\mathbf{R}^n$ is a retractable manifold.

The homeomorphism relation between the manifolds is an equivalence relation. Topological spaces can be classified by their homeomorphism equivalence classes such that manifolds belonging to one class are homeomorphic to each other and have the same topological properties. To label each topological equivalence class, one needs topological invariants such as dimension, compactness and connectedness of the manifold. The homotopy type of the manifold is a topological invariant and the homotopy relation between manifolds is also an equivalence relation. Topological spaces can be classified by their homotopy type. Because the requirement of homotopy is weaker than that of the homeomorphism, the classification by homotopy is coarser than by homeomorphism. Manifolds which are homeomorphic must have the same homotopy type. But the inverse is not true in general. For instance, the dimension of a manifold is a topological invariant under homeomorphism transformations, but the dimension is not an invariance of the homotopy type, manifolds having the same homotopy type may have different dimension. For instance, the topologically trivial manifolds can be retractable manifolds of any dimension and they have the same homotopy type.

One of the most important homotopy invariants of a manifold is its homology groups, which we shall consider in the following.

7.2 Singular homology group

The study of global invariants of manifolds started comes from combinatorial topology. The basic idea is to cut a manifold into many cells and to study how these cells can be glued together.

For a smooth manifold M, the neighborhood of each point is homeomorphic to an open set of $\mathbf{R}^n$. Therefore, we can use the dissection of $\mathbf{R}^n$ as the dissection of M. We first look at the orientable submanifold in $\mathbf{R}^n$, which is called the standard k-simplex:

$$\triangle_k = \{x = (x^0, x^1, \cdots, x^k) \in \mathbf{R}^{k+1} | \sum_{i=0}^{k} x^i = 1, x^i \geq 0\}.$$

As examples, several simple standard k-simplexes are given in Figure 7.1.

In Figure 7.1, v_i are the vertices of the standard k-simplex $\triangle_k$

$$v_0 = (1, 0, 0, \cdots, 0), \qquad v_1 = (0, 1, 0, \cdots, 0), \cdots.$$

The k-dimensional hyper plane $\triangle_k$ spanned by $(k+1)$ vertices is an orientable k-simplex, which can be represented by their vertices $\{v_i\}$ as

$$\triangle_k = [v_0, v_1, \cdots, v_k] = sgn(p)[v_{p_0}, v_{p_1}, \cdots, v_{p_k}],$$

where $p \in P_{k+1}$, and P_{k+1} is a $(k+1)$ order permutation group. The above definition shows that even permutation of the vertices still represents the same simplex and odd permutation of the vertices represents a simplex with the opposite orientation.

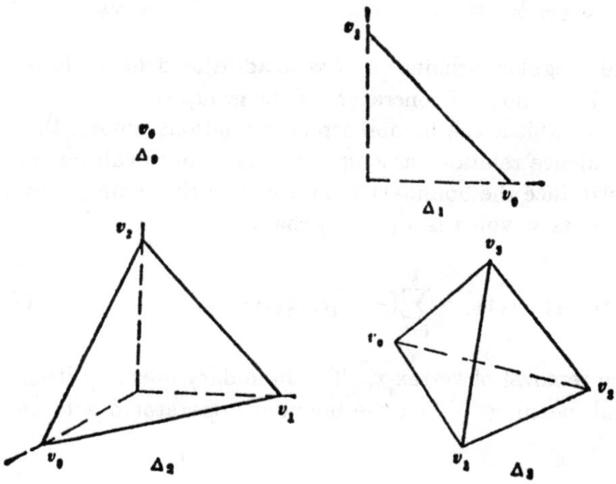

Figure 7.1

Taking homeomorphism mappings from $\mathbf{R}^n$ to manifold M, we obtain the k-dimensional orientable submanifold in M. For the homeomorphism mapping ϕ

$$\phi : \Delta_k \to \sigma_k \subset M,$$

the image σ_k of ϕ is called the singular k-simplex of M , which can be represented by the image (also denoted as v_i) of the vertices as

$$\sigma_k = [v_0, v_1, \cdots, v_k] \subset M.$$

Here, we should emphasize that homology studies the image of the homeomorphism mapping and the k-simplex itself is a k-dimensional submanifold of M. For an n-dimensional manifold M, the highest dimension of its submanifold is n, therefore there is no simplex of dimension higher than n.

On manifold M, the integer coefficients linear combination of the k-simplexes is called the singular k-chain on M:

$$c_k = \sum_i a_i \sigma_k^i, \qquad a_i \in Z.$$

The set $C_k(M) = \{c_k\}$ of all singular k-chains on manifold M is a free Abelian group which is generated by all orientable k-simplexes on M. In other words, to each k-simplex there are two orientation: $\pm\sigma_k$

$$\sigma_k + (-\sigma_k) = 0.$$

For arbitrary integer $a, b \in Z$,

$$a\sigma_k + b\sigma_k = (a+b)\sigma_k, \qquad a(-\sigma_k) = (-a)\sigma_k.$$

The set $C_k(M)$ of all singular k-chains on M with addition defined above is an Abelian group. In general, the number of generators of the group is infinite. To facilitate our analysis, we try to establish the homomorphism relations among the chain groups such that the equivalence relations among the chain groups can be established. For this purpose, we introduce the boundary operator ∂ on the chain group, it maps each k-simplex to its boundary which is a $(k-1)$-chain

$$\partial\sigma_k = \partial[v_0, v_1, \cdots, v_k] = \sum_{i=0}^{k}(-1)^i[v_0, v_1, \cdots, \widehat{v}_i, \cdots v_k] \in C_{k-1}(M),$$

where $\widehat{v}_i$ means the removal of vertex v_i. The boundary operator itself has straightforward geometrical meaning. When the boundary operator ∂ acts on an arbitrary k-chain, we have

$$\partial c_k = \partial(\sum_i a_i\sigma_k^i) = \sum_i a_i(\partial\sigma_k^i) \in C_{k-1}(M).$$

The boundary operator preserves the addition operation of the chain group, so it becomes a homomorphic operator among chain groups with algebraic meaning:

$$\partial : C_k(M) \to C_{k-1}(M),$$

where the boundary of 0-chain is defined as zero. With the boundary operator ∂, two important subgroups of $C_k(M)$ can be obtained:

1). The kernel of the boundary homomorphism:

$$Z_k = \{c_k \in C_k(M) | \partial c_k = 0\}$$

is called k-cycle group and the generator z_k is called k-cycle.

2). The image of the boundary homomorphism:

$$B_k = \{b_k = \partial c_{k+1} | c_{k+1} \in C_{k+1}(M)\}$$

is called k-boundary chain group and the generator b_k is called a boundary chain.

It is easy to prove that the boundary of the boundary chain must be zero

$$\partial \cdot \partial = 0.$$

For example, for 2-simplex $\sigma_2 = [v_0, v_1, v_2]$, one has

$$\partial\sigma_2 = [v_1, v_2] - [v_0, v_2] + [v_0, v_1],$$

$$\partial \cdot \partial \sigma_2 = [v_2] - [v_1] - [v_2] + [v_0] + [v_1] - [v_0] = 0.$$

For an arbitrary k-simplex $\sigma_k = [v_0, v_1, \cdots, v_k]$, we have

$$\begin{aligned}
\partial \cdot \partial \sigma_k &= \partial(\sum_{i=0}^{k}(-1)^i [v_0, v_1, \cdots, \widehat{v}_i, \cdots v_k]) \\
&= \sum_{i=0}^{k}(-1)^i \partial [v_0, v_1, \cdots, \widehat{v}_i, \cdots v_k] \\
&= \sum_{j<i}^{k}(-1)^{i+j}[v_0, v_1, \cdots, \widehat{v}_j, \cdots, \widehat{v}_i, \cdots v_k] \\
&\quad + \sum_{j>i}^{k}(-1)^{i+j-1}[v_0, v_1, \cdots, \widehat{v}_i, \cdots, \widehat{v}_j, \cdots v_k] \\
&= 0.
\end{aligned}$$

Hence, the boundary chain must be a cycle. B_k is a subgroup of Z_k:

$$B_k \subset Z_k.$$

Both B_k and Z_k are Abelian groups so B_k is a normal subgroup of Z_k. Therefore a quotient group can be defined as

$$H_k(M) = Z_k(M)/B_k(M)$$

which is called the k-homology group of manifold M. The homology group H_k is the set of all k-cycle equivalence classes(module boundaries). When two k-cycles $z_k^{(1)}$ and $z_k^{(2)}$ are differ by a boundary chain,

$$z_k^{(1)} - z_k^{(2)} \in B_k(M),$$

we say $z_k^{(1)}$ and $z_k^{(2)}$ are equivalent with each other:

$$z_k^{(1)} \sim z_k^{(2)}.$$

This equivalence relation is called homology. That is to say, $z_k^{(1)}$ and $z_k^{(2)}$ are homologous to each other. Both $z_k^{(1)}$ and $z_k^{(2)}$ belong to one homology class which is written as $< z_k >$.

As the number of chains on M is very large, we will focus on cycles. Cycles, which differ by a boundary chain, belong to the same homology class, so the equivalence class of the nontrivial k-cycles in H_k is the k-dimensional submanifold around a $(k+1)$-dimensional hole. Cycles, which are not around any holes, are boundary chains which belong to zero homology class. In other words, $H_k(M)$ can be used to represent the $(k+1)$-dimensional holes in M and it is a finite generated Abelian group.

For a homeomorphism between two manifolds:

$$f : M \to N,$$

there is an induced mapping of the chain group on M:

$$f_* : C_k(M) \to C_k(N).$$

Since f_* and the boundary homomorphic operation commute

$$\partial f_* = f_* \partial,$$

there exists a homomorphism between homology groups of the two manifolds. It can be proved that for a homeomorphism mapping f, there is an induced mapping f_* between the homology groups of two manifolds:

$$f_* : H_k(M) \to H_k(N).$$

In fact, f_* is an isomorphism for all $k \geq 0$. Therefore, manifolds which are homeomorphic have the same homology group. Furthermore remember that generators of the homology group $H_k(M)$ on manifold M are the equivalence classes of the k-dimensional submanifolds which are around $(k+1)$-dimensional holes, and deformation retraction of the manifold does not change the homology group, so homology groups of manifolds, which have the same homotopy type, are isomorphic to each other.

The topologically trivial retractable manifolds, such as $\mathbf{R}^n$, which are of the same homotopy type as a single point, have the homology group

$$H_k(\mathbf{R}^n) = \begin{cases} \mathbf{Z}, & k = 0, \\ 0, & k \neq 0. \end{cases}$$

A single point v_0 is a zero-dimensional manifold,

$$H_k(v_0) = 0, \qquad k \neq 0.$$

According to the definition, any zero-dimensional chain must be a cycle: $Z_0(v_0) = C_0(v_0)$. For a single point v_0 there is no lower dimensional submanifold, i.e., $B_0(v_0) = 0$. Hence

$$H_0(v_0) = Z_0(v_0) = Z.$$

For path-connected manifolds, any two points can be joined by a one-dimensional submanifold, the zero-dimensional submanifolds are homologous to each other. So there is a theorem:

Theorem 7.1 *For path-connected manifold,*

$$H_0(M) = Z. \tag{7.1}$$

Homology groups of manifold are Abelian groups. The rank of a group is a very important quantity. The rank of the k-dimensional homology group of the manifold is called the k-th Betti number:

$$b_k = \dim(H_k(M)),$$

which is a topological invariant and a homotopy type invariant. The alternative sum of all Betti numbers on the manifold is called the Euler number $\chi(M)$ of M

$$\chi(M) = \sum_{k=0}^{n}(-1)^k b_k. \tag{7.2}$$

The Euler number is a very important topological quantity which characterizes the non-triviality of the manifold. It stems first from combinatorial topology. When a manifold is dissected into a set of simplexes, the number a_k of the k-simplexes depends on the dissection. However, Euler found that despite different triangular dissections of surfaces (two-dimensional manifolds), the alternative sum of the numbers of vertices, edges and faces which are represented by v, e and f respectively

$$\chi = v - e + f$$

is independent of the dissections and is a topological invariant. The Euler formula above can be generalized to any higher dimensional connected manifolds. Taking the triangular dissection of M, the alternative sum of the number a_k of each dimensional simplex is called the Euler number of M:

$$\chi(M) = \sum_{k=0}^{n}(-1)^k a_k. \tag{7.3}$$

It can be proved that the Euler number defined in Eq. 7.3 is just the Euler number defined in Eq. 7.2. Since there exists homomorphism among different dimensional chain groups:

$$\partial_k : C_k(M) \to C_{k-1}(M), \qquad 0 \le k \le n,$$

where C_{-1} is defined as null space. From the theorem on rank in linear algebra, we have

$$\begin{aligned} a_k &= \dim(C_k(M)) = \dim(\ker\partial_k) + \dim(\mathrm{Imag}\partial_k) \\ &= \dim(Z_k(M)) + \dim(B_{k-1}(M)). \end{aligned}$$

The k-th Betti number is

$$b_k = \dim(H_k(M)) = \dim(Z_k(M)) - \dim(B_k(M)).$$

Therefore,

$$\begin{aligned}\sum_{k=0}^{n}(-1)^k b_k &= \sum_{k=0}^{n}(-1)^k \dim(Z_k(M)) + \sum_{k=0}^{n}(-1)^{k+1}\dim(B_k(M)) \\ &= \sum_{k=0}^{n}(-1)^k \dim(Z_k(M)) + \sum_{k=0}^{n}(-1)^k \dim(B_{k-1}(M)) \\ &= \sum_{k=0}^{n}(-1)^k a_k = \chi(M),\end{aligned}$$

where in the second step we used the relation:

$$B_n(M) = B_{-1}(M) = 0.$$

The Betti numbers are topological and homotopic invariants on the manifold. Since the Euler number $\chi(M)$ is the alternative sum of the Betti numbers, so it is also a topological and homotopic invariant on the manifold. The Betti numbers represent a refinement of the Euler number.

7.3 General homology group and universal coefficient theorem

The chain groups discussed in the last section have integer coefficients. The corresponding homology group $H_k(M)$ is called singular homology group $H_k(M, \mathbf{Z})$. The coefficient addition group $\mathbf{Z}$, which is used in the chain group, can be replaced by any other Abelian group G. For instance, G can be $\mathbf{Z}_2$ (mod-2 integer), $\mathbf{Q}$ (rational number), $\mathbf{R}$ (real number) and $\mathbf{C}$ (complex number). Similarly we can have all chain groups with coefficients of the Abelian group elements such as chain group $C_k(M,G)$, cycle group $Z_k(M,G)$, boundary chain group $B_k(M,G)$ and the homology group $H_k(M,G)$. In general we call $H_k(M,\mathbf{Z})$, $H_k(M,\mathbf{Z}_2)$, $H_k(M,\mathbf{Q})$, $H_k(M,\mathbf{R})$ and $H_k(M,\mathbf{C})$ singular homology group, mod-2 homology group, rational homology group, real homology group and complex homology group respectively.

Of all these homology groups for a manifold, the singular homology group $H_k(M,\mathbf{Z})$ is the most fundamental. Other homology groups $H_k(M,G)$ are completely determined by properties of the singular homology group $H_k(M,\mathbf{Z})$ and of the Abelian group G. This is just the universal coefficient theorem.

Here we restate that other homology groups, which provide less information than singular homology group does. The integer homology group in general has the following structure:

$$\begin{aligned} H_k(M,\mathbf{Z}) &= \mathbf{Z}\oplus\cdots\oplus\mathbf{Z}\oplus T_k(M) \\ &= F_k(M)\oplus T_k(M), \end{aligned} \tag{7.4}$$

where $F_k(M)$ is a rank-r free group (the k-th Betti number $b_k = r$), and $T_k(M)$ is a finite group called torsion subgroup. When the integer coefficient $\mathbf{Z}$ is replaced by the rational $\mathbf{Q}$,

$$H_k(M, \mathbf{Q}) = \mathbf{Q} \oplus \cdots \oplus \mathbf{Q}. \tag{7.5}$$

The number r of $\mathbf{Q}$ is called the rank of $H_k(M, \mathbf{Q})$ or the Betti number b_k. The torsion subgroup $T_k(M)$ in Eq. 7.4 does not appear in Eq. 7.5. As the number of element of the torsion group is finite, let $\tau_k \in T_k(M)$ and the order of τ_k be m, i.e.,

$$m\ \tau_k = 0.$$

When $\mathbf{Z}$ is replaced by $\mathbf{Q}$, τ_k is $\frac{1}{m}$ of the identity element, i.e.,

$$\tau_k = \frac{1}{m} \cdot 0 = 0.$$

Hence, $H_k(M, \mathbf{Q})$ has no torsion subgroup.

The integer $\mathbf{Z}$ can be embedded into rational $\mathbf{Q}$, real $\mathbf{R}$ and complex $\mathbf{C}$ and we have

$$\begin{cases} H_k(M, \mathbf{Q}) = H_k(M, \mathbf{Z}) \otimes \mathbf{Q}, \\ H_k(M, \mathbf{R}) = H_k(M, \mathbf{Z}) \otimes \mathbf{R}, \\ H_k(M, \mathbf{C}) = H_k(M, \mathbf{Z}) \otimes \mathbf{C}. \end{cases}$$

they are essentially isomorphic if torsion subgroup is neglected.

In dealing with certain special problems, one needs mod-2 group $\mathbf{Z}_2$ as coefficients. In $\mathbf{Z}_2$, we have $+1 = -1$. Therefore, $C_k(M, \mathbf{Z}_2)$ is the chain group of nonorientable simplexes. For nonorientable manifolds, we often use mod-2 homology group $H_k(M, \mathbf{Z}_2)$. The corresponding Betti number is $b_k^{(2)}$ which is called the k-th mod-2 Betti number of manifold M. In general, b_k of the integer homology group is not equal to the $b_k^{(2)}$ of the mod-2 homology group. But the Euler number $\chi(M)$ of M is independent of the coefficient group of the homology group.

In the following, we should considered some simple manifolds. First, a few theorems are introduced.

Theorem 7.2 *If the manifold M can be decomposed into l connected branches $M_1, M_2, \cdots, M_l$, then*

$$H_k(M, G) = H_k(M_1, G) \oplus H_k(M_2, G) \oplus \cdots \oplus H_k(M_l, G).$$

According to this theorem, it is sufficient to consider the homology group on each connected branch. Unless otherwise specified, we will only consider connected manifolds in the following.

The zero-th homology group and its corresponding structure need special consideration. The zero-th homology group is a criteria of connectivity of the manifold. For a connected manifold, similar to Eq. 7.1, we have

Theorem 7.3 *For a connected manifold M,*

$$H_0(M,G) = G.$$

Theorem 7.4 *For an n-dimensional connected manifold M,*

$$H_n(M,G) = \begin{cases} G, & \textit{for compact orientable manifold,} \\ {}_2G, & \textit{for compact nonorientable manifold,} \\ 0, & \textit{for noncompact manifold.} \end{cases}$$

Here ${}_2G$ is the kernel of the double mapping

$${}_2G = \{g \in G | 2g = 0\}.$$

Now we give a simple explanation. For an n-dimensional connected manifold M, the boundary of an n-dimensional region is an $(n-1)$-dimensional boundary chain. So the highest boundary chain is $B_{n-1}(M,G)$, and $B_n(M,G)$ must be zero. If there is an orientable region in n-dimensional manifold M and it is an n-dimensional cycle $\mathbf{Z}_n$, it must be a substantial cycle of the n-dimensional homology group. For a compact orientable manifold, the whole manifold is orientable and without boundary, so it is a generator of the homology group. Hence, we have

$$H_n(M,G) = G.$$

For a nonorientable manifold, the whole manifold is nonorientable but there is an orientable region such that the boundary of the orientable region is twice the $(n-1)$-dimensional closed submanifold (Please refer to the example of the homology group of $RP(N)$ manifold given in §7.5 example 2.). In this case, we have

$$H_n(M,G) = {}_2G.$$

Therefore, for nonorientable manifold M,

$$\begin{cases} H_n(M,\mathbf{Z}) = 0, \\ H_n(M,\mathbf{Z}_2) = \mathbf{Z}_2. \end{cases}$$

For a noncompact manifold, there doesn't exist an n-dimensional cycle, so that

$$H_n(M,\mathbf{Z}) = 0.$$

When $k \neq 0, n$, the calculation of homology group $H_k(M,G)$ is very difficult. Sometimes, the de Rham theorem and the cohomology theory are needed.

Theorem 7.5 (Duality) *For M an orientable connected manifold and G is a field (such as $\mathbf{R}, \mathbf{C}, \mathbf{Z}_2$), there exists Poincareé duality*

$$H_k(M, G) = H_{n-k}(M, G). \tag{7.6}$$

(For nonorientable manifold, the above theorems is still true if one uses $\mathbf{Z}_2$ as coefficient.) Therefore, we have relations for the Betti numbers of the manifold:

$$b_k = b_{n-k}.$$

Integer $\mathbf{Z}$ is not a field, for integer homology group, there is no dual relation Eq. 7.6. The torsion coefficients do not satisfy the dual relation. However, for orientable connected closed manifold of n-dimension, it can be proved that the torsion coefficients have relations (see reference [VIII] Theorem 7.7)

$$T_{k-1} = T_{n-k}.$$

Two-dimensional surface has been studied thoroughly in topology and there is the following classification theorem:

Theorem 7.6 *If the genus of an orientable surface is g, then*

$$H_1(M, \mathbf{Z}) = \oplus_{2g}\mathbf{Z}.$$

Hence,

$$\dim(H_1(M, \mathbf{Z})) = 2g.$$

If the genus of a nonorientable surface is g, then

$$H_1(M, \mathbf{Z}) = \oplus_{g-1}\mathbf{Z} \oplus \mathbf{Z}_2.$$

In the following, we take several examples to show how to calculate the homology groups of manifolds. We emphasize that the key ingredients of homology theory are the chain groups and the boundary operator. We must pay close attention to them.

Example 1. The n-dimensional sphere S^n is orientable, hence

$$H_0(S^n, \mathbf{Z}) = \mathbf{Z}, \qquad H_n(S^n, \mathbf{Z}) = \mathbf{Z}.$$

As all k-dimensional ($1 \leq k < n$) closed submanifolds of S^n are contractible, we have

$$\begin{aligned} H_k(S^n, \mathbf{Z}) &= 0 \quad (\text{when } 1 \leq k < n), \\ \chi(S^n) &= \sum_{k=0}^{n}(-1)^k b_k = 1 + (-1)^n \\ &= \begin{cases} 2, & \text{if } n \text{ is even}, \\ 0, & \text{if } n \text{ is odd}. \end{cases} \end{aligned}$$

Example 2. Two-dimensional torus is orientable:

$$H_0(T^2, \mathbf{Z}) = \mathbf{Z},$$

$$H_2(T^2, \mathbf{Z}) = \mathbf{Z},$$

$$H_1(T^2, \mathbf{Z}) = \mathbf{Z} \oplus \mathbf{Z}.$$

As shown in Figure 7.2, the generators of H_1 are cycles a and b. And the Euler number

$$\chi(T^2) = 1 - 2 + 1 = 0.$$

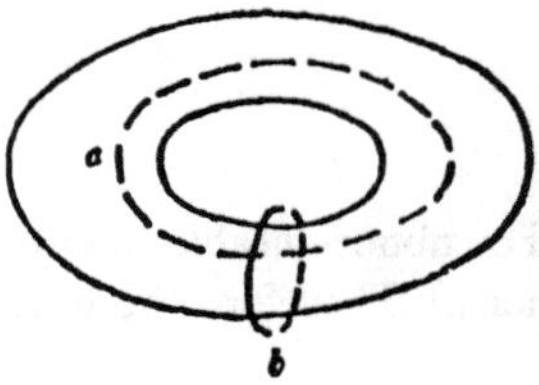

Figure 7.2 Torus

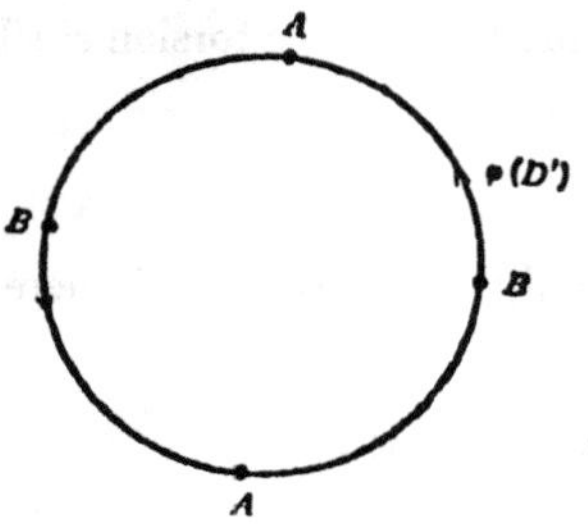

Figure 7.3

Example 3. $M = RP(2) = S^2/\mathbf{Z}_2$ is a nonorientable two-dimensional manifold.

$$H_0(M, \mathbf{Z}) = \mathbf{Z}, \quad H_0(M, \mathbf{Z}_2) = \mathbf{Z}_2$$

$$H_2(M, \mathbf{Z}) = 0, \quad H_2(M, \mathbf{Z}_2) = \mathbf{Z}_2.$$

For H_1, we must pay attention to whether there exists non-boundary cycle. As

$$RP(2) = S^2/\mathbf{Z}_2,$$

let φ be the mapping to identify the antipodal points on S^2:

$$\varphi : S^2 \to RP(2).$$

Let D^k represent the upper hemisphere of $S^k (k = 1, 2)$. $\varphi(D^2)$ a orientable 2 chains on the manifold $RP(2)$, as shown in Figure 7.3. It is easy to prove that

$$\partial\varphi(D^2) = 2\varphi(D^1), \implies \varphi(D^1) = \frac{1}{2}\partial\varphi(D^2).$$

$\varphi(D^1)$ is a one-dimensional cycle on $RP(2)$ and is not a boundary in integral homology, so is a substantial 1-cycle. Hence,

$$H_1(M, \mathbf{Z}) = \mathbf{Z}_2, \quad \text{and} \quad H_1(M, \mathbf{Z}_2) = \mathbf{Z}_2.$$

For the singular homology group, $b_0 = 1$, $b_1 = 0$, $b_2 = 0$. The only one-dimensional torsion coefficient is $\tau_1 = 2$. For the mod-2 homology group, we have

$$b_0^{(2)} = 1, \quad b_1^{(2)} = 1, \quad b_2^{(2)} = 1.$$

Therefore,

$$\chi(M) = \sum_0^2 (-1)^k b_k = \sum_0^2 (-1)^k b_k^{(2)} = 1.$$

Example 4. $M = RP(3) \simeq SO(3) \simeq S^3/\mathbf{Z}_2$ is an orientable three-dimensional manifold. Similar to the last example, let φ be the mapping to equate the antipodal points on S^3:

$$\varphi : S^3 \to RP(3).$$

Let D^k represent the upper hemisphere of $S^k (k = 1, 2, 3)$ and $\varphi(D^k)$ the orientable k chains on the manifold $RP(3)$. It is easy to prove that

$$\begin{cases} \partial\varphi(D^3) = 0. \\ \partial\varphi(D^2) = 2\varphi(D^1). \\ \partial\varphi(D^1) = 0. \end{cases}$$

Therefore, $\varphi(D^3)$ is a cycle and is not a boundary chain. We have

$$H_3(M, G) = G.$$

$\varphi(D^2)$ is not a cycle unless $G = \mathbf{Z}_2$. We have

$$H_2(M, \mathbf{Z}) = 0, \qquad H_2(M, \mathbf{Z}_2) = \mathbf{Z}_2.$$

$\varphi(D^1)$ is a cycle and $\varphi(D^1) = \frac{1}{2}\partial\varphi(D^2)$. We have

$$H_1(M, \mathbf{Z}) = \mathbf{Z}_2, \quad H_1(M, \mathbf{Z}_2) = \mathbf{Z}_2,, \quad H_0(M, G) = G.$$

For the singular homology group, $b_0 = 1, \quad b_1 = 0, \quad b_2 = 0, \quad b_3 = 1$. The only torsion coefficient is $\tau_1 = 2$, and

$$\chi(M) = 0 \text{ (independent of the choice of homology group).}$$

The homology groups of $RP(3)$ are listed in following table:

k	integer homology $H_k(M, \mathbf{Z})$	real homology $H_k(M, \mathbf{R})$	mod-2 homology $H_k(M, \mathbf{Z}_2)$
0	$\mathbf{Z}$	$\mathbf{R}$	$\mathbf{Z}_2$
1	$\mathbf{Z}_2$	0	$\mathbf{Z}_2$
2	0	0	$\mathbf{Z}_2$
3	$\mathbf{Z}$	$\mathbf{R}$	$\mathbf{Z}_2$
$\chi =$	0	0	0

From this table we see that there exists Poincareé duality except for the singular homology group. For integer homology group, there may exist torsion subgroups which does not appear in other homology groups.

7.4 Cohomology theory

In this section we begin to discuss the cohomology theory which is dual to the homology theory. In Appendix D, we point out that the set of homomorphism mappings from a vector space V to real field $\mathbf{R}$, forms a dual vector space of the original vector space V. For convenience, we adopt the Dirac symbol which is more familiar to physicists. The ket $|a>$ represents vector in the vector space and the bra $<b|$ represents vector in the dual vector space. Hom($V,\mathbf{R}$) represents homomorphism mapping from the vector space V to the real field $\mathbf{R}$

$$\begin{array}{rlcl} \mathrm{Hom}(V,\mathbf{R}) : V & & \rightarrow & R, \\ <b| \in \mathrm{Hom}(V,\mathbf{R}) : & |a> \in V & \rightarrow & <b|a> \in \mathbf{R}. \end{array}$$

The set of Hom($V,\mathbf{R}$) forms a vector space V^* which is the dual space of the vector space V.

$H_k(M,\mathbf{Z})$ is an Abelian group with its algebraic structure analogous to a linear space but over integer ring. So we can analyse its dual space as a vector space (only replacing real number by integer number), i.e., the set of homomorphism mappings Hom($A,\mathbf{Z}$) forms the dual Abelian group A^* of A.

Similarly, we can analyse the set of homomorphism mappings from an Abelian group A to the integer addition group $\mathbf{Z}$

$$\begin{array}{rll} \mathrm{Hom}(A,\mathbf{Z}) : & A \rightarrow & \mathbf{Z} \\ f : & a \rightarrow & <f,a> \in \mathbf{Z} \end{array}$$

$<f,a>$ is a bilinear integer-valued function of f and a. Within the set of homomorphism mappings, the addition operator can be defined. For two mappings f and g, the addition $f+g$ can be defined as

$$<(f+g),a> = <f,a> + <g,a> .$$

It can be proved that if A is an n-dimensional free group and has bases $X = \{x_1, x_2, \cdots x_n\}$, then the dual Abelian group A^* =Hom($A,\mathbf{Z}$) is also an n-dimensional free group and has bases $\Xi = \{\xi_1, \xi_2, \cdots, \xi_n\}$, such that

$$<\xi_i, x_j> = \delta_{ij}.$$

The bases Ξ is called the dual bases of X.

For a homomorphism mapping φ between two Abelian groups A and B, there is a dual homomorphism mapping φ^* between the dual groups. As shown in the following diagram:

$$\begin{array}{ccc} A & \xrightarrow{\varphi} & B \\ a & \longrightarrow & \varphi(a) \\ \Downarrow & & \Downarrow \\ \mathrm{Hom}(A,\mathbf{Z}) & \xleftarrow{\varphi^*} & \mathrm{Hom}(B,\mathbf{Z}) \end{array}$$

$$<\varphi^*(\beta), a> = <\beta, \varphi(a)> \in \mathbf{Z},$$

where $a \in A$, $\varphi(a) \in B$, $\beta \in \mathrm{Hom}(B, \mathbf{Z})$, $\varphi^*(\beta) \in \mathrm{Hom}(A, \mathbf{Z})$.

The essential ingredients of homology group are the sequence of chain groups and the boundary homomorphism ∂ among them. The image and kernel of the boundary homomorphism form the homology group.

The set of homomorphism mappings $\mathrm{Hom}(C_k(M), \mathbf{Z})$ forms the dual chain group $C^k(M)$ of the chain group $C_k(M)$. We call $C^k(M)$ the k-cochain group with the label k at upper right corner and $C_k(M)$ the k-chain group. From the boundary homomorphism mapping between k-chain groups

$$\partial : C_k(M) \to C_{k-1}(M),$$

the dual coboundary homomorphism mapping $\bar{\partial} \equiv \partial^*$ between k-cochain groups can be defined

$$\bar{\partial} : C^{k-1}(M) \to C^k(M).$$

As $\partial_{k-1} \cdot \partial_k = 0$, there are relations between the dual homomorphism mappings

$$\bar{\partial}_k \cdot \bar{\partial}_{k-1} = (\partial_{k-1} \cdot \partial_k)^* = 0.$$

Subgroups of the cochain group $C^k(M)$ can be defined by coboundary operation

$$\begin{array}{ll} \text{cocycle}: & Z^k(M, \mathbf{Z}) = \text{kernel of the } \bar{\partial}_k, \\ \text{coboundary}: & B^k(M, \mathbf{Z}) = \text{image of the } \bar{\partial}_{k-1}, \\ \text{cohomology}: & H^k(M, \mathbf{Z}) = Z^k(M, \mathbf{Z})/B^k(M, \mathbf{Z}). \end{array}$$

The cohomology group H^k is the set of equivalence classes of k-cocycle (module the coboundary) and is also an Abelian group. It can be proved that

Theorem 7.7 *For an n-dimensional connected manifold M,*

$$\begin{aligned} H^0(M, \mathbf{Z}) &= \mathbf{Z}, \\ H^n(M, \mathbf{Z}) &= \begin{cases} \mathbf{Z}, & \textit{for manifold orientable,} \\ \mathbf{Z}_2, & \textit{for manifold not orientable.} \end{cases} \end{aligned}$$

Between the homology group and the cohomology group there exists the following theorem:

Theorem 7.8

$$b^k = b_k, \quad T^k(M) = T_{k-1}(M) \qquad (k = 0, 1, \cdots, n; \quad T_{-1}(M) \equiv 0),$$

where $b^k = \dim H^k(M, \mathbf{Z})$ is the upper Betti number corresponding to the cohomology group.

From this theorem we see that the singular homology group $H_k(M, \mathbf{Z})$ completely determines not only the homology group $H_k(M, G)$ but also the cohomology group $H^k(M, G)$. We are convinced that $H_k(M, \mathbf{Z})$ is the fundamental homology group on manifold M.

Cohomology group A^* is the set of $\mathrm{Hom}(A, \mathbf{Z})$. By using the skewsymmetric tensor product on A^*, we can define the cup product on A^*. For an orientable manifold, the product among cochains can be defined according to the rank of the cochains. If $x^p \in C^p(M)$, $\beta^q \in C^q(M)$, then

$$\alpha^p \cup \beta^q \in C^{p+q}(M).$$

Product so defined is a bilinear functional of the homology group. And it is distributive and associative, so the cochain group becomes a cochain ring. The coboundary operation of the product satisfies

$$\bar{\partial}(\alpha^p \cup \beta^q) = \bar{\partial}\alpha^p \cup \beta^q + (-1)^p \alpha^p \cup \bar{\partial}\beta^q.$$

It can further be proved that

$$\begin{array}{rcccl} \text{cocycle} & \cup & \text{cocycle} & = & \text{cocycle.} \\ \text{coboundary} & \cup & \text{cocycle} & = & \text{coboundary.} \\ \text{cocycle} & \cup & \text{coboundary} & = & \text{coboundary.} \end{array} \tag{7.7}$$

It can be proved that the cohomology class of the product of two is completely determined by the cohomology classes of the two factors. Hence, the product of cohomology classes can be defined, which is also distributive and associative. The direct product mapping $\cup$ of the cohomology groups is defined as:

$$\cup : H^k(M) \times H^l(M) \to H^{k+l}(M). \tag{7.8}$$

If we define the direct sum H^* of cohomology groups

$$H^*(M) = \oplus \sum_{k=0}^{n} H^k(M),$$

then the direct product operation in Eq. 7.8 makes $H^*(M)$ a cohomology ring which has ring structure. But the homology group itself is only an Abelian group, with no ring structure. The direct product mapping is

$$\cup : H^*(M) \times H^*(M) \to H^*(M).$$

The dual operation U^* of the direct product $\cup$ inverses the direction of the mapping such that in the direct sum of homology groups

$$H_*(M) = \oplus \sum_{k=0}^{n} H_k(M),$$

there is a dual mapping U^*

$$U^* : H_*(M) \to H_*(M) \times H_*(M).$$

That is to say U^* does not provide any ring structure in homology theory. In general, the homology ring of the manifold does not exist. However, the cohomology ring must exist, which makes cohomology theory more important. When we study the topological invariants on the manifold, cohomology theory is a more effective and simpler tool than homology theory.

7.5 de Rham cohomology theory

When the coefficient group G of the homology group $H_k(M,G)$ is a field, the $H_k(M,G)$ itself is a linear space over G and the dual linear space is just the cohomology group $H^k(M,G)$. When concerned only with the rank (Betti number) of the homology group and the Euler number, we can neglect the torsion subgroup temporarily and study only the linear space $H_k(M,\mathbf{R})$ and the dual linear space $H^k(M,\mathbf{R})$ over real field (The same discussions can be carried out for complex field.). In this case, calculus can be used. We know the Stokes theorem:

$$\int_c d\omega = \int_{\partial c} \omega,$$

which can be written as

$$< c, d\omega > = < \partial c, \omega > .$$

The integral domain (chain) is related to differential forms (cochain). The boundary operator is related to the exterior differential operator. We have

$$< \partial^2 c, \omega > = < \partial c, d\omega > = < c, d^2\omega >= 0.$$

Making use of differential forms and exterior differential operator d, the de Rham cohomology theory is dual to the homology theory. By using differential operator d, one defines two subspaces in differential forms: $(\omega_k \in \Lambda^k(M))$

$$\begin{aligned} Z^k_{dR} &= \{\omega_k; d\omega_k = 0\}, && \text{set kernel of } d_k && \text{(closed } k\text{-forms)}, \\ B^k_{dR} &= \{\omega_k; \omega_k = d\alpha_{k-1}\}, && \text{set of image of } d_{k-1} && \text{(exact } k\text{-forms)}, \end{aligned}$$

two closed k-forms are taken as equivalent if they differ by an exact form, and they belong to the same cohomology class. The set of cohomology classes constitutes the cohomology group

$$H^k_{dR}(M,\mathbf{R}) = Z^k_{dR}/B^k_{dR} = \mathrm{Ker}(d_k)/\ \mathrm{Imag}(d_{k-1}).$$

Definition 7.4 *The k-th cohomology group $H^k_{dR}(M,\mathbf{R})$ of a manifold M is the quotient group of the k-cocycle group, consisting of all closed k-forms and its subgroup of exact k-forms. Thus elements of $H^k_{dR}(M,\mathbf{R})$ are equivalence classes of closed k-forms.*

If $c_k \in Z_k$(cycle), and $\omega_k \in Z^k_{dR}$(closed forms), then

$$< c_k, \omega_k >= \int_{c_k} \omega_k \in \mathbf{R}.$$

It can be proved that this above integral depends only on the equivalence classes, i.e.,

$$\int_{c_k} (\omega_k + d\alpha_{k-1}) = \int_{c_k + \partial c_{k+1}} \omega_k = \int_{c_k} \omega_k,$$

which defines a bi-linear mapping

$$H_k(M, \mathbf{R}) \oplus H^k_{dR}(M, \mathbf{R}) \to \mathbf{R}.$$

This shows that the de Rham cohomology group is dual to general homology group. In other words, $H^k_{dR}(M, \mathbf{R})$ is isomorphic to cohomology group $H^k(M, \mathbf{R})$. This is just the de Rham theorem.

Theorem 7.9 (de Rham) *For a compact manifold M, $H^k_{dR}(M, \mathbf{R})$ is isomorphic to cohomology group $H^k(M, \mathbf{R})$ of M. Therefore, the space $H^k_{dR}(M, \mathbf{R})$ which is defined through differential forms is actually a topological invariant.*

For simplicity, in the following discussions all $H^k(M, \mathbf{R})$ means the de Rham cohomology group with the subscript dR omitted.

As exterior product preserves the equivalence relation of cohomology, a similar relation analogous to Eq. 7.7 can be obtained:

$$\begin{aligned} \text{closed forms} \ \wedge \ \text{closed forms} &= \text{closed forms}, \\ \text{closed forms} \ \wedge \ \text{exact forms} &= \text{exact forms}. \end{aligned}$$

For instance, if α_k is a closed form, we have

$$\alpha_k \wedge d\beta_j = (-1)^k d(\alpha_k \wedge \beta_j) \qquad (\alpha_k \in Z^k).$$

In terms of exterior product, direct product of cohomology groups can be defined

$$H^k(M, \mathbf{R}) \otimes H^j(M, \mathbf{R}) = H^{k+j}(M, \mathbf{R}),$$

i.e., de Rham cohomology ring is isomorphic to the general cohomology ring.

For manifold $M = M_1 \times M_2$, using the product of cohomology groups defined above, we have

$$H^k(M, \mathbf{R}) = \oplus \sum_{i+j=k} H^i(M_1, \mathbf{R}) \otimes H^j(M_2, \mathbf{R}). \tag{7.9}$$

Therefore, the Betti number and the Euler number satisfy the following relations

$$\begin{aligned} b^k(M) &= \sum_{i+j=k} b^i(M_1) b^j(M_2), \\ \chi(M) &= \chi(M_1)\chi(M_2). \end{aligned}$$

This is called the Kunneth formula.

Example 1. For the n-dimensional torus $T^n = S^1 \times S^1 \times \cdots \times S^1$, we have

$$H^k(T^n) = \oplus \begin{pmatrix} n \\ k \end{pmatrix} \mathbf{R}.$$

So the Betti number and the Euler number are

$$b^k(T^n) = \begin{pmatrix} n \\ k \end{pmatrix}, \qquad \chi(T^n) = \sum_{k=0}^{n} (-1)^k \begin{pmatrix} n \\ k \end{pmatrix} = 0.$$

In terms of differential forms, it is easy to study the Poincareé duality. For a compact orientable manifold M, we can define the scalar product of k-form α_k and $(n-k)$-form β_{n-k}:

$$< \alpha_k, \beta_{n-k} > \equiv \int \alpha_k \wedge \beta_{n-k} \in \mathbf{R}.$$

Through this scalar product, duality between $H^k(M, \mathbf{R})$ and $H^{n-k}(M, \mathbf{R})$ is easily established and they are algebraically isomorphic. Hence,

$$b^k = b^{n-k}.$$

The structure of the zero-th homology group needs special consideration. As de Rham cohomology group has no (-1)-forms, i.e., the zero forms (functions) f can not be represented as exact forms. Hence, the closed 0-forms are generators of the zero-th cohomology group:

$$H^0 = \{\text{constant function space}\}.$$

For a connected manifold, we have

$$H^0(M, \mathbf{R}) = \mathbf{R}.$$

Theorem 7.10 *For a manifold with r connected branches, we have*

$$H^0(M, \mathbf{R}) = \mathbf{R} \oplus \mathbf{R} \oplus \cdots \oplus \mathbf{R},$$

i.e.,

$$b^0 = \dim H^0(M, \mathbf{R}) = \textit{number of connected branches}.$$

de Rham cohomology group of the linear space $\mathbf{R}^n$ is very simple, because $\mathbf{R}^n$ is homotopic to a single point. In $\mathbf{R}^n$, except for the zero-forms(functions), any closed k-form can be represented as exterior differentiation of a $(k-1)$-form. So they are homologous to zero, therefore,

$$\dim H^k(\mathbf{R}^n) = 0, \qquad (k > 0)$$
$$\dim H^0(\mathbf{R}^n) = 1.$$

Manifolds satisfying these are usually called topologically trivial manifolds. And there is the Poincareé lemma:

Theorem 7.11 (Poincareé lemma) *In the neighborhood of any point on a manifold M, every closed k-form($k \geq 1$) is exact locally.*

Given a continuous mapping f between two manifolds

$$f : N \to M,$$

for differential forms on the manifolds it induces a pull back mapping f^*

$$f^* : \Lambda^k(M) \to \Lambda^k(N),$$

satisfying

$$\begin{aligned} f^*(d\alpha) &= df^*(\alpha), \\ f^*(\alpha \wedge \beta) &= f^*(\alpha) \wedge f^*(\beta). \end{aligned}$$

The induced pull back mapping of the cohomology classes is

$$f^* : H^k(M) \to H^k(N).$$

Hence, homeomorphic manifolds have the same cohomology group, which is not only topological-invariant, but also homotopical-invariant.

Finally, we summarize de Rham cohomology theory and homology theory in the following table. In terms of the Stokes theorem,

$$< c, d\omega > = < \partial c, \omega >$$

it is easy to prove the dual correspondence relations given in the table 7.2.

De Rham cohomology theory	Singular homology theory
$d : \Lambda^k \to \Lambda^{k-1}$	$\partial : c_k \to c_{k-1}$
$d^2 = 0$	$\partial^2 = 0$
$d\omega = 0$, ω is called closed form	$\partial c = 0$, c is called cycle
$\omega = d\alpha$, ω is called exact form	$c = \partial a$, c is called boundary chain
$\omega = d\alpha \ \underset{\not\Leftarrow}{\Longrightarrow}\ d\omega = 0$	$c = \partial a \ \underset{\not\Leftarrow}{\Longrightarrow}\ \partial c = 0$
$Z^k = \{\omega_k \| d\omega_k = 0\}$	$Z_k = \{c_k \| \partial c_k = 0\}$
$B^k = \{\omega_k \| \omega_k = d\alpha_{k-1}\}$	$B_k = \{c_k \| c_k = \partial c_{k+1}\}$
$H^k = Z^k / B^k = \mathrm{Ker}(d_k)/\mathrm{Imag}(d_{k-1})$	$H_k = Z_k / B_k = \mathrm{Ker}(\partial_k)/\mathrm{Imag}(\partial_{k+1})$
$b^k = \dim H^k$	$b_k = \dim H_k$
$\omega_1 \sim \omega_2$, if $\omega_1 - \omega_2 = d\alpha$	$c_1 \sim c_2$, if $c_1 - c_2 = \partial a$

In the following, we take n-dimensional sphere S^n, n-dimensional projective space $RP(n) = S^n/Z_2$ and the group manifold $U(n)$ as examples to demonstrate how to calculate de Rham cohomology groups.

Example 2. S^n is a unit sphere in $(n+1)$-dimensional Euclidean space E^{n+1}, and is a compact orientable n-dimensional manifold without boundary.

1).

$$H^0(S^n, \mathbf{R}) = \mathbf{R} \qquad (S^n \text{ is connected}).$$

2).

$$H^n(S^n, \mathbf{R}) = \mathbf{R}. \tag{7.10}$$

Now we state the reason. As S^n is compact orientable manifold, a nonzero n-form, i.e., the volume element, can be defined everywhere. The integral of the volume element ω on the whole manifold S^n is not zero:

$$\int_{S^n} \omega \neq 0. \tag{7.11}$$

ω is a closed form (any n-form on an n-dimensional manifold must be a closed form.) and is not exact, because If ω is exact, i.e., $\omega = d\alpha$, since S^n has no boundary, we would have

$$\int_{S^n} \omega = \int_{S^n} d\alpha = \int_{\partial S^n} \alpha = 0.$$

This contradicts with Eq. 7.11. Hence, this n-form is the generator of the cohomology class and Eq. 7.10 is proved. Similarly, we can prove

Theorem 7.12 *An orientable closed n-dimensional manifold M has nontrivial n-cohomology group:*

$$H^n(M, \mathbf{R}) = \mathbf{R}.$$

3).

$$H^k(S^n, \mathbf{R}) = 0 \qquad (0 < k < n).$$

Since any lower dimensional submanifolds in S^n are contractible manifolds, so any k-cycle c_k $(0 < k < n)$ must be boundary chain:

$$c_k = \partial c_{k+1} \qquad (0 < k < n).$$

Therefore, if $\alpha_k \in Z^k$ is a closed form, its integral over any cycle c_k in S^n is

$$\int_{c_k} \alpha_k = \int_{\partial c_{k+1}} \alpha_k = \int_{c_{k+1}} d\alpha_k = 0,$$

as the integral of α_k over any cycle gives zero, α_k must be an exact form. Any closed k-form $(0 < k < n)$ in S^n is an exact form so there is no nontrivial k-dimensional closed forms and this completes our proof.

Example 3. The n-dimensional real projective space: $RP(n) = S^n/\mathbf{Z}_2$.

$RP(n)$ is a manifold of identifying the antipodal points on S^n which is the covering manifold of $RP(n)$. There is a projective mapping which is continuous surjective from S^n to $RP(n)$. We can use the properties of S^n to study the properties of $RP(n)$.

1). As S^n is connected, so is $RP(n)$. Therefore we have

$$H^0(RP(n)),\mathbf{R}) = \mathbf{R}.$$

2). Take S^n as the unit sphere in $\mathbf{R}^{n+1}$, Let

$$E = \{dx^1, \cdots\cdots, dx^n\}$$

be the coframe on S^n. We translate the co frame at a given point to its antipode by two different ways. One is through the parallel transport of the coframe along dx^1 geodesic curve, and we get

$$E' = \{-dx^1, dx^2, \cdots\cdots, dx^n\}.$$

The other is through the antipode map, i.e.,

$$i : dx^k \to -dx^k \quad \Longrightarrow i^*E = \{-dx^1, \cdots, -dx^n\}.$$

For RP(n) $E' \sim i^*E = (-1)^{n-1}E'$, thus it is orientable if n is odd, and is not orientable if n is even.

So in the case if n is even, $\chi(S^n) = 2$, there is no vector field which is nonzero everywhere on S^n. When projected to $RP(n)$ manifold, the left hand frame and the right hand frame on S^n projects to each other, hence, $RP(n)$ is a nonorientable manifold.

When n is odd, $\chi(S^n) = 0$, on S^n there exists a vector field which is nonvanishing everywhere. When two orientable frames (left and right hand frames) are projected to $RP(n)$ manifold, there still exist two orientable frames. So $RP(n)$ is orientable. Hence,

$$H^n(RP(n),\mathbf{R}) = \begin{cases} 0 & (\,n \text{ is even}), \\ \mathbf{R} & (\,n \text{ is odd}). \end{cases}$$

3). The local properties of $RP(n)$ is the same as that of S^n. When $k \neq 0$ or n, any closed k-form on the covering manifold S^n must be an exact form. Therefore, any closed k form on $RP(n)$ must be an exact form. We have

$$H^k(RP(n),\mathbf{R}) = 0 \qquad (0 < k < n).$$

Example 4. Group manifold $U(n) \simeq SU(n) \times U(1)/\mathbf{Z}_n$.

Let g be an $n \times n$ unitary matrix which is an n-dimensional linear representation of the group $U(n)$. The Maurer-Cartan form $g^{-1}dg$ is a 1-form which takes value on the Lie algebra of $U(n)$. Let

$$\omega_k = \mathrm{Tr}(g^{-1}dg)^k,$$

it is easy to prove that $\omega_k = 0$ when k is even and when k is odd, ω_k is the generator of the cohomology group $H^k(U(n),\mathbf{R}) = \mathbf{R}$ (when $k = 1, 3, \cdots, 2n-1$). For instance,

$$U(2) = U(1) \times SU(2)/\mathbf{Z}_2 = S^1 \times S^3/\mathbf{Z}_2.$$

From the Kunneth formulas Eq. 7.9, we have

$$\begin{aligned}
&H^0(U(2),\mathbf{R}) = \mathbf{R} \quad (U(n) \text{ connected}),\\
&H^1(U(2),\mathbf{R}) = \mathbf{R} \quad (\text{generator is } \omega_1),\\
&H^2(U(2),\mathbf{R}) = 0,\\
&H^3(U(2),\mathbf{R}) = \mathbf{R} \quad (\text{generator is } \omega_3),\\
&H^4(U(2),\mathbf{R}) = \mathbf{R} \quad (\text{generator is } \omega_1 \wedge \omega_3).
\end{aligned}$$

The Euler number is zero:

$$\chi(U(2)) = 1 - 1 + 0 - 1 + 1 = 0.$$

The Euler number of any simple Lie group manifold is zero. The above results are consistent with previous analysis.

7.6 Harmonic forms

Given a metric tensor field on a manifold, the Hodge $*$ operation can be defined. Combining exterior differential operator d and the Hodge $*$ operator, we can define the co-differential operator δ and an elliptic differential operator (Laplacian) $\triangle$:

$$\triangle = d\delta + \delta d.$$

Kernel ω_k ($\triangle\omega_k = 0$) of the Laplace operator are called harmonic forms. The set $\{\omega_k\}$ of all harmonic k-forms is a linear space $\mathrm{Harm}^k(M,\mathbf{R})$ over real field. There exists the Hodge decomposition theorem[35].

Theorem 7.13 (Hodge Decomposition Theorem) *For a compact Riemannian manifold M without boundary, an arbitrary k-form on the manifold can be decomposed as the sum of an exact form, a co-exact form and a harmonic form:*

$$\omega_k = d\alpha_{k-1} + \delta\beta_{k+1} + \gamma_k, \tag{7.12}$$

where γ_k is a harmonic k-form. The three terms in the right hand side are mutually orthogonal.

Proof: We apply the Green function operator $\hat{G}$ to this case.
As is known, for a linear differential operator $\hat{L}$, the differential equation

$$\hat{L}\phi = f, \tag{7.13}$$

where f is a given function and ϕ is the function to be determined. When operator $\hat{L}$ is not singular, i.e., equation $\hat{L}\phi = 0$ has no nontrivial solutions, there exists a Green function $\hat{G}$ of the operator $\hat{L}$ satisfying

$$\hat{L}\hat{G} = I,$$

where I is the identity operator, such that the solution of Eq. 7.13 is

$$\phi = \hat{G} f.$$

However, equation $\hat{L}\phi = 0$ has nontrivial solutions when operator $\hat{L}$ is singular. Let $\hat{P}$ be a projection operator which projects arbitrary functions to the solution space of $\hat{L}\phi = 0$. In this case, one can defined the generalized Green function in the orthogonal complement subspace of the solution space of $\hat{L}\phi = 0$. The generalized Green function $\hat{G}$ satisfies the following equation:

$$\hat{L}\hat{G} = I - \hat{P}.$$

For the Laplace operator Δ, let $\hat{P}$ be a projection operator which projects arbitrary differential forms to harmonic forms. The generalized Green function of Δ satisfies the following equation

$$\Delta \hat{G} + \hat{P} = I. \tag{7.14}$$

As the Laplace operator is self-adjoint and positive definite, the generalized Green function which satisfies Eq. 7.14 must exist. Applying Eq. 7.14 to arbitrary k-forms, we have

$$\omega_k = d\delta\hat{G}\omega_k + \delta d\hat{G}\omega_k + \hat{P}\omega_k,$$

which is just the Hodge decomposition theorem.

Furthermore, we can prove that there is only one harmonic form for each cohomology class. That is to say, the set $\mathrm{Harm}^k(M, \mathbf{R})$ of k-forms is isomorphic to $H^k(M, \mathbf{R})$.

Theorem 7.14 *For a compact manifold M without boundary, $Harm^k(M, \mathbf{R})$ is isomorphic to $H^k(M, \mathbf{R})$:*

$$Harm^k(M, \mathbf{R}) \simeq H^k(M, \mathbf{R}).$$

Proof : If $\omega_k \in Z^k$, i.e., $d\omega_k = 0$, taking exterior differentiation of Eq. 7.12, we have

$$d\delta\beta_{k+1} = 0.$$

Therefore,

$$<\beta, d\delta\beta> = 0 \implies <\delta\beta, \delta\beta> = 0 \implies \delta\beta = 0,$$

then

$$\omega = d\alpha + \gamma,$$

i.e., ω and γ are homologous to each other: $\omega \sim \gamma$. The theorem is proved.

It is due to this theorem that we can apply harmonic forms to study the corresponding results of cohomology. For example:

1). On a compact manifold M without boundary, $\mathrm{Harm}^k(M,\mathbf{R})$ is of finite dimension, then $H^k(M,\mathbf{R})$ is also of finite dimension. The dimension of the harmonic k-forms is just the Betti number b^k.

2). The Hodge * operator and the Laplace Δ operator commute so we have

$$*\mathrm{Harm}^k(M,\mathbf{R}) \simeq \mathrm{Harm}^{n-k}(M,\mathbf{R}).$$

Thus the Poincareé duality in cohomology theory can be proved:

$$H^k(M,\mathbf{R}) \simeq H^{n-k}(M,\mathbf{R}),$$

$$b^k = b^{n-k}.$$

3). If $\omega^{(1)}$ is harmonic on M_1 and $\omega^{(2)}$ is harmonic on M_2, then $\omega^{(1)} \wedge \omega^{(2)}$ is harmonic on $M_1 \times M_2$. Hence,

$$\mathrm{Harm}^k(M,\mathbf{R}) = \oplus \sum_{i+j=k} \mathrm{Harm}^i(M_1,\mathbf{R}) \otimes \mathrm{Harm}^j(M_2,\mathbf{R}).$$

The Kunneth formula in cohomology theory is thus proved:

$$\begin{aligned} H^k(M,\mathbf{R}) &= \oplus \sum_{i+j=k} H^i(M_1,\mathbf{R}) \otimes H^j(M_2,\mathbf{R}). \\ b_k(M) &= \sum_{i=0}^{k} b_i(M_1) b_{k-i}(M_2). \end{aligned}$$

7.7 Bi-invariant form on group manifold and invariant form on symmetric space

Now we introduce the bi-invariant forms on group manifold G. A form on Lie group G is called bi-invariant if it is both left and right invariant. Clearly, a left-invariant 1-form θ on G will also be right invariant if and only if it is invariant under the inner automorphism. So an s-form ω is a bi-invariant form if and only if it is Ad-invariant, i.e.,

$$\sum_{i=1}^{s} \omega(X, \cdots, [X, X_i], X_{i+1}, \cdots, X_s) = 0, \quad X_i \in \mathbf{g}. \tag{7.15}$$

By this equation, we can prove the following theorem

Theorem 7.15 *Every bi-invariant form is closed.*

Proof: Let ω be a left invariant r-form on group G, $\{X_i\}$ be the left invariant vector fields on G, then

$$d\omega(X_0, X_1, \cdots, X_r) = \sum_{i<j}(-1)^{i+j}\omega([X_i, X_j], X_0, \cdots, \hat{X}_i, \cdots, \hat{X}_j, \cdots, X_r). \quad (7.16)$$

If ω is also right invariant, the right invariance of ω implies Eq. 7.15

$$\sum_{i=1}^{r}\omega(X_1, \cdots, [X_0, X_i], X_{i+1}, \cdots, X_r) = 0 \quad (7.17)$$

So from Eq. 7.16 and Eq. 7.17, we obtain $d\omega(X_0, X_1, \cdots, X_r) = 0$ for any left invariant vector fields $\{X_i\}$, so $d\omega = 0$.

Since Hodge $*$ commutes with both the left and right translations, $\delta\omega = 0$, so a bi-invariant form must be a harmonic form, i.e., $\Delta\omega = 0$. Conversely, it can also be proved that a harmonic form ω is a bi-invariant form, and there exists a theorem:

Theorem 7.16 (Hodge) *The bi-invariant forms on a connected Lie group manifold G are precisely the harmonic forms. So the cohomology ring of a connected compact Lie group G is isomorphic to the ring of bi-invariant differential forms, which is the ring of Ad-invariant multilinear alternating maps on the Lie algebra* $\mathbf{g}$.

For example, Killing form K on the Lie algebra $\mathbf{g}$ of a Lie group G is a symmetric bilinear mapping

$$\begin{aligned} K:\ & \mathbf{g}\times\mathbf{g} \to \mathcal{F}(G) \\ & X, Y \to K(X,Y) \equiv \mathrm{tr}(ad_X \circ ad_Y) = K(Y,X). \end{aligned}$$

In terms of Killing form K, we can introduce a tri-linear alternating mapping Ω

$$\begin{aligned} \Omega : \mathbf{g}\times\mathbf{g}\times\mathbf{g} &\to \mathcal{F}(G) \\ X\ ,\ Y\ ,\ Z &\to \Omega(X,Y,Z), \end{aligned}$$

where $\Omega(X,Y,Z) \equiv K([X,Y],Z) = K(X,[Y,Z])$.

Every compact Lie group manifold G is a symmetric space of the group $G\times G$. The action $T_{(g_1,g_2)}$ of $G\times G$ on G can be defined as

$$T_{(g_1,g_2)}(g) = g_1 g g_2^{-1},$$

so

$$T_{(h_1,h_2)}T_{(g_1,g_2)} = T_{(h_1g_1,h_2g_2)}.$$

The involution σ is given by $\sigma(g_1, g_2) = (g_2, g_1)$, so the isotropy group $H = G_V = \{(g,g)\}$, i.e.,

$$G = G_L \times G_R/G_V$$

$(G_L \times G_R, G_V, \sigma)$ is a symmetric space. The action $T_{(g_1,g_2)}$ is a transformation of symmetric space $G = G_L \times G_R/G_V$, the bi-invariant form on group G is an invariant form on symmetric space G. Invariant forms play a particularly important role among differential forms on symmetric space. There is a theorem:

Theorem 7.17 *If M is a symmetric space of a compact, connected Lie group, then every invariant form on M is closed, the ring of invariant forms coincides (cohomologous) with the cohomology ring of symmetric space M.*

(Its proof can be seen in reference[VII]B. A. Dubrovin, A. T. Fomenko, S. P. Novikov, Modern Geometry. Part III. P. 10).

7.8 G-structure of manifold and its restriction to the homology group of manifold

Let M be a smooth compact n-manifold without boundary. A vector field $X \in \mathcal{X}(M)$ is a section of tangent bundle $T(M)$, and a covector field $\sigma \in \Lambda^1(M)$ is a section of cotangent bundle $T^*(M)$. In terms of local coordinate $x = (x^1, \cdots, x^n)$ on a neighborhood $U \subset M$, the vector field X and the covector field σ can be expressed as

$$X = \xi^i(x)\partial_i, \qquad \sigma = \sigma_i(x)dx^i,$$

for a general coordinate transformation $x^i \to y^i(x)$,

$$det\left(\frac{\partial y^i}{\partial x^j}\right) \neq 0,$$

there exists its inverse transformation $y^i \to x^i(y)$. Correspondingly the natural frame $\frac{\partial}{\partial x^i}$ and dx^i transformed as

$$\frac{\partial}{\partial y^i} = \frac{\partial x^k}{\partial y^i}\frac{\partial}{\partial x^k}, \qquad dy^i = \frac{\partial y^i}{\partial x^k}dx^k,$$

the matrices of transformation between frames is

$$L = (L_{ij}) = \left(\frac{\partial x^k}{\partial y^i}\right) \in GL(n, \mathbf{R}),$$

the set of all this matrices forms a group $GL(n, \mathbf{R})$, called the structure group of the bundle of frame. $GL(n, \mathbf{R})$ is not a compact group, but it can be contracted to its Lie subgroup.

In the following we consider a differentiable manifold equipped with a G-structure, where G is a Lie subgroup of $GL(n, \mathbf{R})$. G-structure on M is a reduction of transformation group between frames to the subgroup G. A rank-n general linear group $GL(n, \mathbf{R})$ consists of all $n \times n$ nonsingular matrices. For any nonsingular matrix $A \in GL(n, \mathbf{R})$, there is a decomposition

$$A = B \cdot C,$$

where $B \in O(n)$ is an orthogonal matrix and C a positive definite symmetric matrix. Hence the group manifold $GL(n, \mathbf{R})$ can be written as a Cartesian product of two spaces

$$GL(n, \mathbf{R}) = O(n) \times PS,$$

where $O(n)$ is a rank-n orthogonal group manifold and PS is the set of all positive definite symmetric matrices. PS is not a group and is not a vector space either, it is homeomorphic to a vector space S which consists of all rank-n symmetric matrices. PS can be obtained from S via an exponential mapping

$$\begin{aligned} \exp:\ S &\rightarrow PS, \\ s \in S &\rightarrow \exp(s) \in PS. \end{aligned}$$

This exponential mapping is a continuous onto mapping. An arbitrary positive definite matrix is an exponential of a symmetric matrix. The correspondence relation above is a homeomorphism. The space S is contractible as it is a vector space. Therefore the space PS is also contractible. Hence group $GL(n, \mathbf{R})$ and group $O(n)$ are of the same homotopy type, $GL(n, \mathbf{R})$ can be deformation contracted to $O(n)$, i.e, transformation group between frames may be reduced to the compact subgroup $O(n)$. A $O(n)$-structure on M is nothing but a Riemannian metric on M. On Riemannian manifold one can choose orthogonal frame, transformation group between orthogonal frames is $O(n)$. Just because $O(n)$ has the same homotopy type as $GL(n, \mathbf{R})$, so that every smooth manifold M always admits a Riemannian metric. Analogously for any complex manifold of complex dimension m, its structure group $GL(m, \mathbf{C})$ can be reduced to the maximal compact subgroup $U(m)$ of $GL(m, \mathbf{C})$, so that every complex manifold M always admits a Hermitian metric.

An orientable manifold has a nowhere vanish n-form (volume element). The group $SL(n, \mathbf{R})$ provide an oriented frame at each point of M, and thus endow M with an orientation.

A symplectic manifold of dimension $n = 2m$ has a symplectic frame, the group leaving the antisymmetric quadratic form invariant is noncompact group $Sp(n, \mathbf{R})$. A symplectic manifold is a manifold with a $Sp(n, \mathbf{R})$-structure.

As the above analysis, the G-structure of manifold give restricted to the homology group of the manifold, following table give some examples for the compact n-manifold without boundary.

Name	G-structure	Necessary condition
Riemannian	$O(n)$	None
Lorentzian	$O(n-1,1)$	Euler number $= 0$
Orientable	$SL(n,\mathbf{R})$	$H^n(M,\mathbf{R}) \neq 0$
Orient. Riem.	$SO(n)$	$w^1 \in H^1(M,\mathbf{Z}_2) = 0$
Spin	$Spin(n)$	$w^2 \in H^2(M.\mathbf{Z}_2) = 0$
Symplectic	$SP(n,\mathbf{R})$	$H^n(M,\mathbf{R}) \neq 0, H^2(M,\mathbf{R}) \neq 0, n = 2m$
Complex	$GL(m,\mathbf{C})$	$H^{2m}(M,\mathbf{R}) \neq 0,\ H^2(M,\mathbf{R}) \neq 0$
Hermitian	$U(m)$	$H^{2m}(M,\mathbf{R}) \neq 0,\ H^2(M,\mathbf{R}) \neq 0$
Kähler	$U(m)$	$H^{2k}(M,\mathbf{R}) \neq 0, k = 1,2,\cdots,m$
Calabi-Yau	$SU(m)$	$H^{2k}(M,\mathbf{R}) \neq 0, k = 1,2,\cdots,m$ the first Chern class $= 0$

For example, $S^{2n}(n > 1)$ are even orientable manifold, they have neither symplectic nor complex structure. But S^2 is a complex manifold. About the orientation and spin structure ref. §10.5.

Notations and Formulae

- Standard k-simplex in $\mathbf{R}^n$:

$$\begin{aligned}\Delta_k &= \{x = (x^0, x^1, \cdots, x^k) \in \mathbf{R}^{k+1} | \textstyle\sum_{i=0}^k x^i = 1, x^i \geq 0\} \\ &= [v_0, v_1, \cdots, v_k].\end{aligned}$$

- Singular k-simplex on M:

$$\sigma_k = [v_0, v_1, \cdots, v_k] \in C_k(M).$$

Boundary operator ∂:

$$\partial\sigma_k = \sum_{i=0}^{k}(-1)^i[v_0, v_1, \cdots, \widehat{v}_i, \cdots v_k] \in C_{k-1}(M).$$

k-cycle:

$$Z_k = \{c_k \in C_k(M) | \partial c_k = 0\}.$$

k-boundary chain:

$$B_k = \{b_k = \partial c_{k+1} | c_{k+1} \in C_{k+1}(M)\}.$$

k-homology class:

$$H_k(M) = Z_k(M)/B_k(M).$$

- de Rham cohomology:

$$\begin{aligned} Z^k_{dR} &= \{\omega \in \Lambda^k | d\omega = 0\} \quad \text{(set of closed } k\text{-forms)}, \\ B^k_{dR} &= \{\omega \in \Lambda^k | \omega = d\alpha\} \quad \text{(set of exact } k\text{-forms)}, \\ H^k_{dR}(M, \mathbf{R}) &= \mathbf{Z}^k_{dR}/B^k_{dR} \quad = \mathrm{Ker}(d_k)/\ \mathrm{Imag}(d_{k-1}). \end{aligned}$$

- The k-th Betti number $b_k = dim(H_k(M)) = dim(H^k(M))$.

- The Euler number: $\chi(M) = \sum_{k=0}^n (-1)^k b_k$.

- For an n-dimensional connected manifold M,

$$H_n(M, G) = \begin{cases} G, & \text{for compact orientable manifold,} \\ {}_2G, & \text{for compact nonorientable manifold,} \\ 0, & \text{for noncompact manifold.} \end{cases}$$

Duality: $b_k = b_{n-k}$, $\quad T_{k-1} = T_{n-k}$.

- For manifold $M = M_1 \times M_2$, there exists Kunneth formula:

$$H^k(M, \mathbf{R}) = \oplus \sum_{i+j=k} H^i(M_1, \mathbf{R}) \otimes H^j(M_2, \mathbf{R}).$$

$$\begin{aligned} b^k(M) &= \sum_{i+j=k} b^i(M_1) b^j(M_2), \\ \chi(M) &= \chi(M_1)\chi(M_2). \end{aligned}$$

Exercises

1.Find the singular homology group of $SO(3)$ manifold and its Betti numbers and Euler number.

2.Find the de Rham cohomology groups, the Betti numbers and the Euler numbers of the following manifolds:

1).S^6 and S^7.

2).$RP(2)$ and $RP(3)$.

3).$U(3)$.

Chapter 8

Homotopy of Manifold, Fibre Bundle, Classification of Fibre Bundles

8.1 Homotopy group of manifold

First, we introduce path and the closed path. A path is a continuous mapping f from a segment $I = [0,1]$ of real axis $\mathbf{R}$ to a manifold M

$$\begin{aligned} f : I &\rightarrow M \\ \lambda &\rightarrow f(\lambda). \end{aligned}$$

Note that mapping f is called the path, the image $f(I)$ of f is a curve on M.

There are many paths for a given manifold. We are to study the homotopy equivalence classes of paths. We say that two paths are homotopic if they can be continuously deformed to each other.

If the image of mapping f at the end points of I coincide on M

$$f(0) = f(1) = x_0,$$

the corresponding path is called a closed path or loop with base point x_0. If one loop can be continuously deformed to another loop (keeping the base point x_0 invariant), then the two loops are called homotopically equivalent. Homotopy relation is an equivalence relation. Therefore, the space of loops with base point x_0 can be classified by this equivalence relation. Loops in a given class are homotopic to each other. We use $< f >$ to represent the loop class in which all loops are homotopic to loop f.

In Figure 8.1, for instance, R^2 space has two regions D_1 and D_2 where region D_2 has a hole while region D_1 has not. All loops in D_1 with base x_1 can be retracted to point x_1, that is to say, these loops and the identity mapping belong to the same homotopy class. However, in D_2, loops f_1 and f_2 with base point x_2 do not belong to the same homotopy class. This example shows that to distinguish the topological properties of two-dimensional manifold and to see whether there are holes in two-dimensional space, one needs to look at the loop equivalence class and not the closed path itself.

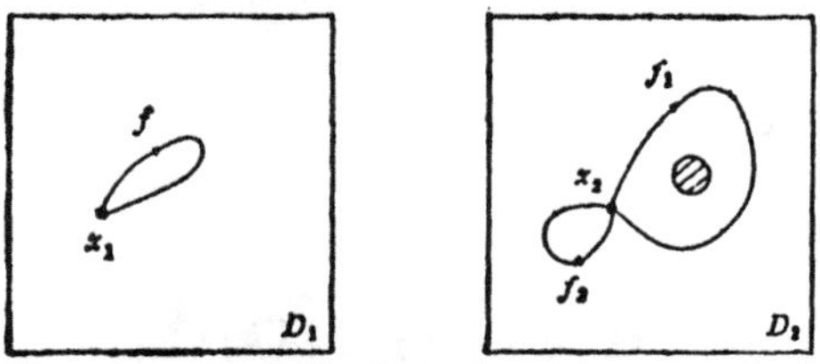

Figure 8.1 Sketch map of the loop homotopy equivalence class

Both the starting point and the end point of a loop are the same point x_0. Hence, one can define the multiplication of two loops with the same base point. For loops f and g with base point x_0, the product $h = f \cdot g$ is defined as

$$h(\lambda) = \begin{cases} f(2\lambda), & 0 \le \lambda \le \frac{1}{2}, \\ g(2\lambda - 1), & \frac{1}{2} \le \lambda \le 1, \end{cases}$$

h is still a loop with base x_0. This operation preserves homotopy, namely, if $f \sim f'$, $g \sim g'$, then $f \cdot g \sim f' \cdot g'$.

For each loop f the inverse loop f^{-1} can be defined as

$$f^{-1}(t) = f(1-t), \qquad 0 \le t \le 1,$$

such that $f \cdot f^{-1} \simeq e$, i.e., the identity mapping

$$\begin{aligned} e : I &\rightarrow M \\ \lambda &\rightarrow e(\lambda) = x_0 \end{aligned}$$

is homotopic to the product of f and f^{-1}.

For the set of homotopy equivalence loop classes with base point x_0, product of loop classes can be defined

$$< f_1 > < f_2 > = < f_1 \cdot f_2 >, \tag{8.1}$$

i.e., product of homotopy classes is defined as the class determined by the product of their representative elements in the corresponding homotopy classes. The homotopy class of identity mapping which maps interval I to a point x_0 is taken as the unit element. Then the set of loop homotopy classes with base x_0 and product defined in Eq. 8.1 is a group, called the homotopy group $\Pi_1(M, x_0)$ with base point x_0 on manifold M. $\Pi_1(M, x_0)$ is also called the fundamental group of the manifold.

Here, we emphasize that it is the set of loop homotopy classes and not the set of loops that has the group structure.

For homotopy group $\Pi_1(M, x_0)$, there is a theorem:

Theorem 8.1 *For two path connected manifolds M and N, if they have the same homotopy type, then their homotopy groups are isomorphic to each other*

$$\Pi_1(M, x_0) \sim \Pi_1(N, y_0), \qquad x_0 \in M, \quad y_0 \in N.$$

A natural corollary of this theorem is that if M is a path connected manifold, then for any two points $x_0, y_0 \in M$, groups $\Pi_1(M, x_0)$ and $\Pi_1(M, y_0)$ are isomorphic and $\Pi_1(M, x_0)$ depends only on M and is independent of the choice of the base point x_0, so we can talk about the abstract fundamental group $\Pi_1(M)$. $\Pi_1(M)$ is called the fundamental group of the manifold. When dealing with the abstract fundamental group $\Pi_1(M)$, we still need to study the homotopy group $\Pi_1(M, x_0)$ for a given base point x_0.

A manifold is called simply connected if the fundamental group of a path connected manifold is trivial (The only group element is unit.). For example, $\mathbf{E}^n$ and $S^n (n \geq 2)$ are simply connected manifolds.

The fundamental group is an effective tool in studying the classification of two-dimensional surfaces. The nontrivial fundamental group of a two dimensional surface indicates the existence of holes on the surface. For multi-dimensional manifolds, higher-order homotopy groups are introduced which are similar to the fundamental group. The higher order homotopy group $\Pi_k(M, x_0)$ is the set of homotopy equivalence classes of mappings of the k-loop with base point x_0. To highlight the point, we generalize the one-loop to the k-loop. Let I_k be a closed unit interval in $\mathbf{E}^k$

$$I_k = \{\lambda = (\lambda^1, \cdots, \lambda^k) \in E^k | 0 \leq \lambda^i \leq 1, \quad i = 1, \cdots, k\}.$$

If the continuous mapping is

$$\begin{aligned} f : I_k &\rightarrow M \\ \lambda^1, \cdots, \lambda^k &\rightarrow f(\lambda^1, \cdots, \lambda^k) \in M \end{aligned}$$

such that the boundary ∂I_k of I_k is mapped to a fixed point x_0 in M:

$$f(\lambda^1, \cdots, \lambda^k) = x_0, \text{ when } \lambda^i = 0, 1 \text{ for } i = 1, \cdots, k,$$

then mapping f is called a k-loop with base point x_0. This mapping is equivalent to the continuous mapping from S^k to M which maps a given point (for instance, north pole) in S^k to a fixed point x_0 in M. The product of two k-loops f and g with the same base point x_0 can be defined as

$$h = f \cdot g :$$

$$h(\lambda^1, \cdots, \lambda^k) = \begin{cases} f(2\lambda^1, \lambda^2, \cdots, \lambda^k), & 0 \leq \lambda^1 \leq \frac{1}{2}, \\ g(2\lambda^1 - 1, \lambda^2 \cdots, \lambda^k), & \frac{1}{2} \leq \lambda^1 \leq 1. \end{cases}$$

Figure 8.2 illustrates the product of two 2-loops.

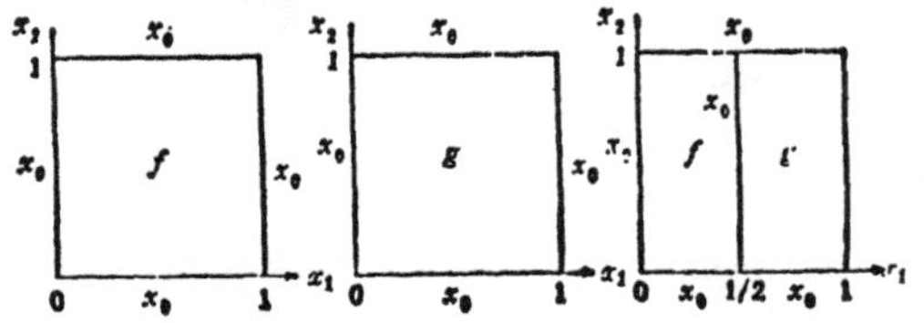

Figure 8.2 Sketch figure of the product of two 2-loops.

If one k-loop can be continuously deformed to another k-loop with the base point x_0 unchanged, we call the two loops are homotopically equivalent and use $< f >$ to represent the homotopy loop class in which any k-loop is homotopically equivalent to k-loop f. Product of homotopy loop classes with the same base point x_0 is defined as the class determined by the product of any element in each class

$$< f > \cdot < g > = < f \cdot g > .$$

The set of homotopy classes with the above defined multiplications is a group and is called the k-th homotopy group $\Pi_k(M, x_0)$ of the manifold M with base point x_0.

The fundamental group $\Pi_1(M, x_0)$ could be a non Abelian group. The higher homotopy group $\Pi_k(M, x_0)$ for $k \geq 2$ must be Abelian. The reason is that the boundary $S^0 = \{0, 1\}$ of $I_1 = [0, 1]$ are two distinct points, but the boundary $\partial I_k \sim S^{k-1}$ of I_k for $k \geq 2$ is connected. Any k-loop f with base point x_0 can be continuously deformed to a homotopy loop such that the boundary $\partial I_k \sim S^{k-1}$ of I_k is thickened. As S^{k-1} $(k \geq 2)$ is connected, the product of closed homotopy $k(\geq 2)$-paths commute. For example, the product of two 2-loop classes f and g, as shown in Figure 8.3, are permutable.

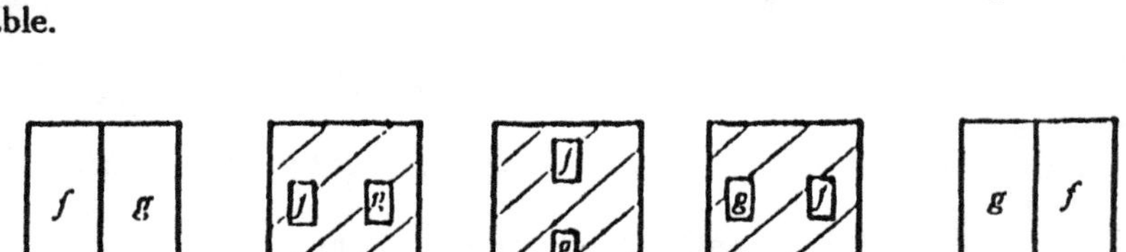

Figure 8.3 Permutation of the product of two 2-loop classes

It is easy to show that for path connected manifold M, homotopy groups $\Pi_k(M, x_0)$ and $\Pi_k(M, x_1)$ with different base point are isomorphic. For path connected manifold M, the homotopy group is simply written as $\Pi_k(M)$.

When calculating homotopy groups of a manifold, it is sometimes necessary to study the homotopy group of the product manifold of manifold M and N. There is a very useful theorem:

Theorem 8.2 *The homotopy group of the product manifold of two manifolds M and N is isomorphic to the direct product of the homotopy groups of M and N:*

$$\Pi_k(M \times N, p \times q) \simeq \Pi_k(M,p) \oplus \Pi_k(N,q), \qquad p \in M, q \in N, \tag{8.2}$$

where the right hand side is the direct product of two groups. Here $\oplus$ denotes direct product of groups G_1 and G_2 which is the set of all ordered pairs

$$G_1 \oplus G_2 = \{(g_1, g_2) | g_1 \in G_1, g_2 \in G_2\}.$$

The product of the ordered pairs is induced from the product of G_1 and G_2:

$$(g_1, g_2) \cdot (g_1', g_2') = (g_1 g_1', g_2 g_2').$$

The proof of Eq. 8.2 is very easy, because the k-loop on manifold $M \times N$ with base point (p,q) is the set of k-closed path pairs of (M,p) and (N,q).

Example: For the two-dimensional torus

$$T_2 = S^1 \times S^1,$$

$$\Pi_1(T_2) = Z \oplus Z, \qquad \Pi_k(T_2) = 0 \ (k \geq 2).$$

8.2 Relative homotopy group and exact homotopy sequence

It is very difficult to calculate homotopy groups of a manifold. Even for the simplest manifold S^2, the calculation of homotopy groups of all orders is not solved completely yet. In this section, we introduce the relative homotopy group and the exact homotopy sequence, which can be used to calculate the homotopy groups of manifolds.

Let manifold N be a closed submanifold of manifold M and point q belong to N:

$$q \in N \subset M.$$

If there exists a mapping f from I_k to M, where one surface I_{k-1} of ∂I_k is mapped to N:

$$f: \quad \begin{array}{ll} I_k = \{(\lambda^1, \cdots, \lambda^k) \in I^k | 0 \leq \lambda^i \leq 1\} & \to M, \\ I_{k-1} = \{(\lambda^1, \cdots, \lambda^k) \in I^k | \lambda^k = 0\} & \to N. \end{array} \tag{8.3}$$

Moreover, other boundary surfaces $\partial I_k \backslash I_{k-1}$ of I_k and ∂I_{k-1} are mapped to the given point q. Mapping f then is called a k-loop in M relative to N with base point q. For instance, the 2-loop with base point q in M relative to N is illustrated in Figure 8.4.

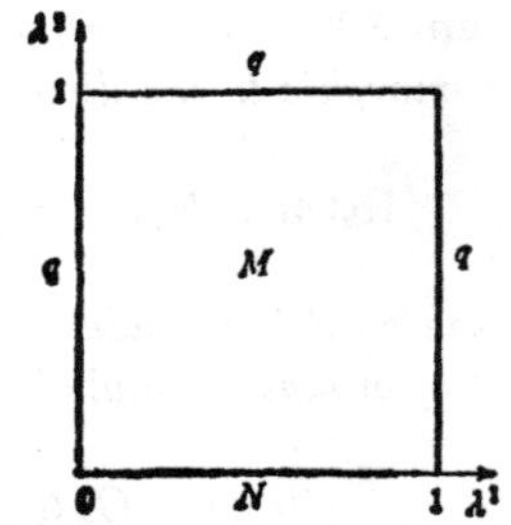

Figure 8.4.

Product of two relative k-loops can be defined similar to Eq. 8.2. For example, product $f \cdot g$ of two relative 2-loops is shown in Figure 8.5.

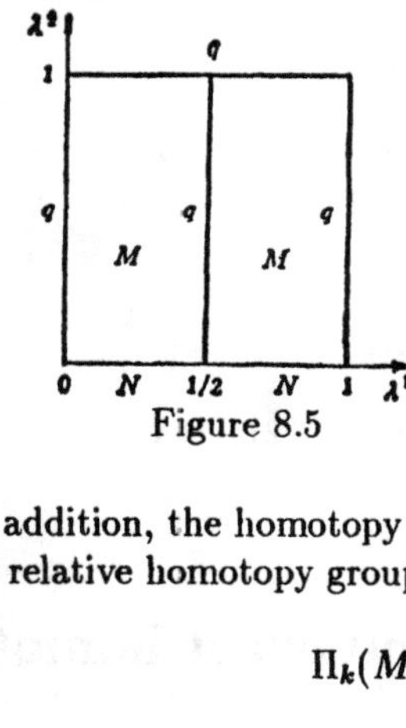

Figure 8.5

The product of homotopy equivalence classes of relative k-loop can be defined as the homotopy equivalence class of the product of the representative elements of respective classes. The set of homotopy equivalence classes of relative k-loop with this product constructed the relative homotopy group $\Pi_k(M, N; q)$ of a pair of manifolds (M, N).

In addition, the homotopy equivalence classes of $\Pi_k(N, q)$ is the identity element of the relative homotopy group $\Pi_k(M, N; q)$ and

$$\Pi_k(M, q; q) = \Pi_k(M, q) \qquad (k \geq 2). \tag{8.4}$$

Now, we look at the relationship between the relative homotopy group $\Pi_k(M, N; q)$ and the homotopy groups $\Pi_k(M, q)$ and $\Pi_k(N, q)$.

As $N \subset M$, there exists an inclusion mapping i

$$i : N \to M,$$

such that there is an induced mapping i_* among the k-loop classes. Correspondingly, there is a homomorphism mapping of the homotopy group:

$$i_* : \Pi_k(N, q) \to \Pi_k(M, q).$$

Similarly, there exists an inclusion mapping j between the pair of manifolds (M, q) and (M, N)

$$j : (M, q) \to (M, N)$$

and the induced mapping j_*

$$j_* : \Pi_k(M, q; q) \rightarrow \Pi_k(M, N; q).$$

From Eq. 8.4, we have the homomorphism sequence

$$\Pi_k(N, q) \xrightarrow{i_*} \Pi_k(M, q) \xrightarrow{j_*} \Pi_k(M, N; q). \tag{8.5}$$

The image of the mapping i_* includes all k-loops with base point q which are homotopic to the identity element of $\Pi_k(M, N; q)$ under mapping j_*, and *vice versa*, i.e., the homomorphism sequence Eq. 8.5 is an exact sequence (refer to Appendix F):

$$Imag\ i_* = Ker\ j_*.$$

In addition, if restricting the mapping f in Eq. 8.3 to the open surface I_{k-1} defined by $\lambda^k = 0$, for k-loops with base point q of the pair of manifold (M, N), we can obtain a homomorphism mapping ∂ from $\Pi_k(M, N; q)$ to $\Pi_{k-1}(N, q)$:

$$\partial : \Pi_k(M, N; q) \rightarrow \Pi_{k-1}(N, q)$$

which is called the boundary homomorphism. It can be proved that the following homomorphism sequence is exact when the pair of manifolds (M, N) are path connected

$$\cdots \longrightarrow \Pi_k(N, q) \xrightarrow{i_*} \Pi_k(M, q) \xrightarrow{j_*} \Pi_k(M, N; q) \xrightarrow{\partial} \Pi_{k-1}(N, q) \longrightarrow \cdots. \tag{8.6}$$

One of the basic properties of manifold pair is that the exact sequence of group homomorphism can be formed.

Fiber bundle, which will be introduced in §8.4, is a powerful tool in calculating the homotopy groups of various manifolds. A fiber bundle E locally can be taken as the direct product manifold of the base manifold B and the fiber F with a projection mapping

$$\pi : E \rightarrow B,$$

at point $x \in B$, the inverse image $\Pi^{-1}(x)$ of the projection mapping is the fiber space F_x. The fiber F_x at each point $x \in B$ is isomorphic to the standard fiber F. Taking arbitrary point y_0 on the fiber $F_{x_0} = \pi^{-1}(x_0)$, among $\Pi_k(E, y_0)$, $\Pi_k(F, y_0)$ and $\Pi_k(E, F; y_0)$ there exists the following exact group homomorphism sequence (similar to exact sequence Eq. 8.6):

$$\cdots \longrightarrow \Pi_k(F) \xrightarrow{i_*} \Pi_k(E) \xrightarrow{j_*} \Pi_k(E, F) \xrightarrow{\partial} \Pi_{k-1}(F) \longrightarrow \cdots. \tag{8.7}$$

The projection mapping of the bundle $\pi : E \rightarrow B$ preserves the fiber such that fiber $F_{x_0} = \pi^{-1}(x_0)$ is mapped to a single point x_0. Therefore, there is a homomorphism mapping

$$\pi_* : \Pi_k(E, F, y_0) \rightarrow \Pi_k(B, x_0).$$

This homomorphism is actually an isomorphism: $\Pi_k(E, F, y_0) \simeq \Pi_k(B, x_0)$ such that the exact sequence Eq. 8.7 leads to an exact homomorphism sequence of the homotopic groups of fiber bundle $E(B, F, \pi)$

$$\cdots \longrightarrow \Pi_k(F) \xrightarrow{i_*} \Pi_k(E) \xrightarrow{\pi_*} \Pi_k(B) \xrightarrow{\partial} \Pi_{k-1}(F) \to \cdots. \qquad (8.8)$$

Making use of this exact sequence, homotopy groups of many homogeneous spaces can be obtained.

Example 1. We know that $S^1 = E^1/Z$. Similar to Eq. 8.8, we have exact sequence

$$\to \Pi_k(Z) \to \Pi_k(E^1) \to \Pi_k(S^1) \to \Pi_{k-1}(Z) \to \cdots \quad (k \geq 2),$$

i.e.,

$$0 \to \Pi_k(E^1) \to \Pi_k(S^1) \to 0 \qquad (k \geq 2).$$

As E^1 is a contractible manifold, the homotopy group of any order is zero, hence

$$\Pi_k(S^1) = \Pi_k(E^1) = 0 \qquad (k \geq 2).$$

Example 2. For the coset space G/H of group G, similarly, there exists an exact sequence

$$\cdots \to \Pi_2(G) \to \Pi_2(G/H) \to \Pi_1(H) \to \Pi_1(G) \to \cdots.$$

For any simple Lie group, one has

$$\Pi_1(G) = \Pi_2(G) = 0.$$

Hence, for any simple Lie group one has

$$\Pi_2(G/H) = \Pi_1(H).$$

Example 3. For the unit tangent vector bundle on S^n

$$\begin{array}{lll} S^{n-1} \to & V_{n+1,2} & \simeq SO(n+1)/SO(n-1) \\ & \downarrow & \\ & S^n & \simeq SO(n+1)/SO(n), \end{array}$$

from Eq. 8.8, there exists the following exact sequence:

$$\cdots \to \Pi_k(S^{n-1}) \xrightarrow{i_*} \Pi_k(V_{n+1,2}) \xrightarrow{\pi_*} \Pi_k(S^n) \to \cdots.$$

When $k < n-1$, we have

$$0 \xrightarrow{i_*} \Pi_k(V_{n+1,2}) \xrightarrow{\pi_*} 0.$$

Hence,

$$\Pi_k(V_{n+1,2}) = 0 \qquad (k < n-1). \qquad (8.9)$$

Finally, we introduce some basic terms. For a manifold M, if the homotopy group satisfies the following condition

$$\Pi_k(M) = 0, \qquad 1 < k \leq n,$$

M is called k-order connected manifold. For $k = 1$, i.e., the fundamental group of the manifold is trivial ($\Pi_1(M) = 0$), the manifold is called simply connected manifold. From Eq. 8.9, one sees that manifold $V_{n+1,2}$ is an $(n-2)$-order connected manifold.

In view of the importance of the fundamental group, the manifold whose fundamental group is an order-m cyclic group is called m-connected manifold, where m is the number of generators of the fundamental group. Please do not confuse m-order connected manifold with m-connected manifold. For instance, $SO(n)(n \geq 3)$ is a double connected manifold, some-times the manifold $SU(1,1)$ is called as an ∞ connected manifold.

In the following, some manifolds and the corresponding homotopy groups are given:

connected manifold	$\Pi_0(M) = 0,$
simply connected manifold	$\Pi_1(M) = 0,$
k-order connected manifold	$\Pi_k(M) = 0 \;\; (k \geq 2),$
double connected manifold	$\Pi_1(M) = Z_2.$

For a Lie group of transformation acting on simply connected manifold M, if the action of Lie group of transformation is effective, i.e., for isotropy subgroup G_x corresponding to arbitrary point x on manifold M, there is an exact sequence

$$0 \to G_x \to G.$$

Consider further the fundamental group $\Pi_1(M/G, x)$ of orbit space M/G, there is a homomorphism mapping from G to fundamental group $\Pi_1(M/G, x)$ and the kernel of the mapping is the isotropy group G_x. Therefore there is the following group exact sequence

$$0 \to G_x \to G \to \Pi_1(M/G, x) \to 0.$$

Hence,

$$\Pi_1(M/G, x) \simeq G/G_x.$$

If the action of the group of transformation is freely, G_x contains only the identity element, in which case

$$\Pi_1(M/G, x) \simeq G \;\; (\text{when } G_x = e). \tag{8.10}$$

Example 4.

$$\begin{cases} \Pi_1(S^1) = \Pi_1(E^1/Z) = Z, \\ \Pi_1(T^2) = \Pi_1(E^2/Z \times Z) = Z \oplus Z, \\ \Pi_1(RP(n)) = \Pi_1(S^n/Z_2) = Z_2, (n > 1). \end{cases}$$

As is known from the theory of Lie groups, for connected Lie group G, there exists a universal covering group $\tilde{G}$ which is a simply connected Lie group and

$$G = \tilde{G}/D,$$

where D is the discrete invariant subgroup of $\tilde{G}$. From Eq. 8.10, we have

$$\Pi_1(G) = \Pi_1(\tilde{G}/D) = D.$$

Example 5.

$$\begin{cases} \Pi_1(SO(n)) = \Pi_1 \quad (Spin(n)/Z_2) = Z_2, \\ \Pi_1(U(1)) = \Pi_1 \quad (R^1/Z) = Z, \\ \Pi_1(SU(n)) = 0 \quad (n \geq 2), \\ \Pi_1(SU(1,1)) = Z. \end{cases}$$

In the following we study the stable homotopy groups of $O(n)$, $U(n)$ and $SP(n)$ by using homotopy group sequences.

First we look at the n orthogonal frame bundle on sphere S^n

$$SO(n) \to SO(n+1) \to S^n \simeq SO(n+1)/SO(n).$$

There is a group homomorphism sequence

$$\cdots \to \Pi_{k+1}(S^n) \xrightarrow{\partial} \Pi_k(SO(n)) \xrightarrow{i_*} \Pi_k(SO(n+1)) \xrightarrow{\pi_*} \Pi_k(S^n) \to \cdots.$$

As

$$\Pi_{k+1}(S^n) = \Pi_k(S^n) = 0 \quad (k < n-1),$$

the group homomorphism sequence above is reduced to

$$0 \to \Pi_k(SO(n)) \to \Pi_k(SO(n+1)) \to 0 \quad (k < n-1).$$

Hence it can be derived that there exist stable homotopy groups for $SO(n)$ group:

$$\Pi_k(SO(n)) = \Pi_k(SO(n+1)) \quad (n > k+1). \tag{8.11}$$

Next we look at the complex n-dimensional projective space

$$CP(n) \simeq U(n+1)/U(n) \times U(1),$$

and notice that the unit sphere in $C^n \simeq R^{2n}$ is

$$S^{2n-1} \simeq U(n)/U(n-1) \simeq SU(n)/SU(n-1),$$

there is a Hopf bundle with fiber $U(1)$ and base space $CP(n)$

$$U(1) \to S^{2n+1} \to CP(n).$$

There exists a homotopy group exact sequence

$$\cdots \to \Pi_{k+1}(S^{2n-1}) \xrightarrow{\partial} \Pi_k(U(n-1)) \xrightarrow{i_*} \Pi_k(U(n)) \xrightarrow{\pi_*} \Pi_k(S^{2n-1}) \to \cdots .$$

When $k < 2n-2$, the exact sequence above is reduced to

$$0 \to \Pi_k(U(n-1)) \simeq \Pi_k(U(n)) \to 0 \quad (k < 2n-2).$$

Hence there exist stable homotopy groups for $U(n)$ group

$$\Pi_k(U(n+1)) \simeq \Pi_k(U(n)) \quad (k < 2n). \tag{8.12}$$

Similarly we can analyse the quaternion Hopf bundle

$$SP(1) \to S^{4n+3} \to HP(n),$$

where $HP(n)$ is quaternion projective space

$$HP(n) \simeq S^{4n+3}/SP(1) \simeq SP(n+1)/SP(n) \times SP(1).$$

It can be proved that

$$\Pi_k(SP(n)) = \Pi_k(SP(n+1)) \quad \text{when } k < 4n+2. \tag{8.13}$$

It is clear from Eqs. 8.11-8.13 that when n is large enough the homotopy groups of orthogonal group, unitary group and symplectic group are independent of n and are called stable homotopy groups. To these Lie groups, the stable homotopy groups of all orders have their corresponding Bott periods. The detail is given in Table 8.1.

Table 8.1 Stable homotopy groups of U, O and SP groups

Π_k	U(n) $(n > \frac{k}{2})$	O(n) $(n > k+1)$	SP(n) $(n > \frac{1}{4}(k-2))$
0	0	$\mathbf{Z}_2$	0
1	$\mathbf{Z}$	$\mathbf{Z}_2$	0
2	0	0	0
3	$\mathbf{Z}$	$\mathbf{Z}$	$\mathbf{Z}$
4	0	0	$\mathbf{Z}_2$
5	$\mathbf{Z}$	0	$\mathbf{Z}_2$
6	0	0	0
7	$\mathbf{Z}$	$\mathbf{Z}$	$\mathbf{Z}$
8	0	$\mathbf{Z}_2$	0
Bott period	2	8	8

Some remarks here:
1). $\Pi_0(M)$ shows the connectivity of manifold M.

$$\Pi_0(U(n), SU(n), SO(n)) = 0,$$

shows that $U(n)$, $SU(n)$ and $SO(n)$ are connected manifolds.

$$\Pi_0(O(n)) = \mathbf{Z}_2,$$

shows that $O(n)$ is not a connected manifold and has a parity operation.
2). For higher order unitary group $U(n)$, the Bott period is 2. When $n > \frac{k}{2}$,

$$\Pi_k(U(n)) = \begin{cases} \mathbf{Z}, & k \text{ is odd}, \\ 0, & k \text{ is even}. \end{cases}$$

For $k = 2n$,

$$\Pi_{2n}(U(n)) = \mathbf{Z}_n \quad (n > 1).$$

3). For higher order orthogonal and symplectic groups, the Bott period is 8. And

$$\Pi_k(O(n)) = \Pi_{k+4}(SP(n)),$$

$$\Pi_3(O(4)) = \mathbf{Z} \oplus \mathbf{Z}.$$

4). For all three types of groups in Table 8.1, $\Pi_2 = 0$, and $\Pi_3 = \mathbf{Z}$. This is related to the fact that all compact semisimple Lie group has $SU(2) \simeq S^3$ as its Lie subgrop, so for compact simple Lie group G

$$\Pi_2(G) = 0, \ \Pi_3(G) = \mathbf{Z}.$$

5). For unitary group, $U(n) \simeq U(1) \times SU(n)$, when $n \geq 2$.

$$\begin{aligned} \Pi_1(U(n)) &= \quad \mathbf{Z} \qquad (n \geq 1), \\ \Pi_1(SU(n)) &= \quad 0 \qquad (n \geq 2), \\ \Pi_k(U(n)) &= \ \Pi_k(SU(n)) \ (k, n \geq 2). \end{aligned}$$

6). For orthogonal group, $O(n) = SO(n) \times \mathbf{Z}_2$.

$$\Pi_1(O(n)) = \Pi_1(SO(n)) = \mathbf{Z}_2 \ \ (n \geq 3).$$

This shows that the orthogonal group has spin representations.

$$\Pi_k(O(n)) = \Pi_k(SO(n)), \quad \text{when } k \geq 1, n \geq 2.$$

7). For general linear group, it can be proved that

$$\Pi_k(GL(n, C)) = \Pi_k(U(n)),$$

$$\Pi_k(GL(n,R)) = \Pi_k(O(n)).$$

From the section 7.8 group $GL(n,R)$ and group $O(n)$ are of the same homotopy type and so they have the same homotopy groups, i.e.,

$$\begin{aligned}\Pi_1(GL(2,R)) &= \Pi_1(O(2)) = \mathbf{Z},\\ \Pi_1(GL(n,R)) &= \Pi_1(O(n)) = \mathbf{Z}_2 \quad (n \geq 3),\\ \Pi_k(GL(n,R)) &= \Pi_k(O(n)).\end{aligned}$$

In a similar way, it can be shown that

$$\Pi_k(GL(n,C)) = \Pi_k(U(n)).$$

Finally we give homotopy groups of some homogeneous manifolds needed in the gauge field theory(Ref. to Table 8.1) as follows.

Table 8.2 Homotopy groups of several homogeneous manifolds.

	SU(2n)/SP(n)	SU(n)/O(n)
Π_2	0	$\mathbf{Z}_2$ $(n > 3)$, $\mathbf{Z}$ $(n = 2)$
Π_3	0	$\mathbf{Z}_2$ $(n > 4)$, $\mathbf{Z}_4$ $(n = 3)$
Π_5	$\mathbf{Z}$ $(n > 2)$	$\mathbf{Z}$ $(n > 3)$

8.3 Relation between homotopy group and homology group

Both homotopy group and homology group are algebraic functors on manifolds. Both are tools used in the study of topological properties of manifolds by algebraic methods and are the bases of algebraic topology (application of algebraic methods such as groups and homomorphism to study the topological properties of manifolds). Homotopy group is conceptually intuitive and has direct links with global properties of manifolds. However, the calculation of homotopy group is very difficult. On the other hand, the calculation of homology group is relatively easy. Particularly, in de Rham cohomology theory and harmonic forms, by using calculus the local properties and the global properties of manifolds are combined. Both the homotopy group and the homology group are homotopy type invariants of manifolds. They are closely related to each other. There exists homomorphism mapping from homotopy group to homology group. There is a theorem:

Theorem 8.3 *If manifold M is path connected, then homomorphism mapping*

$$\psi : \Pi_1(M) \to H_1(M)$$

is surjective. The kernel of the mapping is the commutator subgroup of $\Pi_1(M)$:

$$H_1(M) = \Pi_1(M)/\left[\Pi_1(M), \Pi_1(M)\right],$$

where $[\Pi_1(M), \Pi_1(M)]$ *is the commutator subgroup of* $\Pi_1(M)$, *which is the smallest subgroup that makes the quotient group Abelian. We may say that* $H_1(M)$ *is the Abelization of* $\Pi_1(M)$.

Theorem 8.4 *(Hurewicz isomorphism theorem): For a simply connected and path connected manifold* M $(H_1(M) = \Pi_1(M) = 0)$, *the first nontrivial homotopy group and homology group appear at the same dimension and are equal. That is to say, if*

$$\Pi_k(M) = 0 \qquad (\text{ for } 1 \leq k < m),$$

then

$$H_k(M) = 0 \qquad (\text{ for } 1 \leq k < m),$$

and

$$H_m(M) = \Pi_m(M).$$

For instance, the homology group of the n-dimensional sphere S^n is:

$$H_0(S^n) \simeq H_n(S^n) \simeq Z,$$
$$H_k(S^n) = 0 \qquad (\text{ for } k \neq 0, n).$$

S^n is a nontrivial manifold with the simplest homology group. However, the homotopy group of S^n is quite complicate. From the Hurewicz theorem, when $k \leq n$, the homotopy group and the homology group are isomorphic

$$\Pi_k(S^n) \simeq H_k(S^n) \qquad (k \leq n).$$

But when $k > n$, the homotopy group of S^n is so complicated that even the homotopy groups of S^2 has not be solved completely yet. In the following, some lower order homotopy groups of S^n are given in table 8.3 and table 8.4.

Table 8.3. Homotopy groups of the sphere S^n

$\Pi_k \backslash S^n$	1	2	3	4	5	6	7	8	9	10	11
1	Z	0	0	0	0	0	0	0	0	0	0
2	0	Z	0	0	0	0	0	0	0	0	0
3	0	Z	Z	0	0	0	0	0	0	0	0
4	0	Z_2	Z_2	Z	0	0	0	0	0	0	0
5	0	Z_2	Z_2	Z_2	Z	0	0	0	0	0	0
6	0	Z_{12}	Z_{12}	Z_2	Z_2	Z	0	0	0	0	0
7	0	Z_2	Z_2	$Z + Z_{12}$	Z_2	Z_2	Z	0	0	0	0
8	0	Z_2	Z_2	$Z_2 + Z_2$	Z_{24}	Z_2	Z_2	Z	0	0	0
9	0	Z_3	Z_3	$Z_2 + Z_2$	Z_2	Z_{24}	Z_2	Z_2	Z	0	0
10	0	Z_{15}	Z_{15}	$Z_3 + Z_{24}$	Z_2	0	Z_{24}	Z_2	Z_2	Z	0
11	0	Z_2	Z_2	Z_{15}	Z_2	Z	0	Z_{24}	Z_2	Z_2	Z

Note: 1).

$$\Pi_n(S^n) = \mathbf{Z},$$
$$\Pi_k(S^n) = 0 \qquad (k < n),$$
$$\Pi_k(S^1) = 0 \qquad (k > 1).$$

2).

$$\Pi_{n+k}(S^n) \simeq \Pi_{n+k+1}(S^{n+1}) \qquad (n \geq k+2),$$

which is called the stable homotopy group of the sphere. Stable homotopy groups for $k \leq 19$ of the sphere is given in table 8.4 [36].

Table 8.4 The stable homotopy group of the sphere S^n

k	$\Pi_{n+k}(S^n)$, $n \geq k+2$	k	$\Pi_{n+k}(S^n)$, $n \geq k+2$
0	Z	10	Z_6
1	Z_2	11	Z_{504}
2	Z_2	12	0
3	Z_{24}	13	Z_3
4	0	14	$Z_2 + Z_2$
5	0	15	$Z_{480} + Z_2$
6	Z_2	16	$Z_2 + Z_2$
7	Z_{240}	17	$Z_2 + Z_2 + Z_2 + Z_2$
8	$Z_2 + Z_2$	18	$Z_8 + Z_2$
9	$Z_2 + Z_2 + Z_2$	19	$Z_{264} + Z_2$

Note: In table 8.3, except for $\Pi_n(S^n) = \mathbf{Z}$ and $\Pi_{2n-1}(S^n)$ which might be the direct sum of Z and a finite group, all other homotopy groups are finite groups.

S^n is a nontrivial manifold with the simplest homology group. On the other hand, the Eilenberg-Maclane space $K(G,n)$ is a nontrivial manifold with the simplest homotopy group where G is an Abelian group. $K(G,n)$ is a space which has the following homotopy group:

$$\begin{cases} \Pi_n(K(G,n)) \simeq G, \\ \Pi_k(K(G,n)) = 0 \qquad (\text{ for } k \neq n). \end{cases}$$

For instance, $K(\mathbf{Z},1) = S^1$. The homotopy group of $K(G,n)$ is very simple while the homology group of $K(G,n)$ is in general complicated. But since the homology theory itself is simpler than the homotopy theory, the calculation of homology groups of $K(G,n)$ is still simpler than that of the homotopy groups of the sphere S^n.

In superstring theories, grand unification and super grand unification models, one frequently encounters the 248-manifold E_8. The manifold is quite complicated, but the lower-order homotopy groups are relatively simple[37]:

$$\begin{cases} \Pi_3(E_8) = \mathbf{Z}, \\ \Pi_k(E_8) = 0 \qquad (\text{ for } 1 \leq k \leq 14, \quad k \neq 3). \end{cases} \tag{8.14}$$

So in analyzing phenomena on the manifold with dimension lower than 14, E^8 is equivalent to $K(\mathbf{Z},3)$ which has simple homotopy group and complicated homology group.

For a manifold M, the classification of vector bundles which are endowed with given structure group G is often determined by the homotopy group of the manifold about which we will study in §8.7. For instance, the classification of vector bundles on S^n, S^n can be represented as the union of the northern and the southern hemispheres. The overlapping of the northern and the southern hemispheres is homotopic to the equator S^{n-1}. The gauge transformation at the equator is labelled by the homotopy group $\Pi_{n-1}(G)$. From Eq. 8.14, when $n \leq 15$, except for $n = 4$, E_8 bundle is trivial, and the E_8 bundle on S^4 can be labelled by integer Z, because $\Pi_3(E_8) = \mathbf{Z}$.

8.4 Fibre bundle

Even at the beginning of this book, we have already used the fibre bundle language. For example, tangent bundle $T(M)$, cotangent bundle $T^*(M)$, tensor bundle $T^{r,s}(M)$, frame bundle $L(M), \cdots$. In modern theoretical physics fibre bundles are extensively and effectively used in Yang-Mills theory, Kaluza-Klein theory and string theory, reflecting the global topological properties of gauge field theory. Now, we further highlight some basic concepts about fibre bundles.

A bundle is a triple (E, M, π), where E and M are topological spaces and π is a projection mapping

$$\pi: \quad E \rightarrow M.$$

Space M, which is always a smooth manifold, is called the base manifold. The projection mapping π is surjective and continuous, and for each point $x \in M$, the set of points $\Pi^{-1}(x)$ in E is called the fiber of the bundle over x, denoted as F_x. Space E is called the total space, which is the union of all fibers, $E = \bigcup_{x \in M} F_x$. We shall restrict ourselves to situations in which topological spaces F_x for all $x \in M$ are homeomorphic to a space F, which is called the typical fiber.

A fiber bundle $E(M, F, \pi, G)$ is a bundle with an additional structure, the structure group G, which is a topological group acting on the fibre. Fiber bundle is locally like Cartesian product $U \times F$, where $U \in M$ is an open neighborhood of base manifold M, but in general, fibre bundle is not a trivial Cartesian product globally. If the base manifold M is covered by a set of open neighborhood $\{U_\alpha\}$

$$M = \bigcup_\alpha U_\alpha.$$

In each neighborhood U_α the fiber bundle can be described locally by the Cartesian product $U_\alpha \times F$. The projection $\pi : E \rightarrow M$ satisfies the requirement that there exists a diffeomorphism

$$\begin{array}{llll} \Psi_\alpha: & \pi^{-1}(U_\alpha) & \to & U_\alpha \times F \\ & u & \to & (\pi(u), \alpha_{\pi(u)}(u)) = (x, \xi_\alpha(x)) \end{array} \tag{8.15}$$

where α_x is a homomorphism from fiber F_x to the typical fiber F. Mapping Ψ_α shows that fiber bundle can be trivialized locally.

If $U_\alpha \cap U_\beta \neq 0$,there exists another diffeomorphism

$$\begin{array}{llll} \Psi_\beta: & \pi^{-1}(U_\beta) & \to & U_\beta \times F \\ & u & \to & (\pi(u), \beta_{\pi(u)}(u)) = (x, \xi_\beta(x)). \end{array}$$

Then for a given point $x \in U_\alpha \bigcap U_\beta$, $\Psi_{\alpha\beta}(x) = \alpha \cdot \beta^{-1}(x)$ is an automorphism of F, and belonging to the automorphism group $G(F)$ of the space F. The continuous mapping $\Psi_{\alpha\beta}$:

$$\begin{array}{llll} \Psi_{\alpha\beta}: & U_\alpha \bigcap U_\beta & \to & G(F) \\ & x & \to & \Psi_{\alpha\beta}(x) \end{array}$$

such that

$$\xi_\alpha(x) = \Psi_{\alpha\beta}(x)\xi_\beta(x)$$

is called the transition function between U_α and U_β, which is fibre preserving.

For a given point x of manifold M, the transition function of the fiber bundle E satisfies the following compatibility conditions:

$$\begin{array}{ll} \Psi_{\alpha\alpha}(x) = 1, & x \in U_\alpha; \\ \Psi_{\alpha\beta}(x)\Psi_{\beta\alpha}(x) = 1, & x \in U_\alpha \bigcap U_\beta; \\ \Psi_{\alpha\beta}(x)\Psi_{\beta\gamma}(x)\Psi_{\gamma\alpha} = 1, & x \in U_\alpha \bigcap U_\beta \bigcap U_\gamma. \end{array}$$

where the repeated indices are not summed over.

The set $\Psi_{\alpha\beta}(x)$ constitutes a group $G \subset GL(F)$ called the structure group of fiber bundle E. A fiber bundle is completely determined by its transition functions. Transition functions $\Psi_{\alpha\beta}(x)$ are G-valued functions on base manifold. They are independent of the choose of coordinate systems of base manifold, but if changing the diffeomorphism Ψ_α, i.e., the typical fibre has a transformation $g_\alpha(x) \in G$

$$\begin{array}{llll} g_\alpha(x): & F & \to & F' \\ & \xi_\alpha(x) & \to & \xi'_\alpha(x) = g_\alpha(x)\xi_\alpha(x) \end{array}$$

and the transition functions $\Psi_{\alpha\beta}(x)$ will be changed as

$$\Psi'_{\alpha\beta}(x) = g_\alpha(x)\Psi_{\alpha\beta}(x)g_\beta(x)^{-1}.$$

The set $\Psi'_{\alpha\beta}(x)$ also constitutes the original structure group G. When element $\Psi_{\alpha\beta}(x)$ run over all G, so does $\Psi'_{\alpha\beta}(x)$. The transformation $\Psi_{\alpha\beta}(x) \to \Psi'_{\alpha\beta}(x)$ is called the gauge transformation which does not change the global structure of fiber bundle.

Example 1) The cylinder L is globally a product bundle

$$L = S^1 \times I,$$

where S^1 is the base manifold, $I = [0, 1]$ is the fiber which is an interval of R. The cylinder is a trivial bundle, its structure group has only one element: the identity.

Example 2) The Möbius band (see Figure 1.9 of Chapter 1) has the same base space S^1 and the same fiber I, a Möbius band is locally like cylinder, i.e., it is locally a Carterian product and is called locally trivializable. But globally Möbius band is not trivializable, twisting occurs somewhere. So it is a twisted product and its structured group is Z_2. Although the local topology of the fibre bundle is trivial, the global topology determined by the transition function may be very complicated due to the relative twisting of neighboring fibres. Since there are nontrivial twisted product like this globally, so we need to study the theory of fibre bundles.

Let M be an n-dim smooth manifold, as we have seen in §3.1, tangent bundle $T(M)$, cotangent bundle $T^*(M)$ and tensor bundle $T^{r,s}(M)$ are all fiber bundles $E(M, F, \pi, G)$ and their typical fibers are all vector spaces, acted upon by structure group $GL(n, R)$. A fiber bundle, whose typical fiber is a k-dim vector space V, which is the k-dim representation space of the structure group G, is called a rank-k vector bundle over M, denoted as $E(M, V, \pi, G)$.

A vector field $X(x)$ of a manifold M is an assignment to each point $x \in M$ an element $X(x)$ of $T_x(M)$. Vector field is also called a section of the tangent bundle. This idea can be generalized to fiber bundle as follows:

Definition 8.1 *A section of the fiber bundle $E(M, V, \pi, G)$ is an injective mapping*

$$\sigma: \quad M \to E$$

such that

$$\pi \cdot \sigma = identity$$

i.e., mapping σ is fiber preserving. So we can say a section σ of a fiber bundle E is a rule which assigns a fiber preserving point $\sigma(x)$ on each fiber to each point $x \in M$. But the existence of global sections depends upon the global properties of the fiber bundle E. There are fiber bundles which have no global sections.

Now let us look at the space $\Gamma(E)$ of sections of a vector bundle E:

$$\Gamma(E) = \{\sigma; \quad \sigma \text{ are sections of vector bundle E}\}.$$

For example

$$\mathcal{X}(M) = \{X(x); \quad X(x) \text{ are sections of tangent bundle } T(M)\}.$$

$$\begin{aligned} \Lambda^1(M) &= \{\omega; \ \omega \text{ are sections of cotangent bundle } T^*(M)\} \\ &= \text{Hom}(\mathcal{X}(\mathcal{M}), \mathcal{F}(M)). \end{aligned}$$

The fibers of rank-k vector bundle E inherit the structures of a k-dim vector space.

One can use the vector space structure on fibers of a vector bundle to define the pointwise addition of sections or scalar multiplication of sections by functions, i.e., if $\xi(x)$ and $\eta(x)$ are two local sections of $\Gamma(E)$, then $\xi(x)+\eta(x)=[\xi+\eta](x)$ is defined by adding values in the fibres. If $f(x)$ is a smooth continuous function on M, the new section $f(x)\xi(x)=[f\xi](x)$ can be defined by pointwise scalar multiplication in the fiber. So the set $\Gamma(E)$ of sections is a vector space, which is a rank-k $\mathcal{F}(M)$-module. At each point x of an open neighborhood $U \subset M$, we can choose a bases $\{e_a(x)\}_1^k$ for the k-dim fiber over x, such a collection of bases is called a frame, every vector field $\xi(x)$ can be expressed in terms of it as

$$\xi(x)=\xi^a(x)e_a(x). \tag{8.16}$$

For the rank-k vector bundle, transition functions $\Psi_{\alpha\beta}(x)$ belong to $GL(k,R)$ rather than to the full group $Diff(R^k)$ of diffeomorphisms of the typical fiber R^k. The origin 0 of the fiber is preserved by $GL(k,R)$ and represents a distinguished element of the fiber of a vector bundle, so a global section called the zero-section of vector bundle can be defined. A rank-k vector bundle is locally trivializable, i.e., in any open neighborhood U, one can always choose k linearly independent non-zero smooth sections $\{e_a(x)\}_1^k$ everywhere, and every smooth section can be expressed as the linear combination of these k smooth sections with smooth function coefficients as Eq. 8.16. But the global section which is non-zero everywhere may be not exist.

Example 3) In quantum mechanics, the Schrödinger wave function $\Psi(x,y,z)$ is a section of complex line bundle on base space R^3. When we study a charged particle around a magnetic monopole which is a topologically nontrivial spherical symmetric field, we must consider Schrödinger wave function $\Psi(x,y,z)$ on base space R^3-0, one has to introduce at least two open neighborhoods $U_\pm$:

$$U_+=\{(x,y,z)\in R^3-0, z>-\epsilon\} \qquad U_-=\{(x,y,z)\in R^3-0, z<+\epsilon\}.$$

The complex line bundle is trivializable on each neighborhood. Let $\Psi_\pm$ be sections in the corresponding neighberhoods, in the overlapping region $U_+\bigcap U_-$, which is homotopically equivalent to the $x-y$ plane at $z=0$ with 0 removed,

$$\Psi_+=e^{in\phi}\Psi_- \quad (\phi \text{ is the azimuthal angle}).$$

The transition function $e^{in\phi}\in U(1)$, so the complex line bundle is a nontrivial twisted bundle when $n\neq 0$.

Example 4) Tangent bundle and cotangent bundle of S^2. Since the Euler number $\chi(S^2)=2$, which is not zero, the everywhere non-zero global sections of the tangent bundle and cotangent bundle of S^2 do not exist.

Let S^2 be the sphere of unit diameter, which is covered by at least two open neighborhoods $U_\pm$

$$U_+=S^2-(0,0,-\frac{1}{2}), \qquad U_-=S^2-(0,0,\frac{1}{2}),$$

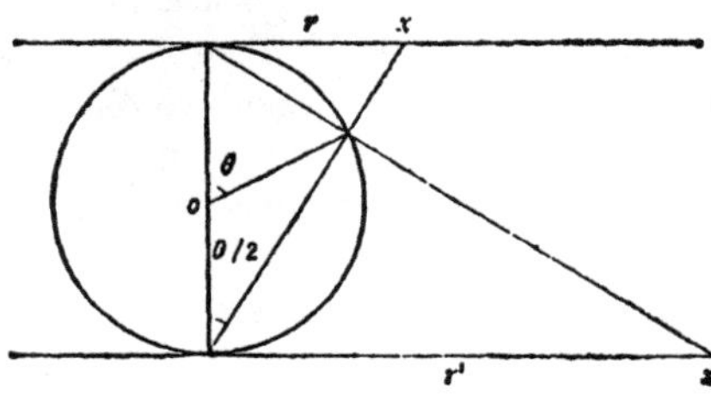

Figure 8.6

these two neighborhoods can be stereographically projected to the plane (see Figure 8.6), so we can obtain the coordinates of points on M:

In U_+,

$$x = (x^1, x^2), \quad x^1 = r\cos\varphi,\ x^2 = r\sin\varphi, \quad r = \tan\frac{\theta}{2}.$$

In U_-,

$$x' = (x'^1, x'^2), \quad x'^1 = r'\cos\varphi,\ x'^2 = r\sin\varphi, \quad r' = \cot\frac{\theta}{2}.$$

U_- coordinates are related to the U_+ coordinates through the inversion

$$\vec{x}' = \vec{x}/r^2, \quad d\vec{x}' = \frac{1}{r^2}d\vec{x} - 2\frac{dr}{r^2}\vec{x} = (r^2 d\vec{x} - 2\vec{x}(\vec{x}\cdot d\vec{x}))/r^4.$$

The transition function $\phi = \frac{\partial(x'^1, x'^2)}{\partial(x^1, x^2)}$ for $T^*(S^2)$ is nothing but the Jacobian matrix for the change of coordinates, is given by

$$\phi_{ij}(x^1, x^2) = (\delta_{ij}r^2 - 2x^1x^2)/r^4$$

restricted to equator of the sphere, $r = r' = 1$, the transition function matrix is

$$\phi(\cos(\theta), \sin(\theta)) = -\begin{pmatrix} \cos 2\theta & \sin 2\theta \\ \sin 2\theta & -\cos 2\theta \end{pmatrix},$$

it represents a non-trivial mapping from S^1 to $GL(2, R)$.

The transition function matrix for the tangent bundle $T(S^2)$ is just the transpose of the inverse of the transition matrix for the cotangent bundle $T^*(S^2)$, which is also a representation matrix of $GL(2, R)$, this leads to $T(S^2)$ and $T^*(S^2)$ being equivalent bundle, both are non-trivial.

In the following, we shall consider the frame bundle $L(E)$. A section σ of vector bundle E is also called a vector field $\sigma(x)$ on base manifold M. At each point x of a neighborhood $U \subset M$, we can choose k linearly independent vector fields $\{e_a(x)\}_1^k$

as bases, such a collection of basis is called a frame. For a given frame, its k linearly independent combination constitutes a new frame,

$$e'_a(x) = B_a^b(x)e_b(x), \quad \det B_a^b(x) \neq 0, \quad x \in U.$$

The entire frame is a matrix-valued mapping $B_a^b(x)$ from U to $GL(k,R)$. The set of all frames at a point $x \in M$ constructed the group $GL(k,R)$. The frame bundle $L(E)$ is a fiber bundle over the same base manifold M, with its fibre at x the set of all frames. The typical fiber of $L(E)$ is the structure group G of E, and the structure group of $L(E)$ is also G. Both $L(E)$ and its associated vector bundle E have the same structure group.

A fiber bundle in which the typical fiber F and the structure group G are identical and G acts on F by left multiplication is called a principal bundle, and is always denoted as $P(M,G,\pi)$ or simply as $P(M,G)$,

$$\begin{array}{ccc} G & \to & P \\ & & \downarrow \pi \\ & & M \end{array}$$

Frame bundle $L(E)$ is a principal bundle associated with vector bundle E. As we have seen in the §3.1, tangent frame bundles play an important role for various tensor bundles. The twists in a vector bundle are determined by twists in the associated principal bundle, so the topological information of fiber bundle are essentially encoded in the theory of principal bundles.

Example 5) When H is a closed Lie subgroup of a Lie group G, we can get a principal H-bundle $G(G/H,H,\pi)$ whose fiber is the group H, the structure group is also H.

For example, the Hopf bundle on sphere:

$$\begin{array}{ccc} U(1) & \to & SU(2) \\ & & \downarrow \pi \\ & & SU(2)/U(1) \end{array} \quad \sim \quad \begin{array}{ccc} S^1 & \to & S^3 \\ & & \downarrow \pi \\ & & S^2 \end{array}$$

We conclude this section with some remarks about the principal bundle and the vector bundle, they have distinctly different characteristics:

(1) The typical fiber of vector bundle is vector space, there are addition of sections and scalar multiplication of section with functions. These operations preserve fiber. Meanwhile, there is G-acting on the fibre and the G-action is also fiber preserving, which preserves the linear structure of vector space.

There is a global zero-section, since the origin 0 of fiber is preserved by the group G.

(2) The fiber of principal bundle is group manifold, not vector space, has no linear structure. There is no addition between sections of principal bundle, but there is the group multiplication on fiber which is fiber preserving. There is no zero-section, since the identity of the fiber is changed by the G-action.

In the next section we look at in detail the principal bundle and its associated vector bundles, and analyse their relations.

8.5 Principal bundle and associated bundle

A principal bundle $P = P(M, G)$ is a fiber bundle whose typical fiber is a group manifold G and the structure group which acts on fiber by left translation is also G.

Since the left and right translations commute, one can define a right action of G on P. This action is a mapping from $P \times G \to P$, it can be denoted by r_g as the transformation of P:

$$\begin{aligned} r_g : P &\to P \\ u &\to ug \in P, u \in P, g \in G. \end{aligned}$$

This G-action r_g does not change the structure group G of P, since the fiber itself is a group manifold, the G-action r_g on P is smooth and free. The freedom of G action implies that each orbit is homeomorphic to the group G. Taking each orbit as a point, hence the space P/G of orbits of the G-action r_g is just the base manifold of $P(M, G)$, and there is projection mapping from P to M:

$$\pi : P \to M$$

This canonical projection is fiber preserving, i.e.,

$$\pi(r_g u) = \Pi(u), \forall u \in P, \quad g \in G.$$

Hence we obtain the principal bundle $P = P(M, G)$.

Now we construct vector bundles associated with principal bundle P. Let ρ be a representation of G on a finite dimensional vector space V.

$$\rho : G \to GL(V)$$

where $GL(V)$ is the general linear transformation group of vector V. Construct the product bundle $P \times V$, and define the G-action on $P \times V$ as

$$(u, y) \to (r_g u, \rho(g) y), \forall u \in P, \quad y \in V, \quad g \in G,$$

denote the orbit under G-action of (u, y) as $[u, y]$, and take every orbit as a point,we obtain the space of orbits, denoted as:

$$P \times_\rho V = \{[u, y]\},$$

There is projection mapping π_E

$$\begin{aligned} \pi_E : P \times_\rho V &\to M \\ [u, y] &\to \pi_E[u, y] = \pi(u) = x. \end{aligned}$$

Note that the G-action on $P \times V$ is free, so $\pi_E^{-1}(X)$ is isomorphic to V. Hence the space of orbits $P \times_\rho V$ is a vector bundle, denote as

$$\begin{array}{ccc} V \to & P \times_\rho V \\ & \downarrow \\ & M \end{array}$$

its transition functions are given by the representation $\rho(G)$. This vector bundle is said to be the vector bundle associated with P for the representation ρ, and denote as

$$E(M, V, G, \pi_E, P) = P \times_\rho V,$$

its topological structure is determined by its associated principal bundle.

Principal bundle and its associated vector bundle have the same structure group, which acts on fiber by left multiplication. On the other hand, the right action of group G on the principal bundle P changes a frame to a new frame in the same fibre, it corresponds to the change of bases in the fiber of vector bundles, and the associated vector bundle is just the equivalence classes on $P \times V$, it can be expressed as:

$$(r_g u, y) = (u, \rho(g^{-1})y) \quad \forall u \in P, y \in V, g \in G.$$

This transformation of frame is called gauge transformation. On each neighborhood U_α of M if the bases can be chosen such that the set of transition functions contains the identity only, then the fiber bundle is trivial. There is the following theorem

Theorem 8.5 *A principal bundle P(M,G) is trivial if and only if it has a global continuous section.*

Proof: If the principal bundle $P(M, G)$ has a global continuous section, let $\{U_\alpha\}$ be the open covering of manifold M, on the neighborhood U_α there exists the map

$$\begin{array}{rcl} \varphi_\alpha : U_\alpha & \to & G \\ x & \to & \varphi_\alpha(x) \in G. \end{array}$$

In the overlapping region $U_\alpha \bigcap U_\beta \neq 0$,

$$\varphi_\alpha(x) = \psi_{\alpha\beta}(x)\varphi_\beta(x),$$

here $\psi_{\alpha\beta}(x) = \varphi_\alpha(x)\varphi_\beta(x)^{-1}$ is the transition function. Now one can introduce gauge transformation, i.e., on every neighborhood U_α, multiply P by $\phi_\alpha(x)$ from right, then the transition function can be expressed as

$$\psi'_{\alpha\beta}(x) = \varphi_\alpha^{-1}(x)\psi_{\alpha\beta}(x)\varphi_\beta(x) = identity,$$

so the principal bundle $P(M, G)$ is trivial. Obviously the reverse is also true.

Principal bundle is trivial implies that its associated bundles are trivial. If the principal bundle has a global continuous section, i.e., there exists the global mapping $\varphi(x)$ from M to G, in terms of the G-action on the associated bundle one obtains the global continuous section in the form

$$\begin{aligned} \sigma : M &\rightarrow P \times_G V \\ x &\rightarrow \rho(\varphi(x))\xi(x) \end{aligned}$$

where $\xi(x)$ is a mapping from M to V, and the number of linearly independent sections is the same as the dimension of the typical fiber V.

Since the zero representation always exists, vector bundle will always admit the global zero-section without being trivial. But when there exists everywhere non-zero global section, and the number of linearly independent global sections is the same as the dimension of the typical fiber, then the associated vector bundle is trivial.

8.6 Induced bundle, reduction of fibre bundle

There are many way to construct new bundle with old one, a way which is often used is the so called pullback. For a fiber bundle E with base manifold M and fiber F, let f: $N \rightarrow M$ be a mapping from a manifold N to M, then the induced bundle of E, denoted by f^*E, can be obtained as follows:

$$\begin{aligned} f : N &\rightarrow M \\ x &\rightarrow y = f(x) \\ f^* : E &\rightarrow f^*E \\ F_y &\rightarrow F_{f(x)} \end{aligned}$$

i.e., copying the fiber F_y of E at each point $y \in M$ to the fiber at $x \in N$ of f^*E. The neighborhoods and transition functions of f^*E can be defined analogously. So one can obtain the induced bundle f^*E over base manifold N.

There are the following theorems (Their proof can be found in reference [1].)

Theorem 8.6 *If two mapping f and g from N to M are homotopic, E is a fiber bundle over M, then the induced bundles f^*E and g^*E are equivalent.*

This result indicates that the topological property of a bundle is unchanged by contracting the base manifold. A bundle over a point is trivial, so one has

Any fibre bundle over a contractible base manifold is trivial.

Proof: Contractibility of base manifold M to point x_0 means that the identity mapping f and collapsed mapping g are homotopic to each other.
Identity mapping f:

$$\begin{aligned} M &\rightarrow M \\ x &\rightarrow x \end{aligned}$$

collapsed mapping g:

$$\begin{array}{ccc} M & \rightarrow & M \\ x & \rightarrow & x_0 \end{array}$$

i.e., g is the mapping which collapses all points of M to a point x_0.

$$g(x) = x_0, \forall x \in M.$$

Since f and g are homotopic, bundles f^*E and g^*E are equivalent, i.e., $f^*E \cong g^*E$. However, g^*E is a bundle over a single point x_0, so it is trivial. $g^*E = M \times F$. On the other hand, as f is the identity ($f^*E = E$), so $E \cong M \times F$, i.e., bundle E itself is trivial.

From this theorem we can see that any bundle over R^n is trivial. Nontrivial fiber bundles can only be constructed when the base manifold is non-trivial. The simplest non-contractible space is the sphere S^m. Even taking it as the base manifold, the bundle may still be trivial. For example, the tangent frame bundle over S^1, S^3 and S^7 are trivial. In terms of homotopy, an analogous result can be obtained.

Theorem 8.7 *If the fiber F of bundle E is contractible, then E always has a (global) section.*

From the previous paragraph we know that the existence of global sections in a principal bundle implies the triviality of the principal bundle. Thus Theorem 7 can be generalized as:

Theorem 8.8 *Fibre bundle E is trivial if either the base manifold M or the structure group G is contractible.*

If the structure group G has a maximal compact closed subgroup H, and the coset G/H is contractible, then the principal bundle $P(M,G)$ can always be reduced to the principal bundle $P(M,H)$.

Next we consider fiber bundles related to the tangent bundle of n-dimension manifold, or its associated bundle, the frame bundle $L(M)$ over a given n-dimension manifold. The structure group is taken to be $GL(n,R)$. Though this group is not contractible, it does have a contractible part, i.e., it has maximal compact closed subgroup $O(n,R)$.

Then the structure group $GL(n,R)$ of the frame bundle $L(M)$ can be reduced to the compact subgroup $O(n)$. This result indicates that every manifold M always admits a Riemannian metric, so we can restrict the tangent frames to be orthonormal with respect to this metric, and one obtains an $O(n)$ frame bundle over M.

In addition, if manifold M is orientable, then by further restricting the permissible frames belong to a particular orientation class, one obtains an $SO(n)$ frame bundle, i.e., for an n-dimensional orientable Riemannian manifold, the structure group $GL(n,R)$ of tangent frame bundle can be reduced to $SO(n)$.

Similarly if M is an n-dimensional complex manifold, the structure group of the complex frame bundle is $GL(n,C)$

$$GL(n,C) = U(n) \times PH(n)$$

where $U(n)$ is the usual unitary group, which is the maximal compact subgroup of $GL(n,C)$, and $PH(n)$ is the set of all positive definite Hermitian matrices, which is contractible, therefore, the structure group $GL(n,C)$ can be reduced to the maximal subgroup $U(n)$. This result shows that every complex manifold M always admits an Hermitian metric, and the frame preserving Hermitian metric generates a $U(n)$ principal bundle. This results had been given in the table 7.3.

Next we describe briefly a number of basic constructions involving vector bundles. We can do all of linear algebra pointwise and glue together globally. For instance:

Whitney sum Consider two vector bundles E_1 and E_2 over the same base manifold M. Let f denote the diagonal coproduct of M.

$$\begin{aligned} f: \quad M &\rightarrow M \times M \\ x &\rightarrow (x, x). \end{aligned}$$

Bundle $f^*(E_1 \times E_2)$ over M is called the Whitney sum of E_1 and E_2, and is denoted by $E_1 \oplus E_2$.

$$E_1 \oplus E_2 = f^*(E_1 \times E_2).$$

Note that fibre $F_x(E_1 \oplus E_2)$ is canonically isomorphic to the direct sum $F_x(E_1) \oplus F_x(E_2)$.

Analogously, we can consider the tensor product bundle $E_1 \otimes E_2$, whose fibre $F_x(E_1 \otimes E_2)$ is canonically isomorphic to the direct product $F_x(E_1) \otimes F_x(E_2)$.

The dual vector bundle E^*, has as its fibre $F_x(E^*)$ the dual vector space of $F_x(E)$,

$$F_x(E^*) = Hom(F_x(E), \mathbf{R}).$$

But if a vector bundle E is endowed with a Euclidean metric, then E is isomorphic to its dual bundle E^*.

Now we study the very important sub-bundle and its orthogonal complements.

Consider two vector bundles E and E' over the same base manifold M, if each fibre $F_x(E)$ is a subvector space of the corresponding fibre $F_x(E')$, then E is a sub-bundle of E', denoted as $E \subset E'$.

Now if E' is endowed with a Euclidean metric g on its fibre, $E \subset E'$, then there exists an orthogonal complementary sub-bundle $E^\perp$, $F_x(E^\perp)$ denotes the subspace of $F_x(E')$ consisting of all vectors v with $g(v,w) = 0$, for all $w \in F_x(E)$. Such that

$$E \oplus E^\perp = E'.$$

As an example, let $M \subset N$ be a smooth Riemannian manifold, then the tangent bundle $T(M)$ is a sub-bundle of the restriction $T(N)|_M$, and the orthogonal complement $T(M)^\perp \subset T(N)|_M$ is called the normal bundle of M in N. And

$$T(M) \oplus T(M)^\perp = T(N)|_M.$$

At the last of this section, we consider the natural (canonical) line bundle and the tangent bundle of $RP(n)$ and $CP(n)$.

Let $\mathcal{L}$ be a line through the origin in $\mathbf{R}^{n+1}$, intersecting S^n at points $\pm x$. Let γ be the subset of $RP(n) \times \mathbf{R}^{n+1}$ consisting of all pairs $(\{\pm x\},\ v)$ with $v \in \mathcal{L}$. Define

$$\begin{array}{rccc} \pi: & \gamma & \to & RP(n) \\ & (\{\pm x\},\ v) & \to & \{\pm x\}. \end{array}$$

Thus each fibre $\pi^{-1}(\{\pm x\})$ can be identified with $\mathcal{L}$, i.e., the real multiple of x. The resulting vector bundle γ is called the natural (or canonical) line bundle over $RP(n)$.

Let $\mathcal{E}^{n+1} = T(\mathbf{R}^{n+1})$ be the tangent bundle of $\mathbf{R}^{n+1}$, $\gamma^\perp$ be the orthogonal complement of γ in $\mathcal{E}^{n+1}$.

$$\gamma \oplus \gamma^\perp = \mathcal{E}^{n+1}.$$

Let $\mathcal{L}^\perp \subset \mathbf{R}^{n+1}$ denote the fibre F_x of $\gamma^\perp$, which is the orthogonal complementary n-plane of L.

Proposition 1. The tangent bundle $T(RP(n))$ is isomorphic to bundle $Hom(\gamma, \gamma^\perp)$.

Proof: The real projective space $RP(n)$ can be defined as the set of all unordered pairs $\{x, -x\}$. Then the tangent manifold can be identified with the set of all pairs $\{(x, v),\ (-x, -v)\}$ satisfying

$$(x, x) = 1, \quad (x, v) = 0.$$

But each such pair determines and is uniquely determined by a linear mapping $\mathcal{L} \to \mathcal{L}^\perp$. Thus the tangent space of $RP(n)$ at $\pm x$ is isomorphic to the vector space $Hom(\mathcal{L}, \mathcal{L}^\perp)$, so that

$$T(RP(n)) \cong Hom(\gamma, \gamma^\perp).$$

Analogously, let L be a complex line through the origin in $\mathbf{C}^{n+1}$ and $L^\perp \subset \mathbf{C}^{n+1}$ be the complementary complex n-plane. The natural (or canonical) complex line bundle λ over $CP(n)$ is a line bundle whose fibre at a point $x \in CP(n)$ is the complex multiple L of $x \in \mathbf{C}^{n+1}$.

Let $\mathcal{E}^{n+1} = T(\mathbf{C}^{n+1})$ be the tangent bundle of $\mathbf{C}^{n+1}$, the normal bundle ν of $T(CP(n))$ in $\mathcal{E}^{n+1}$ is a trivial bundle

$$\mathcal{E}^{n+1} = T(CP(n)) + \nu.$$

Let $\lambda^\perp$ be the orthogonal complement bundle of λ in $\mathcal{E}^{n+1}$, by using the standard Hermitian metric on $\mathbf{C}^{n+1}$, so the Whitney sum

$$\lambda \oplus \lambda^\perp = \mathcal{E}^{n+1}.$$

Proposition 2. The tangent bundle $T(CP(n))$ is isomorphic to the complex vector bundle $Hom_{\mathbf{C}}(\lambda, \lambda^\perp)$.

Its proof is analogous with the proof of Proposition 1.

8.7 The homotopy classification of fibre bundles, universal fibre bundle

For a given base manifold and given typical fiber, how many different types of fiber bundles can be constructed? Since every vector bundle can be associated with a specific principal bundle, twists in an associated bundle are determined by twists in the principal bundle, i.e., the topological structure of associated bundle is completely determined by the topological structure of the principal bundle. Hence we first study the classification of principal bundles, and find out how many different types of principal bundles $P(M, G)$ can be found with given M and G.

In the last section, we see that non-trivial fibre bundles can be constructed only when the base manifold is non-trivial. The simplest noncontractible manifold is the sphere S^m. Now we consider the classification of G-bundle over the sphere S^m. The open covering of S^m has at least two neighborhoods $U_\pm$. Locally the principal bundle is trivializable and can be described by Cartesian product as

$$P|_{U_\alpha} = U_\alpha \times G.$$

In the overlapping region $x \in U_+ \cap U_-$, which is homotopic to S^{m-1}, there is the transition function $\psi_{\alpha\beta}(x)$ between U_+ and U_-,

$$\psi_{\alpha\beta} : U_\alpha \cap U_\beta \to G$$

which is homotopic to mapping from S^{m-1} to G. Hence, we conclude
Proposition 3: The principle G-bundles $P(S^m, G)$ over S^m are determined up to equivalence class by elements of $\Pi_{m-1}(G)$.

In the following we consider the classification of general principal bundle $P(M, G)$. For this propose we introduce the classifying spaces B_G and universal bundle $\xi_P(G)$.

Definition 8.2 *For a given principal bundle $P(M, G)$, bundle $\xi_P(G)$*

$$\begin{array}{ccc} G & \rightarrow & \xi_P(G) \\ & & \downarrow \\ & & B_G \end{array}$$

is called universal bundle for group G if all its homotopy groups are zero,

$$\Pi_k(\xi_P(G)) = 0, \tag{8.17}$$

the base manifold B_G of a universal G-bundle is called the classifying space.

If there exists a continuous mapping from M to B_G

$$f : M \to B_G$$

then the induced bundle $f^*(\xi_P(G))$ is a principal G-bundle over M. There is the following theorem:

Theorem 8.9 (**Classification Theorem**): *There is a natural one-to-one correspondence between the set of equivalence classes of principal fiber bundles with given base M and structure group G, and the set $[M, B_G]$ of homotopy classes of mappings from M to B_G of a universal bundle $\xi_P(G)$.*

For example, consider again the problem of classifying $P(S^m, G)$, which can be reduced to determining the set $[S^m, B_G]$, i.e., to the determination of the homotopy group $\Pi_m(B_G)$. From the homotopy exact sequence of the universal bundle $\xi_P(G)$:

$$\to \Pi_k(\xi_P(G)) \to \Pi_k(B_G) \to \Pi_{k-1}(G) \to \Pi_{k-1}(\xi_P(G)) \to .$$

By Eq. 8.17 one has

$$\Pi_k(B_G) \simeq \Pi_{k-1}(G) \qquad (k \geq 1).$$

Hence this result coincides with Proposition 2.

For an n-dimensional manifold if we weaken Eq. 8.17, replacing it by

$$\Pi_k(\xi_P(G)) = 0, \qquad k < n,$$

then this G-bundle is said to be n-universal for group G, denoted usually by $\xi_P(n, G)$, and

$$\lim_{n\to\infty} \xi_P(n, G) = \xi_P(G).$$

It can be shown that the classifying space B_G is unique up to homotopy equivalence, which is the base manifold of n-universal bundle ($n >$dim M).

From Eq. 8.9, the Stiefel manifold

$$V_{n.k} \simeq O(n)/O(n-k)$$

satisfies

$$\Pi_l(V_{n,k}) = 0 \qquad (l < n-k). \tag{8.18}$$

Manifold $V_{n,k}$ can also be expressed as the $O(k)$-principal bundle over Grassmann manifold $G_r(n, k)$

$$\begin{array}{ccccc} O(k) & \to & V_{n,k} & \simeq & O(n)/O(n-k) \\ & & \downarrow & & \\ & & G_r(n.k) & \simeq & O(n)/O(n-k) \times O(k) \end{array}$$

Eq. 8.18 shows that $V_{n,k}$ is an $(n-k-1)$-universal $O(k)$-bundle $\xi_P(n-k-1, O(k))$.

Similarly the complex Stiefel manifold

$$V^c_{n,k} \simeq U(n)/U(n-k)$$

satisfies

$$\Pi_l(V^c_{n,k}) = 0 \qquad (l < 2(n-k)). \tag{8.19}$$

The complex Stiefel manifold is also a $U(k)$-principal bundle over complex Grassmann manifold $G_r(n,k,\mathbf{C})$.

$$\begin{array}{cccc} U(k) & \rightarrow & V^c_{n,k} & \simeq & U(n)/U(n-k) \\ & & \downarrow & & \\ & & G_r(n,k,C) & \simeq & U(n)/U(n-k)\times U(k) \end{array}$$

Eq. 8.19 shows that $V^c_{n,k}$ is a $(2n-2k-1)$-universal $U(k)$-bundle $\xi_P(2n-2k-1,U(k))$.

The classification of principal bundles determines the classification of the associated vector bundle. In terms of classifying space, one can also introduce the universal vector bundle and consider directly the classification of vector bundles. For example, the complex projective space $CP(n)$ is a complex Grassmann manifold $G^c_r(n+1,1)$, whose points are lines in C^{n+1} passing through the origin. Now we introduce the natural complex line bundle L over $CP(n)$, which can be used to classify other line bundles, so it is also called the universal complex line bundle $\xi_L(C)$.

First we introduce the trivial bundle over $CP(n)$

$$I^{n+1} = CP(n)\times C^{n+1},$$

its points are denoted by the pair (p,z). Addition and scalar multiplication are performed on the second factor while leaving the first factor unchanged. Let $L = L(n+1,1,C)$ be the sub-bundle of I^{n+1}, defined by

$$\begin{array}{cccc} C\rightarrow & L(n+1,1,C) & =(p,z)\in I^{n+1}; & z\in p \\ & \downarrow & & \\ & CP(n) & & \end{array}$$

i.e., the fiber of $L(n+1,1,C)$ over a point p of $CP(n)$ is just the set of points in C^{n+1} which lie in the line joining p and the origin. By using this natural line bundle over $CP(n)$, one can classify other complex line bundles. For example, let E be a complex line bundle over M, one can always find a complementary bundle $\bar{E}$ such that

$$E\oplus\bar{E} = M\times C^{n+1}$$

where $\oplus$ denotes the Whitney sum (or direct sum) of vector bundles. The fiber of E is the set of lines in C^{n+1}.

Let $E_1(M,F_1,\pi_1)$ and $E_2(M,F_2,\pi_2)$ be fiber bundles over the same base manifold M. Then the Whitney sum $(E_1\oplus E_2)\subset(M,F_1\oplus F_2,\Pi)$ is the subspace of all $(u_1,u_2)\in E_1\times E_2$ of which $\pi_1(u_1)=\pi_2(u_2)$ and $\pi(u_1,u_2)=\pi_1(u_1)=\pi_2(u_2)$.

The fiber $\pi^{-1}(x)$ of $E_1\oplus E_2$ is $\pi_1^{-1}(x)\times\pi_2^{-1}(x)$, which is isomorphic to the direct sum $F_1\oplus F_2$ of vector bundles.

Let f be a mapping from M to $CP(n)$

$$f: M\rightarrow CP(n).$$

Then the pullback of the natural line bundle over $CP(n)$ is isomorphic to the line bundle E

$$f^* : L \to f^* L \simeq E$$

and the set $[M, CP(n)]$ of homotopy classes of mappings from M to $CP(n)$ can be denoted by equivalence classes of complex line bundle over manifold M.

The construction above can be generalized. We know that each point of the Grassmann manifold $G_r^c(n,k)$ corresponds to a k-plane in C^n through the origin. The natural k-plane bundle $L(n,k,C)$ over $G_r^c(n,k)$ has as its fiber the k-plane in C^n over the corresponding point in $G_r^c(n,k)$. Therefore, it can be proved that the set of k-dim vector bundles over m-dimensional manifold M is isomorphic to the set of homotopy classes of maps from M to $G_r^c(n,k)$ for $n \geq N$ for some fixed N determined by the dimension m of M.

By taking the natural inclusion map of $G_r^c(n,k)$ into $G_r^c(n+1,k)$, one obtains the universal Grassmann manifold $G_r^c(\infty,k)$. Its cohomology is simpler than that of $G_r^c(n,k)$, and it is a polynomial algebra with generators

$$c_i(L(\infty,k,C)) = c_i(\xi_P(U(k))) = c_i(\xi_P(GL(k,C))$$

which will be studied in the next chapter as Chern cohomology class (characteristic class).

Given a k-dimensional vector bundle E over M and a map

$$f : M \to G_r^c(\infty,k)$$

the pullback bundle $f^* L(\infty,k,C) = E$ and

$$f^* c_i(L(\infty,k,C)) = c_i(E) = H^k(M, \Pi_{k-1}(G))$$

is determined uniquely up to homotopy. So the cohomology classes are all well-defined and depend upon the bundle E only.

G Structure group of fiber bundle	B_G Universal classifying space	$\xi_P(G)$ Universal principal bundle	ξ_E Universal k-dim vector bundle
$GL(k,\mathbf{R})$, $O(k)$	$G_r(\infty,k)$	$V_{\infty,k}$	$L(\infty,k,\mathbf{R})$
$SL(k,\mathbf{R})$, $SO(k)$	$\hat{G}_r(\infty,k)$	$\hat{V}_{\infty,k}$	$\hat{L}(\infty,k,\mathbf{R})$
$GL(k,\mathbf{C})$, $U(k)$	$G_r^c(\infty,k)$	$V_{\infty,k}^c$	$L(\infty,k,\mathbf{C})$
$U(1)=SO(2)$	$CP(\infty)$	$S^\infty = \lim_{\infty} S^{2n+1}$	$L(\infty,1,\mathbf{C})$

Where $\hat{G}_r(\infty,k) = \lim_{n\to\infty} \hat{G}_r(n,k)$ and $\hat{G}_r(n,k)$ is the manifold whose points are oriented k-dim plane in $\mathbf{R}^n$ passing through the origin. So manifold

$$\hat{G}_r(n,k) \simeq O(n)/O(n-k) \times SO(k)$$

forms a 2-sheeted covering space for $G_r(n,k)$.

It is worth noticing that although $\xi(G)$ is contractible trivial manifold, it is a nontrivial bundle over B_G. its twisting can be measured in terms of the cohomology $H^*(B_G)$. The elements of $H^*(B_G)$ are called the characteristic classes. In the next chapter we shall look in detail the characteristic classes of fibre bundle, which is the pullback of the elements of $H^*(B_G)$.

Finally, we consider simply the case of the boundle with infinite dimension. Note that both the set of Yang-Mills potentials (affine connection) and the set of Riemannian metrics are contractible spaces. Thus connection bundle

$$\begin{array}{ccc} \mathcal{G} & \rightarrow & \mathcal{A} \\ & & \downarrow \pi \\ & & \mathcal{A}/\mathcal{G} \end{array}$$

are the universal bundle for the gauge group G. and analogously, the Riemannian metric bundle

$$\begin{array}{ccc} Diff(M) & \rightarrow & Riem(M) \\ & & \downarrow \pi \\ & & Riem(M)/Diff(M) \end{array}$$

are the universal bundle for the diffeomorphism group $Diff(M)$.

Notations and Formulae

- Exact homomorphism sequence of a pair of manifolds (M,N)

$$\cdots \longrightarrow \Pi_k(N,q) \xrightarrow{i_*} \Pi_k(M,q) \xrightarrow{j_*} \Pi_k(M,N;q) \xrightarrow{\partial} \Pi_{k-1}(N,q) \longrightarrow \cdots .$$

- Exact homomorphism sequence of fiber bundle $E(B,F,\pi)$

$$\cdots \longrightarrow \Pi_k(F) \xrightarrow{i_*} \Pi_k(E) \xrightarrow{j_*} \Pi_k(B) \xrightarrow{\partial} \Pi_{k-1}(F) \longrightarrow \cdots .$$

- Some formule of homotopy group:

$$\Pi_2(G/H) = \Pi_1(H) \quad \text{(for simple Lie group G)}.$$

$$\Pi_1(G) = \Pi_1(\tilde{G}/D) = D \quad \text{(for connected Lie group } G = \tilde{G}/D).$$

$$\Pi_1(M/G) \simeq G \quad \text{(for M simple connected, and the action of G is freely)}.$$

- Relation between homotopy group and homology group:

$$H_1(M) = \Pi_1(M)/\left[\Pi_1(M), \Pi_1(M)\right],$$

Hurewicz isomorphism theorem: if $\Pi_k(M) = 0$ (for $1 \le k < m$),

$$\text{then} \qquad H_k(M) = 0 \quad (\text{ for } 1 \le k < m),$$

$$\text{and} \qquad H_m(M) = \Pi_m(M).$$

- Principal bundle $P(M,G)$:

$$\begin{array}{ccc} G & \to & P \\ & & \downarrow \pi \\ & & M \end{array}$$

- Associated bundle $E(M,V,\pi,G,P) = P \times_\rho V$

$$\begin{array}{ccc} V & \to & E \\ & & \downarrow \pi \\ & & M \end{array}$$

Exercises

1. Prove that for semisimple Lie group G and its subgroup H,

$$\Pi_2(G/H) = \Pi_1(H).$$

2. Using Hopf mapping

$$S^3 \to S^2 \simeq SU(2)/U(1),$$

show that

$$\Pi_k(S^3) = \Pi_k(S^2) \quad (k \geq 2).$$

3. Find the topological classification of $SU(2)$ principal bundle on S^4.
4. Is the vector bundle trivial if it has global cross sections? Find the criterial to decide if a vector bundle is trivial or not.
5. Give the definition of universal bundle $\xi(G)$ associated with the structure. How to classify principal bundle $P(M,G)$? Take $U(k)$ as the structure group to highlight the point.
6. As $S^2 \simeq CP(1)$, let L be the natural line bundle on $CP(1)$,

$$L = \{(p,z) \in CP(1) \times C^2 | z \in p\},$$

prove

$$T^*(S^2) \simeq L \otimes L.$$

Chapter 9

Differential Geometry of Fibre Bundle, Yang-Mills Gauge Theory

Field theory is one of the major topics of modern physics. From the point of view of differential geometry, field is the section of fibre bundle which is the generalization of function on manifold. In order to carry out differentiation of sections on fibre bundle, one has to specify the relation among neighboring fibres, thus the introduction of connection is necessary. Covariant derivative is a linear map from one section to another.

There are several equivalent definitions of connection on a fibre bundle, which are generalizations of affine connection we encountered in Chapter 3. Modern treatment of connection usually starts with connection on principal bundle, it is an abstract formal formulation about connection which we will study in §9.1. A connection on principal bundle induces a connection on its associated bundle, and we will study it in §9.2. Then in §9.3 we look at connection on arbitrary vector bundle, which can be induced by connection on the principal bundle of frames. In this case, one can usually give an equivalent definition of connection through the covariant derivative on sections of fibre bundle, analogous to the introduction of affine connection of tangent bundle on the manifold. In the last part of this chapter we study the Yang-Mills theory, which is nothing but the connection on principal bundle, with the gauge field strength as the curvature corresponding to the connection. In terms of the metric structure, gauge theory action and Yang-Mills equation can be obtained, and we have a look at its special solutions: instanton and monopole.

9.1 Connection and curvature on principal bundle

A principal bundle $P(M, G, \pi)$ is a fibre bundle whose fibre and transition function both belong to the same group G, the continuous map

$$\pi :\ P \longrightarrow M$$

is defined such that the right action of G on P is free and smooth.

Let us recall the definition of connection of tangent bundle on manifold. Given a curve $x(\tau)$ on manifold M, the tangent vector at each point of the curve is

$$X = \frac{d}{d\tau} = \frac{dx^i}{d\tau}\frac{\partial}{\partial x^i}.$$

A vector field Y on M is called parallel transported along $x(\tau)$ if

$$D_X Y = 0. \tag{9.1}$$

Namely connection defines the parallel transport of any vector Y along a given direction X.

In fibre bundle language, tangent field is a section of tangent bundle, restricting tangent field Y to the curve $x(\tau)$ on manifold M is equivalent to the intersection between fibre $F(x)$ with section Y, called the lift of $x(\tau)$. This lift is called horizontal if Eq. 9.1 is satisfied.

Generalize the result above to any fibre bundle, connection is defined if horizontal lift $u(\tau)$ of curve $x(\tau)$ on the base manifold is defined on the entire fibre space, and

$$\pi(u(\tau)) = x(\tau).$$

A connection on fibre bundle leads to a correspondence between any two fibres along the curve $x(\tau)$ on base manifold.

Next we look at connection on principal bundle. As the m-dimensional base manifold and k-dimensional Lie group G are both smooth manifolds, and principal bundle is locally a direct product of base manifold and group manifold, which is also a smooth manifold, and we can study the tangent bundle $T(P)$ and cotangent bundle $T^*(P)$, thus obtain different expressions of connection. First we look at the tangent space approach of connection.

Tangent space $T_u(P)$ at a point $u \in \pi^{-1}(x) = F_x$ on bundle P is an $(m+k)$-dimensional space. The right action of structure group G on principal bundle P, which is an automorphism of fibre F_x, can yield k right invariant tangent vector fields V_a $(a = 1, 2, \cdots, k)$ on group manifold F_x, they constitute the vertical subspace V_u tangent to fibre F_x at each point of the fibre F_x:

$$V_u \in T_u(P).$$

The complement subspace H_u in $T_u(P)$ relative to V_u is called the horizontal subspace:

$$T_u(P) = V_u \oplus H_u.$$

H_u can not be determined by equation above alone. Horizontal subspace is determined if connection is defined on the principal bundle, namely when the parallel transport of sections on the bundle is defined.

Principal bundle $P(M,G)$ can locally be trivialized, any point $u \in \pi^{-1}(x)$ can be written as:

$$u = (x, g).$$

A local section of P is a smooth map from a neighborhood U to G. Connection on principal bundle provides a rule for the parallel transport of sections. Let $x(\tau)$ be a curve on M. Using connection, this curve can be horizontally lifted to a curve on the bundle as

$$u(\tau) = (x(\tau), g(\tau)),$$
$$\pi(u(\tau)) = x(\tau).$$

Tangent field of curve $u(\tau)$ belongs to the horizontal subspace H_u. One can define differentiation along the horizontal lifted curve $u(\tau)$ analogous to Eq. 9.1 as:

$$D_\tau g(\tau) \equiv (\frac{d}{d\tau} + A_\mu(x)\frac{dx^\mu}{d\tau})g(\tau) = 0, \tag{9.2}$$

where $A_\mu(x)dx^\mu = A_\mu^a(x)\frac{\lambda_a}{2i}dx^\mu \in T^*(M)$ is a Lie algebra valued 1-form, and $\frac{\lambda_a}{2i} \equiv \tau_a$ are the generators of the Lie algebra.

For simplicity, group element $g \in G$ can be represented by a matrix (g_{ij}). Differentiation along the horizontal lifted curve can then be written as

$$\frac{d}{d\tau} = \frac{dx^\mu}{d\tau}\frac{\partial}{\partial x^\mu} + \frac{dg_{ij}}{d\tau}\frac{\partial}{\partial g_{ij}}.$$

Using the parallel transport equation Eq. 9.2 for $g(\tau)$, one obtains

$$\begin{aligned}\frac{d}{d\tau} &= \frac{dx^\mu}{d\tau}\frac{\partial}{\partial x^\mu} - (A_\mu(x)g)_{ij}\frac{dx^\mu}{d\tau}\frac{\partial}{\partial g_{ij}} \\ &= \frac{dx^\mu}{d\tau}(\frac{\partial}{\partial x^\mu} - A_\mu^a(x)(\frac{\lambda_a}{2i})_{ik}g_{kj}\frac{\partial}{\partial g_{ij}}) = \frac{dx^\mu}{d\tau}\mathcal{D}_\mu.\end{aligned}$$

Where
$$\mathcal{D}_\mu = \frac{\partial}{\partial x^\mu} - A_\mu^a \tilde{V}_a, \tag{9.3}$$

and
$$\tilde{V}_a = Tr(\frac{\lambda_a}{2i}g\frac{\partial}{\partial g^t}),$$

are the right-invariant fundamental vector fields on manifold G, $\mathcal{D}_\mu$ is the covariant derivative, which is the horizontal lift of tangent vector $\frac{\partial}{\partial x^\mu}$. The set $\{\mathcal{D}_\mu\}$ constitutes the basis of the horizontal space H_μ of dimension m, so one obtains a splitting of $T_u(P)$:

$$T_u(P) = H_u(P) \oplus V_u(P),$$
$$\pi_* H_u = T_{\pi(u)}(M),$$
$$\pi_* V_u = 0,$$

which is invariant under the right multiplication by the group:

$$R_{a*}H_u = H_{ua},$$

and the family of horizontal spaces $\{H_u\}$ is right invariant:

$$\{R_{a*}H_u\} = \{H_u\}.$$

The covariant derivatives do not commute in general:

$$[\mathcal{D}_\mu,\ \mathcal{D}_\nu] = -F^a_{\mu\nu}\hat{V}_a, \tag{9.4}$$

where $F^a_{\mu\nu} = \partial_\mu A^a_\nu - \partial_\nu A^a_\mu + f^a_{bc}A^b_\mu A^c_\nu$ is called the curvature corresponding to connection A.

Eq. 9.4 shows that the commutator of covariant derivatives has vertical components only. Furthermore it is easy to show that covariant derivatives satisfy the Jacobi identity:

$$[\mathcal{D}_\mu, [\mathcal{D}_\nu, \mathcal{D}_\lambda]] + [\mathcal{D}_\nu, [\mathcal{D}_\lambda, \mathcal{D}_\mu]] + [\mathcal{D}_\lambda, [\mathcal{D}_\mu, \mathcal{D}_\nu]] = 0.$$

An alternative definition of connection is through the cotangent space approach. Take a Lie algebra valued 1-form ω in $T^*(P)$ whose vertical component is Maurer-Cartan form $g^{-1}dg$, namely in the local coordinates,

$$\omega = g^{-1}Ag + g^{-1}dg \ \in \ \Omega^1(P, \mathbf{g}), \tag{9.5}$$

where $A = A^a_\mu(x)\frac{\lambda_a}{2i}dx^\mu \ \in \ T^*(M)$, and Ω^1(P,$\mathbf{g}$) denote the Lie algebra $\mathbf{g}$-valued 1-form on P.

It is easy to show that ω satisfies:

$$< \omega,\ \mathcal{D}_\mu >= 0,$$
$$< \omega,\ \hat{V}_a > |_e = \frac{\lambda_a}{2i}.$$

So ω annihilates the horizontal basis of $T(P)$ and maps the vertical space basis to a constant. Remember that

$$\omega|_{\text{fiber}} = g^{-1}dg,$$

and under the right action of group G on the principal bundle P:

$$\begin{aligned} R_a: &\quad u \to ua, \\ R^*_a: &\quad \omega \to a^{-1}\omega a. \end{aligned}$$

Namely ω transforms as a tensor while A remains invariant. So given a connection 1-form ω on a principal bundle P ($\omega \in \Omega^1(P, \mathbf{g})$), equation $\omega = 0$ defines a G-invariant family of horizontal direction on P.

Given a connection ω, the curvature Ω be defined as

$$\Omega = d\omega + \omega \wedge \omega \ \in \Omega^2(P, \mathbf{g}),$$

which is a Lie algebra valued 2-form having no vertical component. Namely its contraction with vertical tangent field is zero:

$$<\Omega,\ X,Y>=0,\ \forall X,Y\in V(P).$$

Ω satisfies the following Bianchi identity

$$d\Omega+\omega\wedge\Omega-\Omega\wedge\omega=0,$$

and it transforms as a tensor under the right action of group G on principal P:

$$R_{a*}:\ \Omega\rightarrow a^{-1}\Omega a,\ \forall a\in G.$$

Pull back of Ω to the base manifold can be written as:

$$\begin{aligned}\Omega &= g^{-1}Fg.\\ \text{where}\qquad F &= dA+A\wedge A\\ &= \frac{1}{2}F^a_{\mu\nu}\frac{\lambda_a}{2i}dx^\mu\wedge dx^\nu\ \in\Omega^2(M,\mathbf{g})\end{aligned}$$

is a Lie algebra valued 2-form on M, satisfying the following Bianchi identity:

$$dF+A\wedge F-F\wedge A=0.$$

Now we take a look at the third approach to connection of principal bundle, using differential forms on base manifold. Let $\{U_\alpha\}$ be an open covering of M, principal bundle can be trivialized locally and has smooth sections g_α on U_α:

$$\begin{aligned}g_\alpha:\ U_\alpha &\rightarrow P,\\ x &\rightarrow g_\alpha(x).\end{aligned}$$

Connection 1-form ω and curvature 2-form Ω can be pulled back to the base manifold as

$$g^*_\alpha\omega=A^{(\alpha)},\quad g^*_\alpha\Omega=F^{(\alpha)}.$$

In the overlapping region $U_\alpha\cap U_\beta\neq 0$, we have

$$g_\alpha(x) = g_{\alpha\beta}g_\beta(x). \tag{9.6}$$

$$\begin{aligned}\text{Then}\qquad \Omega &= g_\alpha^{-1}A^{(\alpha)}g_\alpha+g_\alpha^{-1}dg_\alpha\\ &= g_\beta^{-1}A^{(\beta)}g_\beta+g_\beta^{-1}dg_\beta.\end{aligned} \tag{9.7}$$

From Eq. 9.6 we have:

$$A^{(\beta)}=g^{-1}_{\alpha\beta}A^{(\alpha)}g_{\alpha\beta}+g^{-1}_{\alpha\beta}dg_{\alpha\beta}. \tag{9.8}$$

The corresponding curvature 2-form is

$$F^{(\beta)}=g^{-1}_{\alpha\beta}F^{(\alpha)}g_{\alpha\beta}, \tag{9.9}$$

so the curvature 2-form Ω can be defined consistently on $P(M,G)$:

$$\Omega = g_\alpha^{-1} F^{(\alpha)} g_\alpha = g_\beta^{-1} F^{(\beta)} g_\beta. \tag{9.10}$$

Eq. 9.7 and Eq. 9.10 determines how local connection A and curvature F are to be glued together to give globally defined connection ω and curvature Ω, i.e., a connection ω on principal bundle P can be given locally (in each neighborhood U_α) in terms of a 1-form $A(x)$ defined on the base manifold U_α, which takes value in the Lie algebra $\mathbf{g}$ of G.

From the analysis above, we can see that we can use Eq. 9.7 to define the connection 1-form $\omega \in T^*(P)$ on the principal bundle if we can find a Lie algebra valued 1-form $A \in T^*(M)$ on the open set U_α of base manifold, satisfying Eq. 9.8 in the overlapping regions. This is the third approach to the definition of connection on principal bundle, and Eq. 9.8 is equivalent to the gauge transformation in gauge field theory. Gauge potential A is equivalent to the pull back of connection on the principal bundle to the base manifold through local sections on the principal bundle, and gauge field strength F is equivalent to the pull back of curvature 2-form Ω to the base manifold.

In summary, there are three equivalent definitions of connection on principal bundle $P(M,G)$:

1. Tangent space approach. A connection on a principal bundle $P(M,G)$ is a family of smoothly varying horizontal subspaces H_u, which are invariant under the right action of group G on P. Namely, for each point $u \in P$, one define a splitting of the tangent bundle $T_u(P)$ into horizontal component H_u and vertical component V_u,

$$T_u = H_u \oplus V_u,$$

where V_u is consisted of right-invariant fundamental vector fields on vertical fibre G manifold. Given the connection and the horizontal space H_u, which satisfies

$$\pi_* H_u = T_{\pi(u)}(M),$$
$$\{H_{R_a u}\} = \{H_u\},$$

i.e., the splitting is smooth and invariant under right multiplication by group G. The basis $\{\mathcal{D}_\mu\}$ of H_u can be expressed explicitly as Eq. 9.3.

2. Cotangent space approach. Pick a 1-form ω in $T^*(P)$ satisfying:

$$\omega|_{\text{fibre}} = g^{-1}dg, \quad \text{or} \quad <\omega, \tilde{V}_a>|_e = \frac{\lambda_a}{2i},$$
$$R_a^*\omega = a^{-1}\omega a.$$

3. Base manifold differential form approach. Given a connection 1-form A on each neighborhood U_α of base manifold M, which transforms as Eq. 9.8 in the overlapping region.

9.2 Connection on associated vector bundle

We recall that an associated bundle

$$E(M, V, \pi, G, P) = P \times_G V,$$

where V is a vector representation space ρ of G. A trivial G action on the trivial bundle $P \times V$ is:

$$g(u, v) = (u \cdot g,\ \rho(g^{-1})v), \quad \forall u \in P,\ v \in V.$$

The quotient by this G-action is the associated bundle $E = P \times_G V$, which is a vector bundle over $M = P/G$ with fibre V, and is called the associated bundle of P for representation ρ. A connection on the principal bundle P induces a connection on the associated bundle $E = P \times_G V$.

Associated bundle E has the same structure group as that of the principal bundle. Let ρ be the representation of group G acting on vector space V, the representation $\rho(g_{\alpha\beta})$ of transition function $g_{\alpha\beta}$ of principal bundle P is then the transition function on the associated bundle E. P and E have the same base manifold M. The horizontal lift $u(\tau)$ on the principal bundle of a curve $x(\tau)$ on the base manifold is:

$$u(\tau) = (x(\tau), g(\tau)).$$

It determines completely the horizontal lift to the associated bundle $E = P \times_G V$ of curve $x(\tau)$,

$$z(\tau) = (x(\tau), \psi(\tau)),$$
$$\psi(\tau) = g^{-1}(\tau)\psi.$$

For simplicity, we use g^{-1} here in place of its representation matrix $\rho(g^{-1})$ and let $g(0) = 1$. We can obtain the horizontal curve $z(\tau)$ on bundle E from the horizontal lift $g(\tau)$ of curve $x(\tau)$ on base manifold M. The covariant derivative D_τ acting on tangent field ψ is the tangent vector of the horizontal $z(\tau)$ on E, which determines the horizontal subspace $H(E)$ of tangent space $T(E)$ on bundle E:

$$\begin{aligned} D_\tau \psi(\tau)|_{\tau=0} &= \frac{d}{d\tau}(g^{-1}(\tau)\psi(x(\tau)))|_{\tau=0} \\ &= \frac{dx^\mu}{dt}(\frac{\partial}{\partial x^\mu}\psi + A_\mu \psi) \\ &= \frac{dx^\mu}{dt} D_\mu \psi, \end{aligned}$$

where

$$D_\mu = \frac{\partial}{\partial x^\mu} + A_\mu$$

is the covariant derivative operator acting on section $\psi(x)$ of bundle E. Under the right action of G on principal, the corresponding section on E behaves as:

$$\psi \to g^{-1}\psi.$$

And it is easy to prove then that:

$$D_\mu\psi \to g^{-1}D_\mu\psi.$$

Namely covariant derivative operator is a differential operator mapping one section of E to another section on E.

The pull back of curvature Ω on principal bundle P to base manifold M gives the 2-form $F = dA + A \wedge A$, which can be regarded as Lie algebra $\mathbf{g}$-valued rank-2 skew-symmetric contravariant tensor (section of tensor bundle) on base manifold M. From Eq. 9.9 we know this tensor bundle is the associated bundle of principal P, the covariant derivative of its section F is:

$$\begin{aligned} D_\tau F|_{\tau=0} &= \frac{d}{d\tau}(g^{-1}(\tau)F(x(\tau_1)g(\tau)|_{\tau=0} \\ &= \frac{dx^\mu}{dt}(\frac{\partial}{\partial x^\mu}F + [A_\mu, F]) \\ &= \frac{dx^\mu}{dt}D_\mu F. \end{aligned}$$

Alternatively define

$$DF \equiv dx^\mu D_\mu F = dF + [A, F],$$

called the covariant exterior differentiation of F. Under the right action of structure group G, DF transforms in the same way as F,

$$DF^{(\beta)} = g_{\alpha\beta}^{-1}\ DF^{(\alpha)}\ g_{\alpha\beta}.$$

Vector field $\psi(x)$ can be called vector (elementary representation space of group G) valued 0-form, $F(x)$ is called matrix valued (adjoint tensor) 2-form. For adjoint tensor valued k-form Σ

$$\Sigma^{(\beta)} = g_{\alpha\beta}^{-1}\ \Sigma^{(\alpha)}\ g_{\alpha\beta},$$

one may define its covariant exterior differentiation as:

$$D\Sigma = d\Sigma + A \wedge \Sigma - (-1)^k \Sigma \wedge A.$$

And it is easy to prove that $D\Sigma$ is an adjoint tensor type $(k+1)$-form:

$$D\Sigma^{(\beta)} = g_{\alpha\beta}^{-1}\ D\Sigma^{(\alpha)}\ g_{\alpha\beta}.$$

Namely covariant exterior differential operator D maps one gauge covariant form to another gauge covariant form.

Next we give two examples to highlight the significance of connection and covariant derivative.

Example 1. Electromagnetic field is a $U(1)$ gauge field. Wave function of a charged particle moving in electromagnetic field is a section of $U(1)$ associated bundle, which is a complex valued function. Its norm $|\psi(x)|$ is the probability amplitude of finding the partical at x, and $\psi(x) = |\psi(x)|e^{i\theta}$ with the phase factor θ having no particular physical significance. The physical state is unchanged by a global phase factor change of the wave function. But for a local phase factor transformation:

$$\psi(x) \to \psi'(x) = e^{i\theta\alpha(x)}\psi(x),$$

as the phase factor is coordinate dependent, the differential of the wave function is no longer a section of the bundle, covariant differentiation has to be introduced:

$$D_\mu = \partial_\mu + A_\mu(x),$$

$D_\mu\psi(x)$ then is still a section of the associated bundle, which has the same transformation rule as $\psi(x)$:

$$D_\mu\psi(x) \to D'_\mu\psi'(x) = e^{-i\theta\alpha(x)}D_\mu\psi(x),$$

where $D'_\mu = \partial_\mu + A'_\mu(x)$. While under a gauge transformation, gauge potential $A(x)$ transforms as:

$$\begin{aligned} A_\mu(x) \to A'_\mu(x) &= e^{-i\theta\alpha(x)}(A_\mu(x) + \partial_\mu)e^{i\theta\alpha(x)} \\ &= A_\mu(x) + ie\partial_\mu\alpha(x). \end{aligned}$$

Covariant differential operator is a linear operator which maps one section of the associated vector bundle to another section. Covariant derivative specifies the interaction between charged particle and gauge field, and is called minimal coupling.

Example 2. $SU(2)$ Yang-Mills gauge field. The physics background for its introduction is the nuclear isospin symmetry in elementary particle physics. If we neglect the electromagnetic interaction, proton and neutron are a doublet of isospin, written as:

$$\psi_N(x) = \begin{pmatrix} \psi_p(x) \\ \psi_n(x) \end{pmatrix}.$$

The $SU(2)$ gauge covariant Dirac equation satisfied by $\psi_N(x)$ is:

$$\gamma^\mu D_\mu\psi_N(x) = -im_N\psi_N(x),$$

where D_μ is the covariant derivative operator, which includes the Yang-Mills field $A(x)$:

$$D_\mu\psi_N(x) = (\partial_\mu + A_\mu(x))\psi_N(x).$$

Of which $A(x)$ is an $SU(2)$ Lie algebra valued 1-form, which is a pull back to the base manifold of 1-form ω on principal bundle P with structure group $SU(2)$. The local section on the principal bundle P

$$\sigma_\alpha : U_\alpha \to P$$

is called the local gauge, and

$$A(x) = \sigma_\alpha^* \omega = A_\mu^a(x) \tau^a dx^\mu.$$

From the commutator of covariant derivative, we obtain the Yang-Mills gauge field strength:

$$F = dA + A \wedge A.$$

The relationship between gauge field strength F and gauge potential A, is exactly as that between curvature Ω and connection ω of fibre bundle, with gauge potential corresponding to the connection and gauge field strength corresponding to curvature.

9.3 Connection on general vector bundle

A general rank-k vector bundle is also an associated bundle of its frame bundle (the $GL(k, \mathbf{R})$ principal bundle). General vector bundle is very important in theoretical physics because the space of sections of such a bundle carries a natural structure of a vector space. As vector bundles arise so often, here we look at an approach that no direct reference to an underlying principal bundle is needed.

Let $E = E(M, V, \pi, G)$ be a general rank-k vector bundle over an m-dimensional manifold M. Vector bundle E can be trivialized locally. On each neighborhood $U \in M$, one can take a local frame $\{e_\alpha(x)\}$, and the local smooth section S of bundle E is a smooth mapping from U to $\pi^{-1}(U)$:

$$\begin{aligned} S : U &\to \pi^{-1}(U) \\ x &= \to S(x) = \sum_{a=1}^{k} e_a(x) z^a(x). \end{aligned}$$

$S(x)$ could be regarded as a vector valued function, $\{z^a(x)\}$ are its components with respect to the basis $\{e_a(x)\}$. Let $\Gamma(E)$ be the set of sections, the covariant derivative operator ∇ is a map between sections:

$$\begin{aligned} \nabla : \Gamma(E) &\to \Gamma(E \otimes T^*(M)), \\ S(x) &\to \nabla S(x), \\ \nabla S(x) &= \nabla(e_a(x) z^a(x)) \\ &= e_a(x) \otimes dz^a(x) + e_b(x) \Gamma^b{}_a(x) \otimes z^a(x) \\ &= e_a(x) \otimes \omega^a(x), \end{aligned} \qquad (9.11)$$

where

$$\omega^a = dz^a + {\Gamma^a}_b z^b \tag{9.12}$$

is a vector valued 1-form. And

$$\Gamma^b{}_a(x) = dx^\mu \Gamma^b_{\mu a}(x) \ \in T^*(M) \ \ (a, b = 1, \cdots, k)$$

are called the connection 1-forms of vector bundle E, they are independent of the coordinate system $\{x_\mu\}$ on the base manifold. Now let $X = \xi^\mu(x)\frac{\partial}{\partial x^\mu}$ be a tangent vector on base manifold M, we can get the directional derivative of section $S(x)$ in a given direction X on M with the help of connection,

$$\begin{aligned}
\nabla_X S &= < \nabla S, X > = \xi^\mu(x) \nabla_\mu S. \\
\text{where} \ \ \nabla_\mu S &= < \nabla S, \partial_\mu > \\
&= e_a\Big(\frac{\partial z^a}{\partial x^\mu} + \Gamma^a_{\mu b}(x) z^b(x)\Big) \equiv e_a z^a_{;\mu}. \\
\nabla S &= dx^\mu \nabla_\mu S.
\end{aligned}$$

Parallel transport of section S along X can be written as:

$$\nabla_X S = \xi^\mu e_a\Big(\frac{\partial z^a}{\partial x^\mu} + \Gamma^a_{\mu b} z^b\Big) = 0.$$

Thus one can lift a curve $x(\tau)$ on base manifold M to a horizontal curve $u(\tau)$ on vector bundle E:

$$u(\tau) = (x(\tau), z(\tau)),$$

where $z(\tau)$ satisfies the following parallel transport equation:

$$\begin{aligned}
\frac{d}{d\tau} &= \frac{dx^\mu}{d\tau}\frac{\partial}{\partial x^\mu} + \frac{dz^a}{d\tau}\frac{\partial}{\partial z^a} \\
&= \frac{dx^\mu}{d\tau}\Big(\frac{\partial}{\partial x^\mu} - \Gamma^a_{\mu b} z^b \frac{\partial}{\partial z^a}\Big) \\
&= \frac{dx^\mu}{d\tau} D_\mu, \\
\text{where} \ \ D_\mu = \frac{\partial}{\partial x^\mu} &- \Gamma^a_{\mu b} z^b \frac{\partial}{\partial z^a}
\end{aligned}$$

is the covariant derivative operator in the tangent bundle $T(E)$, which is the horizontal lift of tangent vector $\frac{\partial}{\partial x^\mu}$ on M. Hence one can split $T(E)$ into a vertical component $V(E)$ with basis $\{\frac{\partial}{\partial z^\mu}\}$ lying strictly in the fibre and a horizontal component $H(E)$ with basis $\{D_\mu\}$:

$$T(E) = V(E) \oplus H(E).$$

Unlike $\{\frac{\partial}{\partial x^\mu}\}$, the commutator of D_μ is non-vanishing:

$$[D_\mu, D_\nu] = -R^a_{b\mu\nu} z^b \frac{\partial}{\partial z^a}.$$

Note here the right hand side has vertical component only, and the coefficient:

$$R^a_{b\mu\nu} = \partial_\mu \Gamma^a_{\nu b} - \partial_\nu \Gamma^a_{\mu b} + \Gamma^a_{\mu c}\Gamma^c_{\nu b} - \Gamma^a_{\nu c}\Gamma^c_{\mu b}$$

is called the curvature tensor of the connection, which is the obstruction to the integrability of the horizontal subspace. It measures the non-commutation of covariant differentiation and the path dependence of parallel transport.

So far we used the tangent space approach to connection, now we turn to the cotangent space approach. From Eq. 9.12:

$$\omega^a = dz^a + \Gamma^a{}_b z^b$$

is a vector valued 1-form on the cotangent bundle $T^*(E)$, satisfying:

$$\begin{aligned} &< \omega^a, D_\mu >= 0, \\ &< \omega^a, \frac{\partial}{\partial z^b} >= \delta^a_b. \end{aligned}$$

Namely, connection 1-form annihilates the horizontal subspace of $T(E)$. Equation $\omega = 0$ defines a horizontal subspace of $T(E)$. So from Eq. 9.11 for the covariant derivative:

$$\nabla S = e_a \otimes \omega^a,$$

we have the second order derivative of S:

$$\begin{aligned} \nabla^2 S &= \nabla(e_a \otimes dz^a + e_a \otimes \Gamma^a{}_b z^b) \\ &= e_b \otimes \Gamma^b{}_a \wedge dz^a + e_b \otimes \Gamma^b{}_a \wedge \Gamma^a{}_c z^c \\ &\quad + e_a \otimes (d\Gamma^a{}_b z^b - \Gamma^a{}_b \wedge dz^b) \\ &= e_a \otimes \Omega^a{}_b z^b. \\ \text{Where } \Omega^a{}_b &= d\Gamma^a{}_b + \Gamma^a{}_c \wedge \Gamma^c{}_b \\ &= \frac{1}{2} R^a_{b\mu\nu} dx^\mu \wedge dx^\nu \end{aligned}$$

is the tensor valued curvature 2-form.

Under a change of frames on the fibre of E, we have

$$e'_a(x) = e_b(x)\Lambda^b{}_a(x) \quad ((\Lambda^b{}_a) \in GL(k)).$$

The section $S(x)$ is a geometric object and is independent of the choice of frames, so

$$S(x) = e_a z^a = e'_a z'^a.$$

Thus its components transform as:

$$z'^a(x) = (\Lambda^{-1})^a{}_b(x) z^b(x).$$

Next we look at the transformation of the connection. It is easy to show that

$$\nabla e'_a = e'_b \Gamma'^b{}_a = \nabla(e_b \Lambda^b{}_a) = e_b \Gamma^b{}_c \Lambda^c{}_a + e_b d\Lambda^b{}_a,$$
$$\text{so} \qquad \Gamma'^a_b = (\Lambda^{-1})^a{}_c \Gamma^c_d \Lambda^d{}_b + (\Lambda^{-1})^a{}_c d\Lambda^c{}_b.$$

So under a change of frames in the fibre, the connection 1-form $\Gamma^a{}_b(x) \in T^*(M)$ transforms as a gauge potential rather than as a tensor. It is easy to prove that the curvature 2-form $\Omega^a{}_b(x)$ transforms as a tensor:

$$\Omega'^a_b = (\Lambda^{-1})^a{}_c \Omega^c_d \Lambda^d{}_b.$$

Take the exterior differentiation of 1-form $\omega^a \in T^*(E)$ gives:

$$\begin{aligned} d\omega^a &= d\Gamma^a{}_b z^b - \Gamma^a{}_b \wedge dz^b \\ &= \Omega^a{}_b z^b - \Gamma^a{}_b \wedge \omega^b. \end{aligned}$$

So $\Omega^a{}_b z^b$ is the covariant differential of the 1-form ω^a,

$$\Omega^a{}_b z^b = \nabla\omega^a = d\omega^a + \Gamma^a{}_b \wedge \omega^b,$$

and under the change of frames in the fibre,

$$\omega'^a = (\Lambda^{-1})^a{}_b \omega^b.$$

For a given vector bundle E,

$$\pi^{-1}(x) = V$$

is a k-dimensional vector space. On which the set of linear function constitute the space V_x^* dual to V_x, which is also a k-dimensional vector space. Vector bundle E^* dual to E is obtained by taking the dual vector space V_x^* as fibre at each point of the base manifold.

Let $\{e_a\}$ be a local frame of the vector bundle E. One can choose the dual frame $\{\theta^a\}$ such that:

$$< \theta^a, e_b >= \delta^a_b.$$

So we can define the inner product between section S and S^* of the dual vector bundle:

$$< S^*, S >=< \theta^a z_a, e_b z^b >= z_a z^a.$$

If the transition matrix of E in the overlapping region is a $k \times k$ non-singular matrix $\lambda \in GL(k)$, then the corresponding transition matrix for E^* is $(\lambda^{-1})^t$.

For a connection ∇ defined on E,

$$\nabla e_a = e_b \Gamma^b{}_a,$$

the connection ∇^* on E^* is then defined by requiring that the inner product has the following property:

$$d < S^*, S > = < \nabla^* S^*, S > + < S^*, \nabla S >,$$

$$\text{i.e.,} \quad \nabla^* \theta^a = -\theta^b \Gamma^a{}_b.$$

Furthermore if vector bundle E has a fibre metric, then one may obtain a 1-1 correspondence (linear isomorphism) between E and E^*. And if we require the connection to be metric preserving, i.e.,

$$\begin{aligned} \nabla &= \nabla^*, \\ \text{or} \quad \Gamma^a{}_b &= -\Gamma^b{}_a, \end{aligned}$$

in an orthogonal basis, then this connection is called Riemannian . The curvature of a Riemannian connection in an orthogonal basis is skew-symmetric:

$$\Omega^a{}_b = -\Omega^b{}_a.$$

Note here that for general vector bundle E, Riemannian connection is metric preserving. Torsion is of no relevance here, it arises in the tangent bundle only. For the general principal bundle and vector bundle, the torsion of connection in general is not well defined. Torsion is a property of connection on the tangent bundle $T(M)$ (or cotangent bundle $T^*(M)$, frame bundle $L(M)$ and its associated bundle). The Levi-Civita connection on tangent bundle $T(M)$ is the only torsion free Riemannian connection.

Example The natural complex line bundle L of $CP(1)$.

The complex projective space $CP(1)$ is the set of lines passing through the origin in C^2.

Let $I^2 = CP(1) \times C^2$ be the trivial bundle over $CP(1)$, a point which is denoted by coordinate $(x; z^0, z^1)$.

The natural complex line bundle L is a sub-bundle of I^2 defined by:

$$L = \{(x; z^0, z^1) \in I^2 \mid (z^0, z^1) \text{are homogeneous coordinate of } x \in CP(1)\}.$$

The natural fibre metric on L can be induced from the canonical metric of C^2 as:

$$((x; z^0, z^1),\ (x; w^0, w^1)) = z^0 \bar{w}^0 + z^1 \bar{w}^1.$$

Then the length h of each point $u = (x; z^0, z^1)$ is defined as:

$$h(x; z^0, z^1) = |z^0|^2 + |z^1|^2.$$

This manifold can be covered by two open sets $\{U_0, U_1\}$. The coordinate of U_0 $(x^0 \neq 0)$ can be chosen as $(x, 1, \varsigma^1_{(0)})$, where

$$\varsigma^1_{(0)} = z^1/z^0 = x + iy \equiv z.$$

The length is:

$$h(x,z) = 1 + |z|^2.$$

The metric preserving connection 1-form can be written as:

$$\omega = h^{-1}\partial h = \frac{\bar{z}dz}{1+|z|^2} \in T^*(L).$$

And the corresponding curvature is:

$$\begin{aligned}\Omega &= d\omega + \omega\wedge\omega = (\partial+\bar{\partial})(h^{-1}\partial h)\\ &= -\partial\bar{\partial}\ln\, h = -\frac{dz\wedge d\bar{z}}{(1+|z|^2)^2}\\ &= \frac{2idx\wedge dy}{(1+x^2+y^2)^2}.\end{aligned} \tag{9.13}$$

Note here that the connection 1-form $\omega = h^{-1}\partial h$ is formally a pure gauge with respect to a curvature involving ∂ only, the curvature Ω is non-trivial however, as $d = \partial + \bar{\partial}$.

9.4 Gauge theory, action and Yang-Mills equation

Symmetry plays an important role in quantum field theory. From the Noether theory we know that conservation laws correspond to continuous symmetries of the Lagrangian. On the other hand, gauge field is introduced when we require the Lagrangian to be invariant under local symmetry transformations (gauge transformations). Strong and electro-weak interactions all have gauge fields as intermediate bosonic fields, while gravitation has close links with local Lorentz transformation. Topology is ever increasingly being used in gauge theory and gravitation. Gauge field corresponds to the connection on the principal bundle while matter field corresponds to the section of the associated vector bundle, these physical quantities have direct and explicit counterparts in geometry.

As is well known, Maxwell's electromagnetic theory is an Abelian ($U(1)$) gauge theory, and Yang-Mills theory is the non-Abelian gauge theory. Gauge potential $\mathcal{A}_\mu(x)$ is a vector field taking value in the Lie algebra $\mathbf{g}$ of gauge group G. In the matrix representation of $\mathbf{G}$, $\mathcal{A}_\mu(x)$ can be written as:

$$\mathcal{A}_\mu(x) = T_a\mathcal{A}_\mu^a(x),$$

where T_a are the anti-Hermitian matrix generators of $\mathbf{g}$:

$$[T_a, T_b] = f_{ab}^c T_c, \quad T_a^+ = -T_a,$$

where f_{ab}^c are the structure constants of G. For simplicity the normalization of the metric of group G can be chosen such that:

$$Tr(T_a T_b) = -\frac{1}{2}\delta_{ab}.$$

So the components of $\mathcal{A}_\mu^a(x)$ are given as:

$$\mathcal{A}_\mu^a(x) = -2Tr(T_a \mathcal{A}_\mu(x)).$$

The corresponding gauge field strength is:

$$\mathcal{F}_{\mu\nu}^a = \partial_\mu \mathcal{A}_\nu^a - \partial_\nu \mathcal{A}_\mu^a + e f_{ab}^c \mathcal{A}_\mu^b \mathcal{A}_\nu^c,$$

where e is the coupling constant. Note here that the last term of above equation imply that the non-Abelian gauge field has self interaction.

To make all the formulas look simple, we introduce the following notations:

$$A_\mu^a(x) = e\mathcal{A}_\mu^a(x), \quad F_{\mu\nu}^a(x) = e\mathcal{F}_{\mu\nu}^a(x).$$

Then $\quad F_{\mu\nu}^a = \partial_\mu A_\nu^a - \partial_\nu A_\mu^a + f_{ab}^c A_\mu^b A_\nu^c$. Or in matrix form:

$$F_{\mu\nu} = \partial_\mu A_\nu - \partial_\nu A_\mu + [A_\mu A_\nu],$$

in which

$$F_{\mu\nu} = T_a F_{\mu\nu}^a = eT_a \mathcal{F}_{\mu\nu}^a, \quad A_\mu = T_a A_\mu^a = eT_a \mathcal{A}_\mu^a.$$

And we can further express them in differential forms:

$$\begin{aligned} A &= A_\mu dx^\mu, \\ F &= \frac{1}{2} F_{\mu\nu} dx^\mu \wedge dx^\nu \\ &= dA + \frac{1}{2}[A, A] = dA + A \wedge A. \end{aligned}$$

Covariant derivative D_μ is defined as

$$D_\mu = \partial_\mu + [A_\mu, \ \ ],$$

then $\quad F_{\mu\nu} = [D_\mu, D_\nu]$. And the following Jacobi identity is satisfied:

$$[D_\mu, [D_\nu, D_\lambda]] + [D_\nu, [D_\lambda, D_\mu]] + [D_\lambda, [D_\mu, D_\nu]] = 0,$$

which is just the Bianchi identity satisfied by the curvature:

$$D_\mu F_{\nu\lambda} + D_\nu F_{\lambda\mu} + D_\lambda F_{\mu\nu} = 0. \tag{9.14}$$

Or in terms of differential forms:

$$DF = dF + [A, F] = 0.$$

Comparing with formulas in §9.1, it is obvious that the gauge potential A is the pull back of connection ω on principal bundle P back to the base manifold M, and the gauge field strength F is the pull back of curvature Ω on $P(M, G)$ to M.

Up to now we have not introduced the metric on the base manifold. However, metric is needed if we want to write down an action for a system of gauge fields.

In what follows we consider the dynamics of Yang-Mills field on 4-dimensional Minkowski space, where the metric has the diagonal form $\eta_{ab} = (1, -1, -1, -1)$ with $x_0 = t$ is the time. The action S for pure gauge (sourceless) can be written as an integral of Lagrangian density:

$$S = \int_M d^4x \, \mathcal{L},$$
$$\mathcal{L} = \frac{1}{2e^2} Tr \, F^{\mu\nu} F_{\mu\nu} = -\frac{1}{4} \mathcal{F}^{\mu\nu}_a \mathcal{F}^a_{\mu\nu} = \frac{1}{2}(E^2_a - H^2_a),$$

where

$$\begin{cases} E^i_a = \mathcal{F}^{i0}_a, & \text{or } \mathbf{E}_a = -\dot{\mathbf{A}}_a - \nabla\phi_a - e f_{abc}\phi_b \mathbf{A}_c, \\ H^i_a = -\frac{1}{2}\epsilon^{ijk}\mathcal{F}^a_{jk}, & \text{or } \mathbf{H}_a = \nabla \times \dot{\mathbf{A}}_a - \frac{1}{2} e f_{abc} \mathbf{A}_b \times \mathbf{A}_c, \end{cases}$$

where $i = 1, 2, 3$, and $a = 1, \cdots, dim \, G$.

And $\phi_a = \mathcal{A}^0_a$, $\mathbf{A}_a = (\mathcal{A}^1_a, \mathcal{A}^2_a, \mathcal{A}^3_a)$. In terms of differential forms the action can be written as:

$$S = \frac{1}{e^2} \int_M Tr(F \wedge *F), \tag{9.15}$$

where $*$ is the Hodge star operator which acts on differential forms.

From the variation of A for action S, one obtains the Euler-Lagrange equation of motion, i.e. the Yang-Mills equation:

$$\partial_\mu \mathcal{F}^{\mu\nu}_a + e f_{abc} \mathcal{A}^b_\mu \mathcal{F}^{\mu\nu}_c = 0.$$

Or in matrix form:

$$D_\mu F^{\mu\nu} = 0,$$

and in terms of differential forms:

$$D * F = 0.$$

For 4-dimensional Yang-Mills field it is sometimes helpful to introduce the dual field ${}^*F^{\mu\nu}$ (super index *) defined by:

$${}^*F^{\mu\nu} = \frac{1}{2}\epsilon^{\mu\nu\lambda\rho} F_{\lambda\rho}.$$

$$\text{Then} \quad *F = *(\frac{1}{2}F_{\mu\nu}dx^\mu \wedge dx^\nu)$$
$$= \frac{1}{2}F_{\mu\nu}\frac{1}{2}\epsilon^{\mu\nu}{}_{\lambda\rho}dx^\lambda \wedge dx^\rho$$
$$= \frac{1}{2}{}^*F_{\mu\nu}dx^\mu \wedge dx^\nu \equiv {}^* F.$$

Bianchi identity Eq. 9.14 can be rewritten in terms of dual field as

$$D_\mu{}^*F^{\mu\nu} = 0. \tag{9.16}$$

Matrix and differential form expressions of Yang-Mills field

	Matrix tensor expression	In differential form
Gauge potential	$A_\mu(x) = eA_\mu^a(x)T_a$	$A = eA_\mu^a T_a dx^\mu$
Field strength	$F_{\mu\nu} = \partial_\mu A_\nu - \partial_\nu A_\mu + [A_\mu, A_\nu]$	$F = dA + A \wedge A$
Bianchi identity	$D_\mu{}^*F^{\mu\nu} = 0$	$DF = 0$
Field equation	$D_\mu F^{\mu\nu} = 0$	$D * F = 0$

Note here that the introduction of a metric (through index contraction or Hodge * operation) is inevitable for Yang-Mills equations. But the Bianchi identity is independent of the metric. Eq. 9.16 formally involves index contraction, but it is a contraction with the dual field, so is in fact a cyclic sum of three terms as Eq. 9.14, and consequently independent of the metric.

Quite often we have to consider systems of gauge field A interacting with matter field ϕ, the interaction between them is described by covariant derivative:

$$D\phi = (D_\mu\phi)dx^\mu = d\phi + A\phi,$$

or in its components:

$$(D_\mu\phi)^i = \partial_\mu\phi^i + A_\mu^a(T_a)^i_j\phi^j$$

Under a local gauge transformation,

$$\phi \to g^{-1}\phi, \quad D\phi \to g^{-1}D\phi.$$

The Lagrangian density takes the form:

$$\frac{1}{e^2}tr(F \wedge *F) \;+\; (D\phi)^+ \wedge *D\phi - m^2\phi^+ \wedge *\phi$$
$$= \{-\frac{1}{4e^2}F^a_{\mu\nu}F^{\mu\nu}_a + \overline{(D_\mu\phi)}^i(D^\mu\phi)_i - m^2\bar{\phi}^i\phi_i\} * 1.$$

Equations of motion is obtained by variation of A and ϕ:

$$D^\mu F^a_{\mu\nu} = J^a_\nu, \tag{9.17}$$
$$D_\mu D^\mu\phi^i - m^2\phi^i = 0,$$

where $J^a_\mu = \frac{e^2}{2}[\phi^+_i T_a D_\mu\phi^i - (D_\mu\phi)_i T_a\phi^i]$ is the current of matter, which is not conserved.

9.5 Local gauge symmetry and current conservation

Conserved quantities in a physical system is closely related to symmetry transformations of the system. The action in Eq. 9.15 has local gauge symmetry. when there exists a gauge transformation:

$$A_\mu \to A_\mu^g = g^{-1}A_\mu g + g^{-1}\partial_\mu g = A_\mu + g^{-1}D_\mu g,$$
$$F_{\mu\nu} \to F_{\mu\nu}^g = g^{-1}F_{\mu\nu}g,$$

Lagrangian density L and action S are gauge invariant as a result of the trace operation, so gauge transformation is a symmetry transformation of the Yang-Mills equation.

For an infinitesimal gauge transformation

$$g(x) = exp(\Theta(x)) = I + \Theta(x) + \cdots,$$

where the anti-Hermitian matrix $\Theta(x)$ can be written as:

$$\Theta(x) = \theta^a(x)T_a.$$

Here $\theta^a(x)$ are infinitesimal real parameters. Then to the first order, one get

$$\begin{aligned} \delta A_\mu &= D_\mu\Theta, \\ \delta F_{\mu\nu} &= [F_{\mu\nu}, \Theta]. \end{aligned} \tag{9.18}$$

Or in component form
$$\begin{aligned} \delta A_\mu^a &= \partial_\mu\theta^a + f_{abc}A_\mu^b\theta^c, \\ \delta F_{\mu\nu}^a &= f_{abc}F_{\mu\nu}^b\theta_c. \end{aligned}$$

Continuous symmetry transformation of a physical system can induce the conservation of Noether current. Usually one considers the conserved current corresponding to the "first kind" gauge transformation only, of which the parameters θ^a are space-time independent constants.

The variation of A in the action gives the field equation Eq. 9.17:

$$D^\mu F_{\mu\nu} = J_\nu,$$

where the "external current" contribution from the matter field is not necessarily conserved in general. Conservation of current reflects a gauge symmetry of the physical system, thus it should be obtainable through Noether's theorem by varying the gauge invariant Lagrangian. Let us take a sourceless gauge field system (external current $J_\mu = 0$), and assume the gauge parameters θ to be constant, then from Eq. 9.18 we have

$$\delta A_\mu^a = f_{bc}^a A_\mu^b\theta^c,$$

and the Noether current is

$$j_a^\mu = -\frac{\partial \mathcal{L}}{\partial(\partial_\mu A_\nu^b)} \frac{\delta A_\nu^b}{\delta \theta^a} = \frac{2}{e^2} Tr\ F^{\mu\nu}[A_\nu, T_a] = -f_{bc}^a F^{\mu\nu b} A_\nu^c.$$

It is easy to prove that this current j_a^μ is conserved:

$$\partial_\mu j_a^\mu = 0,$$

which is the self-induced current for the non-Abelian gauge theory and can be non-vanishing and conserved by itself. Note here that the self-induced current contains the gauge potential and is not gauge covariant, consequently it has no explicit physical significance.

To obtain the gauge covariant self-induced current[38], of which physical significance is well defined, we may start from comparing the case with gravitation theory. The space-time symmetry transformation group (its generators are Killing vectors) leads to the conservation of energy and momentum. For space-time with constant curvature, there is a ten-parameter group of motion, corresponding to ten locally conserved quantities. But for more general curved space-time, there is no well defined locally conserved energy and momentum, though the stress-energy tensor $T_{\mu\nu}$ has a vanishing covariant divergence, $\nabla^\mu T_{\mu\nu} = 0$, while the ordinary divergence $\partial^\mu T_{\mu\lambda}$ is in general not zero. Only when there exists Killing vector ξ_μ with $(\xi_{\mu;\nu} + \xi_{\nu;\mu} = 0)$ that will not lead to changes in the metric, then will there exist in the corresponding direction locally conserved current $\xi^\mu T_{\mu\lambda}$ satisfying $\partial_\lambda(\xi^\mu T_{\mu\lambda}) = 0$. Similarly in gauge theory, only when gauge invariant potential exists, will there be well defined locally conserved current, i.e., from Eq. 9.18 it is required that:

$$D_\mu \Theta = 0.$$

Let $\Theta(x) = \theta(x)\mathbf{n}(x)$, here $\theta(x)$ is a real scalar parameter, and $\mathbf{n}(x) = n^a(x)T_a$ is the normalized representation matrix, satisfying:

$$(\mathbf{n}(x), \mathbf{n}(x)) = n^a(x)n_a(x) = 1,$$

namely $\mathbf{n}(x)$ represses any preferred direction in isospin space, and $\theta(x)$ is the gauge rotation angle around $n(x)$. Assuming the gauge rotation angle to be constant, then only the orientation of $n(x)$ is variable with respect to the space-time position x but satisfies

$$D_\mu \mathbf{n}(x) = 0. \tag{9.19}$$

Thus $D_\mu \Theta = 0$ is still satisfied. As a result the gauge field is reducible and there is the conserved current:

$$J_\nu = \partial^\mu(\mathbf{n} \cdot F_{\mu\nu}).$$

Eq. 9.19 is the reducible condition of a gauge field, $\mathbf{n}(x)$ is the generator (Killing vector) of gauge symmetry, and is also the generator of the holonomy group.

In gravitation theory, globally conserved quantities are possible if there is no Killing vector but space-time is asymptotically flat at infinity. The energy-momentum density arising from gravitation field itself alone which depends on the choice of coordinates is not a tensor. Sometimes we can choose a preferred set of coordinates according to the motion of matter or to the symmetry of the system, so that an energy-momentum tensor with certain physical significance is obtained. Similarly, there may exist globally conserved current if a gauge field is asymptotically reducible, but locally conserved quantities are not well defined. In a non-Abelian gauge field, there is also a charge current arising from the non-Abelian field itself, this current is gauge dependent, then how can we find a current which has a definite physical significance? When external fields sources are present, the orientation of the charge current operator $\mathbf{n}(x)$ can be determined from the external matter fields (or Higgs fields), and the corresponding conserved current can be derived. For simplicity, we take a look at an irreducible $SU(2)$ gauge field, assuming that the system has an intrinsic "preferred direction $\mathbf{n}(x)$." The gauge potential $A_\mu(x)$ can then be further divided into an $\tilde{A}_\mu(x)$ of the $U(1)$ field created by $\mathbf{n}(x)$ and a vector partial $B_\mu(x)$ carrying this $U(1)$ charge:

$$A_\mu(x) = \tilde{A}_\mu(x) + B_\mu(x),$$

where
$$\tilde{A}_\mu(x) = \mathbf{n} \cdot A_\mu(x) - [\mathbf{n}(x), \partial_\mu \mathbf{n}(x)],$$

and
$$B_\mu(x) = [\mathbf{n}(x), D_\mu \mathbf{n}(x)].$$

It is easy to prove that $\tilde{A}_\mu(x)$ satisfies the following:

$$\tilde{D}_\mu \mathbf{n}(x) \equiv \partial_\mu \mathbf{n}(x) + [\tilde{A}_\mu(x), \mathbf{n}(x)] = 0,$$

where $\tilde{D}_\mu$ means that in taking the covariant derivative only the parallel potential $\tilde{A}_\mu(x)$ is taken as the connection, corresponding to which $\mathbf{n}(x)$ is a translation invariant field, $\tilde{D}_\mu \mathbf{n}(x) = 0$. But for an irreducible gauge field, if the whole gauge potential is taken as connection, $\mathbf{n}(x)$ is no longer translation invariant, i.e.,

$$D_\mu \mathbf{n}(x) \equiv \partial_\mu \mathbf{n}(x) + [A_\mu(x), \mathbf{n}(x)] = [B_\mu(x), \mathbf{n}(x)] \neq 0.$$

This implies that for an irreducible $SU(2)$ gauge field, there exists charged vector particles $B_\mu(x) = [\mathbf{n}(x), D_\mu \mathbf{n}(x)]$, which are gauge covariant. Under a local gauge transformation

$$A_\mu \to A_\mu + D_\mu \Theta.$$

Let $\Theta(x) = \theta \mathbf{n}(x)$, where θ is a space-time independent constant. Then variation of the sourceless Lagrangian yields the conserved current:

$$\begin{aligned} J_\mu &= -\frac{\partial \mathcal{L}}{\partial(\partial_\mu A_\nu)} \frac{\partial A_\nu}{\partial \theta} = -F_{\mu\nu} \cdot D^\nu \mathbf{n} \\ &= -\mathbf{n} \cdot [F^\perp_{\mu\nu}, B^\nu], \end{aligned}$$

in which $F^{\perp}_{\mu\nu} \equiv [\mathbf{n}, [F_{\mu\nu}, \mathbf{n}]] = \tilde{D}_\mu B_\nu - \tilde{D}_\nu B_\mu$ is the transverse field strength, while the longitudinal field strength is:

$$F^{||}_{\mu\nu} = (F_{\mu\nu} \cdot n)\mathbf{n} = F_{\mu\nu} - F^{\perp}_{\mu\nu}.$$

It is easy to show, via the field equation $D^\mu F_{\mu\nu} = 0$, that

$$J_\mu = -\mathbf{n} \cdot [F_{\mu\nu}, B^\nu] = (D^\nu \mathbf{n}) \cdot F_{\nu\mu} = D^\nu(\mathbf{n} \cdot F_{\nu\mu}) = \partial^\nu(\mathbf{n} \cdot F_{\nu\mu}),$$

so $$\partial^\mu J_\mu = 0.$$

Namely the self-induced current is a conserved gauge covariant physical quantity.

Here we have to pay attention to the difference between non-Abelian and Abelian gauge theories. The later has no self-induced current and the equation of motion, which is linear, can be fully described by gauge field strength without the need of gauge potential. While Yang-Mills theory is a non-linear theory, its equation of motion involves gauge potential explicitly, there is self-interaction and the self- induced current. For the Maxwell theory, under a gauge transformation

$$\begin{aligned} &g(x) = e^{i\epsilon(x)} \in U(1) \\ &A_\mu \to A^g_\mu = A_\mu + \partial_\mu \epsilon(x), \\ &F_{\mu\nu} \to F_{\mu\nu}. \end{aligned}$$

The field strength is gauge invariant, and the equation of motion can cotained field strength only. It was thought that the gauge potential is no more than an auxiliary quantity with no physical significance. Until the second half of this century when Buhm-Aharanov effects[39] have been observed, that people began to realize the importance of gauge potential. While for Yang-Mills theory, the equation of motion has gauge potential explicitly and the field strength is no longer gauge invariant, but gauge covariant instead.

The infinitesimal gauge transformation and finite gauge transformation take the same form for Maxwell theory, so a finite gauge transformation can be obtained by summing up infinite number of infinitesimal transformations. While for Yang-Mills theory, the infinitesimal transformation takes a different form from the finite gauge transformation, and not all finite gauge transformations is obtainable from summing up infinite number of infinitesimal transformations. Namely there could be topologically non-trivial finite gauge transformations which could not be deformed continuous to the identity transformation, and this peculiarity again points to gauge potential as a fundamental physical quantity.

Field strength is a tensor, but gauge potential is an affine quantity which has no definite zero point; field strength is equivalent to curvature on principal bundle which has definite geometric significance, gauge potential is equivalent to connection on the principal bundle and also has definite geometric significance, it can even be said that it has more fundamental geometric significance than curvature. Zero curvature means that parallel transport around a closed path is possible, while there could still be large scale non- integrability if the region in which curvature is zero is not simply connected. Gauge potential is a geometric quantity and has physical significance, but the selection of gauge has no physical significance anyhow.

9.6 Instanton

In the path integral quantization of quantum field theory, we often take the analytic extrapolation of time to make it a virtual number, and the 4-dimension Minkowski space time becomes a Euclidean E^4. Then action S is a virtual number iI, so the integrand in the path integral

$$e^{iS} \to e^{-I}$$

decays exponentially. So we have to consider the classical field configuration, which makes the Euclidean action I minimum. In this section we look at solutions obtained when we take certain asymptotic boundary conditions, these are called the instanton solutions.

To make the Euclidean action I finite, we require that the field strength to be square integrable, i.e.,

$$F_{\mu\nu}(x) \to 0(1/|x|^2), \quad \text{as} \quad |x| \to \infty, \tag{9.20}$$

hence the gauge potential $A_\mu(x)$ has to decay to a pure gauge at infinity:

$$A_\mu(x) \to g^{-1}(x)\partial_\mu g(x) + 0(1/|x|) \quad \text{for} \quad |x| \to \infty.$$

The boundary of E^4 at infinity is a 3-dimensional sphere S^3 of infinite diameter. For $SU(2)$ Yang-Mills theory, the group element at infinity corresponds to the continuous mapping:

$$g: \; S^3 \to SU(2).$$

As $\pi_3(SU(2)) = Z$, so there exist homotopically inequivalent mappings and the inequivalent class of g can be denoted by integer $k \in Z$.

Remember that $E^4 \cup \{\infty\} \simeq S^4$, the stereographic projection from E^4 to S^4 is a conformal mapping. It is easy to prove that for 4-dimensional manifold, under the conformal mapping of metric:

$$g_{\mu\nu}(x) \to f(x)g_{\mu\nu}(x), \quad f(x) > 0,$$

the Hodge * operator is conformal invariant, so action $I = -\frac{1}{e^2}\int_M Tr(F \wedge *F)$ and the equation of motion are also conformal invariant. And E^4 can be compactified conformally by adding the point at infinity, solving the equation of motion on S^4, which is topologically non-trivial and on which there exist topologically non-trivial gauge fields. On the principal bundle $P(S^4, SU(2))$ there exists the topologically invariant Chern characteristic number:

$$c_2[S^4] = \frac{1}{8\pi^2}\int_{S^4} Tr(F \wedge F) = -k \ \in Z,$$

where the integrand $\frac{1}{8\pi^2}Tr(F \wedge F) \ \in H^4(M)$ is called the Chern class. The integer k here denotes the topological character of principal bundle $P(S^4, SU(2))$, i.e., it denotes the non-trivial asymptotic character of gauge potential $A_\mu(x)$. Here k labels the equivalence class of $\pi_3(S^3)$, to which g belongs, and thus in effect classifies principal bundle $P(S^4, SU(2))$. Sphere S^4 is topologically non-trivial and its open covering needs at least two open sets:

$$S^4 = U_+ \cap U_-,$$

the overlap region $U_+ \cap U_-$ is homotopic to S^3. The transition function $g(x)$ determines a map from $U_+ \cap U_-$ to $SU(2)$.

It is easy to check that the Chern class c_2 is closed, and from Poincare Lemma, c_2 can be written as an exact form on each open set:

$$c_2 = \frac{1}{8\pi^2}Tr(F \wedge F) = d\omega,$$

where

$$\omega = \frac{1}{8\pi^2}Tr(F \wedge A - \frac{1}{3}A \wedge A \wedge A)$$

is called Chern-Simons form. On the overlapping region $U_+ \cap U_-$ two gauge potentials differ by a gauge transformation:

$$\begin{aligned} A \to A' &= A^g = g^{-1}Ag + g^{-1}dg. \\ \text{So} \quad -8\pi^2 k &= \int_{S^4} Tr(F \wedge F) \\ &= \int_{U_+ \cap U_-} Tr(F \wedge F) \\ &= \int_{S^3} Tr(F^g \wedge A^g - \frac{1}{3}A^g \wedge A^g \wedge A^g - F \wedge A + \frac{1}{3}A \wedge A \wedge A). \end{aligned}$$

Since

$$F^g = g^{-1}Fg,$$

so we have

$$\begin{aligned} -8\pi^2 k &= -\frac{1}{3}\int_{S^3} Tr[g^{-1}dg \wedge g^{-1}dg \wedge g^{-1}dg - d(g^{-1}Ag \wedge dg)] \\ &= -\frac{1}{3}\int_{S^3} Tr(g^{-1}dg \wedge g^{-1}dg \wedge g^{-1}dg) \\ \text{i.e.,} \quad k &= \frac{1}{24\pi^2}\int_{S^3} Tr(g^{-1}dg \wedge g^{-1}dg \wedge g^{-1}dg) \end{aligned} \tag{9.21}$$

The right hand side is just the Kronecker index of the mapping $S^3 \to SU(2)$, i.e., the multiplicity of group $SU(2)$ over S^3.

To see it more explicitly, use the fundamental representation matrix of $SU(2)$:

$$g(x) = n_4(x) + i\mathbf{n}(x) \cdot \sigma,$$

where $\sigma = (\sigma_1, \sigma_2, \sigma_3)$ are the usual Pauli matrices, and $\{n_i(x)\}$ is a unit vector in E^4, satisfying:

$$n_4^2 + \mathbf{n}^2 = 1.$$

It is easy to check that $g^+g = 1$. Substitute g into Eq. 9.21 gives:

$$k = \frac{1}{12\pi^2} \int_{S^3} d^3x \epsilon_{\mu\nu\lambda\rho} \epsilon^{ijk} (n^\mu \partial_i n^\nu \partial_j n^\lambda \partial_k n^\rho).$$

The integrand is just the area element of the unit sphere S^3 in E^4. Here $n(x)$ corresponds to the continuous mapping from the equator S^3 of S^4 (the boundary of E^4 at infinity) to S^3 (the $SU(2)$ group manifold) with the integrand as the Jacobi of the mapping, k is the multiplicity of the mapping.

On a 4-dimensional manifold, the dual of 2-form is still a 2-form. The Yang-Mills equation $D * F = 0$ is a second order non-linear equation, its solution is exceedingly difficult. Note here the similarity between field equation and the Bianchi identity $DF = 0$, and it is easy to show that the solution to the following first order equation:

$$F = \lambda * F$$

are also solutions to the Yang-Mills equation as a result of the Bianchi identity. Take the Hodge * of the equation above gives:

$$F = \lambda^2 * *F = \begin{cases} \lambda^2 F & \text{on } E^4, \\ -\lambda^2 F & \text{on } M^4. \end{cases}$$

So we have $\lambda = \pm$ for E^4 and $\lambda = \pm i$ for M^4.

For 4-dimensional Minkowski space-time, we need

$$*F = \pm iF.$$

Since both F and $*F$ are Lie algebra valued differential forms, this equation can not be satisfied if G is a compact group. It can be satisfied only if G is non-compact, for example $SL(2, C)$

Now we look at 4-dimensional Euclidean space E^4, the equation $*F = \pm F$ is called self-dual and anti-self-dual respectively. Any 2-form can be decomposed as a sum of a self-dual part and an anti-self-dual part:

$$F = F^+ + F^-,$$

in which

$$F^{\pm} = \frac{1}{2}(F \pm *F)$$

satisfying

$$*F^{\pm} = \pm F^{\pm}.$$

Substitute this into the action gives:

$$\begin{aligned} I &= -\frac{1}{e^2}\int_{S^4} Tr(F \wedge *F) \\ &= -\frac{1}{e^2}\int_{S^4} Tr[(F^+ + F^-) \wedge (F^+ - F^-)] \\ &= -\frac{1}{e^2}\int_{S^4} [Tr(F^+ \wedge F^+) - Tr(F^- \wedge F^-)] \\ &= \frac{1}{e^2} \parallel F^+ \parallel^2 + \frac{1}{e^2} \parallel F^- \parallel^2 \geq 0. \end{aligned}$$

Here $\parallel \cdot \parallel$ denotes the norm of differential forms. For any Lie algebra valued form $\alpha \in \Omega^*(M, \mathbf{g})$, the norm is defined as:

$$\begin{aligned} \parallel \alpha \parallel^2 &= -Tr < \alpha, \alpha > \\ &= -\int_M Tr(\alpha \wedge *\alpha) \geq 0. \end{aligned}$$

Applying the same decomposition to the Chern class gives:

$$8\pi^2 k = -\int_{S^4} Tr[(F^+ + F^-) \wedge (F^+ + F^-) = \parallel F^+ \parallel^2 - \parallel F^- \parallel^2 .$$

Since for any two real numbers a and b,

$$a^2 + b^2 \geq |a^2 - b^2|,$$

so we have

$$I \geq \frac{8\pi^2}{e^2}|k|.$$

As $k \in Z$ are topological invariants, so for each given k, the absolute minimum I_0 of I is given by $\frac{8\pi^2}{e^2}|k|$. For example:

1. If $k = 0$, then $I_0 = 0 \Longrightarrow F = F^+ = F^- = 0$.
2. If $k > 0$, $e^2 I_0 = 8\pi^2 k \Longrightarrow F^- = 0$, so $F = *F$.
3. If $k < 0$, $e^2 I_0 = -8\pi^2 k \Longrightarrow F^+ = 0$, so $F = - * F$.

The analysis above shows that the bound $e^2 I \geq 8\pi^2|k|$ is saturated by the self-dual and anti-self-dual fields, which give the absolute minima of S at a given boundary condition.

Belavin et al [40] use a spherical symmetry ansatz for $k = \pm 1$ solution (BPST solution):

$$A_\mu(x) = f(r) g^{-1}(x) \partial_\mu g(x),$$

where $g(x) = (x_4 - ix_i\sigma_i)/r$ with σ_i the Pauli matrices, note here that the manifold is E^4 with the boundary condition given by Eq. 9.20 instead of S^4, and f is a function of $r = \sqrt{x_1^2 + x_2^2 + x_3^2 + x_4^2}$ only and satisfies the boundary condition:

$$f(r) \to 1, \quad \text{for} \quad r \to \infty.$$

Substitute this $A_\mu(x)$ into the self-dual equation $*F = F$ gives a simple soluble equation, which gives

$$f(r) = \frac{(r - r_0)^2}{(r - r_0)^2 + \lambda^2},$$

which means that the instanton solution is specified by five parameters. One may take $r_0 = 0$ as the origin is taken arbitrarily, then the solution is as follows:

$$\begin{aligned} A(x) &= \frac{r^2}{r^2 + \lambda^2} g^{-1}(x) dg(x) = \frac{i}{r^2 + \lambda^2} \sum_{j=1}^{3} \sigma_j \eta^j_{\mu\nu} x^\mu dx^\nu, \qquad (9.22) \\ F(x) &= -\frac{\lambda^2 i}{(r^2 + \lambda^2)^2} \sum_{j=1}^{3} g^{-1}(x) \sigma_j \eta^j_{\mu\nu} dx^\mu \wedge dx^\nu, \end{aligned}$$

here $\eta^j_{\mu\nu}$, $(j = 1, 2, 3; \mu, \nu = 1, \cdots, 4)$ are defined as:

$$\eta^i_{jk} = \epsilon_{ijk}, \quad \eta^i_{j4} = \delta^i_j, \quad \eta^i_{\mu\nu} = -\eta^i_{\nu\mu}.$$

And it is easy to check:

$$\eta^i_{\mu\nu} \eta^j_{\mu\lambda} = \delta_{ij} \delta_{\nu\lambda} + \epsilon_{ijk} \eta^k_{\nu\lambda}.$$

They can be represented by the so called t'Hooft η matrices:

$$\eta^1 = \begin{pmatrix} 0 & 0 & 0 & 1 \\ 0 & 0 & 1 & 0 \\ 0 & -1 & 0 & 0 \\ -1 & 0 & 0 & 0 \end{pmatrix}, \quad \eta^2 = \begin{pmatrix} 0 & 0 & -1 & 0 \\ 0 & 0 & 0 & 1 \\ 1 & 0 & 0 & 0 \\ 0 & -1 & 0 & 0 \end{pmatrix}, \quad \eta^3 = \begin{pmatrix} 0 & 1 & 0 & 0 \\ -1 & 0 & 0 & 0 \\ 0 & 0 & 0 & 1 \\ 0 & 0 & -1 & 0 \end{pmatrix}.$$

Take $\bar{\eta}^i_{j4} = -\eta^i_{j4}$ and $\bar{\eta}^i_{4j} = -\eta^i_{4j}$, then give us another set of t'Hooft matrices $\bar{\eta} = (\bar{\eta}^1, \bar{\eta}^2, \bar{\eta}^3)$. It is easy to prove

$$\begin{aligned} \left[\eta^i, \eta^j\right] &= -2\epsilon_{ijk} \eta^k, \\ \left[\bar{\eta}^i, \bar{\eta}^j\right] &= -2\epsilon_{ijk} \bar{\eta}^k. \end{aligned}$$

Both η and $\bar{\eta}$ are isomorphic to algebra $so(3)$. Recall that

$$SO(4) = SO(3) \times SO(3).$$

Any skew symmetric tensor $f_{\mu\nu}$ of $so(4)$ can be decomposed into the direct sum of two $so(3)$ representations, the matrix $\eta(\bar{\eta})$ maps the $so(3)$ vector covariantly to the self-dual (anti-self-dual) $so(4)$ skewsymmetric tensor. Solution Eq. 9.22 containing the η matrices is called the self-dual instanton solution, replacing η with $\bar{\eta}$ gives the anti-self-dual instanton solution.

Note that the product of elements of a group generates the multiplicity of homotopy group $\pi_3(G)$, replacing $g(x)$ in Eq. 9.22 with $g(x)^k$ will give one the k-instanton solution. The general solution of $*F = F$ can be written as:

$$A_\mu^j(x) = \eta_{\mu\nu}^j \partial^\nu \ln\ \rho(x),$$
$$\rho(x) = \sum_{j=1}^{k+1} \frac{\lambda_j^2}{(x - a_j)^2}.$$

This solution has $5(k+1)$ parameters (λ_j, a_j), not all of them are independent. There are no more than $(5k+4)$ parameters, as any constant multiple of ρ does not change connection.

The number of independent parameters of instanton solution can be determined by Atiyah-Singer index theorem. For $SU(2)$ gauge theory, there are $8k-3$ parameters, $4k$ of which denote the position of the instanton, another k parameters denote its length, and $3k-3$ parameters denote its orientation relative to the 3-dimensional $SU(2)$ manifold.

Finally we list the different expressions of action and Chern number in the table below.

Table 9.2 Various expressions of action and the Chern number

	Matrix tensor	Differential form
Action I	$-\frac{1}{2e^2}\int d^4x Tr(F^{\mu\nu}F_{\mu\nu})$	$-\frac{1}{e^2}\int Tr(F\wedge *F)$
Chern number $c_2[M]$	$\frac{1}{16\pi^2}\int d^4x Tr(\ {}^*F^{\mu\nu}F_{\mu\nu})$	$\frac{1}{8\pi^2}\int Tr(F\wedge F)$
Chern-Simons	$\frac{1}{16\pi^2}\epsilon^{\mu\alpha\beta\gamma}Tr(F^{\alpha\beta}A_\gamma - \frac{2}{3}A_\alpha A_\beta A_\gamma)$	$\frac{1}{8\pi^2}Tr(F\wedge A$ $-\frac{1}{3}A\wedge A\wedge A)$

Chern number c_2 is a topological invariant. It is unchanged under local gauge transformations. For c_2, as the index contraction in its tensor expression involves contraction with dual field, it is independent of the metric on the base manifold, while action I depends on the choice of the metric.

9.7 Yang-Mills-Higgs monopole

Yang-Mills potential $A(x)$ corresponds to the connection on principal bundle, the matter field $\phi(x)$ corresponds to section of the associated bundle. One can construct a dynamic system of matter field interacting with Yang-Mills field by further introducing metric on the base manifold, and the corresponding Lagrangian and equation of motion can be found.

Now we consider an $SU(2)$ gauge theory with spontaneous broken symmetry, i.e., coupled to a Higgs field with a potential allowing for non-zero vacuum expectation values. The Lagrangian is:

$$\mathcal{L} = -\frac{1}{4}F^a_{\mu\nu}F^{\mu\nu}_a + \frac{1}{2}D_\mu\phi^a D^\mu\phi_a - \lambda^2(\phi^a\phi_a - c^2)^2.$$

By rescaling and taking the Coulomb gauge $A_0 = 0$, the static energy density $\mathcal{E}(\partial_0\mathcal{E} = 0)$ can be written as:

$$\begin{aligned}\mathcal{E} &= \frac{1}{2}\int d^3x[H^a_i H^i_a + D_i\phi^a D^i\phi_a + \lambda^2(\phi^a\phi_a - 1)^2] \\ &= \mp\int d^3x\partial_i(H^i_a\phi^a) + \int d^3x[\frac{1}{2}|H^a_i \pm D_i\phi^a|^2 + \lambda^2(\phi^a\phi_a - 1)^2].\end{aligned}$$

The first term on the right hand side is gauge invariant, it is an integer self-induced conserved charge n, called the topological charge. As a result of the positive definite nature of the last two terms on the right hand side we have:

$$\mathcal{E} \geq n.$$

Saturation of this inequality requires $\lambda = 0$, and this requirement is called the Prasad-Sommerfield limit[41,42], which amounts to introducing a massless Higgs particle with a non-zero vacuum expectation value:

$$\phi^2 \to 1.$$

In this simple case

$$\mathcal{L} = -\frac{1}{4}F^a_{\mu\nu}F^{\mu\nu}_a + \frac{1}{2}D_\mu\phi^a D^\mu\phi_a,$$
$$\mathcal{E} = n + \frac{1}{2}\int d^3x|H^a_i \pm D_i\phi^a|^2.$$

Now instead of the second order field equation, which gives the minima of the action, we have the simpler first order equation:

$$H^a_i = \pm D_i\phi^a$$

called the Bogomolny equation[42], which gives the absolute minimum of the static energy.

To compare with the 4-dimensional sourceless Yang-Mills theory, let the potential (A^a, ϕ^a) corresponds to a 4-dimensional connection W^a with ϕ^a as the 4-th component of W^a, then the Bogomolny equation corresponds to the self-dual equation:

$$F^a_{\mu\nu} = \pm\frac{1}{2}\epsilon_{\mu\nu\lambda\rho}F^a_{\lambda\rho},$$

where $F_{\mu\nu} = \partial_\mu W_\nu - \partial_\nu W_\mu + [W_\mu, W_\nu]$, $F^a_{ij} = \frac{1}{2}\epsilon_{ijk}H^a_k$, and $F^a_{i4} = D_i\phi^a$.

Thus a BPS monopole (Bogomolny[42], Prasad-Sommerfield[41]) corresponds to a (anti-) self-dual static sourceless gauge field. If we want the solution to be non-singular and of finite energy, we have to look for them in the irreducible situation. As a result, irreducibility in spherical symmetry leads to synchronosymmetry, and one can take the following ansatz:

$$W^a_i(x) = \epsilon_{iaj}\frac{(\phi(r)-1)x^j}{r^2},$$
$$W^a_0(x) = G(r)\frac{x^a}{r^2},$$

in which $r = \sqrt{x_1^2 + x_2^2 + x_3^2}$, i, j and $a = 1, 2, 3$. Then

$$E^a_i(x) = F^a_{i4}(x) = E_r(r)\frac{x_i x^a}{r^2} + E_t(r)\frac{(r^2\delta^a_i - x_i x^a)}{r^2},$$
$$H^a_k(x) = F^a_{ij}\epsilon_{ijk} = H_r(r)\frac{x_k x^a}{r^2} + H_t(r)\frac{(r^2\delta^a_k - x_k x^a)}{r^2}.$$

where $$\begin{cases} E_r(r) = \frac{(rG'(r)-G(r))}{r^2}, & E_t(r) = \phi(r)\frac{G(r)}{r^2}, \\ H_r(r) = \frac{(\phi^2(r)-1)}{r^2}, & H_t(r) = \frac{\phi'(r)}{r^2}, \end{cases}$$

with the subscript r denotes the radial component and t denotes the transverse component. Through the separation of variables the sourceless field equation is reduced to:

$$r^2 G" = 2\phi^2 G,$$
$$r^2\phi" = \phi(\phi^2 - 1 - G^2).$$

To make the fields non-singular at the origin, we have:

$$\phi^2 = 1 + 0(r^2), \quad G = (r^2) \quad \text{as} \quad r \to \infty.$$

In order to obtain a solution of finite energy, assume $\phi(r)$ to be a real function, then at infinity:

$$\begin{matrix} \frac{G}{r} \to i\beta \\ \phi \to P(r)e^{-\beta r} \end{matrix} \qquad \text{as} \quad r \to \infty$$

where $P(r)$ is some finite polynomial of r.

From the asymptotic behavior above, we can see that the radial component of the field approaches Coulomb field as $e^{-2\beta r}$, while the transverse component approaches zero as $e^{-\beta r}$.

Making use of the asymptotic behavior, it is easy to prove that the solution of the sourceless equation exists only in the self-dual form[38], i.e., one has to solve the following equation:

$$\begin{matrix} F^-_r \equiv H_r - iE_r = 0 & \Longrightarrow & \phi^2(r) - 1 = i(rG'(r) - G(r)), \\ F^-_t \equiv H_t - iE_t = 0 & \Longrightarrow & r\phi'(r) = i\phi(r)G(r). \end{matrix} \tag{9.23}$$

From this solution it is easy to get the solution analytically as follows:

Let $\phi = \frac{r}{\psi}$ in Eq. 9.23, we have $iG = 1 - r\frac{\psi'}{\psi}$, and from Eq. 9.23 we have

$$(\frac{\psi'}{\psi})' = -\frac{1}{\psi^2}.$$

And the solution of this ordinary differential equation is:

$$r = \int \frac{d\psi}{\pm\sqrt{1 + c_1\psi^2}} = \pm\frac{1}{\sqrt{c_1}} \sinh^{-1}(\sqrt{c_1}\psi) + c_2.$$

The boundary condition at the origin gives us $c_2 = 0$, and from the boundary condition at infinity we see that c_1 is the same as β chosen before, so finally:

$$\phi = \pm\frac{\beta r}{\sinh \beta r},$$
$$G = i(\beta r \coth \beta r - 1),$$

which is the same as that obtained by T.S.Wu, C.N.Yang,[43]

As a result, there exists a continuous distribution of magnetic charge, which falls off exponentially as $e^{-2\beta r}$ with a distribution radius of $\sim (2\beta)^{-1}$. The total magnetic charge and magnetic flux are quantized, but there could be a continuous distribution of magnetic charge, neither is the magnetic flux through the surface of a sphere of radius r quantized. As is well known, in the case of $U(1)$ and irreducible $SU(2)$, Bianchi identity requires that there be either no magnetic charge or only a point magnetic charge, with the non-existence of magnetic charge corresponds to a trivial bundle in $U(1)$, while a point magnetic charge corresponds to a non-trivial bundle in $U(1)$, where the winding number is determined by the magnetic charge present. A reducible $SU(2)$ can have a non-trivial adjoint bundle, since

$$\pi_2(SU(2)/U(1)) = \pi_1(U(1)) = Z,$$

the point magnetic charge and its quantized value are given completely by the topological property, with the magnetic charge located at the singular point of the field. On the other hand, the Bianchi identity of an irreducible $SU(2)$ gauge theory gives a magnetic charge with a continuous distribution, which comes entirely from the contribution of the field itself. Although the distribution is not quantized, the topological properties of the reducible boundary condition still requires that the total magnetic charge at infinity be quantized.

9.8 Seiberg-Witten monopole equation

The last section is devoted to the study of irreducible non-Abelian gauge theories with spontaneous broken symmetry, and the Prasad-Sommerfield limit. Now we look at its supersymmetric extensions, in which the Higgs field is fermionic and the massless is linked to supersymmetry, i.e., we look at $U(1)$ gauge theory with unbroken supersymmetry[44].

Let (M, G) be a compact 4-dimensional Riemannian manifold with local spin structure:

$$S = S^+ \oplus S^-,$$

Let $\Gamma(S^\pm)$ denote the set of sections of bundles $S^\pm$. The Dirac operator $D = \sum e_i \cdot \nabla_{e_i}$ is a mapping from $\Gamma(S^+)$ to $\Gamma(S^-)$, here $\{e_i\}$ are the basis of the Clifford algebra (see App.M), satisfying

$$e_i \cdot e_j + e_j \cdot e_i = -2\delta_{ij}.$$

The product between e_i and ∇_{e_i} is Clifford product, and ∇_{e_i} denotes covariant derivative with Levi-Civita connection of (M, G). It is easy to prove the Weitzenböck formula:

$$D^+ D\psi = -\Delta\psi + \frac{\Re}{4}\psi, \tag{9.24}$$

In which $\Re$ is the scalar curvature of (M, G).
Hint: its proof can be sketched as:

$$\begin{aligned}
D^2\psi &= \sum_{ij} e_i \cdot e_j \cdot (\nabla_{e_i}\nabla_{e_j} - \nabla_{\nabla_{e_i}e_j})\psi \\
&= -\Delta\psi + \sum_{i<j} e_i \cdot e_j \cdot [(\nabla_{e_i}\nabla_{e_j} - \nabla_{\nabla_{e_i}e_j})(\nabla_j\nabla_i - \nabla_{\nabla_{e_j}e_i})]\psi \\
&= -\Delta\psi + \sum_{ij} e_i \cdot e_j R(e_i, e_j)\psi \\
&= -\Delta\psi + \frac{\Re}{4}\psi.
\end{aligned}$$

The spin field may not be well defined globally (see §10.5, the Stiefel-Whitney class $w_2(M) \in H^2(M, Z_2)$ is the topological obstruction). But a variant, a *Spin*^{c} structure always exist. I.e., If there is a complex line bundle L such that there exists globally defined spin field ψ which is the section of the complex vector bundle W^+:

$$W^+ = S^+ \oplus L^{1/2}, \quad L^{1/2} \otimes L^{1/2} = L,$$

then this is called the Spin^c structure associated to the complex line bundle L. Let $D_A = \sum e_i \cdot \nabla^A_{e_i}$ denote the Dirac operator, where

$$\nabla^A_{e_i}(S \otimes L^{1/2}) = (\nabla^{LC}_{e_i} S) \otimes L^{1/2} + S \otimes \nabla^A_{e_i} L^{1/2}.$$

Here $\nabla^{LC}_{e_i}$ denotes covariant derivative with Levi-Civita connection of (M, G), and

$$\nabla^A_{e_i} l = A_i l, \quad \nabla^A_{e_i} l^{\frac{1}{2}} = \frac{A_i}{2} l^{\frac{1}{2}}, \quad l^{1/2} \in L^{1/2},$$

where A is a $U(1)$ valued connection 1-form on complex line bundle L, and $L^{1/2}$ is another complex line bundle on M with $L^{1/2} \otimes L^{1/2} = L$. A is pure imaginary and the curvature $F(A) = dA = iF_A$. F_A is a real 2-form, its self-dual part is

$$F_A^+ = \frac{1}{2}(F_A + *F_A),$$

which is also a 2-form on 4-dimensional manifold. Now we look at the system in which spin field $\psi \in \Gamma(W^+)$ is coupled to the $U(1)$ gauge field. The action is called the Seiberg-Witten functional:

$$SW(A,\psi) = \int_M (|D_A\psi|^2 + |F_A^+|^2 + \frac{\Re}{4}|\psi|^2 + \frac{1}{8}|\psi|^4)d^4x.$$

In which $\Re$ is the scalar curvature of (M, G), $|\psi|^2 = <\psi, \psi>$ is the invariant Hermitian product on $\Gamma(W^+)$. As $Spin(4) = SU(2) \times SU(2)$, we only have to consider the self-dual part, and the corresponding field equation is called the monopole equation in spin representation with the monopole carrying spin.

The Euler-Lagrange equations of motion is:

$$-\Delta_A\psi + \frac{\Re}{4}\psi + \frac{1}{4}|\psi|^4\psi = 0.$$

$$\delta F_A^+ + \frac{1}{2}Im < \nabla_i\psi, \psi > e^i = 0.$$

The first equation shows that spin field ψ satisfies non-linear Schrödinger equation coupled with $U(1)$ gauge potential. As

$$\delta = - * d*, \qquad \text{and} \qquad ** = 1,$$

$$\text{so} \qquad \delta F_A^+ = \frac{1}{2}(- * d*)(F_A + *F_A) = \frac{1}{2}\delta F_A,$$

and the second Euler-Lagrange can be rewritten as:

$$\delta F_A = -Im < \nabla_{e^i}^A\psi, \psi > e^i,$$

i.e., spin field ψ is the source carrying the $U(1)$ charge.

In this case, The Weitzenböck formula must be rewritten as:

$$D_A^+ D_A\psi = -\Delta_A\psi + \frac{\Re}{4}\psi + \frac{i}{2}F_A\psi \tag{9.25}$$

By using this formula, Seiberg-Witten functional can be rewritten as:

$$SW(A,\psi) = \int_M (|D_A\psi|^2 + |F_A^+ - \frac{i}{4} < e_i \cdot e_j\psi, \psi > e^i \wedge e^j|^2) \tag{9.26}$$

from which the Euler-Lagrange equation can be obtained as:

$$\begin{cases} D_A^+ D_A\psi - \frac{i}{2}F_A^+\psi - \frac{1}{8} < e_i e_j\psi, \psi > e^i \wedge e^j\psi = 0, \\ \delta(F_A^+ - \frac{i}{4} < e_i e_j\psi, \psi > e^i \wedge e^j) + \frac{1}{2}Im < D_A\psi, e_i\psi > e^i = 0. \end{cases}$$

They are pretty complicated, fortunately just like the case with Bogomolny equations, the absolute minima of the Seiberg-Witten functional can be get from Eq. 9.26, which satisfy a much simpler set of equations, the famous Seiberg-Witten monopole equations:

$$D_A\psi = 0,$$
$$F_A^+ = \frac{i}{4} < e_i e_j\psi, \psi > e^i \wedge e^j.$$

These equations are gauge invariant but not conformal invariant.

These new classical equations correspond to the self-dual Yang-Mills equations, they play an important role in 4-dimensional manifold geometry. They have simpler analytic properties because which is Abelian U(1) gauge theory with spin, and are far more powerful than the self-dual Yang-Mills equations.

Notations and Formulea

- The connection on the principal bundle $P(M, G)$.

  1. Tangent space approach.
  Given the horizontal space H_u satisfying

$$\pi_* H_u = T_{\pi(u)}(M),$$
$$\{H_{R_a u}\} = \{H_u\},$$

The basis D_μ of H_u can be expressed as:

$$\begin{aligned} \mathcal{D}_\mu &= \frac{\partial}{\partial x^\mu} - A_\mu^a \tilde{V}_a \\ \tilde{V}_a &= Tr(\frac{\lambda_a}{2i} g \frac{\partial}{\partial g^t}) \\ [\mathcal{D}_\mu, \mathcal{D}_\nu] &= -F_{\nu\mu}^a \tilde{V}_a, \end{aligned}$$

where $\frac{\lambda_a}{2i} \equiv \tau_a$ are the anti-Hermitian elements of the Lie algebra $\mathbf{g}$.

2. Cotangent space approach.
Pick a 1-form ω in $T^*(P)$ satisfying

$$\omega|_{\text{fibre}} = g^{-1}dg, \quad \text{or} \quad < \omega, \tilde{V}_a > |_e = \frac{\lambda_a}{2i}$$
$$R_a^*\omega = a^{-1}\omega a$$

expressed by local coordinates

$$\omega = g^{-1}Ag + g^{-1}dg \ \in \ \Omega^1(P, g), \qquad A = A_\mu^a(x)\frac{\lambda}{2i}dx^\mu.$$

$$\begin{aligned} \Omega &= d\omega + \omega \wedge \omega \ \in \Omega^2(P, \mathbf{g}) \\ &= g_{-1}Fg \\ F &= dA + A \wedge A. \end{aligned}$$

3. Base manifold differential form approach.
Given a connection 1-form A on each neighborhood U_α of base manifold M, in the overlapping region $U_\alpha \bigcap U_\beta \neq 0$,

$$A^{(\beta)} = g_{\alpha\beta}^{-1}A^{(\alpha)}g_{\alpha\beta} + g_{\alpha\beta}^{-1}dg_{\alpha\beta},$$

$$F^{(\beta)} = g_{\alpha\beta}^{-1}F^{(\alpha)}g_{\alpha\beta},$$

- The connection on associated bundle:

$$\mathcal{D}_\mu = \partial_\mu + A_\mu, \qquad A_\mu = A_\mu^a\tau_a, \qquad F = dA + A \wedge A.$$

- Connection on general vector bundle:

$$S(x) = e_a(x)z^a(x) \ \in \ \Gamma(E): \ \text{ the space of section of vector bundle } E.$$

$$\nabla z^a = \omega^a = dz^a + \Gamma^a{}_b z^b, \qquad \Gamma^a{}_b = dx^\mu \Gamma^a_{\mu b}.$$

$$\nabla^2 z^a = \Omega^a{}_b z^b.$$

$$\Omega^a{}_b = d\Gamma^a{}_b + \Gamma^a{}_c \wedge \Gamma^c{}_b = \frac{1}{2}R^a_{b\mu\nu}dx^\mu \wedge dx^\nu.$$

- The action S of Yang-Mills theory:

$$S=\frac{1}{e^2}\int_M Tr(F\wedge *F)=\frac{1}{2e^2}\int_M d^4x\; Tr\; F^{\mu\nu}F_{\mu\nu}.$$

The Yang-Mills equation:

$$\partial_\mu F_a^{\mu\nu}+ef_{abc}A_\mu^b F_c^{\mu\nu}=0.$$

The self-dual equation:

$$^*F^{\mu\nu}=\frac{1}{2}\epsilon^{\mu\nu\lambda\rho}F_{\lambda\rho}=F^{\mu\nu}.$$

- The Seiberg-Witten functional:

$$SW(A,\psi)=\int_M(|D_A\psi|^2+|F_A^+|^2+\frac{\Re}{4}|\psi|^2+\frac{1}{8}|\psi|^4)d^4x.$$

The equations of motion is:

$$-\Delta_A\psi+\frac{\Re}{4}\psi+\frac{1}{4}|\psi|^4\psi=0.$$
$$\delta F_A^+ +\frac{1}{2}Im<\nabla_i\psi,\psi>e^i=0.$$

The Seiberg-Witten monopole equations:

$$D_A\psi=0,$$
$$F_A^+=\frac{i}{4}<e_ie_j\psi,\psi>e^i\wedge e^j.$$

Exercises

1. Derive the Levi-Civita connection on S^2 in terms of spherical coordinate system.
2. To verify natural linear bundle L over $CP(n)$ being holomorphic bundle. From the canonical metric on C^{n+1}, the induced metric of the fiber on L can be gotten, then to derive the Riemannian connection and curvature preserving this metric.
3. For non-Abelian gauge field, $F=dA+A\wedge A$, to show that

$$\mathcal{D}F=dF+[A,F]=0.$$

4. The energy-moment tensor of Yang-Mills fields can be repressed as

$$\theta^{\mu\nu}=2Tr(F^{\mu\rho}F_\rho^\nu-\frac{1}{4}g^{\mu\nu}F^{\lambda\rho}F_{\lambda\rho}),$$

here F satisfy the Yang-Mills equation. To verify $\theta^{\mu\nu}$ is conserved, i.e., satisfying following equation:

$$\partial_\mu\theta^{\mu\nu}=0.$$

Chapter 10

Characteristic Classes

10.1 Introduction, Weil homomorphism

Locally fibre bundle is the Cartesian product of the base manifold and fibre, while globally it can not be taken as a product space as there could be global nontrivial structures. The issue is how to decide the deviation from triviality.

As pointed out in §8.7, classification of fibre bundles is usually accomplished through the analysis of homotopy groups. And the deviation from trivial bundle is often characterized by certain topologically invariant integers (called characters). With the introduction of connections, which defines the relationship among neighboring fibres, characters can be obtained through the integral of characteristic classes over the base manifold. The characteristic class is defined in terms of connection, but the resulting integral is independent of the choice of connections. Namely, different characteristic classes given by different choice of connections differ only by an exact form on the base manifold, and the resulting integral is a topological invariant, independent of the choice of connections.

In general we assume the base manifold M to be a topologically nontrivial manifold without boundary, and can be covered by certain number of open sets $\{U_\alpha\}$. The global topological properties of the fibre bundle is determined completely by the transition functions on the overlapping region, which in turn requires the characteristic classes to be closed forms on the base manifold, and locally could be written as an exact form on any open set U_α (Poincare Lemma), such that its integral over open sets depends upon the transition functions on the overlapping region only, as a result of Stoke's theorem. Characteristic class is a closed form on the base manifold, and is not an exact form globally for non-trivial bundles, so its integral over the entire closed manifold is nonvanishing. Thus characteristic class $\chi(E)$ corresponds to the cohomology class on base manifold M

$$\chi(E) \in H^*(M).$$

Characteristic class of fibre bundles is a class of closed, scalar-valued forms on the base manifold M of the bundles, which is defined in terms of the connection of the bundles and having the property that if one connection is replaced by another, the corresponding form is altered by the addition of an exact form only, so the integral of the characteristic class over a closed orientable submanifold of M is a topological invariant of the bundle. Characteristic classes thus distinguish inequivalent fibre bundles.

From §8.7, we know that each principal bundle $P(M,G)$ can be induced by a smooth map

$$f:\ M \to B_G$$

unique up to homotopy. Here B_G is the base manifold of the universal principal G-bundle ξ_G. There is a one to one correspondence between the set of principal bundles $P(M,G)$, with given base manifold M, and the set $[M, B_G]$ of homotopy class of map from M to B_G. So each element χ_G of the cohomology group $H^*(B_G, \mathbf{Z})$ or $H^*(B_G, \mathbf{Z}_2)$ determines a topological characteristic class $\chi(P)$ of $P(M,G)$:

$$\chi(P) = \chi(f^*\xi_G) = f^*\chi_G.$$

Analogously we can consider the vector bundle $E(M, V, \pi, G)$. One defines the characteristic classes of E as the pull back of the universal characteristic classes, the cohomology class of classifying space. For example,

1. k-$\dim_C$ complex vector bundle. Its structure group $GL(k, \mathbf{C})$ can be reduced to $U(k)$, and its classifying space is $G_r^C(\infty, k)$, we can obtain the classification of vector bundle $E(M, V, \pi, U(k))$ through a smooth map f

$$f:\ M \to G_r^C(\infty, k),$$

and $E = f^*\xi_E$. Here ξ_E is the universal bundle with base $G_r^C(\infty, k)$ and structure group $U(k)$. In terms of the characteristic class $\chi(\xi_E) \in H^*(G_r^C(\infty, k), \mathbf{Z})$, one can obtain the characteristic class $C(E)$ of vector bundle E

$$C(E) = f^*\chi(\xi_E) \in H^*(M, \mathbf{Z}),$$

called the Chern class.

2. k-dim real vector bundle. The structure group $GL(k, \mathbf{R})$ can be reduced to $O(k)$, the classifying space is $G_r(\infty, k)$. Through a smooth map

$$f:\ M \to G_r(\infty, k),$$

we can obtain the characteristic class in terms of the characteristic class $\chi(\xi_E) \in H^*(G_r(\infty, k), \mathbf{Z})$

$$P(E) = f^*\chi(\xi_E) \in H^*(M, \mathbf{Z}),$$

and is called the Pontrjagin class.

3. Sometimes for vector bundle E as in the example above, we also consider the cohomology group with coefficients in $\mathbf{Z}_2$. In terms of $H^i(G_r(\infty,k),\mathbf{Z}_2)$, one can obtain

$$w_i(E) = f^*(H^i(G_r(\infty,k),\mathbf{Z}_2)) \in H^i(M,\mathbf{Z}_2),$$

which is called the Stiefel-Whitney class of vector bundle E.

4. For even-dimensional real orientable vector bundle E with structure group $SO(k)$, its classifying space is $\hat{G}_r(\infty,k)$, except the Pontrjagin class, we can obtain another characteristic class $\chi(E)$ called Euler class

$$\chi(E) = f^*(H^k(\hat{G}_r(\infty,k),\mathbf{Z})) \in H^k(M,\mathbf{Z}).$$

The four characteristic classes above are the major sources of characteristic classes of fibre bundles.

Next we outline the Chern-Wril description of characteristic class in terms of curvature forms. For any compact Lie group G with Lie algebra $\mathbf{g}$, the cohomology $H^*(B_G,\mathbf{R})$ of the classifying space is isomorphic to the algebra consisting of all polynomial functions

$$P:\ \mathbf{g}\to R$$

which are invariant under the adjoint of G.

This is the general Chern-Weil result. The Chern, Pontrjagin, and Euler class can all be obtained by Weil Homomorphism, and can be expressed in terms of invariant symmetric polynomials of curvature 2-form on the principal bundle.

Let G be a Lie group with Lie algebra $\mathbf{g}$, P be a symmetric multi-linear mapping, i.e.,

$$\begin{aligned} P:\ \mathbf{g},\cdots,\mathbf{g} &\to \mathbf{R} \\ a_1,\cdots,a_k &\to P(a_1,\cdots,a_k)\in\mathbf{R},\quad a_i\in\mathbf{g}, \end{aligned}$$

where $P(a_1,\cdots,a_k)$ is

(1). Multi-linear.

$$P(ca_1,\cdots,a_k) = P(a_1,\cdots,ca_i,\cdots,a_k) = cP(a_1,\cdots,a_k),\ c\in\mathbf{R}.$$

(2). Symmetric. $P(a_{\sigma_1},\cdots,a_{\sigma_k}) = P(a_1,\cdots,a_k)$, where σ is an element (permutation) of symmetry group S_n.

(3). Invariant with respect to G: $P(a_1^g,\cdots,a_k^g) = P(a_1,\cdots,a_k)$, where $a_i^g = ad_g a_i \equiv g^{-1}a_i g,\ g\in G$

Let $I^k(G)$ denote the set of all invariant symmetric multi-linear mappings, which is a vector space over $\mathbf{R}$, and furthermore let

$$I(G) = \sum_{k=0}^{\infty} I^k(G).$$

For any two elements $P_1 \in I^k(G), P_2 \in I^l(G)$ one can introduce the inner product as follows:

$$P_1 \cdot P_2(a_1, \cdots, a_{k+l}) = \frac{1}{(k+l)!} \sum_\sigma P_1(a_{\sigma_1}, \cdots, a_{\sigma_k}) P_2(a_{\sigma_{k+1}}, \cdots, a_{\sigma_{k+l}}),$$

where the summation is taken over all permutations of S_{k+l}. $I(G)$ with this inner product operation forms a commutative algebra over $\mathbf{R}$.

Taking a connection 1-form ω on the principal bundle $P(M, G)$, the curvature 2-form is

$$\Omega = d\omega + \omega \wedge \omega.$$

If the Lie algebra valued 2-form is substituted for the elements a_i in an invariant polynomial $P \in I^k(G)$, one obtains a $2k$-form $P(\Omega)$ on base manifold M

$$P(\Omega) = P(\Omega, \cdots, \Omega)$$

satisfying

(1). $P(\Omega)$ is a closed $2k$-form on M.

$$\begin{aligned} dP(\Omega) &= kP(D\Omega, \Omega, \cdots, \Omega) \\ &= 0. \end{aligned}$$

As $D\Omega = d\Omega + [\omega, \Omega] = 0$, this is the Bianchi identity which the connection must satisfy.

(2). $P(\Omega)$ has topologically invariant integrals, independent of the choice of the connection.

So the closed 2-form $P(\Omega)$ defines an element of de Rham cohomology $H^k(M, \mathbf{R})$. There exists an algebraic Homomorphism W

$$W: \ I(G) \to H^*(M, \mathbf{R}).$$

Prove: Let ω_0 and ω_1 be two connection 1-forms, Ω_0 and Ω_1 be the corresponding curvature 2-forms. Let

$$\omega_t = \omega_0 + t\eta, \quad \eta = \omega_1 - \omega_0 \quad (0 \le t \le 1),$$

be the interpolation between ω_0 and ω_1, then

$$\Omega_t = d\omega_t + \omega_t \wedge \omega_t,$$
$$\frac{\partial \Omega_t}{\partial t} = d\eta + \{\omega_t, \eta\} = D_t\eta.$$

Let $P(\alpha_1, \cdots, a_k)$ be an invariant symmetric polynomial

$$\begin{aligned} \frac{\partial}{\partial t} P(\Omega_t^k) &= kP(D_t\eta, \Omega_t^{k-1}) \\ &= kD_tP(\eta, \Omega_t^{k-1}) \\ &= kdP(\eta, \Omega_t^{k-1}), \end{aligned}$$

so

$$P(\Omega_1^k) - P(\Omega_0^k) = dQ(\omega_0, \omega_1), \tag{10.1}$$

where

$$Q(\omega_0, \omega_1) = \int_0^1 P(\omega_1 - \omega_0, \Omega_t^{k-1})dt.$$

This equation shows that the difference between $P(\Omega_1^k)$ and $P(\Omega_0^k)$ is an exact form, thus the cohomology class of $P(\Omega^k)$ is independent of the choice of connection.

So given a G-invariant symmetric k-linear polynomial on vector bundle E, one can obtain a characteristic class

$$\chi_P(E) = P(\Omega, \cdots, \Omega),$$

and the mapping

$$\begin{aligned} W: I(G) &\rightarrow H^*(M, \mathbf{R}) \\ P(a_1, \cdots, a_k) &\rightarrow P(\Omega, \cdots, \Omega) \end{aligned}$$

is an algebraic Homomorphism, called Weil Homomorphism.

Let B_G be the classifying space with group G, it could be proved that there exists an isomorphism W_0 between $I(G)$ and $H^*(B_G, \mathbf{R})$. Let f be a map from M to B_G, defined up to homotopy, the following diagram is commutative

$$\begin{array}{ccc} & I(G) & \\ W \swarrow & & \searrow W_0 \\ H^*(M, \mathbf{R}) & \overleftarrow{f^*} & H^*(B_G, \mathbf{R}) \end{array}$$

As a result of the existence of the isomorphism W_0 between $I(G)$ and $H^*(B_G, \mathbf{R})$, the invariant polynomials can be identified with the cohomology classes of B_G. So Weil Homomorphism gives the representation of characteristic classes for fibre bundles.

10.2 Chern class, the splitting principle

At first, we look at the complex vector bundle over M with $GL(k, \mathbf{C})$ transition function, its curvature Ω is a $gl(k, \mathbf{C})$ valued 2-form. Taking the k-dimensional fundamental representation of $gl(k, \mathbf{C})$, then Ω can be expressed by a complex $k \times k$ matrix of 2-forms, $\Omega = (\Omega_j^i)$.

For the invariant polynomial,

$$\begin{aligned} det(1 + a) &= expTr \ \ln(1 + a) \\ &= expTr(a - \frac{a^2}{2} + \frac{a^3}{3} - \cdots) \end{aligned}$$

$$\begin{aligned}
&= 1 + Tr(a - \frac{a^2}{2} + \cdots) + \frac{1}{2}[Tr(a - \frac{a^2}{2} + \cdots)]^2 + \cdots \\
&= 1 + Tra + \frac{1}{2}[(Tra)^2 - Tra^2] + \cdots \\
&= \sum_{j=0}^{\infty} S_j(a),
\end{aligned} \tag{10.2}$$

where $S_k(a)$ is the degree-k symmetric polynomial of a, invariant with respect to G. Substitute a by $\frac{i}{2\pi}\Omega$, then we get the total Chern class

$$C(E) = det(I + \frac{i}{2\pi}\Omega) = 1 + c_1(E) + c_2(E) + \cdots = \sum_{j=0}^{\infty} c_j(E)$$

where c_k is called the k-th Chern class, and can be expressed as

$$\begin{aligned}
c_1(E) &= \tfrac{i}{2\pi} Tr\Omega = \tfrac{i}{2\pi}\delta^j_k \Omega^k_j, \\
c_2(E) &= \tfrac{1}{8\pi^2}[Tr\Omega^2 - (Tr\Omega)^2], \\
c_3(E) &= \tfrac{i}{48\pi^3}[-2Tr\Omega^3 + 3(Tr\Omega^2)Tr\Omega - (Tr\Omega)^3], \\
&\cdots\cdots
\end{aligned}$$

A few remarks here:
(1). As $c_j(E) \in \Lambda^{2j}(M)$, so that

$$\begin{aligned}
&c_j(E) = 0, \quad \text{for } 2j > n = dimM, \\
\text{and} \quad &c_j(E) = 0 \quad , \text{for } j > k = rankE (\text{ the complex dimension of the fibre}),
\end{aligned}$$

thus the total sum is always finite.
(2). From Eq. 9.5, connection 1-form ω satisfies

$$\omega^g = g^{-1}\omega g + g^{-1}dg$$

and curvature Ω satisfies

$$\Omega^g = g^{-1}\Omega g.$$

Since $S_j(a)$ is an invariant symmetric polynomial, $S_j(\Omega^g) = S_j(\Omega)$ is well defined globally on the manifold, satisfying $dS_j(\Omega) = 0$, so

$$S_j(\Omega) \in H^{2j}(M, \mathbf{C})$$

(3). Assuming E be an Hermitian vector bundle, the connection 1-form can be chosen as [1,0] type consistent with the Hermitian metric. Then the curvature 2-form is anti-Hermitian, $i\Omega$ is an Hermitian 2-form, so

$$S_j(i\Omega) \in H^{2j}(M, \mathbf{R}).$$

(4). Further taking the factor $(2\pi)^{-j}$,

$$c_j(E) = (\frac{i}{2\pi})^j S_j(\Omega),$$

such that Chern class $c_j(E)$ actually belongs to the integral class

$$c_j(E) \in H^{2j}(M, \mathbf{Z})$$

and one obtains an integer if $c_j(E)$ is integrated over any $2j$-cycle in M with integer coefficients.

Integrating the characteristic form over the entire manifold M will yield an integral number, called the Chern number of the bundle. For example, if the dimension of M is $n = 4$, the only two Chern numbers are

$$c_2[M] = \int_M c_2(E)$$
$$c_1^2[M] = \int_M c_1(E) \wedge c_1(E)$$

As any invariant polynomial $P(a)$ can be expressed in terms of the elementary symmetric polynomials $S_j(a)$, $P(a)$ can be expressed as a polynomial of Chern classes. Thus Chern classes generate the characteristic ring.

As for any two matrices A and B ,

$$det(A \oplus B) = detA \times detB$$

so for Whitney sum bundle, one get

$$C(E \oplus F) = C(E) \wedge C(F). \tag{10.3}$$

For computational purposes, a complex vector bundle E can always be expressed as direct sum of line bundles. If E does not split as a sum of line bundles, one may lift it to a space N where it does, i.e., there is the splitting principle:
Let E be a complex bundle over compact manifold M, then there exists a smooth map $f: \ N \to M$, such that bundle f^*E over N can be splited into a direct sum of complex line bundles

$$f^*E = l_1 \oplus \cdots \oplus l_k,$$

and the pullback class $C(f^*E) = f^*C(E)$ can be splitted.

As a result of the splitting principle, to prove a polynomial identity, it suffices to prove it under the assumption that the vector bundle is a direct sum of line bundles.

Now let Ω be a $k \times k$ matrix of curvature 2-forms, and assuming that Ω is diagonalizable to k 2-forms $\{\Omega_l\}$, then the total Chern class can be written as

$$C(E) = det(I + \frac{i}{2\pi}\Omega) = \prod_{l=1}^{k}(1 + \frac{i}{2\pi}\Omega_l) = \prod_{l=1}^{k}(1 + \chi_l),$$

where $\chi_l = \frac{i}{2\pi}\Omega_l$. And the Chern classes $c_j(E)$ can be written in terms of $\{\chi_l\}_1^k$ as

$$c_1(E) = \sum_l \chi_l,\ c_2(E) = \sum_{l<m} \chi_l\chi_m,$$
$$\cdots\cdots$$
$$c_k(E) = \sum_l \chi_l\chi_2\cdots\chi_k.$$

Thus the Chern classes $c_j(E)$ are the elementary symmetric functions in $\{\chi_i\}$. Conversely, every symmetric function in $\{\chi_i\}$ can be expressed in terms of the elementary symmetric functions and hence in terms of the Chern classes $\{c_j(E)\}$. This is how that all characteristic classes for complex bundle can be expressed as polynomials of Chern classes. For example, by means of splitting principle, the Todd class

$$\begin{aligned} Td(E) &= \prod_j \frac{\chi_i}{1-e^{\chi_j}} \\ &= 1 + \frac{1}{2}c_1(E) + \frac{1}{12}(c_1(E)^2 + c_2(E)) + \cdots \end{aligned}$$

The Td(E) satisfy

$$Td(E \oplus F) = Td(E)Td(F)$$

The total Chern class $C(E)$ behaves well for direct sums, but not for tensor product of bundles. Note that $Tr(\Omega^j) = 0$ if $2j > dimM$, the following form

$$(exp(\frac{i}{2\pi}\Omega)) \in H^*(M, \mathbf{R})$$

defined on bundle E also projects to a closed form on M, and the cohomology class defined by which, is called the Chern character of vector bundle E, and is denoted by

$$ch(E) = Tr(exp(\frac{i}{2\pi}\Omega)).$$

For any pair of complex vector bundle E and F over M, it is easy to prove that the Chern character satisfies

$$\begin{cases} ch(E \oplus F) = ch(E) + ch(F), \\ ch(E \otimes F) = ch(E)ch(F). \end{cases} \tag{10.4}$$

The importance of the Chern character lies in that it preserves the ring structure on the set of vector bundles. By means of the splitting principle

$$\begin{aligned} ch(E) &= \sum_{j=1}^k e^{\chi_j} = k + \sum_1^k \chi_j + \frac{1}{2}\sum_1^k \chi_j^2 + \cdots \\ &= k + c_1(E) + \frac{1}{2}(c_1^2(E) - 2c_2(E)) + \cdots = \sum_{l=0}^\infty ch_l(E), \\ ch_0(E) &= k, \quad ch_1(E) = c_1(E), \\ ch_2(E) &= \frac{1}{2}(c_1^2(E) - 2c_2(E)) \in H^4(M, \mathbf{Q}), \end{aligned}$$

and all the Chern character $ch_l(E)$ can be written as a polynomial of the Chern classes with rational numbers as coefficients. So

$$ch_l(E) \in H^{2l}(M, \mathbf{Q}).$$

For any two complex line bundle L and L', its tensor product bundle is also a complex line bundle and by Eq. 10.4 one has

$$1 + c_1(L \otimes L') = (1 + c_1(L)(1 + c_1(L')) = 1 + c_1(L) + c_1(L')$$

$$\text{so} \quad c_1(L \otimes L') = c_1(L) + c_1(L'). \tag{10.5}$$

In summary, with the help of equations 10.3 and 10.5 and the fundamental properties of characteristic classes, one can give the axiomatic definitions of Chern classes as follows:

Axiom 1. For each complex vector bundle E over M, there is a sequence of cohomology classes

$$c_j(E) \in H^{2j}(M, \mathbf{Z}), \quad j = 0, 1, 2, \cdots$$

called the Chern classes of E. Of which $c_0(E) = 1$ and $c_j(E) = 0$ for $j > k =$ *rank* of E.

Let $C(E) = 1 + c_1(E) + \cdots + c_k(E)$, and is called the total Chern class.

Axiom 2. (Naturalness) Let $f: M' \to M$ be a differential map, $E' = f^*E$, then we have

$$C(E') = f^*(C(E)) \in H^*(M', \mathbf{Z}).$$

Axiom 3.(Whitney sum formula) Let E and F be two complex vector bundles over the same base manifold, one has

$$C(E \oplus F) = C(E)C(F),$$

$$i.e., \qquad c_l(E \oplus F) = \sum_{j=0}^{l} c_j(E)c_{l-j}(F).$$

Axiom 4.(Normalization) For natural complex line bundle L over $CP(1)$, Chern class $c_1(L)$ is non-zero, and is the given generator of $H^2(CP(1), \mathbf{Z})$. $c_1(L)$ evaluated on the fundamental 2-cycle $CP(1)$ (i.e., the Chern number) is equal to -1,

$$c_1[CP(1)] = \int_{CP(1)} c_1(L) = -1.$$

Making use of Axiom 2 and the inclusion map

$$CP(1) \to CP(2) \to \cdots$$

one can rewrite Axiom 4 as

Axiom 4'. For the natural complex line bundle λ over $CP(\infty)$, element $c_1(\lambda)$ is the generator of the cohomology ring $H^*(CP(\infty, \mathbf{Z}))$.

$CP(\infty)$ is the universal classifying space of complex line bundles. From the splitting principle and Whitney sum formula in terms of the universal classifying space $G^c_r(\infty, k)$ of the complex k-dim vector bundle, one can obtain the Chern classes of all complex vector bundles.

It is easy to show that the cohomology class obtained above by Weil Homomorphism Eq. 10.3 satisfies the four axioms. These results can also be obtained through the cohomology of classifying space.

Example 1. The natural complex line bundle L over $CP(1)$. The first Chern class

$$c_1(L) = \frac{i}{2\pi} Tr\Omega.$$

By using Eq. 9.2, the Chern number

$$\begin{aligned} c_1[CP(1)] &= \int_{S^2} \frac{-dx \wedge dy}{\pi(1 + x^2 + y^2)} = -\frac{1}{\pi}\int_{S^2} \frac{\rho d\rho \wedge d\phi}{(1+\rho^2)} \\ &= -\int_0^\infty \frac{\rho d\rho}{(1+\rho^2)} \\ &= -1. \end{aligned}$$

It is easy to see from this example that due to the inclusion of factor $\frac{i}{2\pi}$ in front of Ω that makes the topological constant an integer, called the Chern number.

Example 2. The tangent bundle of complex projective space $CP(n)$. From §8.6 we know that

$$T(CP(n) \simeq Hom_C(\lambda, \lambda^\perp), \tag{10.6}$$

where λ is the natural complex line bundle over $CP(n)$. $\lambda^\perp$ is the orthogonal complement of λ obtained through the standard Hermitian metric in C^{n+1}. So that the Whitney sum

$$\lambda \oplus \lambda^\perp = \epsilon^{n+1},$$

is a trivial complex $(n+1)$-dim vector bundle over $CP(n)$. Now, adding the trivial bundle

$$\epsilon^1 \simeq Hom_C(\lambda, \lambda)$$

to both sides of Eq. 10.6, on has

$$\begin{aligned} T(CP(n)) \oplus \epsilon^1 &\simeq Hom_C(\lambda, \lambda \oplus \lambda^\perp) \\ &= Hom_C(\lambda, \epsilon^1 \oplus \epsilon^1 \oplus \cdots \oplus \epsilon^1) \\ &= \lambda^* \oplus \lambda^* \oplus \cdots \oplus \lambda^* \\ &\equiv \oplus_{n+1}\lambda^*. \end{aligned}$$

So the total Chern class of the tangent bundle is

$$C(T(CP(n))) = (1+\chi)^{n+1},$$

where $\chi = -c_1(\lambda)$ is the generator of $H^2(CP(n), \mathbf{Z})$.

10.3 Pontrjagin class

Now we look at real vector bundle E of rank k over M, with its transition function belonging to $GL(k, \mathbf{R})$. The transition function could be reduced to $O(k)$, which is the maximum compact subgroup of $GL(k, \mathbf{R})$, if a fibre metric is given. The structure group can always be reduced to $O(k)$. Taking a Riemannian connection on the bundle, the curvature 2-form is Lie algebra $o(k)$ valued, satisfying

$$\Omega = -\Omega^t$$

where Ω^t denotes the transposed matrix of Ω.

From Weil homomorphism and substitute the curvature 2-form Ω into the invariant polynomial Eq. 10.2, one has the total Pontrjagin class as follows

$$P(E) = det(I - \frac{1}{2\pi}\Omega) = 1 + p_1(E) + p_2(E) + \cdots \ .$$

As $\Omega = -\Omega^t$, we have

$$det(I - \frac{1}{2\pi}\Omega) = det(I - \frac{1}{2\pi}\Omega)^t = det(I + \frac{1}{2\pi}\Omega)$$

$$\text{so} \qquad \sum_j (-1)^j S_j(\frac{\Omega}{2\pi}) = \sum_j S_j(\frac{\Omega}{2\pi})$$

which implies that $S_j(\frac{\Omega}{2\pi})$ is zero for odd j. Thus one get

$$p_j(E) \in H^{4j}(M, \mathbf{R})$$

and $p_j(E) = 0$ if $4j > n = dimM$ or if $2j > k =$ rank E. The normal factor $(2\pi)^{-j}$ can be chosen such that $p_j(E)$ belong to integral classes,

$$p_j(E) = \frac{1}{(2j)!(2\pi)^{2j}} \delta^{k_1 \cdots k_{2j}}_{l_1 \cdots l_{2j}} \Omega^{l_1}_{k_1} \wedge \cdots \wedge \Omega^{l_{2j}}_{k_{2j}} \in H^{4j}(M, \mathbf{Z}).$$

For example

$$\begin{aligned} p_1(E) &= -\frac{1}{8\pi^2} Tr\Omega^2, \\ p_2(E) &= \frac{1}{(2\pi)^4}[\frac{1}{8}(Tr\Omega^2)^2 - \frac{1}{4}Tr\Omega^4], \end{aligned}$$

$$\begin{aligned}
p_3(E) &= \frac{1}{(2\pi)^6}[-\frac{1}{48}(Tr\Omega^2)^3 + \frac{1}{8}Tr\Omega^2 Tr\Omega^4 - \frac{1}{6}Tr\Omega^6],\\
p_4(E) &= \frac{1}{(2\pi)^8}[\frac{1}{384}(Tr\Omega^2)^4 - \frac{1}{32}(Tr\Omega^2)^2 Tr\Omega^4 + \frac{1}{12}Tr\Omega^2 Tr\Omega^6,\\
&\quad +\frac{1}{32}(Tr\Omega^4)^2 - \frac{1}{8}Tr\Omega^8],\\
p_5(E) &= \frac{1}{(2\pi)^{10}}[-\frac{1}{3840}(Tr\Omega^2)^5 + \frac{1}{192}Tr(\Omega^2)^3 Tr\Omega^4 - \frac{1}{48}(Tr\Omega^2)^2 Tr\Omega^6\\
&\quad -\frac{1}{64}Tr\Omega^2(Tr\Omega^4)^2 + \frac{1}{16}Tr\Omega^2 Tr\Omega^8 + \frac{1}{24}Tr\Omega^4 Tr\Omega^6 - \frac{1}{10}Tr\Omega^{10}].
\end{aligned}$$

It is easy to prove that the total Pontrjagin class satisfies

$$P(E \oplus F) = P(E)P(F).$$

For any real bundle E, its complexfication is denoted by

$$E_C = E \otimes \mathbf{C},$$

and the total Chern class of the complexfication of E can be written as

$$C(E_C) = det(I + \frac{i}{2\pi}\Omega) = 1 - p_1(E) + p_2(E) - \cdots .$$

Thus one obtains

$$p_j(E) = (-1)^j c_{2j}(E \otimes \mathbf{C}).$$

Conversely, given any complex bundle E of rank k, we can simply forget about the complex structure and think of each fibre as a real space of dimension $2k$ and denote this rank-$2k$ real vector bundle by E_r. Note that both E and E_r have the same total space, base space and the projective map π.

From Eq.(6.5), we know that for any complex vector bundle, the complexfication of the underlying real vector bundle is isomorphic to Whitney sum $E \oplus \bar{E}$, i.e.,

$$E_r \otimes \mathbf{C} = E \oplus \bar{E},$$

so
$$\begin{aligned}
(E_r \otimes \mathbf{C}) &= 1 - p_1(E_r) + p_2(E_r) - \cdots\\
&= C(E)C(\bar{E})\\
&= [1 + c_1(E) + c_2(E) + \cdots][1 - c_1(E) + c_2(E) - \cdots].
\end{aligned}$$

The proper identification gives

$$(-1)^j p_j(E_r) = \sum_{l=0}^{j}(-1)^l c_l(E) c_{2j-l}(E).$$

For instance

$$p_1(E_r) = c_1^2(E) - 2c_2(E),$$
$$p_2(E_r) = c_2^2(E) - 2c_1(E)c_3(E) + 2c_4(E),$$

and as a result of splitting principle for the complexfication of real vector bundle, one get

$$E_C \simeq L_1 \oplus \bar{L}_1 \oplus \cdots \oplus L_k \oplus \bar{L}_k,$$
$$C(E_C) = \prod_{j=1}^{k}(1-\chi_j^2) = 1 - p_1(E) + p_2(E) - \cdots \quad ,$$

and $P(E) = 1 + p_1(E) + p_2(E) + \cdots = \prod_{j=1}^{k}(1+\chi_j^2)$. And other multiplicative characteristic classes can be obtained by using generating function. All characteristic classes for real vector bundle can be expressed as polynomials of Pontrjagin classes. For example,

1. Hirzebruch L-sequence associated to the formal power series

$$\frac{\chi}{tanh\chi} = (1 - \frac{1}{3}\chi^2 + \frac{2}{15}\chi^4 - \frac{17}{315}\chi^6 + \cdots)^{-1}.$$

The total L-class of vector bundle E is

$$\begin{aligned} L(E) &= \prod_{j=1}^{k}\frac{\chi_j}{tanh\chi_j} \\ &= 1 + \frac{1}{3}p_1 + (-\frac{1}{45}p_1^2 + \frac{7}{45}p_2) + (\frac{2}{945}p_1^3 - \frac{13}{945}p_1p_2 + \frac{62}{945}p_3) \\ &\quad +(-\frac{1}{4725}p_1^4 + \frac{22}{14175}p_1^2p_2 - \frac{71}{14175}p_1p_3 - \frac{19}{14175}p_2^2 + \frac{127}{14175}p_4) \\ &\quad +(\frac{2}{93555}p_1^5 - \frac{83}{467775}p_1^3p_2 + \frac{79}{155925}p_1^2p_3 + \frac{127}{467775}p_1p_2^2 + \frac{919}{467775}p_1p_4 \\ &\quad -\frac{16}{22275}p_2p_3 + \frac{146}{13365}p_5) + \cdots \quad . \end{aligned}$$

2. A-roof sequence associated to the formal series

$$\begin{aligned} \frac{\chi/2}{\sinh\chi/2} &= [1 + \frac{1}{3!}(\frac{\chi^2}{2} + \frac{1}{5!}\frac{\chi^4}{2} + \cdots]^{-1} \\ &= (1 - \frac{1}{24}\chi^2 + \frac{7}{2^7 3^2 5}\chi^4 - \cdots \frac{17}{315}\chi^6 + \cdots)^{-1}, \end{aligned}$$

the total A-roof class of vector bundle E is

$$\hat{A}(E) = \prod_{j=1}^{k}\frac{\chi_j}{\sinh\chi_j/2}$$

$$\begin{aligned}
&= 1+\frac{1}{2^2}(-\frac{1}{6}p_1)+\frac{1}{2^4}(\frac{7}{360}p_1^2-\frac{1}{90}p_2)\\
&+ \frac{1}{2^6}(\frac{-31}{15120}p_1^3+\frac{11}{3780}p_1p_2-\frac{1}{945}p_3)\\
&+ \frac{1}{2^8}(\frac{127}{604800}p_1^4-\frac{113}{226800}p_1^2p_2+\frac{4}{14175}p_1p_3+\frac{13}{113400}p_2^2-\frac{1}{9450}p_4)\\
&+ \frac{1}{2^{10}}(\frac{-73}{3421440}p_1^5+\frac{1073}{14968800}p_1^3p_2+\frac{-61}{1247400}p_1^2p_3+\frac{-31}{7484400}p_1p_2^2\\
&+ \frac{53}{1871100}p_1p_4+\frac{1}{44550}p_2p_3-\frac{1}{93555}p_5)+\cdots .
\end{aligned}$$

10.4 Euler class

Now we look at orientable real vector bundle E of even rank $k=2l$, its transition function could be reduced to $SO(k)$, whose adjoint representation is given by $2l\times 2l$ real antisymmetric matrices. The determinant of an antisymmetric $2l\times 2l$ matrix is the square of a polynomial of its entries, this polynomial is invariant and is called the Pfaffian of the matrix concerned. Thus for even rank orientable real vector bundle, there exists an $SO(k)$ invariant polynomial in addition to the characteristic determinant. This polynomial is not invariant under the orientation preserving group $GL^+(k,\mathbf{R})$, so the corresponding characteristic class can only be computed by means of a Riemannian connection, but not a general linear connection.

The $so(2l)$ valued curvature 2-form Ω can be written as a $2l\times 2l$ skew symmetric matrix $\Omega=(\Omega_{ij})$ in the adjoint representation. There exists one $Pf(\Omega)$, closed and $SO(2l)$ invariant, whose square is the determinant of Ω. As a result of Weil homomorphism, $Pf(\Omega)$ is a characteristic class of bundle E, called the Euler class of bundle E, and is denoted by $e(E)=Pf(\Omega)$.

As a result of the splitting principle, Euler class $e(E)$ can be written as

$$e(E)=\chi_1\chi_2\cdots\chi_l$$

whose square is

$$e(E)^2=det(\Omega)=\chi_1^2\chi_2^2\cdots\chi_l^2=p_l(E).$$

Obviously the Euler class is multiplicative for Whitney sum

$$e(E\oplus F)=e(E)e(F).$$

For a complex vector bundle E of rank l, its real counterpart is a real rank-$2l$ orientable vector bundle. And it is easy to prove that

$$e(E_r)=(p_l(E_r))^{\frac{1}{2}}=c_l(E), \qquad (10.7)$$

i.e., the top Chern class of a complex vector bundle is just the Euler class of its underlying real vector bundle.

The tangent bundle $T(M)$ of a differentiable manifold M contains much of the geometric data of the base manifold. The characteristic classes of the tangent bundle are usually called the characteristic classes of the base manifold, and the Euler class of $T(M)$ is also called the Euler class of the base manifold, simply denoted as $e(M)$.

Example. Let M be a k-dimensional Riemannian manifold, its Euler classes are

$$\begin{aligned}
k=2, &\quad e(M)=\frac{1}{4\pi}\epsilon_{ab}\Omega^{ab}=\frac{1}{2\pi}\Omega^{12},\\
k=4, &\quad e(M)=\frac{1}{32\pi^2}\epsilon_{abcd}\Omega^{ab}\wedge\Omega^{cd},\\
k=2l, &\quad e(M)=\frac{1}{2^l(2\pi)^l l!}\epsilon_{a_1a_2\cdots a_{2l}}\Omega^{a_1a_2}\wedge\cdots\Omega^{a_{2l-1}a_{2l}}.
\end{aligned}$$

Remark:

1. Formally the Euler class is the square root of Pontrjagin class (see Eq. 10.7), but for the tangent bundle of a $2l$-dimensional Riemannian manifold

$$p_l(M)\in H^{4l}(M)=0,$$

while the Euler class could be non-zero

$$e(M)\in H^{2l}(M).$$

2. Chern class and Pontrjagin class can be computed using any linear connection. While in the case of Euler class one has to find a connection compatible with the fibre metric of E, i.e., must computed in terms of Riemannian connection. The Euler class is only defined for oriented bundle.

10.5 Stiefel-Whitney class, orientation and spin structure

For a real vector field bundle $E(M,F,\pi,G)$, the i-th Stiefel-Whitney class of E is denoted by $w_i(E)$

$$w_i\in H^i(M,\mathbf{Z}_2),\quad i=0,1,\cdots$$

which forms a base of the cohomology ring of M with coefficients in $\mathbf{Z}_2$. Unlike other characteristic classes, these characteristic classes can not be expressed in terms of polynomials of the curvature form, thus can not be obtained through the Weil homomorphism as Chern, Pontrjagin and Euler classes. But it can be studied from an axiomatic point of view, just like for the Chern class of a complex vector bundle, Stiefel-Whitney class can be characterized by introducing four axioms. We refer interested readers to other books (e.g., J.W.Milnor and J.D.Stasheff: Characteristic Classes) for the existence and the uniqueness of cohomology classes satisfying these axioms. These results could be obtained through obstruction theory and cohomology of classifying space (see §8.7).

Axiom 1. For each real vector bundle E over M, there corresponds a sequence of cohomology classes

$$w_i(E) \in H^i(M, \mathbf{Z}_2), i = 0, 1, 2, \cdots$$

called the Stiefel-Whitney classes of E, in which $w_0(E) = 1$ and $w_i(E) = 0$ for $i > n$, where n is the dimension of the real vector bundle. In addition, let

$$W(E) = 1 + w_1(E) + \cdots + w_n(E)$$

which is called the total Stiefel-Whitney class.

Axiom 2.(Naturalness) Let $f : M' \to M$ be a differentiable map, $E' = f^*E$, then

$$W(E') = f^*(W(E)) \in H^*(M', \mathbf{Z}_2).$$

Axiom 3.(Whitney sum formula) Let E_1 and E_2 be two real vector bundles over the same base manifold, then

$$W(E_1 \oplus E_2) = W(E_1)W(E_2),$$

i.e.,
$$w_k(E_1 \oplus E_2) = \sum_{i=0}^{k} w_i(E_1)w_{k-i}(E_2).$$

Axiom 4.(Normalization) For the natural real line bundle γ_1^1 over $RP(1)$, the Stiefel-Whitnet class of γ_1^1 is non-zero, and is the given generator of $H^1(RP(1), \mathbf{Z}_2)$.

Furthermore, taking use of Axiom 2 and the inclusion map

$$RP(1) \to RP(2) \to \cdots,$$

axiom 4 can be rewritten as

Axiom 4'. For the natural real line bundle γ_∞^1 over $RP(\infty)$, element γ_∞^1 is the generator of the cohomology ring $H^*(RP(\infty), \mathbf{Z}_2)$.

Axioms 4 and 4' are equivalent. It is clear that the two sets of characteristic classes, Stiefel-Whitney class and Chern class, have many properties in common, they are defined as obstructions to the existence of certain vector fields on real and complex vector bundles respectively.

Let E be a trivial vector bundle, there exists a map from E to a vector bundle over a point, and from axiom 2 we know that

$$W(E) = 1.$$

Now we take a look at the tangent bundle $T(S^n)$ of unit sphere $S^n \subset \mathbf{R}^{n+!}$. From the standard embedding $S^n \to \mathbf{R}^{n+1}$, the normal bundle ν is trivial, and the Whitney sum

$$T(S^n) \oplus \nu = \epsilon$$

is also trivial.

$$W(\epsilon) = W(\nu)W(T(S^n)) = 1, \quad W(\nu) = 1,$$
$$\text{so} \quad W(T(S^n)) = 1.$$

From the point of view of Stiefel-Whitney class, tangent bundle over S^n can not be distinguished from trivial bundle.

From Axiom 4, we see that the total Stiefel-Whitney class of the natural line bundle γ_n^1 over $RP(n)$ is

$$W(\gamma_n^1) = 1 + \chi.$$

The natural line bundle γ_n^1 over $RP(n)$ is a sub-bundle of trivial bundle ϵ^{n+1} over $RP(n)$.

Now we consider the tangent bundle $T(RP(n))$. Let ϵ' be a trivial line bundle over $RP(n)$. The Whitney sum $T(RP(n)) \oplus \epsilon'$ is isomorphic to the $(n+1)$-fold Whitney sum $\gamma_n^1 \oplus \gamma_n^1 \oplus \cdots \oplus \gamma_n^1$. So the total Stiefel-Whitney class of tangent bundle $T(RP(n))$ denoted as $W(RP(n))$ is given as

$$W(RP(n)) = (1 + \chi)^{n+1}.$$

Expanding by the binomial theorem, module all the coefficients by 2 and remember that $\chi^{n+1} = 0$ since $H^{n+1}(RP(n), \mathbf{Z}_2) = 0$, we have for example

$$\begin{aligned} W(RP(2)) &= 1 + \chi + \chi^2, \\ W(RP(3)) &= 1 \\ W(RP(4)) &= 1 + \chi + \chi^4, \\ W(RP(5)) &= 1 + \chi^2 + \chi^4. \end{aligned}$$

Now we look at the orientation and spin structure of a manifold M.

Theorem 10.1 *The following statements are equivalent*

(1). M is orientable.

(2). There exists an atlas $\mathcal{A} = \{(U_\alpha, \phi_\alpha)\}$ such that all the Jacobi matrices of transition functions are positive.

(3). The structure group of tangent bundle $T(M)$ can be reduced to $SO(n)$, here n is the dimension of M.

(4). Line bundle $\Lambda^n(M)$ is trivial, i.e., it has a nowhere vanishing cross section.

(5). $w_1(M) \equiv w_1(T(M)) = 0$

The first Stiefel-Whitney class vanishes if and only if M is orientable. From §1.7 we see that as coordinate transformations the line bundle $\Lambda^n(M)$ (n-form) has the determinant of coordinate transformations of $T(M)$. Therefore $T(M)$ is orientable if and only if line bundle $\Lambda^n(M)$ is trivial, namely $w_1(\Lambda^n(M)) = 0$. But tangent bundle can be considered as a Whitney sum of line bundles

$$T(M) = \oplus_n \lambda, \quad w_1(T(M)) = w_1(\oplus_n \lambda) = w_1(\Lambda^n(M)).$$

Following we considered the spin structure on M. Let M be an n-dimensional orientable manifold, the structure group of tangent bundle $T(M)$ can be reduced to $SO(n)$. The fundamental group of $SO(n)$ is $\Pi_1(SO(n)) = \mathbf{Z}_2$. Now let us consider lifting the structure group of $T(M)$ to $Spin(n)$, which is the 2-fold covering of $SO(n)$. $Spin(n)$ is a compact simple connected Lie group.

Choose an open covering $\{U_\alpha\}$ of M such that $T(M)$ is trivial on each of the $\{U_\alpha\}$, and $T(M)$ is determined by transition functions $\{g^{\alpha\beta}\}$

$$g^{\alpha\beta} : \ U_\alpha \cap U_\beta \to SO(n)$$

on the overlapping region $U_\alpha \cap U_\beta \cap U_\gamma \neq 0$ these transition functions satisfy the cocycle condition

$$g^{\alpha\beta} g^{\beta\gamma} g^{\gamma\alpha} = 1.$$

The existence of spin structure is equivalent to being able to choose an open covering $\{U_\alpha\}$ and trivialization such that transition function $g^{\alpha\beta}$ can be lifted to $h^{\alpha\beta}$

$$h^{\alpha\beta} : \ U_\alpha \cap U_\beta \to Spin(n)$$

in such a way that $h^{\alpha\beta}$ also satisfies the cocycle condition

$$h^{\alpha\beta} h^{\beta\gamma} h^{\gamma\alpha} = 1.$$

We say that the tangent bundle $T(M)$ admits a spin structure if there exists an open covering $\{U_\alpha\}$, and the lift $\{h^{\alpha\beta}\}$ can be chosen as above, i.e., restriction of the tangent bundle T(M) to any immersed surface $\Sigma \subset M$ admits a trivialization. A manifold M is called spin manifold if its tangent bundle T(M) admits a spin structure. In two- or three-dimension, an orientable manifold is a spin manifold. But in four and higher dimensions, there are many manifolds having no spin structure. For spinors to be well defined on $T(M)$, i.e., spin structure exists, if and only if its second Stiefel-Whitney class

$$w_2(T(M)) \in H^2(M, \mathbf{Z}_2) \quad \text{vanishes} .$$

If $w_2(T(M)) \neq 0$, there is a sign ambiguity when spinors are parallel transported around a closed path on M, which means that spinors are not well defined.

Example 1. n-dimensional sphere S^n. Its total Stiefel-Whitney class is

$$W(T(S^n)) = 1 + (1 + (-1)^n)\tau,$$

where τ is the normal volume element n-form. From this equation we have

$$w_1(T(S^n)) = 0, \ w_2(T(S^n)) = 0, \ \text{for all } S^n$$

thus S^n are orientable and have spin structure.

Example 2. Complex projective space $CP(n)$.

Manifold $CP(n)$ are compact Kähler manifold, its Kähler form K and the l-th power $K^l = \Lambda_l K$ (the exterior product of l Kähler forms) $(1 \le l \le n)$ are the nontrivial generators of the cohomology group

$$H^k(CP(n), \mathbf{R}) = \begin{cases} 0, & k \text{ is odd}, \\ \mathbf{R}, & k \text{ is even and } k \le 2n, \end{cases}$$

and

$$H^k(CP(n), \mathbf{Z}_2) = \begin{cases} 0, & k \text{ is odd}, \\ \mathbf{Z}_2, & k \text{ is even and } k \le 2n. \end{cases}$$

Let χ denote the generator of the cohomology group $H^2(CP(n), \mathbf{Z}_2)$, then the total Stiefel-Whitney class can be written as

$$W(T(CP(n))) = (1+\chi)^{n+1} = 1 + w_2 + w_4 + \cdots + w_{2n},$$

where the expansion coefficients are taken in $\mathbf{Z}_2$, so

$$w_2 = (n+1)|_{\text{mod } 2}\chi = \begin{cases} 0, & n \text{ is odd}, \\ \chi, & n \text{ is even}. \end{cases}$$

Thus $CP(n)$ have no spin structure for n even. Spin structure exists for $CP(n)$ only if n is odd.

The major sources of characteristic classes on fibre bundle are the four classes above, its main properties refer table 10.1.

Table 10.1

name of class	vector bundle	struct. group	characteristic class	example
Chern	complex	$U(k)$	$c_i(k) \in H^{2i}(M,R)$	$c_1 = \frac{i}{2\pi} Tr\Omega$ $c_2 = \frac{1}{8\pi}(Tr(\Omega)^2 - (Tr\Omega)^2)$
Pontrj.	real	$O(k)$	$p_i(E) \in H^{4i}(M,R)$	$p_1 = -\frac{1}{8\pi^2} Tr\Omega^2$ $p_2 = \frac{1}{64}\frac{1}{(2\pi)^4}(\epsilon_{abcd}\Omega^{ab} \wedge \Omega^{cd})^2$
Euler	real orient.	$SO(k)$ k=2j	$e(E) \in H^k(M,R)$	$e(E) = \frac{\epsilon_{a_1 \cdots a_{2j}}}{2^j (2\pi)^j j!}\cdot$ $\Omega^{a_1 a_2} \wedge \cdots \wedge \Omega^{a_{2j-1} a_{2j}}$
Stiefel-Whitney	real	$O(k)$	$w_i(E) \in H^i(M, Z_2)$	$W(\gamma_n^1) = 1 + \theta$ $\theta \in H^1(RP(n), \mathbf{Z}_2)$

10.6 Secondary characteristic class (Chern-Simons form)

The characteristic class $\chi(E)$ of vector bundle E is a cohomology class of base manifold M

$$\chi(E) \in H^*(M, \mathbf{R})$$

which can usually be expressed in terms of the symmetric invariant polynomial of curvature 2-form Ω (Weil homomorphism). As curvature Ω is a 2-form, only on even-dimensional manifold does non-zero characteristic class can exists and can be written in terms of polynomials of curvature forms. From now on we try to construct topological invariants for odd-dimensional manifold[45].

Let $P(\Omega^k)$ denote the k-th Chern class

$$P(\Omega^k) \in H^{2k}(M,\mathbf{R}), \ dP(\Omega) = 0.$$

Locally $P(\Omega^k)$ can be written as an exact form, and from Eq. 10.1,

$$P(\Omega^k) = dQ(0,\omega) \equiv dQ(\omega)$$

where ω is the connection 1-form, and

$$Q(\omega) = k\int_0^1 P(\omega, \Omega_t^{k-1})dt,$$

$$\Omega_t = t\omega + t^2\omega \wedge \omega.$$

If the $2k$-form $P(\Omega^k)$ be zero identically, $Q(\omega)$ then is a closed $(2k-1)$-form and defines an element of $H^{2k-1}(M,\mathbf{R})$. The vanishing of the characteristic form leads to other invariants which deserve further studies.

As $Q(\omega)$ depends on connection ω, it is desirable to look at the effect on $Q(\omega)$ under a change of the connection. There is the following lemma:

Lemma 1. Let ω_s be a family of connection 1-forms depending on a parameter s, the corresponding curvature 2-form is denoted by

$$\Omega_s = d\omega_s + \omega_s \wedge \omega_s.$$

Then one obtain

$$\frac{\partial}{\partial s}Q(\omega_s) = k(k-1)dR(s) + kP(\frac{\partial\omega_s}{\partial s}, \Omega_s^{k-1}),$$

where

$$\begin{cases} R(s) = \int_0^1 P(\frac{\partial\omega_s}{\partial s}, \omega_t, \Omega_t^{k-2})dt, \\ \omega_t = t\omega_s, \\ \Omega_t = td\omega_s + t^2\omega_s \wedge \omega_s. \end{cases}$$

Proof: Analogous to the proof for Eq. 10.1, from

$$Q(\omega_s) = k\int_0^1 P(\omega_s, \Omega_t^{k-1})dt,$$

one has

$$\frac{\partial Q(\omega_s)}{\partial s} = k\int_0^1 P(\sigma, \Omega_t^{k-1})dt + k(k-1)\int_0^1 P(\omega_s, tD_t\sigma, \Omega_t^{k-2})dt,$$

where $\sigma = \frac{\partial \omega_s}{\partial s}$, $D_t\,\sigma = d\,\sigma + \{\omega_t, \sigma\}$.

On the other hand

$$dR(s) = \int_0^1 [P(D_t\sigma, \omega_t, \Omega_t^{k-1})dt - P(\sigma, D_t\omega_t, \Omega_t^{k-2})]dt,$$

so

$$\begin{aligned}
&\frac{1}{k}\frac{\partial}{\partial s}Q(\omega_s) - (k-1)dR(s) \\
&= \int_0^1 P[\sigma, k\Omega_t + (k-1)t^2\omega_s \wedge \omega_s, \Omega_t^{k-1}]dt \\
&= k\int_0^1 t^{k-1}P[\sigma, d\omega_s + \frac{2k-1}{k}t\omega_s \wedge \omega_s, (d\omega_s + t\omega_s \wedge \omega_s)^{k-2}]dt \\
&= k\int_0^1 t^{k-1}P[\sigma, \Omega_s + (at-1)\omega_s \wedge \omega_s, (\Omega_s + (t-1)\omega_s \wedge \omega_s)^{k-2}]dt,
\end{aligned}$$

where $a = \frac{2k-1}{k}$. Expanding the right hand side gives

$$\begin{aligned}
&P[\sigma, \Omega_s, (at-1)\omega_s \wedge \omega_s, (\Omega_s + (t-1)\omega_s \wedge \omega_s)^{k-2}] \\
&\quad = P(\sigma, \Omega_s^{k-1}) + \sum_{l=1}^{k-1} b_l P[\sigma, \Omega_s^{k-l-1}, (\omega_s \wedge \omega_s)^2],
\end{aligned}$$

where

$$b_l = (t-1)^l \binom{k-2}{l} + (t-1)^{l-1}(at-1)\binom{k-2}{l-1}.$$

As

$$\int_0^1 t^l (1-t)^k dt = \frac{k!l!}{(k+l+1)!} \quad (1 \le l \le k-1),$$

it is easy to prove

$$\int_0^1 t^{k-1} b_l = 0, \quad (1 \le l \le k-1).$$

So

$$\frac{1}{k}\frac{\partial}{\partial s}Q(\omega_s) - (k-1)dR(s) = k\int_0^1 t^{k-1}P(\sigma, \Omega_s^{k-1})dt = P(\sigma, \Omega_s^{k-1}),$$

and this completes the proof of the lemma.

From this lemma one has the following theorem:

Theorem 10.2 *Let $P \in I(G)$ be an invariant polynomial, ω_s be a family of connection 1-forms depending on a parameter s, with the corresponding curvature $\Omega_s = d\omega_s + \omega_s \wedge \omega_s$. If*

(1). $P(\Omega_s^k) = 0$;

(2). $P(\frac{\partial \omega_s}{\partial s}, \Omega_s^{k-1}) = 0$.

then $Q(\omega_s)$ is a closed form, and the cohomology class is independent of s.

Odd-form $Q(\omega)$ is an invariant polynomial of vector bundle E over odd-dimensional manifold if the conditions above are satisfied, and its cohomology class is independent of the choice of connections in the family $\omega(s)$. $Q(\omega)$ is called secondary characteristic class, and is also called Chern-Simons form.

Example 1. Let M be an n-dimensional Riemannian manifold, so the structure group of the tangent bundle $T(M)$ is $GL(n, \mathbf{R})$. Picking a local coordinate $\{x^i\}_1^n$, the Riemannian metric h can be written by the tensor

$$h_{ij} = h_{ji} = h(\frac{\partial}{\partial x^i}, \frac{\partial}{\partial x^j}),$$

which are elements of a positive definite symmetric matrix $H = (h_{ij})$.

The Levi-Civita connection can be written as

$$\omega = (\Gamma^j{}_i), \quad \Gamma^j{}_i = \Gamma^j_{ki} dx^k \in \Lambda^1(M)$$

which can be determined by the metric compatible and torsion free conditions:

$$dH = \omega H + H^t \omega, \quad \Gamma^j_{ki} = \Gamma^j_{ik}.$$

Exterior differentiation of the first equation yields the equation of the curvature Ω, which has to satisfy:

$$\Omega H + H^t \Omega = 0,$$

which can be written as

$$\Omega_{ik} + \Omega_{ki} = 0,$$

i.e., matrix $\Omega H = (\Omega_{ik})$ is skew symmetric.

There still has the following lemma.

Lemma 2. Let $P \in I(G)$ be an invariant polynomial of odd-degree k, then

$$P(\Omega^k) = 0, \quad (\text{ for degree } k \text{ odd}) \tag{10.8}$$

Its proof is quite straightforward. Since

$$P(\Omega^k) = P(\Omega^k)^t = P(-H^{-1}\Omega^k H) = (-1)^k P(\Omega^k),$$

hence one get the result of the lemma.

Now we look at the conformal family of Riemannian metrics, given as

$$H(s) = e^{2\rho s}H,$$

where $\rho = \rho(x)$ is a scalar function and s is a parameter. One can prove that the one-parameter conformal family $\omega(s)$ of Riemannian connection satisfies

$$\frac{1}{s}(\omega(s) - \omega(0)) = Id\rho + A - B, \tag{10.9}$$

where I is the unit matrix and matrices A and B are

$$A = (\frac{\partial \rho}{\partial x^i}dx^j), \quad B = (h_{il}dx^l \frac{\partial \rho}{\partial x^m}h_{jm}).$$

There is a lemma for the conformal family of Riemannian connections:

Lemma 3. Let $P \in I(G)$ be an invariant polynomial of even degree $2l$, then for the Levi-Civita connections of a conformal family of Riemannian metrics, one has

$$P[\frac{\partial \omega(s)}{\partial s}, \Omega_s^{2l-1}] = 0. \tag{10.10}$$

Proof: From Eq. 10.9, the left hand side of the equation above is

$$d\rho\, P[I, \Omega_s^{2l-1}] + P[A, \Omega_s^{2l-1}] - P[B, \Omega_s^{2l-1}],$$

the first term is zero as a result of Lemma 2 (Eq. 10.8). The second term has a factor $dx^j \wedge \Omega_j^k(s)$ which is zero, as is the result of torsion free and integrable conditions Eq.4.8 (the first Bianchy identity) for Riemannian curvature. Similarly the third term has a factor $h_{il}dx^l \wedge \Omega_k^i(s)$, which also vanishes for the same reason. Thus the lemma is proved.

From the two lemmas above, one obtains the following theorem.

Theorem 10.3 *Let M be a Riemannian manifold of dimension n, $P \in I(GL(n, \mathbf{R}))$ be an invariant polynomial of degree $2l$. Then in terms of the Levi-Civita connections of a conformal family of Riemannian metrics, one has*

$$P[\frac{\partial \omega(s)}{\partial s}, \Omega_s^{2l-1}] = 0.$$

If furthermore, $P[\Omega^{2l}] = 0$, then

$$Q(\omega) \in H^{4l-1}(M, \mathbf{R}),$$

which depends on the underlying conformal structure of the Riemannian manifold only.

Example 2. Let L be the complex line bundle over S^1, the local section of bundle L can be written as

$$\sigma(\theta) = e^{iq\theta} \ (0 \leq \theta < 2\pi \text{ is local coordinate of } S^1),$$

which is well defined when q is an integer. Now choose connection ω such that section σ is parallel transported: $\nabla\sigma = 0$,

$$\omega = -iqd\theta.$$

If the first Chern class

$$c_1(L) = \frac{i}{2\pi} Tr\Omega = 0,$$

then the corresponding secondary form is

$$Q(\omega) = \frac{i}{2\pi} \int_0^1 \omega dt = \frac{q}{2\pi} d\theta \in H^1(S^1, \mathbf{R}).$$

$Q(\omega)$ is a closed 1-form which in general is not exact, due to θ is not well defined globally. Hence $Q(\omega)$ defines a non-trivial element of $H^1(S^1, \mathbf{R})$.

This example corresponds to the Bohm-Aharonov effect in $U(1)$ gauge field (electromagnetism). For a topological non-trivial space, there could be topologically non-trivial electromagnetic potential even if the electromagnetic field is zero, and the effects of which can be observed experimentally. Analogously, though curvature is usually regarded as an intrinsic geometric quantity of the manifold while connection is not, as the transformation of gauge potential (connection) is not homogeneous under a gauge transformation. The example above demonstrates that connection is an even more fundamental geometric quantity than curvature. Zero curvature simply means that the holonomy group is zero when parallel transported around a tiny closed path. Parallel transport is usually described by a non-integrable phase factor, non-vanishing field represents the local non-integrability, while non-integrability on large scale has topological characteristics. Secondary characteristic classes of connections may be needed to describe zero field (curvature). It is important to notice that the connection itself is an intrinsic geometric object but the specific gauge for the connection is non-essential.

Global and local properties of a manifold are closely related. The curvature at a point on the manifold is a local quantity. The curvature of a 2-dimensional surface in E^3 could be a constant locally, while a closed compact surface of constant curvature must be S^2. Whether a curve near a point on the manifold is a geodesic line is a local geometric question, but the number of geodesic lines between two given points on the manifold is a global invariant of the manifold. Characteristic classes (secondary characteristic classes) make use of curvature (and connection) to measure the deviation from triviality of fibre bundles. Characteristic classes (secondary characteristic classes) are local quantities, while their integration over the entire manifold represents the global topological invariants of the manifold.

10.7 Generalized Chern-Simons forms

From Eq.9.5 one known that the connection 1-form ω on the principal bundle $P(M,G)$ can be written as

$$\omega(x) = g^{-1}(x)A(x)g(x) + g^{-1}(x)dg(x),$$

where $g(x) \in G$ and

$$A(x) \equiv A^a_\mu(x)\frac{\lambda_a}{2i}dx^\mu$$

is the Lie algebra $\mathbf{g}$- valued 1-form on the base manifold, called the gauge potential or simply connection. When a covering $\mathcal{U} = \{U_\alpha\}$ of base manifold is given, the connection on the overlapping region $U_\alpha \cap U_\beta$ transforms as

$$A^\beta = g^{-1}_{\alpha\beta}A^\alpha g_{\alpha\beta} + g^{-1}_{\alpha\beta}dg_{\alpha\beta}.$$

Let $\mathcal{U} = \{A\}$ be the set of connections on $P(M,G)$, called the connection space, each point A of the connection space transforms under a gauge transformation as

$$A(x) \to A(x)^g = g^{-1}(x)A(x)g(x) + g^{-1}(x)dg(x). \tag{10.11}$$

The connection space $\mathcal{U}$ is a convex set, i.e., it contains the line segment joining any two points. So any two points $A_0, A_1 \in \mathcal{U}$ can be interpolated as

$$A_t = A_0 + t(A_1 - A_0),\ 0 \le t \le 1. \tag{10.12}$$

It is easy to see that A_t transforms the same way as Eq. 10.11

$$A_t \to A_t^g = g^{-1}A_t g + g^{-1}dg.$$

So A_t can also be chosen as connection, and is a point of connection space $\mathcal{U}$.

Next we look at the homology and cohomology on $\mathcal{U}$. Analogous to the standard k-simplex, let

$$\Gamma_k = \Gamma_k(A_0, A_1, \cdots, A_k)$$

be the k-simplex in $\mathcal{U}$, here $\{A_i\}$ are its vertices. For example

$$\begin{aligned}
&\Gamma_0 = \Gamma_0(A_0),\\
&\Gamma_1 = \Gamma_1(A_0, A_1) = \{A_t = A_0 + t(A_1 - A_0) \in \mathcal{U} | 0 \le t \le 1\},\\
&\Gamma_2 = \Gamma_2(A_0, A_1, A_2) = \{A_{ts} = A_0 + t(A_1 - A_0) + s(A_2 - A_0) \in \mathcal{U} | 0 \le s + t \le 1\}.
\end{aligned}$$

The boundary operator ∂_k acting on Γ_k can be defined as

$$\partial_k \Gamma_k = \sum_i (-1)^i \Gamma_{k-1}(A_0, \cdots, \widehat{A_i}, \cdots, A_k),$$

where $\widehat{A_i}$ represents its omission. It is easy to prove

$$\partial_{k-1} \cdot \partial_k = 0.$$

On other hand one can introduced the cochain sequence on $\mathcal{U}$, let the functional on $\mathcal{U}$

$$Q^k(A_0, A_1, \cdots, A_k)$$

is the k-cochain. The coboundary operator Δ can be defined as the alternating difference as follows

$$(\Delta Q^k)(A_0, \cdots, A_{k+1}) = \sum_{i=0}^{k+1}(-1)^i Q^k(A_0, \cdots, \widehat{A_i}, \cdots, A_k), \tag{10.13}$$

where $\widehat{A_i}$ still represents its omission. This equation can be rewritten as

$$< \Delta Q^k, \Gamma_{k+1}(A_0, \cdots, A_{k+1}) >=< Q^k, \partial\Gamma_{k+1}(A_0, \cdots, A_{k+1}) >,$$

the coboundary operator Δ maps k-cochain to $(k+1)$-cochain, and is dual to the boundary operator ∂. As $\Delta^2 = 0$, we can construct the cohomology of the connection space $\mathcal{U}$. The cohomology of connection space can be combined with the de Rham cohomology of the base manifold. And we can study the topological obstruction of the gauge group of connections on $P(M, G)$.

Step 1. From Chern class to Chern–Simons form.

Let $Q^0_{2n}(A) = Tr(F^n)$ denote the n-th Chern class (up to a constant coefficient) on $P(M, G)$, here

$$F = dA + A^2 \in \Lambda^2(M)$$

be the Lie algebra $\mathbf{g}$- valued 2-form on base manifold M, so $Q^0_{2n}(A) \in \Lambda^{2n}(M)$ is also the $\mathbf{g}$- valued $2n$-form on M.

From Eq. 10.13 and Eq.10.1

$$\begin{aligned}(\Delta Q^0_{2n})(A_0, A_1) &= Q^0_{2n}(A_1) - Q^0_{2n}(A_0) \\ &= dQ^1_{2n-1}(A_0, A_1),\end{aligned}$$

where

$$Q^1_{2n-1}(A_0, A_1) = n\int_0^1 Tr(A_1 - A_0, F_t^{n-1})dt, \tag{10.14}$$

here $F_t = dA_t + A_t^2$, $A_t = A_0 + t(A_1 - A_0)$.

It is easy to show that $Q^1_{2n-1}(A_0, A_1)$ satisfies

$$Q^1_{2n-1}(A_0^g, A_1^g) = Q^1_{2n-1}(A_0, A_1),$$

so it is well defined on $P(M,G)$ and is called the Chern-Simons form.

Step 2. From Chern-Simons form to generalized Chern-Simons forms[46].

Let $A_{ts} = A_0 + t\eta_1 + s\eta_2$, $F_{ts} = dA_{ts} + A_{ts}^2$, here

$$\eta_1 = A_1 - A_0, \quad \eta_2 = A_2 - A_1,$$

then from Eq. 10.13 and Eq. 10.14, we have

$$\begin{aligned}(\Delta Q_{2n-1}^1)(A_0, A_1, A_2) &= Q_{2n-1}^1(A_1, A_2) - Q_{2n-1}^1(A_0, A_2) \\ &\quad + Q_{2n-1}^1)(A_0, A_1) \\ &= n \int\limits_{\Gamma_1(A_1,A_2)-\Gamma_1(A_0,A_2)+\Gamma_1(A_0,A_1)} Tr(dt\eta_1 + ds\eta_2, F_{ts}^{n-1}) \\ &= n \int\limits_{\partial\Gamma_2(A_0,A_1,A_2)} Tr(dt\eta_1 + ds\eta_2, F_{ts}^{n-1}).\end{aligned}$$

So by Stoke's theorem one can obtain

$$(\Delta Q_{2n-1}^1)(A_0, A_1, A_2) = -dQ_{2n-2}^2(A_0, A_1, A_2),$$

where

$$Q_{2n-2}^2)(A_0, A_1, A_2) = -n(n-1)\int_0^1 dt \int_0^{1-t} Str(\eta_1, \eta_2, F_{ts}^{n-2})ds$$

is called the generalized Chern-Simons form. Here $Str(\,)$ denotes the symmetric trace, its definition and properties can be found in Appendix O.

The steps above could be progressed further. From the k-functional Q_{2n-k+1}^{k-1} one can obtain the $(k+1)$- functional Q_{2n-k}^k of connection space $\mathcal{U}$ as follows:

$$Q_{2n-k}^k(A_0, \cdots, A_k) = (-1)^{\frac{1}{2}k(k-1)}\frac{n!}{(n-k)!}\int\limits_{\Delta_k} Str(\eta_1, \cdots, \eta_k, F_{t_1,\cdots,t_k}^{n-k} dt_1 \cdots dt_k, \tag{10.15}$$

where

$$\begin{aligned}F_{t_1,\cdots,t_k} &= dA_{t_1,\cdots,t_k} + A_{t_1,\cdots,t_k}^2, \\ A_{t_1,\cdots,t_k} &= A_0 + \sum_{i=1}^k t_i\eta_i, \ \eta_i = A_i - A_0, \\ \Delta_k &= \{(t_1, \cdots, t_k) | 0 \le \sum_{i=1}^k \le 1\}.\end{aligned}$$

which is symmetric: $Q^k_{2n-k}(A^g_0,\cdots,A^g_k)=Q^k_{2n-k}(A_0,\cdots,A_k)$,
gauge invariant: $Q^k_{2n-k}(A_{i_0},\cdots,A_{i_k})\delta^{i_0,\cdots,i_k}_{0,\cdots,k}=Q^k_{2n-k}(A_0,\cdots,A_k)$,
and satisfies:

$$\Delta Q^{k-1}_{2n-k+1}(A_0,\cdots,A_k)=(-1)^k dQ^k_{2n-k}(A_0,\cdots,A_k),\quad (0\le k\le n),$$
$$\Delta Q^n_n(A_0,\cdots,A_{n+1})=0.$$

So we get the generalized Chern-Simons hierarchy[46]. We can further unveil the deep rooted relation between cohomology of connection space and cohomology of gauge transformations in Chapter 13, and we can analyse the non-Abelian anomaly, Wess-Zumino effective action, and Schwinger term etc., in terms of Chern-Simons hierarchy[47].

Notations and Formulae

- The axiomatic definitions of Chern class and Stiefel-Whitney class.

	Chern class on complex vector bundle	Stiefel-Whitney class on real vector bundle
1	$c_j(E)\in H^{2j}(M,\mathbf{Z})$	$w_i(E)\in H^i(M,\mathbf{Z}_2)$
2	$C(E')=f^*(C(E))$, for $E'=f^*E$	$W(E')=f^*(W(E))$, for $E'=f^*E$
3	$c_l(E\oplus F)=\sum_{j=0}^{l}c_j(E)c_{l-j}(F)$	$w_k(E\oplus F)=\sum_{i=0}^{k}w_i(E)w_{k-i}(F)$
4	$c_1(L)\neq 0$ $(\in H^2(CP(1),\mathbf{Z}))$	$w_1(\gamma^1_1)\neq 0$ $(\in H^1(RP(1),\mathbf{Z}_2))$

- Chern class:

$$C(E)=det(I+\frac{i}{2\pi}\Omega)=1+c_1(E)+c_2(E)+\cdots=\sum_{j=0}^{\infty}c_j(E).$$

$$c_1(E)=\tfrac{i}{2\pi}Tr\Omega=\tfrac{i}{2\pi}\delta^j_k\Omega^k_j,$$
$$c_2(E)=\tfrac{1}{8\pi^2}[Tr\Omega^2-(Tr\Omega)^2],$$
$$c_3(E)=\tfrac{i}{48\pi^3}[-2Tr\Omega^3+3(Tr\Omega^2)Tr\Omega-(Tr\Omega)^3],$$
$$\cdots\cdots$$

- Chern character:

$$ch(E)=Tr(exp(\frac{i}{2\pi}\Omega))=\sum_j chj(E).$$

$$ch_0(E)=k,\quad ch_1(E)=c_1(E),\quad ch_2(E)=\frac{1}{2}(c_1^2(E)-2c_2(E)).$$

- Todd class:

$$Td(E) = \prod_j \frac{\chi_i}{1-e^{\chi_j}} = \sum_j td_j(E).$$

$$td_1(E) = \frac{1}{2}c_1(E), \quad td_2(E) = \frac{1}{12}(c_1(E)^2 + c_2(E)).$$

- Pontrjagin class:

$$P(E) = det(I - \frac{1}{2\pi}\Omega) = 1 + p_1(E) + p_2(E) + \cdots.$$

$$p_j(E) = \frac{1}{(2j)!(2\pi)^{2j}}\delta^{k_1\cdots k_{2j}}_{l_1\cdots l_{2j}}\Omega^{l_1}_{k_1}\wedge\cdots\wedge\Omega^{l_{2j}}_{k_{2j}} \in H^{4j}(M,\mathbf{Z}).$$

- Hirzebruch L-sequence:

$$L(E) = \prod_{j=1}^{k}\frac{\chi_j}{tanh\chi_j}.$$

- A-roof sequence:

$$\hat{A}\,(E) = \prod_{j=1}^{k}\frac{\chi_j}{\sinh\chi_j/2}.$$

- Euler class (rank E = 2l):

$$e(M) = \frac{1}{2^l(2\pi)^l l!}\epsilon_{a_1a_2\cdots a_{2l}}\Omega^{a_1a_2}\wedge\cdots\Omega^{a_{2l-1}a_{2l}}.$$

- generalized Chern Simons form:

$$Q^k_{2n-k}(A_0,\cdots,A_k) = (-1)^{\frac{1}{2}k(k-1)}\frac{n!}{(n-k)!}\int_{\Delta_k} Str(\eta_1,\cdots,\eta_k,F^{n-k}_{t_1,\cdots,t_k}dt_1\cdots dt_k,$$

where

$$F_{t_1,\cdots,t_k} = dA_{t_1,\cdots,t_k} + A^2_{t_1,\cdots,t_k},$$

$$A_{t_1,\cdots,t_k} = A_0 + \sum_{i=1}^{k} t_i\eta_i, \ \eta_i = A_i - A_0$$

$$\Delta_k = \{(t_1,\cdots,t_k)|0 \le \sum_{i=1}^{k} \le 1\},$$

which is symmetric: $Q^k_{2n-k}(A^g_0,\cdots,A^g_k) = Q^k_{2n-k}(A_0,\cdots,A_k)$,
gauge invariant: $Q^k_{2n-k}(A_{i_0},\cdots,A_{i_k})\delta^{i_0,\cdots,i_k}_{0,\cdots,k} = Q^k_{2n-k}(A_0,\cdots,A_k)$
and satisfies:

$$\Delta Q^{k-1}_{2n-k+1}(A_0,\cdots,A_k) = (-1)^k dQ^k_{2n-k}(A_0,\cdots,A_k), \quad (0 \le k \le n),$$
$$\Delta Q^n_n(A_0,\cdots,A_{n+1}) = 0.$$

Exercise

1. Show that $Tr\Omega$ is a closed form, and $(Tr\Omega - Tr\Omega')$ is an exact form.

2. Derive the total Chern class of the tangent bundle $T(S^2)$, and find the Chern number and the Euler number.

3. Derive the total Chern class of CP(n), find its Euler number.

Chapter 11

The Atiyah-Singer Index Theorem

11.1 Introduction, Euler number and the associated theorem

In the quantization of a classical field, one has to consider the spectrum of quantum fluctuations about a stationary point of Euclidean domain. And in mathematical terms, it is the problem about the determination of the eigenvalues of an elliptic differential operator. The eigenvalue distribution (spectrum) reflects the topological and metric features of the underlying system, just as the shape of a drum and the thickness of its membrane determines the sound. But conversely, "Can you hear the shape of the drum?" The relation between the analytic properties of differential operators and the topological and metric features of the underlying fiber bundle on which the operators act, can indeed be studied, and this is the Atiyah-Singer index theorem, which can be put as

$$I_a(D) = I_t(D),$$

$$\text{or} \qquad Ind(D) = Ind(\sigma(D)), \tag{11.1}$$

i.e., the analytic index and the topological index coincide. On the left hand side, the analytic index of the elliptic differential operator is an object of infinite dimensional functional analysis, which is related to the existence and uniqueness of solutions of certain partial differential equations, is well defined and quite stable, in the sense that it remains unchanged during continuous perturbation of the operator, about which we will study in the §11.2. On the right hand side, the topological index of the operator is defined via homotopy theory by the global behavior of the symbol $\sigma(D)$ of operator D. In differential geometry, we consider only the essential operator, which is a geometric object, i.e., a linear map from one smooth cross section of fibre bundle to another section. Symbol $\sigma(D)$ of an operator (see next section) is defined locally by the data of the linear algebra of finite rank vector bundle, which is related to the topology of the base manifold and the winding number of the vector bundle. And the topological index depends on the homotopy class of symbol $\sigma(D)$ only. From the asymptotic expansion of the heat kernel, one can further obtain the topological index in terms of cohomology, which is closely related to the renormalization of quantum field theory. The topological index can be written as the integration of the corresponding characteristic classes of the fibre bundle under consideration, and we will take a look at this in §11.3.

de Rham cohomology and harmonic analysis demonstrate that the topology of a manifold can be studied via analysis. Atiyah-Singer index theorem demonstrates further that the analytic index of differential operators can be obtained via topological method, and can be written as an integration of relevant characteristic classes over the tangent bundle involved. Index theorem has since found many applications in physics. For example, when we study the interaction between fermion and gauge field, the spectrum of fermion (left hand side of Eq. 11.1) is related to the global topological feature of the underlying gauge field (right hand side of Eq. 11.1). The index theorem give a relation between analytical properties of the field theory and topological properties of the field configuration.

First let us recall some well known facts and theorems concerning the Euler number. As is the case, Euler number is the origin and starting point for many problems in geometry. Euler number and the associated theorem may be regarded as a prototype of index theorem, and it can be used as an example to interpret the meaning of the two sides of Eq. 11.1.

(1). The topology of manifold determines the permissible triangulations. There are certain invariants independent of the triangulations, and Euler number is one of the most important one of these. For a triangulated manifold M,

$$\chi = \sum_n (-1)^k d_k,$$

where d_k is the number of k-dimensional cells in the triangulation. This is the combinatorial topological meaning of Euler number.

(2). For a compact Riemann surface, the Euler number

$$\chi = 2 - 2g,$$

where g denotes the genus of the surface, which is a topological invariant.

(3). In terms of homology group of the manifold,

$$H_k(M) = Z_k(M)/B_k(M),$$
$$\chi = \sum_k (-1)^k dim H_k = \sum_k (-1)^k b_k,$$

where $b_k = dim H_k(M)$ is the k-th Betti number of the manifold. This approach gives the algebraic topological meaning of Euler number.

(4). From de Rham theory one has

$$b_k = dim H^k_{dR}(M),$$

$$\chi = \sum_k (-1)^k dim H^k_{dR}(M). \tag{11.2}$$

This is the differential topological meaning of Euler number.

(5). From Hodge theorem

$$dim H^k_{Harm}(M) = dim H^k_{dR}(M),$$

of which the Harmonic forms are the zero modes of Laplacian, so

$$\chi = \sum_k (-1)^k dim\ ker \Delta_k = d_e - d_o, \tag{11.3}$$

where d_e (d_o) is the number of even (odd) harmonic forms. This is the "elliptic topological" meaning of Euler number.

(6). From Gauss-Bonnet theorem

$$\chi(M) = \frac{1}{2\pi}\int_M \Omega,$$

where Ω is the curvature 2-form of surface M.

(7). From Hopf-Poincare theorem

$$\chi = \sum_{x_o \in Zero(v)} j(v, x_o),$$

where v is the continuous tangent field of the 2-dim closed smooth surface, $x_o \in M$ are the isolated zero points ($v(x_o) = 0$). $j(v, x_o)$ are the winding number of v at x_o, i.e., the orientation of a vector field is rotated by an integral multiple $j(v, x_o)$ of 2π when the position of v transported along a closed path around the point x_o.

From (1)-(3), we see that Euler number is a topological invariant; (4)-(5) show that Euler number is related to the analytic index of de Rham operator d and Laplacian Δ; Gauss-Bonnet theorem (6) is the simplest prototype of Atiyah-Singer index theorem, which relates Euler number to an integration of characteric form. Hopf-Poincare theorem is the simplest prototype of the G-index theorem, using the pointwise local index to express the topological index. The G-index theorem is a generalization of the ordinary Atiyah-Singer index theorem.

11.2 Elliptic differential operator, elliptic complex and its analytic index

Let M be a compact smooth manifold without boundary of dimension n, E and F be vector bundles over M, and $\Gamma(E)$ denote the space of sections of rank-m vector bundle E. For point x of M having a neighborhood with local coordinates $x = \{x_i\}_1^n$ and local trivialization, $\Gamma(E)$ can be expressed as

$$\Gamma(E) = \{(x;\ \xi^1(x), \cdots, \xi^m(x))\}$$

analogously, $\Gamma(F) = \{(x;\ \eta^1(x), \cdots, \eta^l(x))\}$ is the space of sections of rank-l vector bundle F.

A differential operator of order ν on M is a linear map between the cross section of vector bundles

$$D : \Gamma(E) \to \Gamma(F)$$

which can be written as

$$\begin{aligned} &D(\xi^1(x), \cdots, \xi^m(x)) = (\eta^1(x), \cdots, \eta^l(x)), \\ &\eta^i(x) = \sum_j a^i_{j;\mu_1,\cdots,\mu_n(x)} \frac{\partial^{\mu_1+\cdots+\mu_n}}{\partial x_1^{\mu_1} \cdots \partial x_n^{\mu_n}} \xi^j(x), \\ &\mu_1 + \cdots + \mu_n = \mu \le \nu. \end{aligned} \tag{11.4}$$

Operator D can be written in an $l \times m$ matrix form

$$D = \sum_{\mu \le \nu} a_{\mu_1\cdots\mu_n}(x) \frac{\partial^{\mu_1+\cdots+\mu_n}}{\partial x_1^{\mu_1} \cdots \partial x_n^{\mu_n}},$$

where $a_{\mu_1\cdots\mu_n}$ are matrix valued. Under Fourier transformation,

$$\xi(x) = \int e^{ikx} \hat{\xi}(k) dk,$$

Eq. 11.4 becomes

$$\eta = D\xi = \int e^{ikx} \sigma(D) \hat{\xi}(k) dk,$$

where $\sigma(D) = \sum_{\mu \le \nu} a_{\mu_1\cdots\mu_n}(x) k_1^{\mu_1} \cdots k_n^{\mu_n} (i)^\mu$, $k = (k_1, \cdots, k_n) \in C^n$ is called the symbol of operator D.

The leading symbol $\hat{\sigma_D}(x,k)$ of D is the highest order part of its Fourier transform (For simplicity, the factor $(i)^\nu$ is omitted.)

$$\hat{\sigma_D}(x,k) = \sum_{\mu=\nu} a_{\mu_1\cdots\mu_n} k_1^{\mu_1} \cdots k_n^{\mu_n},$$

which represents a section of bundle $(\odot T^\nu M) \otimes Hom(E, F)$, here (x, k) is the local coordinate for cotangent bundle $T^*(M)$ and $\odot$ denotes symmetric tensor product. The symmetric tensor space $\odot^\nu V$ is isomorphic to the space of homogeneous polynomial function of degree ν on V^*. The leading symbol represents a linear map from E to F, D is said to be an elliptic operator if $\hat{\sigma_D}(x,k)$ is invertible for $k \ne 0$.

Example 1. Let $M = R^3$. Operators $grad(\nabla)$, $curl(\nabla\times)$, and $Div(\nabla\cdot)$ have symbols:

$$\begin{pmatrix} k_1 \\ k_2 \\ k_3 \end{pmatrix}, \quad \begin{pmatrix} 0 & -k_3 & k_2 \\ k_3 & 0 & -k_1 \\ -k_2 & k_1 & 0 \end{pmatrix}, \quad (k_1, k_2, k_3).$$

None of these operators is elliptic.

Example 2. The linear map between plane bundle over 2-dim manifold

$$\eta^1(x_1,x_2) = \frac{\partial\xi^1}{\partial x_1} - \frac{\partial\xi^2}{\partial x_2},$$
$$\eta^2(x_1,x_2) = \frac{\partial\xi^1}{\partial x_2} + \frac{\partial\xi^2}{\partial x_1},$$

has the symbol

$$\sigma(x,k) = \begin{pmatrix} k_1 & -k_2 \\ k_2 & k_1 \end{pmatrix}$$

with $det\ \sigma \neq 0$ for $k \neq 0$. Hence this is an elliptic operator. It is called Cauchy-Riemann operator.

Example 3. The map between line bundles over 2-dim manifold

$$\eta^1(x_1,x_2) = (\frac{\partial^2}{\partial x_1^2} + \frac{\partial^2}{\partial x_2^2})\xi(x_1,x_2)$$

has symbol

$$\sigma(x_1,x_2) = k_1^2 + k_2^2$$

which is invertible for $k \neq 0$, so this is an elliptic operator, called the Laplacian.

Now let us recall some basic concepts about linear differential operator.

The space of zero mode solution of the operator is called the kernel of the operator

$$Ker\ D = \{\xi \in \Gamma(E) | D\xi = 0\}.$$

The image of the operator is defined as

$$Im\ D = \{\eta \in \Gamma(F) | \eta = D\xi\ \forall \xi \in \Gamma(E)\}.$$

And the cokernel of the operator is

$$Coker\ D = \Gamma(F)/Im\ D.$$

Their relations are illustrated in Fig. 11.1.

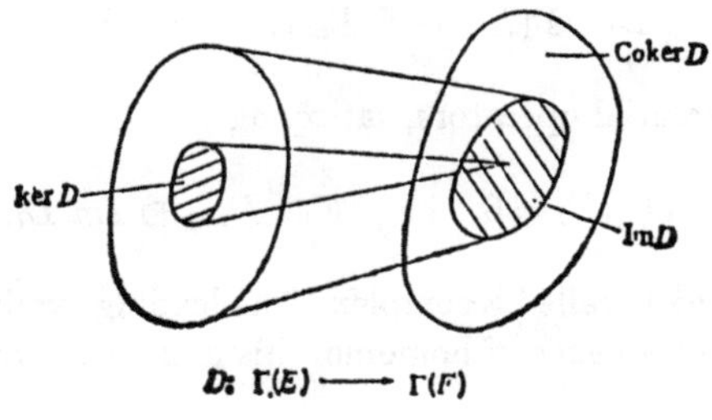

Fig. 11.1

From functional analysis we know that for an elliptic operator D on a compact manifold, the linear space $Ker\ D$ and $Coker\ D$ are finite dimensional, this operator D is called Fredholm operator.

For Fredholm operator D, $Ker\ D$ is finite dimensional, i.e., the homogeneous equation

$$D\xi = 0$$

has finite number of linearly independent solutions. On the other hand, as $Coker\ D$ is finite dimensional, thus the non-homogeneous equation

$$D\xi = \eta$$

has solution if η satisfies finite number of linear conditions. Note that these are two different problem: the former is the problem of uniqueness of the solution, and the later is the problem of existence of the solution.

One can define the analytic index of D as

$$Ind\ D = dim\ Ker\ D - dim\ Coker\ D.$$

$dim\ Ker\ D$ and $dim\ Coker\ D$ are important features of an elliptic operator. When there is Hermitian metric on the space of sections, we can define the adjoint operator D^+ of D, which satisfies

$$(D\xi, \eta) = (\xi, D^+\eta),$$

D^+ is a linear map from $\Gamma(F)$ to $\Gamma(E)$. Linear space $Ker\ D^+$ and $Coker\ D$ are of the same dimension

$$dimKer\ D^+ = dimCoker\ D.$$

So the analytic index can be written as

$$Ind\ D = dimKer\ D - dimKer\ D^+.$$

Frequently one has to consider a finite sequence of differential operators. Let $\{E_l\}$ be a finite sequence of vector bundle over M and

$$D_l:\ \Gamma(E_l) \to \Gamma(E_{l+1}),\quad l = 0, 1, \cdots$$

be a sequence of differential operators, satisfying

$$D_{l+1}D_l = 0,\quad \text{i.e.,}\quad Ker\ D_{l+1} \supset Im\ D_l. \tag{11.5}$$

Then sequence $\{E_l, D_l\}$ is called a complex. The leading symbols associated to operators D_l give rise to a sequence of homomorphism of vector bundles

$$\sigma_l(x, k):\ E_l \to E_{l+1}.$$

Complex $\{E_l, D_l\}$ is called elliptic if the associated symbol sequence is exact for all $k \in R^n - \{0\}$, i.e.,

$$Ker\ \sigma_{l+1}(x,k) = Im\ \sigma_l(x,k), \quad k \neq 0. \tag{11.6}$$

As a result of Eq. 11.5 and Eq. 11.6, we can define cohomology groups for the elliptic complex $(E, D) \equiv \{E_l, D_l\}$ by

$$H^l(E,D) = Ker\ D_l / Im\ D_{l-1}.$$

These cohomology groups are finite dimensional, and the index of the elliptic complex (E, D) is defined as

$$Ind\ (E,D) = \sum_l (-1)^l dim H^l(E,D).$$

As an example, let $E_l = \Lambda^l(M)$ be the l-order differential forms on M, and $D_l = d$ be the exterior differential operator on l-forms. As $d^2 = 0$, complex $(\Lambda(M), d)$ is called de Rham complex, and it is an elliptic complex

$$H^l(E,D) = H^l_{dR}(M,R).$$

Comparing with Eq. 11.2, the index of de Rham complex is the Euler characteristic number $\chi(M)$ of the manifold M

$$Index(\Lambda(M),d) = \sum_l (-1)^l dim\ H^l(\Lambda(M),d) = \chi(M).$$

The adjoint operators D_l^+ can be defined when there is Hermitian metric on the bundle

$$\begin{aligned} (D_l\xi_l, \xi_{l+1}) &= (\xi_l, D_l^+\xi_{l+1}), \\ D_l^+ : \Gamma(E_{l+1}) &\to \Gamma(E_l). \end{aligned}$$

Decomposing the complex into even and odd part

$$E_{even} = \oplus \sum_k E_{2k}, \quad E_{odd} = \oplus \sum_k E_{2k+1},$$

and consider the operators

$$\begin{aligned} \mathcal{D} &= \oplus \sum_k (D_{2k} + D^+_{2k-1}) : \quad \Gamma(E_{even}) \to \Gamma(E_{odd}), \\ \mathcal{D}^+ &= \oplus \sum_k (D^+_{2k} + D_{2k-1}) : \quad \Gamma(E_{odd}) \to \Gamma(E_{even}). \end{aligned}$$

The associated Laplacians are

$$\begin{aligned} \Box_{even} &= \mathcal{D}^+\mathcal{D} = \oplus \sum_k (D^+_{2k}D_{2k} + D_{2k-1}D^+_{2k-1}) = \oplus \sum_k \Delta_{2k}, \\ \Box_{odd} &= \mathcal{D}\mathcal{D}^+ = \oplus \sum_k (D^+_{2k-1}D_{2k-1} + D_{2k}D^+_{2k}) = \oplus \sum_k \Delta_{2k-1}. \end{aligned}$$

Hence

$$\begin{aligned} Index(E, \mathcal{D}) &= dim\ Ker\Box_{even} - dim\ Ker\Box_{odd} \\ &= \sum_k (-1)^k dim\ Ker\Delta_k = ind(E, D). \end{aligned}$$

For instance, in Eq. 11.3, the Euler characteristic number is the sum of even Betti numbers minus the sum of odd Betti numbers, i.e., for de Rham complex $(\Lambda(M), d)$, its analytic index can be expressed as Euler characteristic

$$Ind(\Lambda, d) = \sum_k (-1)^k dim_C H^k_{dR}(T(M) \otimes C) = \sum_k (-1)^k b_k.$$

11.3 Atiyah-Singer index theorem, the symbol bundle and its topological index

Differential operator we encounter is a linear map between smooth cross sections of vector bundle. The leading symbol $\sigma_D(x, k)$ of operator D is defined by the data of linear algebra of finite rank vector bundle, which represents a section of bundle $(\odot T^\nu M) \times Hom(E, F)$. Now let us have a closer look at it. Let (x, k) be the local coordinate for cotangent bundle $T^*(M)$, and B be the unit disk bundle in $T^*(M)$:

$$B = \{(x, k),\ |k|^2 < 1\}.$$

Now take two copies $B^\pm$ of unit disk bundle and glue them together along their common boundary, and we have a unit sphere bundle ψ over M, which is homeomorphic to the one point compactification of cotangent bundle T^*M. Let ρ be the projection:

$$\rho : \ \psi \to M,$$

and $\rho_\pm$ be the restriction of ρ to the hemisphere bundle $B^\pm$:

$$\rho_\pm : \ B^\pm \to M.$$

For the operator

$$\begin{aligned} D : \ \Gamma(E) &\to \Gamma(F), \\ v &\to Dv. \end{aligned}$$

Symbol $\sigma_D(x, k)$ identifies pair $(k(x), v(x))$ with pair $(k(x), \sigma_D(x, k)v(x))$.

Consider the pull back bundles

$$\begin{aligned} \mathcal{F} = \rho^*_+(F) &\quad \text{over } B^+, \\ \mathcal{E} = \rho^*_-(E) &\quad \text{over } B^-, \end{aligned}$$

we want to glue them together and look upon $\sigma_D(x,k)$ as a map from $\mathcal{E}$ to $\mathcal{F}$ over $B^- \cap B^+$. This map is an isomorphism over $B^- \cap B^+$ if D is elliptic. Thus one has a vector bundle $\Sigma(D)$ over Ψ, called the symbol bundle,

$$\Sigma(D) \in K(\Psi) = K_{cpr}(T^*M),$$

i.e., the leading symbol of the operator defines an equivalence class of compact K-theory of cotangent bundle T^*M (see Appendix P).

For any manifold M, the tangent bundle TM is an almost complex manifold, i.e., let

$$\pi : TM \to M,$$
$$T(TM) = \pi^*(TM) \oplus \pi^*(TM) \simeq \pi^*(TM) \otimes C.$$

Atiyah-Singer index theorem is stated as follows:

Theorem 11.1 *(Atiyah-Singer): Let M be a compact smooth manifold and D be an elliptic operator over M, $\Sigma(D)$ is its symbol bundle, then*

$$Ind\ D = (-1)^n (ch(\Sigma(D)) td(TM \otimes C))[TM],$$

here $[TM]$ denotes the integral over the entire TM. Specifically, $[TM]$ indicates that one has to extract the part of degree $dim TM$ and integrate it over the entire tangent bundle TM. Then through the integration along the fibre (see App.P), this formula can also be written as

$$Ind\ D = (-1)^{\frac{n(n+1)}{2}} \frac{(chE - chF) td(TM \otimes C)}{e(TM)} [M].$$

Factor $(-1)^{\frac{n(n+1)}{2}}$ compensates for the difference between the orientation on TM induced by the one on M. Because if $(e_1, \cdots, e_n)$ is a basis of TM, then TM has a canonical orientation, the oriented basis of $T(TM)$ can be chosen as

$$(e_1, e_1', e_2, e_2' \cdots, e_n, e_n'), \quad \text{where} \quad e_k' = Je_k.$$

J carries the horizontal to the vertical factor. In this case, an arbitrary permutation of the frames in M produces an even permutation in TM, so it gives rise to a global orientation.

For elliptic complex (E,D) one get

$$Ind\ (D,E) = (-1)^{\frac{n(n+1)}{2}} \frac{ch(\sum(-1)^i E_i) td(TM \otimes C)}{e(TM)} [M].$$

In the equations above, ch, td and e are the characteristic classes (Chern charecter, Todd and Eular respectively), which are all even order closed differential forms on base manifold M. It shows that the topological index can be expressed by integrating the corresponding characteristic classes of the fibre bundles involved.

Following let us take the de Rham complex $(\Lambda(M), d)$ again as an example, from §11.2 and Eq. 11.2 and Eq. 11.3, we have

$$Ind(\Lambda, d) = \sum_k (-1)^k b_k = \chi(M).$$

Now from Atiyah-Singer index theorem

$$Ind(\Lambda, d) = (-1)^{\frac{n(n+1)}{2}} \int_M \frac{\sum_k (-1)^k ch(\Lambda^k T(M) \otimes C) td(T(M) \otimes C)}{e(TM)}$$

to simplify the right hand side of the equation, use the splitting principle

$$T(M) \otimes C = \oplus \sum_{k=1}^{l} (L_k + \bar{L}_k), \quad dim\ M = n = 2l.$$

Let $\chi_k = c_1(L_k)$, $c_1(\bar{L}_k) = -\chi_k$, then

$$\begin{aligned} td(T(M) \otimes C) &= \prod_{i=1}^{l} \frac{\chi_i}{1 - e^{-\chi_i}} \frac{-\chi_i}{1 - e^{\chi_i}} \\ e(T(M)) &= \prod_{i=1}^{l} \chi_i. \end{aligned}$$

As $ch(L_j) = e^{\chi_j}$, $ch(L_j \otimes L_k) = e^{\chi_j} e^{\chi_k}$, and

$$\begin{aligned} \sum_{k=1}^{n=2l} (-1)^k ch(\Lambda^k T(M) \otimes C) &= 1 - \sum_{j=1}^{n} e^{\chi_j} + \sum_i \sum_j e^{\chi_i} e^{\chi_j} - \cdots \\ &= \prod_{j=1}^{2l} (1 - e^{\chi_j}) = \prod_{j=1}^{l} (1 - e^{\chi_j})(1 - e^{-\chi_j}), \end{aligned}$$

so

$$\begin{aligned} Ind(\Lambda, d) &= (-1)^{\frac{n(n+1)}{2}} \int_M \frac{\sum_k (-1)^k ch(\Lambda^k T(M) \otimes C) td(T(M) \otimes C)}{e(TM)} \\ &= \int_M (-1)^l \frac{\prod_{j=1}^{l} (1 - e^{\chi_j})(1 - e^{-\chi_j})}{\prod_{i=1}^{l} \chi_i} \prod_{k=1}^{l} \frac{\chi_k}{1 - e^{-\chi_k}} \frac{-\chi_k}{1 - e^{\chi_k}} \\ &= \int_M \prod_{i=1}^{l} \chi_i = \int_M e(T(M)). \end{aligned}$$

Hence Atiyah-Singer index theorem for de Rham complex is just the Gauss-Bonnet theorem.

11.4 Other classical elliptic complex

An operator D is called classical if its principal symbol is homogeneous outside of some compact subset of T^*M. There are in general four types of classical elliptic complexes without boundary: de Rham, signature, Dolbault, and spin complexes. In the previous section we look at de Rham complex and the corresponding Gauss-Bonnet theorem, which related the Euler class and the Euler number. In this section we look at the other three classical elliptic complexes and the corresponding index theorems.

(1). The signature complex.

Let M be a compact oriented Riemannian manifold of even dimension $n = 2l$. One can define inner product (,) for middle cohomology group $H^l(M, \mathbf{R})$ as:

$$(\theta, \omega) \equiv \int_M \theta \wedge \omega, \qquad \theta, \omega \in H^l(M, \mathbf{R})$$

which is non-degenerate as a result of Poincare duality, i.e., for any $\theta \neq 0$, there is an ω such that $(\theta, \omega) \neq 0$. This inner product is symmetric if l is even. And the signature $\sigma(M)$ is then defined as the signature of non-degenerate bi-linear symmetric form, i.e., the number of positive eigenvalues minus the number of negative eigenvalues. This inner product is skew symmetric if l is odd, and signature $\sigma(M)$ vanishes automatically.

In what follows we focus our attention on the case when l is even, i.e., the dimension n of the manifold is divisible by 4,

$$n = 4k, \qquad k \in \mathbf{Z}$$

In terms of Hodge * operator, one can define an operator σ acting on p-form as:

$$\sigma = i^{p(p-1)+2k} * .$$

It is easy to show that

$$\sigma d = -\delta\sigma, \quad \sigma\delta = -d\sigma.$$

So σ anticommutes with $(d + \delta)$:

$$\sigma(d + \delta) = -(d + \delta)\sigma$$

and $\sigma^2 = 1$. Let $\Lambda^{\pm}$ be the $\pm$ eigenspaces of σ, so $\Lambda^*(M)$ decomposes into $\Lambda^{\pm}$. As σ anticommutes with $d + \delta$, so $d + \delta$ maps $\Lambda^{\pm}$ to $\Lambda^{\mp}$ and one obtains a signature complex $(\Lambda^{\pm}, d + \delta)$. We consider its analytic index first. It is easy to show that contributions of harmonic forms with eigenvalue $\pm$ of σ cancel each other except in the middle dimension. As $\sigma \sim *$, the eigenspace of σ must be a linear combination of Λ_p and Λ_{n-p}, let

$$\omega = \theta_p + \theta_{n-p} \in \Lambda^+$$

i.e., $\sigma(\theta_p+\theta_{n-p}) = (\theta_p+\theta_{n-p})$ thus

$$\sigma\theta_p = \theta_{n-p}, \quad \sigma\theta_{n-p} = \theta_p,$$

so $\omega' = \theta_p - \theta_{n-p} \in \Lambda^-$. Namely, except for $p = n/2 = 2k$, the eigenvectors of $\pm$ appear in pair, their contributions cancel and consequently

$$\begin{aligned} Ind(\Lambda^\pm, d+\delta) &= dim\ H_+(M,\mathbf{R}) - dim\ H_-(M,\mathbf{R}) \\ &= dim\ H_+^{2k}(M,\mathbf{R}) - dim\ H_-^{2k}(M,\mathbf{R}). \end{aligned}$$

For the middle cohomology one has

$$\sigma = * \quad (p = 2k).$$

Space $H^{2k}(M,\mathbf{R})$ decomposes into $H_\pm^{2k}(M,\mathbf{R})$, which are the eigensapces of $*$ with eigenvalues $\pm$. Let

$$b_{2k}^\pm = dim\ H_\pm^{2k}(M,\mathbf{R}),$$

so the Betti number of middle cohomology $H^{2k}(M,\mathbf{R})$ can be split as

$$b_{2k} = b_{2k}^+ + b_{2k}^-,$$

and the analytic index of signature complex is

$$\begin{aligned} \sigma(M) &= Ind(\Lambda^\pm, d+\delta) \\ &= dim\ H_+^{2k}(M,\mathbf{R}) - dim\ H_-^{2k}(M,\mathbf{R}) \\ &= b_{2k}^+ - b_{2k}^-. \end{aligned}$$

Now the topological index, i.e., the right hand side of the index formula, can be reduced to

$$Ind(\Lambda^\pm, d+\delta) = (-1)^l \int_M \frac{ch(\Lambda^+ - \Lambda^-)}{e(TM)} td(T(M)\otimes C).$$

To simplify it further, use the splitting principle

$$ch(\Lambda^+ - \Lambda^-)(TM\otimes C) = \prod_{i=1}^{l} ch(\Lambda^+ - \Lambda^-)(L_i) = \prod_{i=1}^{l}(e^{-\chi_i} - e^{\chi_i}).$$

Substitute it into the right hand side of index formula above gives

$$\begin{aligned} (-1)^l \frac{ch(\Lambda^+ - \Lambda^-)}{e(TM)} td(T(M)\otimes C) &= (-1)^l \frac{\prod_{i=1}^{l}(e^{-\chi_i} - e^{\chi_i})}{\prod_{i=1}^{l}\chi_i} \frac{\prod_{i=1}^{l}\chi_i(-\chi_i)}{\prod_{i=1}^{l}(1-e^{\chi_i})(1-e^{-\chi_i})} \\ &= \frac{\prod_{i=1}^{l}\chi_i(e^{-\chi_i} - e^{\chi_i})}{\prod_{i=1}^{l}(1-e^{\chi_i})(1-e^{-\chi_i})} \\ &= \prod_{i=1}^{l}\frac{\chi_i}{tanh\frac{\chi_i}{2}} = 2^l \prod_{i=1}^{l}\frac{\chi_i/2}{tanh\frac{\chi_i}{2}}. \end{aligned}$$

So

$$Ind(\Lambda^{\pm}, d+\delta) = \int_M 2^l \prod_{i=1}^{l} \frac{\chi_i/2}{tanh\frac{\chi_i}{2}} = \int_M \prod_{i=1}^{l} \frac{\chi_i}{tanh\frac{\chi_i}{2}} \equiv \int_M L(M).$$

This is the classical Hirzebruch signature theorem. Where the last equality comes from the fact that the cohomology classes in question have the same order for integral over M of dimension $n = 2l$. It is easy to show from Poincare dual that

$$\begin{aligned} \chi(M) &= b_{2k} \quad mod\ 2 \\ &= \sigma(M)\ mod\ 2. \end{aligned}$$

Example For $M = CP(2k)$,

$$\chi(M) = 2k+1, \quad b_{2k} = b_{2k}^{+} = 1,$$

then $$\sigma(M) = 1,$$

2. Dolbault complex.

Let M be a compact complex manifold of real dimension n (complex dimension $l = n/2$), $T = TM$ be its tangent bundle, let T_C denote the complex tangent bundle of M and $\bar{T_C}$ denote the conjugate bundle. There is the following isomorphism

$$T \otimes C \sim T_C \oplus \bar{T_C},$$

and the complex space $\Lambda^k T^* \otimes C$ decomposes into a sum as $\oplus \sum_{p+q=k} \Lambda^{p,q}$, where

$$\Lambda^{p,q} \sim \Lambda^p T_C^* \otimes \Lambda^q \bar{T}_C^*$$

is called the space of smooth forms of type (p, q). The exterior operator on each $\Lambda^{p,q}$ is

$$d = \partial_{p \cdot q} + \bar{\partial}_{p \cdot q},$$

in which $\bar{\partial} = d\bar{z}_i \frac{\partial}{\partial \bar{z}^i}$ is a mapping

$$\begin{aligned} \bar{\partial} : C^{\infty}(\Lambda^{p,q}) &\rightarrow C^{\infty}(\Lambda^{p,q+1}), \\ \bar{\partial}\alpha_{p,q} &= \alpha_{p,q+1}, \quad \alpha_{p,q} \in \Lambda^{p,q}(M). \end{aligned}$$

So the sequence

$$\rightarrow \Lambda^{0,q} \xrightarrow{\bar{\partial}_q} \Lambda^{0,q+1} \xrightarrow{\bar{\partial}_{q+1}} \cdots$$

is called the Dolbault complex, written as $(\Lambda^{0,*}, \bar{\partial})$. Its analytic index is

$$Ind(\Lambda^{0,*}, \bar{\partial}) = \sum_{q=0}^{l} (-1)^q dim\ H^{0,q}(M),$$

where $H^{0,q}(M) = ker\bar{\partial}_q / Im\bar{\partial}_{q-1}$ is the cohomology group of $\bar{\partial}$ on $C^{\infty}(\Lambda^{0,q}(M))$.

Now we turn to the right hand side of the index formula, the topological index, it can be written as

$$(-1)^l \int_M \frac{ch(\sum_{q=1}^{l}(-1)^q \Lambda^{0,q}\bar{T}_C^*)td(T\otimes C)}{e(TM)},$$

and according to the splitting principle

$$ch(\sum_{q=1}^{l}\Lambda^{0,q}) = \prod_{j=1}^{l}(1-e^{\chi_j}),$$

substitute this in the previous equation gives

$$\begin{aligned}(-1)^l \frac{ch(\sum_{q=1}^{l}\Lambda^{0,q})td(T\otimes C)}{e(TM)} &= (-1)^l \frac{\prod_{i=1}^{l}(1-e^{\chi_j})}{\prod_{i=1}^{l}\chi_i} \frac{\prod_{i=1}^{l}\chi_i(-\chi_i)}{\prod_{i=1}^{l}(1-e^{\chi_i})(1-e^{-\chi_i})} \\ &= \prod_{i=1}^{l}\frac{\chi_i}{1-e^{\chi_j}}.\end{aligned}$$

Thus one obtain

$$Ind(\Lambda^{0,*},\bar{\partial}) = \int_M \prod_{i=1}^{l}\frac{\chi_i}{1-e^{-\chi_i}} = \int_M td(T_C M).$$

The analytic index of Dolbault complex is also called the arithmetic genus of the manifold. In the special cases of $n=1$ and $n=4$, as

$$td(T_C(M)) = 1 + \frac{1}{2}c_1 + \frac{1}{12}(c_2 + c_1^2) + \cdots,$$

it is easy to show that the arithmetic genus is related to the Euler characteristic $\chi(M)$ and signature $\sigma(M)$ as follows

$$\begin{aligned} n=2, &\quad Index(\bar{\partial}) = \frac{1}{2}\chi(M), \\ n=4, &\quad Index(\bar{\partial}) = \frac{1}{4}(\chi(M)+\sigma(M)). \end{aligned}$$

Consequently one can show that certain manifolds do not admit complex structures. For instance, for $M = S^4$, $\chi(S^4) = 2$, $\sigma(S^4) = 0$, then $Ind(\bar{\partial}) = \frac{1}{4}(2+0) = \frac{1}{2}$ is not an integer, so S^4 is not a complex manifold.

Manifold	Euler number $\chi(M)$	Signature $\sigma(M)$	$Ind(\delta)$
S^{2k}	2	0	-
S^{2k+1}	0	-	-
$S^1 \times S^1$	0	-	0
$CP(2k)$	$2k+1$	1	1
$CP(2k+1)$	$2k+2$	-	1
Footnote	Sum of index of vector at zeros. $\chi = 0$ for n odd.	Difference of S-D and anti-S-D harm. form $\sigma = 0$ for $n \neq 4k$	Only on complex manifold

3. Spin complex.

Spin complex could very well be the most important classical elliptic complex. One can use Clifford algebra approach to show that spin complex can be interpreted as the square root of de Rham complex.

Dirac operator D is defined, by using the covariant derivative with respect to the basis of orthonomal frame vierbein $\{e^\mu_a(x)\}$ as

$$\begin{aligned} D &= \gamma^a e^\mu_a(x) D_\mu \\ &= \gamma^a e^\mu_a(x)(\frac{\partial}{\partial x^\mu} + \frac{1}{4}[\gamma_b, \gamma_c]\omega^{bc}_\mu(x)). \end{aligned}$$

From now on, we restrict our attention to a 4-dimensional Riemannian spin manifold M, $\{\gamma^a\}$ are the 4×4 Dirac matrices satisfying

$$\{\gamma^a, \gamma^b\} \equiv \gamma^a\gamma^b + \gamma^b\gamma^a = 2g^{ab},$$

$e_a(x) = e^\mu_a(x)\frac{\partial}{\partial x^\mu}$ are the orthonomal frame fields, and $\omega^{bc}(x) = \omega^{bc}_\mu(x)dx^\mu$ are spin connection 1-forms.

It is easy to show that (Weitzenböck formula) (see Eq.9.24)

$$D^+D = DD^+ = -g^{\mu\nu}D_\mu D_\nu + \frac{1}{16}[\gamma_a, \gamma_b][\gamma_c, \gamma_d]R^{abcd},$$

so the leading part of the operator is elliptic for Riemannian metric.

Using $\gamma_5 \equiv \gamma^0\gamma^1\gamma^2\gamma^3$, which anticommutes with the Dirac operator, one may split the space of Dirac spinors into two eigenspaces $\Delta_\pm$ of chirality $\pm$,

$$\begin{aligned} \gamma_5\psi_\pm = \pm\psi_\pm, &\qquad \psi_\pm \in \Delta_\pm, \\ D : \Gamma(\Delta_+) &\rightarrow \Gamma(\Delta_-), \\ D^+ : \Gamma(\Delta_-) &\rightarrow \Gamma(\Delta_+). \end{aligned}$$

The analytic index of spin complex $(\Delta_\pm, D)$ is

$$\begin{aligned} Ind(\Delta_\pm, D) &= dim\ KerD - dim\ KerD^+ \\ &= n_+ - n_-\ , \end{aligned}$$

where $n_{\pm}$ are the numbers of $\pm$ chirality zero-modes. The index itself is the difference between the numbers of $\pm$ chirality spinors in the kernel of Dirac operator.

Now we turn to the topological index of spin complex, it can be expressed as

$$(-1)^l \int_M \frac{ch(\Delta^+ - \Delta^-)}{e(TM)} td(TM) \otimes C).$$

From the splitting principle

$$ch(\Delta^+ - \Delta^-) = \prod_{j=1}^{l} (e^{-x_j/2} - e^{x_j/2}),$$

substitute this into the index formula gives

$$(-1)^l \frac{\prod_{j=1}^{l} (e^{-x_j/2} - e^{x_j/2})}{\prod_{j=1}^{l} x_j} \frac{\prod_{j=1}^{l} x_j(-x_j)}{\prod_{j=1}^{l} (1 - e^{-x_j})(1 - e^{x_j})} = \frac{\prod_{j=1}^{l} x_j}{\prod_{j=1}^{l} (e^{x_j/2} - e^{-x_j/2})}$$

$$= \frac{\prod_{j=1}^{l} x_j/2}{\prod_{j=1}^{l} \sinh x_j/2}.$$

$$\text{So} \quad Ind(\Delta^{\pm}, D) = \int_M \frac{\prod_{j=1}^{l} x_j/2}{\prod_{j=1}^{l} \sinh x_j/2} = \int_M \hat{A}(M),$$

where $\hat{A}(M) = \prod_{j=1}^{l} \frac{x_j/2}{\sinh x_j/2} = 1 - \frac{1}{24}p_1 + \frac{1}{5760}(7p_1^2 - 4p_2) + \cdots$.

For, example when $dim\ M = 4$

$$n_+ - n_- = -\frac{1}{24}\int_M p_1(TM) = \frac{1}{24 \cdot 8\pi^2}\int_M Tr(\Omega \wedge \Omega).$$

11.5 Twisted elliptic complex

The four classical elliptic complexes, we look at in the previous section, are complexes over tangent bundle of closed manifold. For sections taking coefficients in another vector bundle V, we can obtain the twisted elliptic complexes. In the formulas of the topological index, additional factor $ch(V)$ is needed to obtain the nontrivial twisted index. For example,

1. Twisted signature complex. The extended operator

$$(d+\delta)_V : C^\infty(\Lambda^+ \otimes V) \to C^\infty(\Lambda^- \otimes V).$$

And the index becomes

$$\begin{aligned} Ind(\Lambda^\pm \otimes V, (d+\delta)_V) &= \int_M \prod_{j=1}^{l} \frac{\chi_j}{tanh\chi_j/2} chV \\ &= \int_M \prod_{j=1}^{l} \frac{\chi_j}{tanh\chi_j/2} ch_2V = \int_M L(M) \wedge ch_2V, \end{aligned}$$

where $ch_2V = \sum\limits_k \frac{2^k}{k!} Tr(\frac{-F}{2\pi})^k$. Note its difference with $chV = \sum\limits_k \frac{1}{k!} Tr(\frac{-F}{2\pi})^k$, and let $\Omega_V = iF$ be the curvature of the bundle V, here F is a 2-form corresponding to gauge field strength for bundle V.

In particular,

$$\begin{aligned} n=2, \quad index(\tau_V(M),(d+\delta)_V) &= \int_M 2c_1(V) = -\frac{1}{\pi}\int_M Tr\ F, \\ n=4, \quad index(\tau_V(M),(d+\delta)_V) &= dim(V)\frac{1}{3}\int_M P_1 + \int_M (2c_1^2(V) - 2c_2(V)) \\ &= -\frac{dim\ V}{24\pi^2}\int_M Tr\Omega \wedge \Omega \ + \ \frac{1}{2\pi^2}\int_M TrF \wedge F. \end{aligned}$$

2. The twisted Dolbault complex. The extended operator is

$$\bar\partial_V : C^\infty(\Lambda^{0,q} \otimes V) \to C^\infty(\Lambda^{0,q+1} \otimes V),$$

and the index theorem becomes:

$$index(\bar\partial_V) = \int_M td(T_C(M)) \wedge ch\ V.$$

In particular,

$$\begin{aligned} n=2, \quad Index((\bar\partial_V) &= \frac{1}{2} dim\ V \int_M c_1(T_C(M)) + \int_M c_1(V) \\ &= \frac{1}{2} dim\ V \ \ \chi(M) - \frac{1}{2\pi}\int_M Tr\ F. \end{aligned}$$

Let M be a Riemannian surface with genus g, and V be the holomorphic line bundle on M associated with a divisor D of order d, then

$$\tfrac{1}{2} dim\ V \cdot \chi(M) = \tfrac{1}{2}(2-2g) = 1-g,$$
$$\int_M c_1(V) = d,$$

and the index theorem can be written as

$$dim\ H^0(M,D) - dim\ H^1(M,D) = 1 - g + d,$$

where $H^0(M,D)$ is the space of meromorphic functions subordinate to the divisor D, and $H^1(M,D)$ is the space of meromorphic 1-forms subordinate to D.

$$\begin{aligned} H^0(M,D) &= \{f \in K(M) \mid (f) + D \geq 0\}, \\ H^1(M,D) &= \{\omega \in K'(M) \mid (\omega) \geq D\}, \end{aligned}$$

where $K(M)$ is the meromorphic function, $K'(M)$ is the meromorphic 1-form, (f) and (ω) are divisors of f and ω respectively. So this is just the famous Riemann-Roch theorem which is a special case of Atiyah-Singer index theorem.

3. Twisted spin complex.

The extended Dirac operator D_V including the connection on V is:

$$\begin{aligned} D_V &: C^\infty(\Delta_+ \otimes V) \to C^\infty(\Delta_- \otimes V), \\ D_V^+ &: C^\infty(\Delta_- \otimes V) \to C^\infty(\Delta_+ \otimes V), \end{aligned}$$

The index theorem becomes

$$Index(\Delta_\pm \otimes V, D_V) = \int_M \hat{A}(M) \wedge ch(V).$$

In particular,

$$\begin{aligned} n = 2,\quad Index(\Delta_\pm \otimes V, D_V) &= \int_M c_1(V) = -\frac{1}{2\pi}\int_M F, \\ n = 4,\quad Index(\Delta_\pm \otimes V, D_V) &= -\frac{dim\ V}{24\pi}\int_M p_1(T(M))) + \frac{1}{2}\int_M (c_1^2(V) - 2c_2(V)) \end{aligned}$$

$$= \frac{dim\ V}{24 \cdot 8\pi^2}\int_M Tr(\Omega \wedge \Omega) + \frac{1}{8\pi^2}\int_M Tr(F \wedge F).$$

4. Twisted de Rham complex.

$$(d+\delta)_V : C^\infty(\Lambda^{even} \otimes V) \to C^\infty(\Lambda^{odd} \otimes V)$$

$$\text{so}\quad \begin{aligned} Ind(\Lambda \wedge V, d_V) &= \int_M e(T(M)) \wedge ch(V) \\ &= dim\ V \cdot \chi(M). \end{aligned}$$

Namely unlike the other three classical elliptic complexes, the twisting is not detected by de Rham complex.

Classical index theorem	Typical elliptic complex	Analytic index	Topological index
Gauss-Bonnet theorem	de Rham complex (Λ, d)	$\chi(M)$ $=\sum(-1)^p b_p$	$\int_M e(M)$
Hirzebruch theorem	Hirzebruch complex $(\Lambda, d+\delta)$	$\sigma(M)$ $= b_{2k}^+ - b_{2k}^-$	$\int_M L(M)$
	$(\Lambda^\pm \otimes V, (d+\delta)_V)$		$\int_M L(M)\wedge ch(V)$
Riemann-Roch theorem	Dolbault complex $(\Lambda^{0,q}, \bar{\partial})$	$\sum_{q=0}^{n/2}(-1)^q\cdot$ $dim H^{0,q}(M)$	$\int_M td(T_C(M))$
	$(\Lambda^{0,q}\otimes V, \bar{\partial}_V)$		$\int_M td(T_C(M))\wedge ch(V)$
Dirac operator index theorem	spin complex $(\Lambda_\pm, D)$	$n_+ - n_-$	$\int_M \hat{A}(M)$
	$(\Lambda^\pm \otimes V, D_V)$	$\nu_+ - \nu_-$	$\int_M \hat{A}(M)\wedge ch(V)$

11.6 Brief comment on the proof for index theorem, the heat kernel method

Atiyah-Singer index theorem encompasses and generalises many results in geometry, topology, and analysis. There are essentially three kinds of proof, all of them are fairly complicated and difficult. Atiyah once remarked: "These different proofs differ only in their use and presentation of algebraic topology". "The analysis is essentially the same in origin"[48].

1. The cobordism proof.

The first proof sketched by Atiyah-Singer[49,50] and then fully worked out by Brieskom et al[51], Palais[52], $\cdots$. The main idea could be traced back to the Hirzebruch theory. As every elliptic operator on a closed even-dimensional oriented manifold is in the sense of K-theory equivalent to a generalized signature operator, making use of Thom cobordism and for many cases one can compute their generalized signature index and obtain the general formula. This formula can be reduced from the elliptic operator space to $K(TM)$ by using pseudo-differential operator theory, where $K(TM)$ is a ring over $K(M)$. From Bott theorem, one may represent $K(TM)$ approximately by $K(M)$ and obtain the index formula.

2. Imbedding proof.

This was given by Atiyah-Singer in a series of monographs[50,53]. The main idea could be traced back to the proof of Riemann-Roch theorem by Grothendieck, which is quite simple and elegant, that the index is an insensitive quantity, unperturbed by some however drastic changes in the analytic data. As is the usual case in topology, any elliptic operator on a closed manifold could be reduced to a standard form by embedding one manifold into another, and in terms of K-theory one can formulate a topologically defined index endowed with the same transformation properties as the analytic index. Subsequently everything can be reduced to the trivial case, and it is easily seen that the analytic and topological indices coincide, thus it follows that they must coincide in general.

The two approaches above requires a good deal of topology and functional analysis, which are beyond the realm of most physicists. In what follows we look at the heat equation proof, which offers a completely different analytic approach to index theorem.

3. Heat equation proof.

This was worked out by Patodi [54] and Gilky [55]. Its main idea could be traced back to the paper of Rierz on spectral theory of positive definite self-adjoint operators. The heat equation proof utilises more analysis and differential geometry, and is more straightforward and transparent in the relations between classical operators and Riemannian structure of the manifold. Given an elliptic operator, one can get the integral representation of local topological invariant through asymptotic expansion of heat kernel, which intepolates between the identity operator (when $t \to 0^+$) and the projection onto the kernel (when $t \to \infty$). Applications of this method can lead to illuminating proofs of many classical index theorems.

Despite its complexity, this method explicitly associates classical operator with Riemannian structure of the manifold, which is helpful to the understanding of relations between operator spectral and topological and metric properties of the manifold. In particular, it is closely related to the renormalization in quantum field theory and supersymmetry, topics well familiar with physicists.

Let M be a closed compact manifold, E and F be vector bundles over M, $\Gamma(E)$ and $\Gamma(F)$ denote the space of sections on E and F respectively. D is an operator that transforms between $\Gamma(E)$ and $\Gamma(F)$:

$$D:\ \Gamma(E) \to \Gamma(F).$$

Its adjoint $D^+ :\ \Gamma(F) \to \Gamma(E)$, is defined as

$$(\eta, D\xi) = (D^+\eta, \xi), \qquad \xi \in \Gamma(E),\ \eta \in \Gamma(F).$$

Two non-negative self-adjoint elliptic operators can be obtained from D and D^+:

$$\begin{aligned} \Delta_E &= D^+D\ \in End\, \Gamma(E), \\ \Delta_F &= DD^+\ \in End\, \Gamma(F). \end{aligned}$$

As manifold M is closed and without boundary, the spectral of these two operators are discrete. Let

$$\begin{aligned} \Gamma_\lambda(E) &= \{\xi \in \Gamma(E) \mid \Delta_E\xi = \lambda\xi\}, \quad \Gamma_0(E) = Ker\ D, \\ \Gamma_\lambda(F) &= \{\xi \in \Gamma(F) \mid \Delta_F\xi = \lambda\xi\}, \quad \Gamma_0(F) = Ker\ D^+. \end{aligned}$$

It is easy to prove that Δ_E and Δ_F have both the same eigenvalue λ and same multiplicity for $\lambda \neq 0$, thus

$$dim\ \Gamma_\lambda(E) = dim\ \Gamma_\lambda(F),\ \forall\lambda \neq 0$$

only the zero mode may have the different multiplicities.

Proof: Let $\xi_\lambda \in \Gamma_\lambda(E)$, $\lambda \neq 0$, and

$$\Delta_E\xi_\lambda = D^+D\xi_\lambda = \lambda\xi_\lambda.$$

Then $\quad \Delta_F(D\xi_\lambda) = DD^+D\xi_\lambda = \lambda D\xi_\lambda$, so

$$D\xi_\lambda = \eta_\lambda \in \Gamma_\lambda(F),$$

besides $\eta_\lambda \neq 0$, for otherwise $\Delta_E\xi_\lambda = D^+\eta_\lambda = 0$ which contradicts with $\lambda \neq 0$.

Consequently the analytic index is

$$\begin{aligned} Ind\ D &= dim\ Ker D - dim\ Ker D^+ \\ &= dim\Gamma_0(E) - dim\Gamma_0(F) \\ &= \sum_\lambda f(\lambda)(dim\ \Gamma_\lambda(E) - dim\ \Gamma_\lambda(F))\ \ (f(0)=1). \end{aligned} \tag{11.7}$$

One may choose an arbitrary function $f(\lambda)$ with $f(0) = 1$ only need make the sum converge. Note that for the non-negative operator Δ, one can construct the series

$$h(t) = \sum_\lambda e^{-t\lambda} = Tr\ e^{-t\Delta}$$

which converges for all $t > 0$, and has an asymptotic expansion for $t \to 0^+$. Take $f(\lambda) = e^{-\lambda t}$ $(t > 0)$ in Eq. 11.7 gives

$$Ind\ D = h_E(t) - h_F(t),\ \forall t > 0.$$

This equation is independent of t, but is well defined for $t > 0$ only. For $t \to 0^+$, the two terms on the right hand side could be singular, the equation is still well defined however, as the singular parts cancel with each other.

The heat transport operator $e^{-t\Delta}$ has a kernel $K(t, \mathbf{x}, \mathbf{y})$, which is a linear map from the fibre of $\Gamma(E)$ over $\mathbf{y}$ to the fibre over $\mathbf{x}$:

$$e^{-t\Delta}\xi(\mathbf{x}) = \int_M K(t, \mathbf{x}, \mathbf{y})\xi(\mathbf{y})d\mathbf{y}. \tag{11.8}$$

The kernel $K(t, \mathbf{x}, \mathbf{y})$ has a spectral decomposition in terms of the eigenfunvtions of Δ:

$$K_t(\mathbf{x}, \mathbf{y}) = \sum_\lambda e^{-\lambda t}\xi_\lambda(\mathbf{x})\xi^*_\lambda(\mathbf{y}),$$

and it is a solution of heat transport equation for the origin, i.e., the matrix valued kernel satisfies

$$\begin{cases} (\frac{\partial}{\partial t} + \Delta_{\mathbf{x}})K(t,\mathbf{x},\mathbf{y}) = 0, \\ K(0,\mathbf{x},\mathbf{y}) = \delta(\mathbf{x}-\mathbf{y}). \end{cases} \tag{11.9}$$

For $t \to \infty$ it approaches the projector $P_0(\mathbf{x},\mathbf{y})$ to the space of zero modes of Δ.

Now we look at properties of the kernel function. For the simplest case, i.e., let

$$\Delta_0 = -\partial_\mu \partial^\mu,$$

the solution of Eq. 11.9 can then be written as:

$$K_0(t,\mathbf{x},\mathbf{y}) = (4\pi t)^{-\frac{n}{2}} exp(-(\mathbf{x}-\mathbf{y})^2/4t)$$

where n is the dimension of the base manifold M.

Taking the trace of $K(t,\mathbf{x},\mathbf{y})$ and integrating over the manifold gives the series

$$h(t) = \sum_\lambda e^{-\lambda t} = \int_M Tr\ K(t,\mathbf{x},\mathbf{y})d\mathbf{x}. \tag{11.10}$$

Next we try to find the asymptotic expansion of $K(t,\mathbf{x},\mathbf{y})$ as $t \to 0^+$. First we take the Fourier transform of the left hand side of Eq.11.8:

$$\begin{aligned} \int_M K(t,\mathbf{x},\mathbf{y})\xi(\mathbf{y})d\mathbf{y} &= \int e^{i\mathbf{k}\mathbf{x}}\sigma(e^{-t\Delta})\hat{\xi}(\mathbf{k})d\mathbf{k} \\ &= \int\int e^{i\mathbf{k}(\mathbf{x}-\mathbf{y})}\sigma(e^{-t\Delta})\xi(\mathbf{y})d\mathbf{y}d\mathbf{k}. \end{aligned}$$

As it is valid for any section $\xi(\mathbf{y})$, so

$$K(t,\mathbf{x},\mathbf{x}) = \int \sigma(e^{-t\Delta})d\mathbf{k}.$$

Consequently the asymptotic expansion of $K(t,\mathbf{x},\mathbf{x})$ at $t \to 0^+$ can be obtained:

$$\lim_{t\to 0^+} K(t,\mathbf{x},\mathbf{x}) \sim (4\pi t)^{-\frac{n}{2}} \sum_{j=0}^{\infty} t^j \mu_j(\mathbf{x})\sqrt{g},$$

in which there are only finite number of negative power terms and $g = det(g_{ij})$ is the determinant of the metric tensor.

Substitute this into the heat equation through Eq. 11.10 yields the equations satisfied by $\mu_j(\mathbf{x})$. The proof of this asymptotic expansion can be carried out from pseudo-differential operator theory, and in terms of the powers and coefficients of operator D. For differential operators (covariant differential operator etc., independent of the choice of local coordinates), only the curvature and its derivatives appear in the equations satisfied by $\mu_j(\mathbf{x})$, namely $\mu_j(\mathbf{x})$ have differential geometric meaning. And the index becomes

$$Ind\ D = Tr \int_M (\mu_n^E(\mathbf{x}) - \mu_n^F(\mathbf{x}))\sqrt{g}d\mathbf{x},$$

i.e., $(\mu_n^E(\mathbf{x}) - \mu_n^F(\mathbf{x}))$ is identifiable with characteristic class, has topological significance.

Comparing the heat equation with the partition function in statistical physics tells us that index theorem relates the high temperature limit with the low temperature limit. Low temperature limit ($t \to \infty, t \sim \beta = 1/kT$) gives the analytic index of the elliptic operator, which is an index with global topological significance. The asymptotic expansion at high temperature ($t \to 0^+$) is equivalent to the expansion in terms of the symbol of the elliptic operator, its limit is the integral of a local topological invariant, which is like the integral of a topological characteristic class.

In order to get the high temperature asymptotic expansion of the heat kernel, it is helpful to use the Mellin transformation:

$$\zeta_\Delta(s, \mathbf{x}, \mathbf{y}) = \frac{1}{\Gamma(s)} \int_0^\infty dt\ t^{s-1}(K(t, \mathbf{x}, \mathbf{y}) - P_0(\mathbf{x}, \mathbf{y}))$$

where $P_0(\mathbf{x}, \mathbf{y}) = \sum_{\lambda=0} \xi_\lambda(\mathbf{x})\xi_\lambda^*(\mathbf{y}) = lim_{t\to\infty} K(t, \mathbf{x}, \mathbf{y})$ is the projector onto the space of zero modes of Δ.

Taking the trace of $\zeta_\Delta(s, \mathbf{x}, \mathbf{y})$ and integrating over M gives

$$\zeta_\Delta(s) = \sum_{\lambda\neq 0} \lambda^{-s} = \int_M d\mathbf{x}\ Tr\zeta_\Delta(s, \mathbf{x}, \mathbf{x})$$

which is called the generalized Riemann zeta function. It is analytic for sufficiently large s and for $s \to 0$,

$$\frac{d}{ds}\zeta_\Delta(s)|_{s=0} = -\sum_i e^{-s \ln\ \lambda_i} \ln\ \lambda_i|_{s=0} = -\ln\ (\prod_i \lambda_i),$$

i.e.,

$$det\ \Delta = \prod_i \lambda_i = exp(-\zeta_\Delta'(0)).$$

In quantum field theory, ζ function is used extensively in the quantum correction of semi-classical calculations. Using it to calculate the high temperature ($t \to 0^+$) limit, one can get the asymptotic integral formula of the topological index of the operator, namely Atiyah-Singer index theorem.

Generalize the positive definite Laplacian to the non-positive definite Dirac operator, η function can be used instead of ζ function to study the asymmetry of the spectral

$$\eta(s) = \sum_{\lambda\neq 0} Sgn(\lambda)|\lambda|^{-s}.$$

And one can further study the index theorem for manifold with boundary and odd-dimensional manifold.

11.7 Some applications in physics

Dirac's study of magnetic monopole[12], Bhom-Aharonov's study of Fermion in electromagnetic field[39], the introduction of non-integrable phase factor and particularly the study of monopole as well as instanton and quantum anomaly all indicate the relevance and importance of topological analysis in quantum field theory. Atiyah-Singer index theorem lies at the core of global analysis in quantum field theory. And we give two simple examples to illustrate its application.

Example 1. The number of zero energy Fermion solutions around a magnetic monopole[56].

Because of the stability spherical symmetry, we take moving frame to separate the radial and angular variables. So this equivalent to the zero energy Fermion solution with gauge indices on S^2. The corresponding analytic index is the difference between the number of positive and negative chirality zero energy solutions.

$$Ind(\Delta_\pm \otimes V, D) = dim\ \psi^+ - dim\ \psi^-.$$

As the base manifold is S^2, from index theorem, we have

$$dim\ \psi^+ - dim\ \psi^- = \int_M c_1(V) = \text{The first Chern number of the gauge field.}$$

Explicit calculation indeed gives this result.

Example 2. Number of independent parameters of instanton solutions[58].

Let $M = S^4$, the compactification of R^4 is $S^4 = R^4 \cup \{\infty\}$. Instanton is the self-dual solution of Yang-Mills equations on M. The number of independent parameters of instanton solutions is just the dimension of instanton module space $\mathcal{M}$,

$$\mathcal{M} = \mathcal{A}^+/\mathcal{G}, \quad \mathcal{A}^+ = \{A \in \mathcal{A} \mid F_A^- = 0\}$$

which depends on the conformal class of metric of base manifold M and the isomorphism class k of the principal bundle $P(M, G)$.

$$k = c_2(E)[M] = \frac{1}{8\pi^2}\int_M tr\ F_A^2.$$

The topological charge k is an integer, independent of the connection A on $P(M, G)$.

The infinitesimal gauge transformation on A is

$$\delta A = d\Theta + A \wedge \Theta \equiv D_0\Theta \ \in \Omega^1(M, \mathbf{g})$$

where $\Theta \in \Omega^0(M,\mathbf{g})$ is a Lie algebra $\mathbf{g}$ valued function on M. Further

$$\delta F = d\delta A + A \wedge \delta A \equiv D_1 \delta A \in \Omega^2(M,\mathbf{g}).$$

Introducing the projection operators $P_\pm$ on $\Omega^1(M,\mathbf{g})$:

$$P_\pm = \frac{1}{2}(1 \pm *),\ P_\pm^2 = P_\pm,\ P^+P^- = 0.$$

As we consider self-dual solutions only, so

$$(P_- D_1) D_0 \Theta = 0.$$

And we obtain the following complex (instanton deformation complex):

$$0 \to \Omega^0(M,\mathbf{g}) \overset{D_0}{\to} \Omega^1(M,\mathbf{g}) \overset{P_-D_1}{\to} \Omega^2(M,\mathbf{g}) \to 0.$$

As we have quotiented the action of $\mathcal{G}$, this complex is thus elliptic and its cohomology groups are finite dimensional. The analytic index of the complex can be written as:

$$Ind\ D_G = h_A^0 - h_A^1 + h_A^2,$$

$$\text{where} \begin{cases} h_A^0 = dim\ Ker\ D_0, \\ h_A^2 = dim\ (\Omega^2(M,\mathbf{g})/Im\ P_- D_1), \\ h_A^1 = dim\ Ker\ (P_- D_1) - dim\ Im\ D_0. \end{cases}$$

As the connection is irreducible, $Ker\ D_0 = 0$ and so $h_A^0 = 0$. For Riemannian manifold with positive scalar curvature and self-dual Weyl tensor, there is a vanishing theorem which states that $h_A^2 = 0$ if self-dual instanton solution exists. This also guarantees the smoothness of the module space of irreducible instantons.

In terms of Atiyah-Singer index theorem, the number of inequivalent instanton solutions $h_A^1 = -Ind\ D_G$ can be determined as

$$h_A^1 = 4k\ C(G) - \frac{1}{2} dim\ G[\chi(M) - \sigma(M)]$$

where $\chi(M)$ is the Euler characteristic of M, $\sigma(M)$ is the signature of M. $C(G)$ is the dual Coexter number of group G. In the case

$$\begin{array}{lll} G = SU(2), & C(G) = 2, & dim(G) = 3. \\ M = S^4, & \chi(M) = 2, & \sigma(M) = 0. \end{array}$$

One obtains the well known result

$$h_A^1 = 8k - 3.$$

There are $(8k-3)$-parameter family of instantons. These correspond to k-instanton with $5k$ parameters specifying their position and size, and the other $3k - 3 = (k-1) dim\ G$ parameters specifying the relative orientations of instantons in the Lie algebra space $\mathbf{g}$.

Notations and Formulae

- The analytic index of elliptic differential operator D

$$Ind\ D = dimKer\ D - dimCoker\ D.$$

If there is Hermitian metric on the space of sections, then

$$Ind\ D = dimKer\ D - dimKer\ D^{+}.$$

For the elliptic complex (E,D)

$$H^{l}(E, D) = Ker\ D_{l}/Im\ D_{l-1}.$$

$$Ind\ (E, D) = \sum_{l}(-1)^{l} dimH^{l}(E, D).$$

- Atiyah Singer index theorem

$$\begin{aligned} Ind\ D &= (-1)^{n}(ch(\Sigma(D))td(TM \otimes C))[TM] \\ &= (-1)^{\frac{n(n+1)}{2}} \frac{(chE - chF)td(TM \otimes C)}{e(TM)}[M]. \end{aligned}$$

For elliptic complex (E,D)

$$Ind\ (D, E) = (-1)^{\frac{n(n+1)}{2}} \frac{ch(\sum(-1)^{i}E_{i})td(TM \otimes C)}{e(TM)}[M].$$

Exercise

1. Find the Euler Characteristic class for CP(n) and proceed to prove that its integral over the whole manifold, which is the Euler number, is n+1.

2. Take CP(2) as an example, verify the Hirzeruch signature theorem:

$$\sigma(M) \equiv \int_{M} L(M) \equiv -\frac{1}{24\pi^{2}} \int_{M} Tr(\Omega \wedge \Omega),$$

and the Riemann-Roch theorem:

$$Ind(\bar{\partial}) = \int_{M} td(T_{C}(M)).$$

3. Making use of the index theorem for twisted spin complex

$$Index(\Delta_{\pm} \otimes V, D_{V}) = \int_{M} \hat{A}(M) \wedge ch(V),$$

prove that the difference of the numbers of fermions with different chiralities is just the instanton number of the background field (the SU(2) gauge field).

Chapter 12

Index Theorem on Manifold with Boundary and on Open Infinite Manifold

12.1 Introduction

The index theorem described in the previous chapter is valid only for bundle with base manifold M, which are compact without boundary. If the base manifold M is noncompact, or is compact but with boundary, the index theorem must be modified. For an noncompact manifold, just as instanton solutions to the self-dual Yang-Mills equations on Euclidean space R^4 can be obtained by compactifying R^4 to S^4, an open infinite manifold in general can be treated as a limiting case of a manifold with boundary when the boundary approaches infinity. So first we have to study how the index of an elliptic operator depends on the boundary ∂M. Before generalizing the index theorem to the manifold with boundary we first take a look at the following example.

The Taub-Nut metric on R^4:[59]

$$ds^2 = \frac{r+a}{r-a}dr^2 + (r^2 - a^2)[\sigma_x^2 + \sigma_y^2 + (\frac{2a}{r+a})^2\sigma_z^2], \tag{12.1}$$

where $\sigma_x, \sigma_y, \sigma_z$ are the Maurer-Cartan forms on S^4 in terms of Euler angle. They are orthogonal moving frame fields on S^4, i.e., invariant vector fields on the group manifold $SU(2)$,

$$\begin{aligned}
\sigma_x &= \frac{1}{2}(-\cos\psi d\theta - \sin\theta\sin\psi d\psi), \\
\sigma_y &= \frac{1}{2}(\sin\psi d\theta - \sin\theta\cos\psi d\psi), \\
\sigma_z &= \frac{1}{2}(-d\psi - \cos\theta d\phi).
\end{aligned}$$

In which $0 \leq \phi < 2\pi, \quad 0 \leq \theta < \pi, \quad 0 \leq \psi < 2\pi, \quad a \leq r < \infty$.

With this metric we can calculate the Riemannian curvature which is self-dual, i.e., it is a solution of gravitational instanton.

It is worth mentioning that calculating the index of the Dirac operator by the index theorem in the previous chapter will not give an integer:

$$Ind(D) = \int_M \tilde{A} = \frac{1}{24 \cdot 8\pi^2} \int_M Tr\ \Omega \wedge \Omega = -\frac{1}{12}.$$

This is a puzzle! This is due to that there are singularities at $r = a$ and $r = \infty$. The one at $r = a$ is a removable singularity coming from the choice of polar coordinates, but the singularity at $r = \infty$ is an intrinsic one. Therefore one has to restrict E^4 to $a \leq r \leq r_0$ which is a manifold with boundary and with asymmetric boundary. The asymmetry grows as $r_0 \to \infty$. One must consider the effect of the boundary and modify the index theorem in order to get an integer.

However, as the boundary conditions determine the spectrum of the operator, there is a boundary correction to the index. In ordinary physical problems, one usualy take a local boundary condition. But Atiyah and Bott [60] had shown that the spin, signature, and Dolbeault complexes do not admit local boundary condition, there exist topological obstruction to find it. One must consider the nonlocal boundary condition. Atiyah, Patodi, and Singer [61] considered this situation and formulated an index theorem for elliptic complexes on manifold with boundary, i.e., the Atiyah-Patodi-Singer index theorem (APS index theorem). For simplicity, we look at APS index theorem for classical elliptic complexes $(E, D, \partial M)$ with cohomology groups $H^p(E, D, \partial M)$, and the analytic index is defined as:

$$Ind(E, D, \partial M) = \sum_p (-1)^p dim\ H^p(E, D, \partial M).$$

And the APS index theorem can be written in the following form:

$$Ind(E, D, \partial M) = V[M] + S[\partial M] + \xi(\partial M). \tag{12.2}$$

In which $V[M]$ is the integral over M just as in the case without boundary, $S[\partial M]$ is the integral over ∂M of the Chern-Simons form described in \$10.7. This terms arises out of the consideration that the metric near the boundary may not be a product metric, and we will have a good look at this contribution in the next section. $\xi(\partial M)$ is determined by the eigenvalue spectrum of the tangent part H of operator D restricted to the boundary ∂M. For the de Rham complex, $\xi(\partial M)$ vanishes because the local boundary condition (Dirichlet condition) can be used instead. For the other three classical elliptic complexes however, due to the topological obstructions to finding local boundary conditions, one has to give up local boundary conditions in favor of the so called spectral boundary condition determined by the eigenvalue spectrum of H (an operator on ∂M). This is the main content of APS index theorem.

After the general analysis of the APS index theorem in §12.3, as an example, we demonstrate in §12.4 why a spin manifold with boundary does not admit local boundary conditions, and give a brief introduction to the proof of the APS index theorem.

Remember that both the AS and APS index theorems are valid for compact manifold only. To study the topological properties of continuous spectrum of operators, we will give the index theorem on an open infinite manifold in §12.5.

The APS index theorem is related to the non-local boundary conditions which, from the point of view of physics, violates causality. Moreover, the index obtained by the trace identities in quantum field theory is different from that calculated by means of the APS index theorem as the boundary goes to infinity. To get rid of these apparent inconsistencies, we have a close study of the milder boundary condition for the (2+1)-dimensional electrodynamic system in §12.6, and obtain a result consistent with the index theorem over an open infinite manifold from the techniques of both regularization and renormalization in quantum field theory.

12.2 Index theorem for de Rham complex on manifold with boundary

The conventional local boundary condition (Dirichlet condition) can be chosen for the de Rham complex over manifold M with boundary ∂M, and the $\xi(\partial M)$ correction vanishes as a result. Thus we concentrate on the surface correction $S[\partial M]$ in this section.

Let M be a $2n$-dimensional compact manifold with boundary ∂M. If the following product metric can be defined near its boundary($r = r_o$), i.e.,

$$ds^2 = f(r_0, r)dr^2 + g_{ij}(r_0, x)dx^i dx^j$$

and the boundary manifold ∂M is determined by $r = r_0$. Near the boundary, it is easy to calculate the Riemannian connection and curvature corresponding to the metric, and to verify that the Riemannian curvature has no normal component on the boundary. However, the surface correction can be calculated from the connection determined by a suitable choice of the normal to the boundary. Consequently the surface correction $S[\partial M]$ vanishes, if the metric is a product metric near the boundary.

On the one hand, should M have a non-product metric near the boundary, the corresponding Riemannian connection 1-form ω can be decomposed into the tangential and normal components on ∂M according to $\omega = \omega_0 + \theta$, here θ is the normal component called the second fundamental form. Then we can find a product metric on M such that the corresponding connection 1-form is ω_0 near the boundary ∂M, namely the difference between ω and ω_0 is the second fundamental form:

$$\theta = \omega - \omega_0,$$

which is covariant under transformation of frames, and is represented by a matrix valued 1-form.

On the other hand, characteristic classes associated with TM can be constructed from the symmetric polynomial $P(\Omega)$ of the curvature Ω corresponding to the connection ω of manifold M. $P(\Omega)$ is a cohomology class. The characteristic classes measure the deviation from trivial bundle, and is independent of the choice of connection and metric. Hence the difference between characteristic classes $P(\Omega)$ and $P(\Omega_0)$, associated with connection ω and ω_0 respectively, is an exact form $dQ(\Omega, \Omega_0)$ for some differential form $Q(\Omega, \Omega_0)$, namely

$$dQ(\Omega, \Omega_0) = P(\Omega) - P(\Omega_0). \tag{12.3}$$

In fact, $Q(\Omega, \Omega_0)$ is the Chern-Simons form.

As mentioned above, if the metric is a product metric near the boundary, the surface correction $S[\partial M]$ vanishes and the topological index contains only the term $V[M]$, which is an integral of the curvature polynomial $P(\Omega_0)$. However, for the non-product metric near the boundary, the topological index contains both $V[M]$ and $S[\partial M]$. As the topological index is independent of the choice of metric on the manifold, we obtain the following identity :

$$V[M] + S[\partial M] = c \int_M P(\Omega_0).$$

From Eq. 12.3 we have:

$$\begin{aligned} V[M] + S[\partial M] &= c \int_M P(\Omega) - c \int_{\partial M} Q(\Omega, \Omega_0) \\ \text{i.e.,} \quad S[\partial M] &= -c \int_{\partial M} Q(\Omega, \Omega_0). \end{aligned}$$

The surface term $S[\partial M]$ is in general determined by the connection ω, curvature Ω and the second fundamental form θ on the boundary.

From the analysis above, we see that for de Rham complex on 4-dimensional manifold with boundary, the index theorem can be written as:

$$\begin{aligned} \chi &= \frac{1}{32\pi^2}[\int_M \epsilon_{abcd} \Omega^a_b \wedge \Omega^c_d \\ &\quad - \int_{\partial M} \epsilon_{abcd}(2\theta^a_b \wedge \Omega^c_d - \frac{4}{3}\theta^a_b \wedge \theta^c_e \wedge \theta^e_d)]. \end{aligned}$$

Where Ω^a_b are the curvature 2-forms, and θ^a_b are the second fundamental forms.

12.3 APS index theorem

Unlike the de Rham complex, due to the topological obstruction in finding local boundary conditions for the other classical elliptic complexes, non-local boundary conditions have to taken into account, so that the term $\xi(\partial M)$ in Eq. 12.2 does not vanish and is determined by the eigenvalue spectrum of the tangent part H of operator restricted to the boundary ∂M. If the metric is a non-product, we can get the surface correction $S[\partial M]$ in a way similar to that used in the last section. Here we shall concentrate on $\xi(\partial M)$.

First we find a product metric near the boundary ∂M so that the elliptic operator D can be written in the following form:

$$\begin{aligned} D &= A\cdot\partial + B\frac{\partial}{\partial r} \\ &= B(B^{-1}A\cdot\partial + \frac{\partial}{\partial r}) = B(H + \frac{\partial}{\partial r}), \end{aligned}$$

here $A\cdot\partial = A_i\partial_i$ is the tangent part of D, A and B are matrices. Ellipticity of the operator D guarantees that the matrix B is invertible, thus the equation above is well defined. The operator $H = B^{-1}A\cdot\partial|_{\partial M}$ is the tangent part of operator D restricted to the boundary ∂M. H is also an elliptic operator on ∂M if it is self-adjoint, in addition to being independent of r. Consequently, H does have a well defined eigenvalue spectrum $\{\lambda_i\}$ associated with the non-local boundary condition on ∂M, which has a contribution $\xi(\partial M)$ to the topological index of operator D. Now let us look at properties of $\xi(\partial M)$.

Take another manifold M' with the same boundary and product metric near the boundary as manifold M, namely $\partial M = \partial M'$. Glue M and M' together along their common boundary and we have a new closed manifold $M \cup M'$. For simplicity, we take the Hirzebruch signature complex over 4-dimensional manifold M with boundary ∂M as an example. The boundary free index theorem for the signature complex on $M \cup M'$ gives:

$$\begin{aligned} \sigma(M\cup M') &= -\frac{1}{24\pi^2}\int_{M\cup M'} Tr(\Omega\wedge\Omega) \\ &= -\frac{1}{24\pi^2}(\int_M Tr(\Omega\wedge\Omega) - \int_{M'} Tr(\Omega'\wedge\Omega')], \end{aligned}$$

where in the second step we make use of the fact that the orientation of M' with curvature Ω' is opposite to its orientation on $M \cup M'$.

On the other hand, substitute the Novikov formula

$$\sigma(M\cup M') = \sigma(M) - \sigma(M')$$

into the left hand side of the equation above, and we have:

$$\sigma(M) + \frac{1}{24\pi^2}[\int_M Tr(\Omega\wedge\Omega) = \sigma(M') + \int_{M'} Tr(\Omega'\wedge\Omega')].$$

It is obvious that the left hand side of this equation depends on manifold M only while the right hand side depends on M' only. As M and M' are different with only the boundary in common, this equation indicates that there is a quantity, denoted as $\eta_{sg}(\partial M)$, determined by the boundary only, and not by manifold M itself, so

$$\eta_{sg}(\partial M) = \sigma(M) + \frac{1}{24\pi^2}\int_M Tr(\Omega \wedge \Omega).$$

This is the index formula for signature complex over manifold with product metric near the boundary. For non-product metric, the surface correction $S[\partial M]$ should be included and the final result is:

$$\sigma(M) = -\frac{1}{24\pi^2}\int_M Tr(\Omega \wedge \Omega) + \frac{1}{24\pi^2}\int_{\partial M} Tr(\theta \wedge \Omega) + \eta_{sg}(\partial M),$$

where θ is the second fundamental form normal to the boundary, η_{sg} is the so called η-invariant. Next we give an alternative expression of $\eta_{sg}(\partial M)$ in terms of the eigenvalue spectrum $\{\lambda_i\}$ of the tangent part H of the signature operator restricted to the boundary.

For the index theorem over closed maniford, the asymmetry of zero modes play an impotent role. Now for the index over manifold with boundary, the asymmetry of the spectrum $\{\lambda_i\}$ also play an impotent role. So we introduce the generalized $APS\eta$ function $\eta_H(s, \partial M)$ as follows:

$$\eta_H(s, \partial M) = \sum_{\lambda_i \neq 0} \frac{sign(\lambda_i)}{|\lambda_i|^s}. \tag{12.4}$$

The series representing $\eta_H(s, \partial M)$ defines a meromorphic function of s which can be analytic extended to $s = 0$. And the APS η-invariant $\eta_{sg}(\partial M)$ can be written as:

$$\eta_{sg}(\partial M) = \eta_H(s = 0, \partial M).$$

As is clear in Eq. 12.4, $\eta_{sg}(\partial M)$ counts the asymmetry between the number of positive and negative eigenvalues of operator H on ∂M. $\eta_{sg}(\partial M) = 0$, if the metric near the boundary has orientation-reversing isometry.

Analogously, we can study the other three classical elliptic complexes with boundary. As an example, for the spin complex over the 4-manifold with boundary, APS index theorem is:

$$\begin{aligned} Ind(\Delta_\pm, D, \partial M) &= \frac{1}{192\pi^2}\int_M Tr(\Omega \wedge \Omega) - \frac{1}{192\pi^2}\int_{\partial M} Tr(\theta \wedge \Omega) \\ &+ \frac{1}{2}[\eta_H(\partial M) + h(\partial M)]. \end{aligned} \tag{12.5}$$

Where $h(\partial M) = dim\ KerH$. $\eta_H(\partial M) = 0$, if manifold M is simply connected and has orientation-reversing isometry. Particularly, for a manifold with the Taub-Nut metric as Eq.12.1,, the contribution to the index of the Dirac operator from boundary $r = r_0$ can be evaluated as[62]:

$$\eta_D(\partial M) = \frac{1}{12}[1 - (\frac{2a}{r_0 + a})^2]^2.$$

It is obvious that $\eta_D(\partial M) = 0$ if $r_0 = a$. This is the case when the metric has orientation-reversing isometry as we discussed earlier. Moreover,

$$\eta_D(\partial M) \stackrel{r\to\infty}{\longrightarrow} \frac{1}{12}.$$

Substituting this into Eq. 12.5, thus the correct index is: $Ind(D) = -\frac{1}{12} + \frac{1}{12} = 0$. Namely there is no asymmetry between positive and negative chirality Dirac spinors for space with the Taub-Nut metric.

On the other hand, for the twisted spin complex over 4-manifold M with boundary ∂M, APS index theorem is:

$$\begin{aligned} Ind(\Delta_\pm \otimes V, \partial M) &\equiv \nu_+(\partial M) - \nu_-(\partial M) \\ = \frac{1}{192\pi^2}[\int_M Tr(\Omega\wedge\Omega) &- \int_{\partial M} Tr(\theta\wedge\Omega)] - \frac{1}{8\pi^2}\int_M Tr(\Omega_v\wedge\Omega_v) \\ &+\frac{1}{2}[\eta(\Delta_\pm\otimes V,\partial M) + h(\Delta_\pm\otimes V,\partial M)]. \end{aligned}$$

To close this section, we note that for the APS index theorem it is crucial to find the spectral boundary conditions and to calculate APS η-invariant in Eq. 12.5. In the next section, we take the spin complex as an example to study the origin of the topological obstruction, to finding a local boundary condition, and the way to develop the spectral boundary conditions related to the η-function.

12.4 APS index theorem for spin complex, spectral boundary condition

Let M be a n-dimensional spin manifold with boundary ∂M. In the vicinity of the boundary, the metric is a product metric which can be written in terms of the local coordinates as:

$$ds^2 = dr^2 + g_{ij}(r_0, x)dx^i dx^j,$$

where x^i is the local coordinate on $N = \partial M$. $r < r_0$ corresponds to the interior of M. The Dirac operator near the boundary ($r \approx r_0$) can be written as:

$$\not{D} = i\Gamma^\mu D_\mu.$$

D_μ is the covariant derivative including the spin connection and background field. In diagonal representation of Γ_5, Γ_μ have the following form:

$$\Gamma_5 = \begin{pmatrix} I & 0 \\ 0 & -I \end{pmatrix}, \quad \Gamma_\mu = \begin{pmatrix} 0 & \gamma_\mu \\ \gamma_\mu^+ & 0 \end{pmatrix}$$

and in which the radical part

$$\Gamma_r = \begin{pmatrix} 0 & -iI \\ iI & 0 \end{pmatrix}.$$

Thus the Dirac operator in the vicinity of the boundary is:

$$\not{D} = \begin{pmatrix} 0 & \partial_r + iA_r + H \\ -\partial_r + iA_r + H & 0 \end{pmatrix}.$$

Here A_r is the normal component of the gauge field near the boundary, and H is the tangent part of $\not{D}$ restricted to the boundary. Taking the gauge $A_r = 0$, it gives:

$$\not{D} = \begin{pmatrix} 0 & \partial_r + H \\ -\partial_r + H & 0 \end{pmatrix} = \begin{pmatrix} 0 & L \\ L^+ & 0 \end{pmatrix}.$$

$$\text{and} \qquad \{\not{D}, \Gamma_5\} = 0.$$

Thus the spectrum of $\not{D}$ with respect to the chirality is symmetric, i.e.,

$$\not{D}\psi_E = E\psi_E, \qquad \not{D}\Gamma_5\psi_E = -E\Gamma_5\psi_E.$$

Let $\psi = \begin{pmatrix} u \\ v \end{pmatrix}$, we can then write the eigenvalue equations in terms of these two components as:

$$\begin{cases} Lv = Eu \\ L^+u = Ev \end{cases} \rightarrow \begin{cases} L^+Lv = E^2v \\ LL^+u = E^2u \end{cases} \tag{12.6}$$

where L^+L and LL^+ are Hermitian self-conjugate non-negative elliptic differential operators.

It is obvious, from the equations above, that the Dirac operator has the same zero modes as L and L^+. Thus the zero modes of $\not{D}$ can be divided into $\begin{pmatrix} u \\ 0 \end{pmatrix}$ and $\begin{pmatrix} 0 \\ v \end{pmatrix}$, which are the eigenstates of Γ_5 corresponding to the eigenvalues ± 1 respectively. Moreover, the analytic index of $\not{D}$ is:

$$\begin{aligned} Ind\ \not{D} &= dim\ Ker\ L^+ - dim\ Ker\ L \\ &= dim\ Ker\ LL^+ - dim\ Ker\ L^+L. \end{aligned} \tag{12.7}$$

On the other hand, the requirement that $\not{D}$ be Hermitian gives:

$$< u, Lv > = < L^+u, v > .$$

The terms involving H can be eliminated from both sides of this equation as H is a self-adjoint operator in the vicinity of the boundary. Using the Stokes theorem, we get:

$$\int_{N=\partial M} v^+(x,r)u(x,r)|_{r=r_0} = 0. \tag{12.8}$$

LL^+ and L^+L are second order elliptic differential operators, since L and L^+ are first order differential operators. From the theory of differential equation, we known that it is impossible to find solutions to eigenvalue equations Eq. 12.6 with identical local boundary conditions. In other words, the usual local homogeneous boundary conditions can not be used for u and v at the same time when the topological property of background field is non-trivial. Thus, there exist topological obstructions to finding local boundary conditions.

To get rid of topological obstructions, we have to go beyond the usual local boundary conditions. The non-local boundary conditions are often used in mathematics. For instance, on the boundary $r = r_0$, the appropriate boundary condition for u and v is:

$$\begin{cases} Pv(x,r)|_{r=r_0} = 0, \\ (1-P)u(x,r)|_{r=r_0} = 0, \end{cases} \tag{12.9}$$

where P is a projection operator. This equation means that the space including u is an orthogonal complement of the space including v so Eq. 12.8 is always satisfied. Since the difference between L^+ and L under the gauge condition $A_r = 0$ near the boundary is the orientation of $N = \partial M$, and H is a first order differential operator on $N = \partial M$, so the coordinate independent choice of P is a projection operator onto either the positive or the negative spectrum of H.

For brevity, we use the Dirac symbol familiar to physicists. Let $\{|\lambda>\}$ be the complete set of orthonomal eigenstates of H on the boundary $N = \partial M$, i.e.,

$$H|\lambda> = \lambda|\lambda> . \tag{12.10}$$

Without loss of generality, we assume here that H has no zero modes and we leave the case where H has zero modes to the end of this section. Now, we can think of P in Eq. 12.9 as a projection operator onto the positive spectrum of H, namely,

$$P = \sum_{\lambda>0} |\lambda><\lambda|. \tag{12.11}$$

We can expand both u and v near the boundary as:

$$\begin{cases} |u> = \sum_\lambda |\lambda><\lambda|u>, \\ |v> = \sum_\lambda |\lambda><\lambda|v> . \end{cases}$$

Substituting this into Eq. 12.6 gives:

$$\begin{cases} (-\partial_r + \lambda) <\lambda|u> = E<\lambda|v>, \\ (-\partial_r + \lambda) <\lambda|v> = E<\lambda|u> . \end{cases} \tag{12.12}$$

On the other hand, we can write down the boundary conditions Eq. 12.9 on the boundary ∂M:

$$\begin{cases} |u> = \sum_{\lambda>0} |\lambda><\lambda|u>, \\ |v> = \sum_{\lambda<0} |\lambda><\lambda|v> . \end{cases}$$

Namely near the boundary ∂M one get:

$$\begin{cases} <\lambda|u>=0, & \text{for } \lambda<0, \\ <\lambda|v>=0, & \text{for } \lambda>0. \end{cases} \tag{12.13}$$

Substituting them into Eq. 12.12 gives us zero mode solution near the boundary conditions as follows:

$$\begin{cases} \partial_r \ln <\lambda|u>=\lambda, & \text{for } \lambda>0, \\ \partial_r \ln <\lambda|v>=-\lambda, & \text{for } \lambda<0. \end{cases} \tag{12.14}$$

Now we give a brief elaboration on the interpretation of this boundary condition. First, we can attach a semi-infinite cylinder $N \times R^-$ on the boundary $N = \partial M$ to form a manifold $\hat{M} = M \cup (N \times R^-)$, where R^- is the negative half-line and $N \times R^-$ corresponds to the region $r < r_0$. Further, we can extend the Dirac operator onto this semi-infinite cylinder by a constant extension of the background gauge field onto $N \times R^-$ from the boundary $N = \partial M$. This result in the boundary condition holding exactly for the cylinder $N \times R^-$. Second, we look at the zero-modes of L and L^+. Continuity of the background gauge field on $\hat{M}$ gives rise to that the zero-mode solutions of L^+ and L on $\hat{M}$, which are smooth extensions of the original zero-mode solutions on M with continuous first derivatives on the boundary ∂M. Hence we can see from Eq. 12.13 and Eq. 12.14 that zero-mode solutions of L correspond to $\lambda < 0$ and the wave function has a positive logarithmic derivative on the boundary ∂M. Thus, for the region $r < r_0$, $v(x, r)$ vanishes exponentially so that it is square integrable over $\hat{M}$. Similar analysis shows that zero-mode solutions $u(x, r)$ of L^+ associated with $\lambda > 0$ have similar properties to that of L. Finally, the zero-mode solutions of L and L^+ on M, which satisfy the boundary condition Eq. 12.14, are those who having natural square integrable extensions on the boundary ∂M. This is the implication of the boundary conditions. In addition, the operator $\not{D}$ acting on M with boundary conditions Eq. 12.13 and Eq. 12.14 has the same index as the operator which is the extension of $\not{D}$ onto the manifold $\hat{M}$, and the zero-mode solutions of which are restricted to the square integrable function space.

Now, we give a brief proof of the APS index theorem by the heat equation method discussed in §11.7. For the boundary condition Eq. 12.14, both LL^+ and L^+L are self-adjoint non-negative operators as demonstrated above. Since M is a compact manifold with boundary, the spectra of both L^+L and LL^+ are discrete, and the eigenfunctions of L^+L are in one-to-one correspondence with that of LL^+. More precisely, if u is a solution to the equation

$$LL^+u = E^2u,$$

it is easy to verify that

$$v = \frac{1}{|E|}L^+u$$

is the solution to the equation

$$L^+L = E^2 v.$$

In a manner similar to that used in Eq. 11.9, we can introduce the heat equation as follows:

$$\begin{cases} (\frac{\partial}{\partial \beta} + LL_x^+)K_1(x,y;\beta) = 0, \\ (\frac{\partial}{\partial \beta} + L^+L_x)K_2(x,y;\beta) = 0, \end{cases} \tag{12.15}$$

with boundary condition Eq. 12.10 and initial condition:

$$K_1(x,y;0) = K_0(x,y;0) = \delta(x-y). \tag{12.16}$$

the analytic index of the Dirac operator is:

$$\begin{aligned} Ind &= dim\ Ker\ L^+ - dim\ Ker\ L \\ &= dim\ Ker\ LL^+ - dim\ Ker\ L^+L \\ &= \lim_{\beta\to\infty} \Delta(\beta), \\ \text{where} \qquad \Delta(\beta) &= Tr < x|e^{-\beta LL^+} - e^{-\beta L^+L}|y> \\ &= \int dx\ Tr[K_1(x,x;\beta) - K_2(x,x;\beta)] \end{aligned} \tag{12.17}$$

is the heat kernel function corresponding to Eq. 12.15 and Eq. 12.16. The eigenfunctions of LL^+ and L^+L are one-to-one correspondence, so that $\Delta(\beta)$ is in fact independent of β. Thus, taking the limit $\beta \to 0^+$, we get the asymptotic expression of $Tr|K(x,x;\beta)|$,

$$TrK_1(x,x;\beta)|_{\beta\to 0^+} \sim (4\pi\beta)^{-\frac{n}{2}} \sum_{j=0}^{\infty} \beta^j \mu_j(x) + R_1(\beta),$$
$$TrK_2(x,x;\beta)|_{\beta\to 0^+} \sim (4\pi\beta)^{-\frac{n}{2}} \sum_{j=0}^{\infty} \beta^j \nu_j(x) + R_2(\beta).$$

Substituting this into Eq. 12.15 we can get the cyclic formulas for $\mu_j(x)$ and $\nu_j(x)$ which are similar to that on compact manifold without boundary, and independent of the boundary conditions. The information of the contribution from the boundary appears in the last terms of the equations above, i.e.,

$$\begin{aligned} R_1(\beta) - R_2(\beta) &= \frac{1}{2}\sum_{\lambda} Sgn(\lambda) erfc(|\lambda|\beta^{1/2}) \\ \text{where} \qquad erfc(x) &= \frac{2}{\sqrt{\pi}}\int_x^{\infty} dt\ e^{-t^2} \end{aligned}$$

is the so called co-error function, and λ is the eigenvalue of H. The equation above corresponds to the regularization of the asymmetry of the spectrum of H:

$$\begin{aligned} \eta_H &= \lim_{\beta\to 0} \sum_{\lambda} Sgn(\lambda) erfc(|\lambda|\beta^{1/2}) \\ &= \lim_{\beta\to 0} 2[R_1(\beta) - R_2(\beta)]. \end{aligned}$$

Taking the higher temperature limit ($\beta \to 0^+$) for Eq. 12.16. Since $\Delta(\infty) = \Delta(0)$, we can have the APS index theorem:

$$\begin{aligned} Ind\not{D} &= V[M] + \frac{1}{2}\eta_H, \\ \text{here} \quad V[M] &= \int_M dx(\mu_n(x) - \nu_n(x)) = \int_M \hat{A}(M)\, ch(F), \end{aligned}$$

$V[M]$ is the same characteristic form as the case in which the manifold is compact and without boundary. The boundary correction is $\frac{1}{2}\eta_H$.

On the other hand, if the operator H have zero mode, then the projector P [Eq. 12.11] must instead by $\hat{P} = \sum_{\lambda \geq 0} |\lambda><\lambda|$. Then the result above had to be rewritten as:

$$Ind\ \not{D} = V[M] + \frac{1}{2}(h + \eta_H),$$

here h is the dimension of the zero mode of H, and

$$\eta_H = \lim_{s \to 0} \sum_{\lambda \neq 0} \frac{sign(\lambda)}{|\lambda|^s}.$$

Further the surface correction $S[\partial M]$ has to be included on the right hand side of the equation above, if the metric near ∂M is non-product as analyzed above.

12.5 Index theorem on open infinite manifold

Atiyah-Singer index theorem and Atiyah-Patodi-Singer index theorem are both valid on compact manifold, on which the spectrum of Dirac operator are discrete, the positive and negative eigenvectors are in one-to-one correspondence, so the heat kernel function $\Delta(\beta)$ is independent of β, the low temperature limit is the same as the high temperature limit. In physics, however, one frequently has to consider the continuum spectra for Dirac operator, and study the scattering state of background field with non-trivial topology, in which case the heat kernel function may be β dependent, and this dependence needs special attention. As the continuum eigenmodes are normalized, infinities such as $\Delta(0)$ often appear during normal ordering, so we have to be careful to avoid any possible singularities, and the regularization and renormalization are always used as in quantum field theory. All of these can be taken into account by the index theorem on open infinite manifolds[63].

Let M be an infinite n-dimensional open manifold endowed with Euclidean metric

$$ds^2 = g^{\mu\nu} dx^\mu dx^\nu,$$

and ∇_μ denotes the covariant derivative acting on spin field:

$$\nabla_\mu = \partial_\mu + \frac{1}{2}\omega_\mu^{ab}\sigma_{ab}.$$

Here $\omega_\mu^{ab} dx^\mu \equiv \omega^{ab}$ is the spin connection 1-form, and

$$\sigma_{ab} = \frac{1}{2}[\gamma_a, \gamma_b]$$

are the $SO(n)$ representation matrices with the γ_a matrices satisfy:

$$\gamma_a = \gamma_a^+, \quad \tfrac{1}{2}\{\gamma_a, \gamma_b\} = \delta_{ab},$$
$$\text{and} \quad \gamma_\mu(x) \equiv \gamma_a e_\mu^a(x) \qquad \text{satisfy} \quad \tfrac{1}{2}\{\gamma_\mu, \gamma_\nu\} = g_{\mu\nu} \quad .$$

The Dirac operator can be written as:

$$\not{D} = i\Gamma_a e_\mu^a(x)\nabla^\mu + K(x) \qquad ,$$
$$\text{here} \quad K(x) = \begin{pmatrix} 0 & A(x) \\ A^+(x) & 0 \end{pmatrix}$$

is the background field containing all other gauge potential, and Γ_a are constant matrices:

$$\Gamma_a = \begin{pmatrix} 0 & \gamma_a \\ \gamma_a^+ & 0 \end{pmatrix}, \quad \Gamma_5 = \begin{pmatrix} I & 0 \\ 0 & -I \end{pmatrix}.$$

In this representation $\not{D} = \begin{pmatrix} 0 & L \\ L^+ & 0 \end{pmatrix}$ has the property:

$$\{\not{D}, \Gamma_5\} = 0.$$

Just like the situation we analysed in the previous section, the analytic index of operator $\not{D}$ can be given by Eq. 12.7:

$$\begin{aligned} Ind\not{D} &= dim\ Ker\ L^+ - dim\ Ker\ L \\ &= \lim_{\beta\to\infty} \Delta(\beta). \end{aligned}$$

Now we turn to the topological index of $\not{D}$. The asymmetry of the spectrum is a manifestation of the non-trivial topology of the background field. For compact manifold, the spectrum is discrete and only the zero modes has topological contributions. Now that the density of continuum states could be asymmetric, so it can also have topological contributions. In order to overcome the difficulty of working with zero modes, one introduces a mass term:

$$\not{D}_m = \not{D} + m\Gamma_5 = \begin{pmatrix} m & L \\ L^+ & -m \end{pmatrix},$$

and it is easy to prove that:

$$\not{D}_m^2 = \not{D}^2 + m^2 \geq m^2,$$

so all eigenvalues of $\not{D}_m$ are non-zero, i.e.,

$$\not{D}_m\psi_\lambda = \lambda\psi_\lambda, \quad \lambda^2 \geq m^2.$$

And the η function is introduced to denote the asymmetry of the spectrum of operator $\not{D}$:

$$\eta_m(\epsilon) = \int_{-\infty}^{\infty} d\lambda\ \rho_m(\lambda) sgn(\lambda) e^{-\epsilon\lambda}. \tag{12.18}$$

The integral here comes directly from the fact that operator $\not{D}$ has continuum spectrum, and $\rho_m(\lambda)$ is the density of the spectrum, ϵ is the infinitesimal real parameter of regularization.

Recall that solutions u of $L^+u = 0$ constitute the eigenvectors $\begin{pmatrix} u \\ 0 \end{pmatrix}$ of operator $\not{D}_m$ with eigenvalue m, and solutions v of $Lv = 0$ constitute the eigenvectors $\begin{pmatrix} 0 \\ v \end{pmatrix}$ of operator $\not{D}_m$ with eigenvalue $-m$, as $\not{D}_m \to \not{D}$ for $m \to 0$, so the index of $\not{D}$ can be written as:

$$Ind\ \not{D} = sgn(m) \lim_{m\to 0} \lim_{\epsilon\to 0+} \eta_m(\epsilon).$$

Note that $\quad sgn(\lambda)e^{-\epsilon|\lambda|} = \frac{1}{\pi}\int_{-\infty}^{\infty} d\omega\ e^{-i\epsilon\omega}\frac{\lambda}{\lambda^2+\omega^2}, \qquad (12.19)$

so Eq. 12.18 can be written as:

$$\begin{aligned} \eta_m(\epsilon) &= \frac{1}{\pi}\int_{-\infty}^{\infty} d\omega\ e^{-i\epsilon\omega}\int_{-\infty}^{\infty} d\lambda\ \rho_m(\lambda)\frac{\lambda}{\lambda^2+\omega^2} \\ &= \frac{1}{\pi}\int_{-\infty}^{\infty} d\omega\ e^{-i\epsilon\omega}\int_M dx\ |g|^{\frac{1}{2}} Tr < x|\frac{\not{D}_m}{\not{D}_m^2+\omega^2}|x >, \end{aligned}$$

where $\{|x>\}$ are the complete set of eigenmodes of $\not{D}_m$, satisfying:

$$\begin{aligned} &< x|y >= |g|^{\frac{1}{2}}\delta(x-y), \\ &\int_M dx\ |g|^{\frac{1}{2}}|x >< x| = 1. \end{aligned}$$

Making use of the point splitting technique, i.e., regularization which is often used in quantum field theory, we have:

$$\begin{aligned} Tr < x|\frac{\not{D}_m}{\not{D}_m^2+\omega^2}|y > &= m\ Tr < x|\Gamma_5\frac{1}{\not{D}^2+m^2+\omega^2}|y > \\ &= i\frac{m}{\sigma}Tr < x|\Gamma_5\frac{1}{\not{D}+i\sigma}|y >, \end{aligned}$$

where $\sigma = \sqrt{m^2+\omega^2}$. To avoid any possible divergence that might arise from $x = y$, we shall momentarily assume $x \neq y$.

Next we introduce the trace identity which is the generalization of the axial anomaly equation.

Proposition (trace identity):

$$\begin{aligned} Tr < x|\Gamma_5|x > &= \frac{i}{2}\nabla_\mu Tr < x|\sqrt{|g|}\Gamma^\mu \frac{1}{\sqrt{|g|}(\not{D}+i\sigma)}|x > \\ &+ i\sigma Tr < x|\Gamma_5 \frac{1}{\sqrt{|g|}(\not{D}+i\sigma)}|x > . \end{aligned}$$

Proof: By using the regularization technique in quantum field theory (so called quantum regularization):

$$\begin{aligned} Tr < x|\Gamma_5|x > &= \lim_{s\to 0} Tr < x|\Gamma_5 \frac{1}{|\sqrt{|g|}(\not{D}+i\sigma)|^s}|x > \\ &= \lim_{s\to 0}\frac{1}{2}[Tr < x|\Gamma_5\sqrt{|g|}(\not{D}+i\sigma)\frac{1}{|\sqrt{|g|}(\not{D}+i\sigma)|^{s+1}}|x > \\ &+ Tr < x|\frac{1}{|\sqrt{|g|}(\not{D}+i\sigma)|^{s+1}}\sqrt{|g|}(\not{D}+i\sigma)|x > . \end{aligned}$$

Note that $\nabla_\mu |g|^{\frac{1}{2}}\Gamma^\mu = |g|^{\frac{1}{2}}\Gamma^\mu\nabla_\mu$ and

$$\not{D} = i\Gamma^\mu\nabla_\mu + K(x), \quad \{\gamma_\mu, \Gamma_5\} = 0, \quad \{K, \Gamma_5\} = 0.$$

One can obtain:

$$\begin{aligned} Tr < x|\Gamma_5|x > &= \lim_{s\to 0}\{\frac{i}{2}\nabla_\mu Tr < x|\sqrt{|g|}\Gamma^\mu\Gamma_5\frac{1}{|\sqrt{|g|}(\not{D}+i\sigma)|^{s+1}}|x > \\ &\quad i\sigma Tr < x|\Gamma_5\frac{1}{|\sqrt{|g|}(\not{D}+i\sigma)|^{s+1}}|x >\} \\ &= \frac{i}{2}\nabla_\mu Tr < x|\sqrt{|g|}\Gamma^\mu\Gamma_5\frac{1}{\sqrt{|g|}(\not{D}+i\sigma)}|x > \\ &+ \quad i\sigma Tr < x|\Gamma_5\frac{1}{\sqrt{|g|}(\not{D}+i\sigma)}|x > . \end{aligned}$$

This completes the proof of the identity.

From the trace identity, one have:

$$\begin{aligned} \eta_m(\epsilon) &= sgn(m)e^{-\epsilon|m|}\int_M dx\,\sqrt{|g|}Tr < x|\Gamma_5|x > \\ &+\frac{1}{2\pi}\int_{-\infty}^{\infty} d\omega e^{-i\epsilon\omega}\frac{m}{\omega^2+m^2} + \oint d\Sigma_\mu Tr < x|i\Gamma^\mu\Gamma_5\frac{1}{\not{D}+i\sigma}|x > . \end{aligned}$$

Here the first term is the contribution from the anomaly, determined by the topological properties of the background field, and can be written in terms of the integral of the characteristic class over M, i.e.,

$$\int_M dx\sqrt{|g|}Tr < x|\Gamma_5|x >= \int_M \hat{A}(M)ch(F) \equiv V[M].$$

The second term is called the surface term, N is the hypersurface of manifold M at infinity. For simplicity, we assume that the metric of M at infinity can be written as:

$$ds^2 = dr^2 + r^2 g_{ij}dx^i dx^j,$$

and the outer normal of surface N is denoted as:

$$n^\mu(x) = E^\mu_a(x)n^a.$$

Further assume the normal to be geodesic flat asymptotically :

$$n^\mu\omega_\mu \to 0, \quad n^\mu\partial_\mu n_\nu \to 0, \quad \text{for} \quad r \to \infty,$$

and all the background field normal coordinate independent. Then the surface term can be written as:

$$\begin{aligned} &\tfrac{1}{2\pi}\textstyle\int_{-\infty}^{\infty} d\omega e^{-i\epsilon\omega}\tfrac{m}{\omega^2+m^2}\int d\mu_N Tr < x|i\hat{\Gamma}\Gamma_5\tfrac{1}{\not{D}+i\sigma}|x > \\ &= -\tfrac{1}{2}\textstyle\int_N d\mu_N Tr < x|i\Gamma_5\tfrac{1}{i\hat{\partial}+i\hat{\Gamma}\Gamma^T\nabla_T+\hat{\Gamma}K}|x >, \end{aligned}$$

where $\hat{\Gamma} = n^\mu\Gamma_\mu, \Gamma^T_\mu = \Gamma_\mu - \hat{\Gamma}n_\mu.$

Take the limit $m \to 0^+$, let $\mathcal{H} = \hat{\gamma}\gamma^+_T\nabla^T - i\hat{\gamma}A^+$ and remember that:

$$\delta(\omega) = \lim_{m\to 0^+}\frac{1}{\pi}\frac{m}{m^2+\omega^2},$$

then the surface term becomes:

$$\frac{1}{2}\int_N d\mu_N Tr < x|\frac{i}{i\mathcal{H}+i\hat{\partial}} - \frac{i}{-i\mathcal{H}^+ + i\hat{\partial}}|x > .$$

Note here that all the background fields are radial coordinate independent, the Fourier transformation of the formula above is:

$$\frac{1}{4\pi}\lim_{t\to 0^+}\int_{-\infty}^{\infty} d\hat{\omega}e^{-it\hat{\omega}}\int_N d\mu_N Tr < x|\frac{1}{i\mathcal{H}+i\hat{\omega}} + \frac{i}{i\mathcal{H}^+ - i\hat{\omega}}|x > .$$

For simplicity we assume that $\mathcal{H}$ and $\mathcal{H}^+$ have no zero modes, and they have identical spectrum density, then the expression above becomes:

$$\begin{aligned} \frac{-i}{4\pi}\lim_{t\to 0^+}\int d\mu(\lambda)\int_{-\infty}^{\infty} d\hat{\omega}e^{-it\hat{\omega}}(\frac{1}{\hat{\omega}-i\lambda} - \frac{1}{\hat{\omega}+i\lambda^*}) &= \frac{1}{2}\int d\mu(\lambda)sgn(Re(\lambda)) \\ &= \frac{1}{2}\eta(Re(\mathcal{H})), \end{aligned}$$

where Eq. 12.19 is used in the last step. Substitute this in to Eq. 12.18 and we get the index theorem on open infinite manifold M:

$$\begin{aligned} Ind\ \not{D} &= \Delta(\infty) \\ &= \Delta(0) + \frac{1}{2}\eta(Re(\mathcal{H})) \\ &= V[M] + \frac{1}{2}\eta(Re(\mathcal{H})). \end{aligned}$$

This is quite similar to the APS index theorem on manifold with boundary, but with some differences:

1. For APS index theorem, spectral boundary condition (non-local boundary condition) is used. So the heat kernel function $\Delta(\beta)$ is β independent, $\Delta(\beta) = \Delta(0)$ gives the APS index theorem for Dirac operator on manifold with boundary.

2. For open infinite manifold, Dirac operator has continuous spectrum, the spectral density is asymmetric so the heat kernel $\Delta(\beta)$ is a non-trivial function of β, $\Delta(0) \neq \Delta(\beta)$, and trace identity has to be used. These two cases coincide if $\mathcal{H}^+ = \mathcal{H}$ on an infinite hypersurface.

Finally we take a brief look at the vacuum fermion number problem, which is related to the spectrum of Dirac Hamiltonian, and can be expressed in terms of the APS η-invariant. In the model, in which there is charge conjugation symmetry relating to the positive and negative energy eigenmodes of the Dirac Hamiltonian, vacuum fermion number can be evaluated from the zero modes of Dirac operator in topologically no-trivial background field. In the case, where charge conjugation symmetry is absent, vacuum fermion charge evaluation usually requires a non-trivial continuum state contribution.

Note that zero modes are absent when $m \neq 0$, we introduce the second quantized Dirac field with ζ-function regulator:

$$\psi_s(x) = \int dk(b_k u_k(x) + d_k^+ v_k(x))|\lambda(k)|^{-\frac{s}{2}},$$

here $u_k(x)$ and $v_k(x)$ are the positive and negative solutions of the eigenvalue equation: $\not{D}\psi_\lambda(k,x) = \lambda\psi_\lambda(k,x)$, operators b_k, d_k^+ are the annihilation and creation operators for the fermion and anti-fermion bound states respectively. Then the fermion number operator is:

$$Q_s = \int dx\ :\psi_s^+(x)\psi_s(x) := \frac{1}{2}\int dx[\psi_s^+(x), \psi_s(x)],$$

where : : is the normal ordering operation. Substitute the eigenmodes expansion into Q_s and we get:

$$Q_s = \textstyle\int dk[b_k^+ b_k - d_k^+ d_k]|\lambda(k)|^{-s} - \frac{1}{2}\eta_{\not{D}}s,$$

$$\text{where} \quad \eta_{\not{D}}s = \textstyle\int_{-\infty}^{\infty} d\lambda\rho_{\not{D}}(\lambda)sgn(\lambda)|\lambda|^{-s} = 2\int_0^\infty d\lambda\sigma_{\not{D}}(\lambda)\lambda^{-s} \quad ,$$

in which $\rho_{\not{D}}(\lambda)$ is the spectra density matrix, $\sigma_{\not{D}}(\lambda)$ is its odd part:

$$\sigma_{\not{D}}(\lambda) = \frac{1}{2}(\rho_{\not{D}}(\lambda) - \rho(-\lambda)).$$

The fermion number operator changes sign under charge conjugation so the fermion number in ground state is:

$$< N >= \lim_{s\to 0} < Q_s >= -\frac{1}{2}\eta_{\not{D}}(0) = -\lim_{s\to 0}\int_0^\infty d\lambda \sigma_{\not{D}}(\lambda)\lambda^{-s}.$$

Thus the ground state fermion number is the Mellin transformation of the odd spectral density evaluated at $s = 0$. Its value can be obtained by using the trace identity above.

12.6 Weak local boundary condition for Dirac operator

We have mentioned in §12.4 that for Dirac operator, local boundary conditions give rise to topological obstructions, so spectral boundary condition has to be used for fermionic field. But this spectral boundary condition is non-local, it violates causality so is physically undesirable. Let us look at a simple example. Assume that the background field to be time independent, and the space to be flat circular disc of radius R. The Dirac Hamiltonian for fermions interacting with static spherical symmetric Abelian gauge field can be written as:

$$H = \alpha^k(-i\partial_k + A_k) + \beta m.$$

In the representation where

$$\alpha^1 = \begin{pmatrix} 0 & i \\ -i & 0 \end{pmatrix}, \quad \alpha^2 = \begin{pmatrix} 0 & 1 \\ 1 & 0 \end{pmatrix}, \quad \beta = \begin{pmatrix} 1 & 0 \\ 0 & -1 \end{pmatrix},$$

we have

$$H = \begin{pmatrix} m & L \\ L^+ & -m \end{pmatrix}.$$

$$\text{where} \quad \begin{cases} L = \partial_x - iA_x - i\partial_y - A_y, \\ L^+ = -\partial_x + iA_x - i\partial_y - A_y. \end{cases}$$

For the static field, we take the Coulomb gauge:

$$A_0 = 0, \quad A_i = -\epsilon_{ij}\partial_j\chi,$$
$$L = (\partial_x - i\partial_y) - (\partial_x - i\partial_y)\chi \equiv \nabla - \nabla\chi,$$
$$L^+ = -(\partial_x + i\partial_y) - (\partial_x + i\partial_y)\chi \equiv -\bar{\nabla} - \bar{\nabla}\chi.$$

From the symmetry of the problem we choose the plane polar coordinate:

$$x = \rho\cos\theta, \quad y = \rho\sin\theta,$$

then

$$\nabla = \partial_x - i\partial_y = e^{-i\theta}(\partial_\rho - \frac{i}{\rho}\partial\theta),$$

$$\bar{\nabla} = \partial_x + i\partial_y = e^{i\theta}(\partial_\rho + \frac{i}{\rho}\partial\theta).$$

The eigenvalue equation of H is:

$$H\psi_E = \begin{pmatrix} m & L \\ L^+ & -m \end{pmatrix}\psi_E = E\psi_E, \quad \psi_E = \begin{pmatrix} u \\ v \end{pmatrix},$$

i.e.,

$$Lv = (E-m)u,$$
$$L^+u = (E+m)v.$$

The vacuum charge, induced by the background field, is given by the spectral asymmetry of H:

$$< N >= -\frac{1}{2}\eta_H(0) = -\frac{1}{2}\lim_{s\to 0}\sum_n sgn(E_n)|E_n|^{-s}.$$

Note that operator H has an eigenvalue-reversing discrete symmetry when $E \neq \pm m$, the contribution to $\eta_H(0)$ cancel each other except for $E = \pm m$.

1. $E = m, \quad v = 0$. Solutions of $L^+u = 0$ are:

$$u_n = e^{-\alpha}\rho^n e^{-in\theta}, \quad n = 0, 1, 2, \cdots. \tag{12.20}$$

2. $E = -m, \quad u = 0$. Solutions of $Lv = 0$ are:

$$v_n = e^{\alpha}\rho^n e^{-in\theta}, \quad n = 0, 1, 2, \cdots. \tag{12.21}$$

As we require the wavefunctions be square integrable around $\rho = 0$, so only positive values of n are acceptable.

From the Hermiticity of H we have:

$$< u, Lv >=< L^+u, v > . \tag{12.22}$$

So under this boundary condition the Stokes theorem gives:

$$\int u^* v|_{\rho=R} d\theta = 0.$$

If we take the local boundary condition (Dirichlet condition) instead, then

$$\psi|_R = 0 \implies u|_R = 0, \quad v|_R = 0. \tag{12.23}$$

Namely there are no non-trivial solutions.

Usually the spectral boundary condition consistent with Eq. 12.22, is taken to obtain physically sensible results. As an example, take the angular momentum operator

$$J = \rho \times \nabla = -i\partial_\theta - \rho\partial_\rho\alpha + \frac{1}{2}\sigma_3.$$

Assuming that the gauge potential becomes a pure gauge locally at the boundary, then

$$\alpha \rightarrow \phi \ln\rho, \quad \text{for } \rho \rightarrow R. \tag{12.24}$$

$$\text{Where} \qquad \phi = \frac{1}{2\pi}\int_D F = \frac{1}{2\pi}\int_{\partial D} A_\theta d\theta = \rho\alpha'$$

is the integral of the first Chern class. Then

$$J|_R = -i\partial_\theta - \phi + \frac{1}{2}\sigma_3.$$

Thus, near the boundary, solution Eq. 12.21 can be written as:

$$\begin{aligned} u_n|_R &= R^{n-\phi}e^{in\theta}, \\ v_n|_R &= R^{n+\phi}e^{-in\theta}. \end{aligned} \tag{12.25}$$

Let P denote the projector onto the positive eigenvalue of $J|_R$, the spectral boundary condition gives:

$$\begin{aligned} &(1+\sigma_3)P\psi|_R = 0 &&\Longrightarrow \quad n - \phi + \tfrac{1}{2} < 0, \\ &(1-\sigma_3)(1-P)\psi|_R = 0 &&\Longrightarrow \quad -n - \phi - \tfrac{1}{2} > 0. \end{aligned}$$

Number n thus is heavily constrained by the boundary condition, and the analytic index of H has the definite number:

$$\begin{aligned} Ind\ H_0 &= dim\ Ker\ L^+ - dim\ Ker\ L \\ &= [|\phi + \frac{1}{2}|]. \end{aligned}$$

In which [ ] denotes the integer part. The asymmetric part η_J of spectral J is:

$$\eta_J = -2\{|\phi + \frac{1}{2}|\} + 1,$$

where { } denotes the fractional part. Comparing the two equations above gives:

$$Ind\ H_0 = \phi + \frac{1}{2}\eta_J.$$

This is just the result from APS index theorem on manifold with boundary.

The vacuum fermion charge

$$\begin{aligned} < N > &= -\frac{1}{2} sgn(m) Ind\ H_0 \\ &= -\frac{1}{2} sgn(m)[\phi + \frac{1}{2}] \end{aligned} \quad (12.26)$$

is a half integer. For circular symmetric vortex background field having an orientation-reversing isometry, from the index theorem on open infinite manifold and making use of the trace identity in quantum field theory, the tangent direction contribution can be ignored, and

$$\begin{aligned} < N > &= -\frac{1}{2} sgn(m) V[M] \\ &= -\frac{1}{2} sgn(m) \Phi. \end{aligned}$$

The fermionic charge of background field is a fractional charge!

The two results above are not consistent. The reason is that the non-local spectral boundary is taken in Eq. 12.26, hence there is an extra contribution to the charge localized near the boundary.

Polychronakos[64] introduced the weak local boundary condition for spin field: one of the spinor components vanishes pointwise. For example:

$$u|_R = 0, \quad \text{or} \quad v|_R = 0.$$

The first gives solutions Eq. 12.20, while the second gives solutions Eq. 12.21. For concreteness, we shall considered the later one only, i.e., choice $v|_R = 0$. There are infinitely many solutions if $m \to 0$ and all of them are zero modes of H_0, so $Ind\ H_0$ diverges. To have a well defined result, as the regularization and renormalization in quantum field theory, we redefine:

$$Ind\ H_0 = \lim_{\epsilon \to 0^+} [\sum_{n=0}^{\infty} \int_0^{R-\epsilon} (|u_n(A)|^2 - |u_n(0)|^2) d^2x].$$

Here $u_n(0)$ are the solutions of trivial gauge potential $A = 0$,

$$u_n(0) = c_n^0 \rho^n e^{in\theta}, \quad n = 0, 1, \cdots.$$

The Hamiltonian also has a spectral asymmetry, and induced fermionic charge can be obtained form the asymmetric part of $H(A)$.

Since each term in the summation above goes to 1 as $\epsilon \to 0$, so we can neglect any finite number of initial terms from both summations. For sufficiently large $n \gg N$, the zero modes become highly localized near the boundary, and α can be approximated by Eq. 12.23, and u_n by Eq. 12.25. So the integral in the redefined $Ind\ H_0$ above can be calculated and we have:

$$Ind\ H_0 = \lim_{\epsilon \to 0+} \sum_{n=N}^{\infty} [(1 - \frac{\epsilon}{R})^{2(n-\phi+1)} - (1 - \frac{\epsilon}{R})^{2(n+1)}] = \phi.$$

And the vacuum charge is:

$$< N >= -\frac{1}{2} sgn(m)\phi,$$

which is the same as for the open infinite manifold.

From this simple example, we see that spectral boundary condition is needed for fermionic field if the final result is to be well defined, which is the same as in the usual APS index theorem. The APS index theorem can be generalized to open infinite manifold if we apply the regularization and renormalization techniques from quantum field theory and take the weak local boundary condition for spin field, then the final result is physically acceptable.

Notation and Formula

- Atiyah-Patodi-Singer index theorem

$$Ind(E, D, \partial M) = V[M] + S[\partial M] + \frac{1}{2}(\eta(\partial M) + h).$$

- For the spin complex

$$Ind\ \not{D} = V[M] + \frac{1}{2}(h + \eta_H),$$

  here h is the dimension of the zero mode of the tangent part H of operator D, and

$$\eta_H = \lim_{s \to 0} \sum_{\lambda \neq 0} \frac{sign(\lambda)}{|\lambda|^s}.$$

- On open infinite manifold

$$Ind\ \not{D} = V[M] + \frac{1}{2}\eta(Re(\mathcal{H})),$$

  here $\mathcal{H} = \hat{\gamma}\gamma_T^+ \nabla^T - i\hat{\gamma}A^+$ and $\hat{\gamma} = n^\mu \gamma_\mu, \gamma_\mu^T = \gamma_\mu - \hat{\gamma} n_\mu . n_\mu$ is the asymptotically flat unit normal.

Exercise

1. What are the local and nonlocal boundary condition respectively? Try to find out the difficulties taking the local boundary condition for Dirac operator if the background field is endowed with non-trivial topology.

Chapter 13

Family Index Theorem, Topological properties of Quantum Gauge Theory

Let $\mathcal{U} = \{A\}$ be the set of connections on $P(M, G)$, called the connection space or gauge potential space. As disscussed in §10.7, this connection space is a topologically trivial convex set, and is contractible.

One can construct the space of gauge orbit $\mathcal{U}/\mathcal{G}$(orbit space) by identifying points in $\mathcal{U}$ which correspond to gauge potentials differing by gauge transformations. Here $\mathcal{G}$ is the group of gauge transformation, i.e., it maps $g(x)$ from base manifold M into group G. $\mathcal{U}/\mathcal{G}$ is nontrivial simply because $\mathcal{G}$ is topologically nontrivial.

In considering the interaction between gauge field and matter field, the action is functional on connection space $\mathcal{U} = \{A\}$. On the one hand, we keep in mind that the physical system is gauge invariant and the dynamical variables are functionals on orbit space $\mathcal{U}/\mathcal{G}$. On the other hand, gauge potential is field variable and gauge invariance is merely a constraint on the physical system, so we still need to lift the orbit space to the entire connection space to analyse the constrained system on connection space $\mathcal{U}$.

$\mathcal{U}$, $\mathcal{U}/\mathcal{G}$, $\mathcal{G}$ are closely related by the Atiyah-Singer family index theorem[65]. In §13.1, we construct the index bundle of Dirac operator family on connection space in terms of Atiyah-Singer formulation. We study the gauge group cohomology, orbit space cohomology and connection space cohomology, and proceed to analyse their relations[66]. In §13.2 we study the relationship between field theory on orbit space and cohomology of gauge group (BRST cohomology), and relate the family index on orbit space with topological obstruction of gauge group. In §13.3, we give the global topological interpretation of cocycles of gauge group in terms of Čech-de Rham double complex[67]. We can overcome topological obstruction and promote it to a higher order by using the subdivision of regions. In the last part of this chapter, we look at various applications of family index theorm in physics and analyse the global topological properties of quantum field theory and quantum anomaly[68-74].

13.1 Family index theorem of Dirac operator

For fermion field interacting with gauge field, one have the Dirac operator $\not{D}_A$

$$\not{D}_A \equiv \gamma^\mu D_\mu \equiv \gamma^\mu(\partial_\mu + A_\mu),$$

in which A is the gauge potential. The analytic index of $\not{D}_A$ is:

$$Ind\ \not{D}_A = \dim\ Ker \not{D}_A - \dim\ Ker \not{D}_A^+.$$

Both $Ker\ \not{D}A$ and $Ker\ \not{D}_A^+$ change with A, while their difference $Ker\ \not{D}_A - Ker\ \not{D}_A^+$ could be a well defined vector bundle on $\mathcal{U}$ and is called the index bundle:

$$Ind\ \not{D}_A = Ker\ \not{D}_A - Ker\ \not{D}_A^+ \ \in K(\mathcal{U}),$$

where $K(\mathcal{U})$ is a formal difference in K-theory (see Appendix Q). $Ind\not{D}_A$ is covariant under gauge transformation. By identifying points in $\mathcal{U}$ which correspond to gauge potential differing by gauge transformations, $Ind\ \not{D}_A$ becomes a vector bundle on $\mathcal{U}/\mathcal{G}$, i.e.,

$$Ind\ \not{D}_A = Ker\ \not{D}_A - Ker\ \not{D}_A^+ \ \in K(\mathcal{U}/\mathcal{G}). \tag{13.1}$$

Now we look at the characteristic classes of the index bundle. First we consider the connection bundle $\mathcal{U}$, which is a principal bundle on $\mathcal{U}/\mathcal{G}$ with action by the gauge group $\mathcal{G}$:

$$\begin{array}{ccc} \mathcal{G} & \rightarrow & \mathcal{U} \\ & & \downarrow \\ & & \mathcal{U}/\mathcal{G} \end{array}$$

Pick a cross section on $\mathcal{U}$, i.e., pick a representive point $a = a_\mu(x)dx^\mu$ on any orbit, which is a Lie algebra $\mathbf{g}$ valued 1-form, such that each point A on the same orbit (same fibre) is different from a by a local gauge transformation:

$$A = g^{-1}ag + g^{-1}dg, \quad g \in \mathcal{G}.$$

The covector at A on bundle $\mathcal{U}$ can be written as:

$$\begin{aligned} \delta A &= g^{-1}\delta a g - D_A g^{-1}\delta g \equiv \delta_0 A - D_A v, \\ \text{here} \quad v &= g^{-1}\delta g, \quad D_A v = dv + \{A, v\}, \quad \delta_0 A = g^{-1}\delta a g. \end{aligned}$$

For simplicity we assume that D_A has no zero mode, and only D_A^+ has zero modes. The elliptic Hermitian operator $D_A^+ D_A$ is then positive definite and one can define the Green function as:

$$G_A \equiv (D_A^+ D_A)^{-1}.$$

And the connection $\mathcal{A}$ on connection bundle $\mathcal{U}$ is defined as:

$$\mathcal{A} \equiv -G_A D_A^+ \delta A \equiv u + v, \quad u = -G_A D_A^+ \delta_0 A. \tag{13.2}$$

It is easy to prove that

$$\mathcal{A}|_{\text{fibre}} = g^{-1}\delta g = v,$$

satisfies the fundamental criteria for the connection (Eq.9.1). That is, a choice of horizontal tangent bundle in T(M) transverse to the vertical tangent bundle, then give a connection on fibre bundle. So one can decompose δA into vertical and horizontal parts:

$$\delta A = \delta^{\perp} A + \delta^{\|} A.$$

Here the vertical covector

$$\delta^{\perp} A = D_A G_A D_A^+ \delta A = -D_A \mathcal{A}$$

is the image of D_A, and the horizontal covector

$$\delta^{\|} A = (1 - D_A G_A D_A^+)\delta A \equiv \pi \delta A$$

satisfies $$D_A^+ \delta^{\|} A = 0.$$

Namely it is an element of $Ker\ D_A^+$. So connection $\mathcal{A}$ vanishes along the direction of $\delta^{\|} A$. The projection operator $\pi = 1 - D_A G_A D_A^+$ satisfies

$$\pi^2 = \pi = \pi^+.$$

It projects δA to the kernel of D_A^+, i.e., sending $T_A^*(\mathcal{U})$ to its horizontal subspace H_A.

Next let us look at the following product bundle

$$P(M, G) \times \mathcal{U}(\mathcal{U}/\mathcal{G}, \mathcal{G}),$$

on which there is the total connection: $A + \mathcal{A} = A - G_A D_A^+ \delta A$. The corresponding total curvature is

$$\begin{aligned} \mathcal{F} &= (d + \delta)(A + \mathcal{A}) + \frac{1}{2}[A + \mathcal{A}, A + \mathcal{A}] \\ &= F + \delta^{\|} A + \delta \mathcal{A} + \mathcal{A}^2. \end{aligned}$$

Let Λ_j^k denote the set of j-forms on M and k-forms on $\mathcal{U}$, then

$$F = dA + A^2 \in \Lambda_2$$

$$\delta^{\|} A = (1 - D_A G_A D_A^+)\delta A \in \Lambda_1^1$$

$$\delta \mathcal{A} + \mathcal{A}^2 \in \Lambda^2$$

as $$\delta^2 = \delta^{\|2} = \delta^{\perp 2} = 0,$$

so $$\delta^{\perp} D_A \mathcal{A} = -D_A \delta^{\perp} \mathcal{A} + [\delta^{\perp} A, d] = -D_A(\delta^{\perp} \mathcal{A} + \mathcal{A}^2) = 0.$$

As D_A has no zero mode, so

$$\delta^{\perp}\mathcal{A} + \mathcal{A}^2 = 0.$$

Thus $\delta\mathcal{A} + \mathcal{A}^2 = \delta^{\|}\mathcal{A}$ is a horizontal 2-form on $\mathcal{U}$, consequently the total curvature

$$\mathcal{F} = F + \delta^{\|}A + \delta^{\|}\mathcal{A} \tag{13.3}$$

has horizontal component only.

Restricting the connection of $\mathcal{U}$ to the irreducible case, the action of $\mathcal{G}$ on $P \times \mathcal{U}$ given by

$$(u, A) \longrightarrow (ug, A^g)$$

is free (no fix point). Note here that the action of $\mathcal{G}$ commutes with the action of G, one can construct a principal bundle $Q = P \times \mathcal{U}/\mathcal{G}$,

$$\begin{array}{ccc} G & \rightarrow & Q \\ & & \downarrow \\ & & M \times \mathcal{U}/\mathcal{G}. \end{array}$$

The connection on Q can be easily obtained from the connection $A + \mathcal{A}$ on bundle $P \times \mathcal{U}$, its corresponding curvature has horizontal component only (the horizontal lift of $M \times \mathcal{U}/\mathcal{G}$) just like Eq. 13.3.

To study the topological obstruction to the existence of chiral Dirac operator with background field, one starts from studing the characteristic class of index bundle (Eq. 13.1),

$$ch(Ind\not{D}_A) = \int_M \hat{A}(M)ch(Q).$$

Here $\hat{A}(M)$ is the A-roof polynomial on base manifold M. $\hat{A}(M) = 1$ for $M = S^{2n}$. The Chern character $ch(Q) = Tr(\exp \frac{i}{2\pi}\mathcal{F})$. Expand $\mathcal{F}$ in terms of differential forms on $M \times \mathcal{U}$. Integrate the order-n symmetric invariant polynomial of $\mathcal{F}$ over M yields the family index (assuming M to be $(2n-k-1)$-dimensional):

$$Q^{k+1}[M] = \int_M Tr(\mathcal{F}^n).$$

It is easy to see from Eq.13.3 that Q^{k+1} is a $(k+1)$-form on $\mathcal{U}/\mathcal{G}$. This is the result given by Atiyah-Singer.

Theorem 13.1 *(Family index theorem) The characteristic classes for the index bundle of Dirac operator $\not{D}_A$*

$$ch^{k+1}(Ind\not{D}_A) = c_n Q^{k+1}[M] = c_n \int_M Tr(\mathcal{F}^n), \text{ here } c_n = \frac{i^n}{n!(2\pi)^n}, \tag{13.4}$$

can be written as differential forms on $\mathcal{U}/\mathcal{G}$ only.

13.2 Relation among cohomology on orbit space, on connection space and on gauge group

From the family index theorem, $Q^{k+1}[M]$ is a $(k+1)$-form on $\mathcal{U}/\mathcal{G}$ only. Let $\bar{\delta}A$ be the horizontal variation satisfying:

$$D_A^{+}\bar{\delta}A = 0$$

and take the local horizontal gauge

$$\bar{\delta}\mathcal{A} + \mathcal{A}^2 = 0,$$

$$\text{then} \qquad \mathcal{F} = F + \bar{\delta}A.$$

Expand $Tr(\mathcal{F}^n)$ in the powers of $\bar{\delta}A$ yields

$$Tr(\mathcal{F}^n) = \sum_{k=0}^{n} \bar{q}^k_{2n-k}(\bar{\delta}A, F), \tag{13.5}$$

$$\text{here} \qquad \bar{q}^k_{2n-k}(\bar{\delta}A, F) = \tfrac{n!c_n}{(n-k)!k!} Str[(\bar{\delta}A)^k, F^{n-k}] \ \in \Lambda^k_{2n-k}$$

are $(2n-k, k)$-forms on $M \times \mathcal{U}/\mathcal{G}$. Then from

$$(d + \bar{\delta})Tr(\mathcal{F}^n) = 0$$

and compare terms of the same order, one gets the descent equations:

$$\begin{cases} d\bar{q}^0_{2n} = dTr(F^n) = 0, \\ d\bar{q}^k_{2n-k} + \bar{\delta}\bar{q}^{k-1}_{2n-k+1} = 0 \quad (k = 1, \cdots, n), \\ \bar{\delta}\bar{q}^n_n = 0. \end{cases} \tag{13.6}$$

Keep in mind that the base manifold is $(2n-k-1)$-dimensional, and substitute Eq. 13.5 into the family index formula (Eq.13.4), one has

$$\begin{aligned} \bar{\delta}\bar{Q}^{k+1}[M] &= \bar{\delta}\int_M \bar{q}^{k+1}_{2n-k-1} = (-1)^{k+1}\int_M \bar{\delta}\bar{q}^{k+1}_{2n-k-1} \\ &= (-1)^k \int_M d\bar{q}^{k+2}_{2n-k-2} = (-1)^k \bar{Q}^{k+2}[\partial M]. \end{aligned}$$

The second equality is a result of $d\bar{\delta} = -\bar{\delta}d$. For M compact and without boundary, $\partial M = 0$, so

$$\bar{\delta}\bar{Q}^{k+1}[M] = 0.$$

As the orbit space $\mathcal{U}/\mathcal{G}$ is topologically non-trivial, so $\bar{Q}^{k+1}[M] \neq \bar{\delta}(\cdot)$ in general. Next we lift and embed the k-form to orbit space in the bundle $\mathcal{U}$, thus becoming a k-form on $\mathcal{U}$. Expand $Tr(\mathcal{F}^n)$ in the powers of any given variation δA on $\mathcal{A}$, then one has

$$Tr(\mathcal{F}^n) = \sum_{k=0}^{2n} q^k_{2n-k}.$$

From $(d+\delta)Tr(\mathcal{F}^n)=0$, one obtains the descent equations analogous to Eq. 13.6:

$$\begin{cases} dq_{2n}^0 = dTr(F^n) = 0, \\ dq_{2n-k}^k + \delta q_{2n-k+1}^{k-1} = 0 \quad (k=1,\cdot,2n), \\ \delta q_0^{2n} = 0. \end{cases}$$

Then from the family index formula $Q^{k+1}[M] = \int_M q_{2n-k-1}^{k+1}$, one gets $\delta Q^{k+1}[M] = (-1)^k Q^{k+2}[\partial M]$. If M is compact and without boundary, $\partial M = 0$, so

$$\delta Q^{k+1}(M) = 0.$$

Here δ is any given variation on $\mathcal{A}$. As space $\mathcal{U}$ is topologically trivial, one has the following as a result of Poincare lemma:

$$Q^{k+1}[M] = \delta\beta^k[M], \tag{13.7}$$

where $\beta^k[M]$ can be determined as follows: As $Tr(\mathcal{F}^n)$ satisfies

$$(d+\delta)Tr(\mathcal{F}^n) = 0,$$

so locally

$$Tr(\mathcal{F}^n) = (d+\delta)\omega_{2n-1}(A+\mathcal{A}),$$

here

$$\begin{aligned} \omega_{2n-1}(A+\mathcal{A}) &= Q_{2n-1}(0, A+\mathcal{A}) \\ &= n\int_0^1 dtTr(A+\mathcal{A}, \mathcal{F}_t^{n-1}), \\ \mathcal{F}_t &= t\mathcal{F} + t(t-1)(A+\mathcal{A})^2. \end{aligned}$$

Expand $\omega_{2n-1}(A+\mathcal{A})$ in the powers of A:

$$\omega_{2n-1}(A+\mathcal{A}) = \sum_{n=0}^{2n-1} \omega_{2n-k-1}^k(A, \mathcal{A}).$$

Substitute this into the family index formula (Eq.13.4) gives:

$$\begin{aligned} Q^{k+1} &= \int_M \delta\omega_{2n-k-1}^k(A, \mathcal{A}) \\ &= (-1)^{k+1}\delta\int_M \omega_{2n-k-1}^k(A, \mathcal{A}). \end{aligned}$$

Comparing with Eq. 13.7 and one has

$$\beta^k[M] = (-1)^{k+1}\int_M \omega_{2n-k-1}^k(A, \mathcal{A}).$$

In the following we restrict the variation along the fibre, i.e., along the direction of the image of D_A, which is called the vertical variation $\hat{\delta}A$, and require $\hat{\delta}A$ to satisfy

$$\begin{aligned} &\hat{\delta}a = 0, \\ \text{then} \qquad &\hat{\delta}A = -D_A g^{-1}\hat{\delta}g = -D_A v, \end{aligned}$$

where $v = g^{-1}\hat{\delta}g$ satisfies $\hat{\delta}v = -v^2$. This $\hat{\delta}$ corresponds to the BRST operator, and

$$\mathcal{A}|_{\delta A=\hat{\delta}A} = g^{-1}\hat{\delta}g = v, \qquad \mathcal{F} = F.$$

Expand $\omega_{2n-1}(A+\mathcal{A})|_{\delta A=\hat{\delta}A} = \omega_{2n-1}(A+v)$ in the powers of v:

$$\omega_{2n-1}(A+v) = \sum_{k=0}^{2n-1} \omega^k_{2n-k-1}(A,v), \tag{13.8}$$

and substitute this into $Tr(\mathcal{F}^n) = Tr(F^n) = (d+\hat{\delta})\omega_{2n-1}(A+v)$, one arrives at the following descent equations:

$$\begin{cases} d\omega^0_{2n-1}(A) = Tr(F^n), \\ d\omega^1_{2n-2}(A,v) + \hat{\delta}\omega^0_{2n-1}(A) = 0, \\ d\omega^k_{2n-k-1}(A,v) + \hat{\delta}\omega^{k-1}_{2n-k}(A,v) = 0, \\ \hat{\delta}\omega^{2n-1}_0(A,v) = 0. \end{cases} \tag{13.9}$$

Further integrate $\omega^k_{2n-k-1}(A,v)$ along the k-chain Γ^k on fibre $\mathcal{G}_a$ through a, and define the k-cochain density of the gauge group as:

$$\Omega^k_{2n-k-1}[\Gamma^k] \equiv (-1)^{k-1}c_n \int_{\Gamma^k} \omega^k_{2n-k-1}(A,v) \ \in \Lambda^{2n-k-1}(M),$$

then, as a result of the descent equations Eq. 13.9, one arrives at the following:

$$\begin{aligned} d\Omega^k_{2n-k-1}[\Gamma^k] &= -c_n \int_{\Gamma^k} d\omega^k_{2n-k-1}(A,v) \\ &= -c_n \int_{\Gamma^k} \hat{\delta}\omega^{k-1}_{2n-k}(A,v) \\ &= c_n \int_{\partial\Gamma^k} \omega^{k-1}_{2n-k}(A,v) = \Omega^{k-1}_{2n-k}[\partial\Gamma^k]. \end{aligned} \tag{13.10}$$

Integrate the k-cochain density $\Omega^k_{2n-k-1}[\Gamma^k]$ over M and we have the k-cochain of the gauge group,

$$\alpha^k[\Gamma^k, M] = 2\pi \int_M \Omega^k_{2n-k-1}[\Gamma^k].$$

It is linear with respect to both Γ^k and M.

If we extend the $(2n-k-1)$-dimensional compact manifold M to a $(2n-k)$-dimensional disc D such that $\partial D = M$, integrate Eq. 13.10 over the disc and we have:

$$\begin{aligned}\alpha^{k-1}[\partial\Gamma^k, D] &= 2\pi\int_D \Omega^{k-1}_{2n-k}[\partial\Gamma^k] \\ &= 2\pi\int_D d\Omega^k_{2n-k-1}[\Gamma^k] \\ &= \alpha^k(\Gamma^k, \partial D).\end{aligned}$$

This shows clearly the correspondence between the cohomology of gauge group and the cohomology of the base manifold in terms of the functional α. In the proof above, we assumed Eq. 13.10 to be valid on the whole of D, which is true only if D is topologically trivial. For a topologically non-trivial manifold, we can not take it for integrand such that $\alpha^{k-1}(\partial\Gamma^k, S_{2n-k}) = \alpha^k(\Gamma^k, \partial S_{2n-k}) = 0$, it is wrong. We leave the problem where the base manifold is topologically non-trivial to the next section.

Now we choose a $(k+1)$-chain Γ^{k+1} on $\mathcal{U}$ such that $\partial\Gamma^{k+1}$ belongs to the same gauge orbit: $\partial\Gamma^{k+1} \subset \mathcal{G}_a$, as illustrated in Fig.13.1. Here $\mathcal{G}_a$ is the gauge orbit through a. Integrate the family index formula over the $(k+1)$-chain Γ^{k+1} and we have

$$\begin{aligned}Q[\Gamma^{k+1}, M] &= \int_{\Gamma^{k+1}} Q^{k+1}[M] \\ &= \int_{\Gamma^{k+1}} \delta\beta^k[M] \\ &= \int_{\partial\Gamma^{k+1}} \beta^k[M] \\ &= \frac{(-1)^{k+1}}{2\pi c_k}\alpha^k[\partial\Gamma^{k+1}, M].\end{aligned}$$

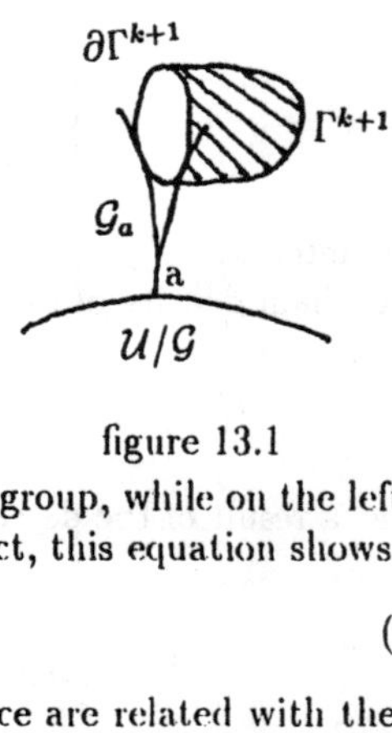

figure 13.1

On the right hand side, we have k-cochain on gauge group, while on the left hand side, it is the $(k+1)$-chain on orbit space. Thus in effect, this equation shows that

$$\pi_{k+1}(\mathcal{U}/\mathcal{G}) = \pi_k(\mathcal{G}). \tag{13.11}$$

Consequently the topological obstructions on orbit space are related with the topological obstruction on gauge group.

13.3 Topological obstruction of variety degree on gauge group and the Čech-de Rham double complex

Let the gauge potential $A = A^a_\mu(x)T_a dx^\mu$ be a Lie algebra g valued 1-form, and gauge transformation $g(x)$ be a Lie group G valued function of base manifold M, i.e., a map from M into G. Then under the gauge transformation $g(x)$,

$$A(x) \longrightarrow A(x)^g = g(x)^{-1}A(x)g(x) + g(x)^{-1}dg(x),$$

the set of gauge potentials $\mathcal{U} = \{A(x)\}$ is a topologically trivial affine space. The set of gauge transformations $\mathcal{G} = \{g(x), x \in M\}$ is an infinite dimensional group, simply called the gauge group and could be topologically non-trivial, so that the orbit space $\mathcal{U}/\mathcal{G}$ could also be topologically non-trivial. Making use of the embedding map i and the projection map π:

$$\mathcal{G} \xrightarrow{i} \mathcal{U} \xrightarrow{\pi} \mathcal{U}/\mathcal{G},$$

we can obtain the following exact sequence of homotopy group:

$$\rightarrow \pi_{n+1}(\mathcal{U}) \rightarrow \pi_{n+1}(\mathcal{U}/\mathcal{G}) \rightarrow \pi_n(\mathcal{G}) \rightarrow \pi_n(\mathcal{U}) \rightarrow$$

As the gauge potential space $\mathcal{U}$ is a topologically convex set, i.e.,

$$\pi_n(\mathcal{U}) = 0, \ n \geq 1.$$

So the exact sequence above is reduced to

$$0 \rightarrow \pi_{n+1}(\mathcal{U}/\mathcal{G}) \rightarrow \pi_n(\mathcal{G}) \rightarrow 0.$$

Consequently we have: $\pi_{n+1}(\mathcal{U}/\mathcal{G}) = \pi_n(\mathcal{G})$. It is just Eq.13.11 .

The topologically non-trivial gauge group $\mathcal{G}$ makes the orbit space $\mathcal{U}/\mathcal{G}$ non-trivial topologically, this is the topological obstruction to the construction of gauge covariant Dirac operator on the orbit space. In what follws we use Čech-de Rham double complex to study the topological obstructions of Dirac operator and end up naturally with the cohomology sequence of gauge group.

The analytic index of Dirac operator can be written in terms of Atiyah-Singer index theorem as:

$$\begin{aligned} Ind\ \not{D} &= \int_M \hat{A}(M)ch(F) = Z, \\ \text{where} \quad ch(F) &= Tr\exp(\frac{iF}{2\pi}) \\ &= r + \frac{i}{2\pi}Tr\ F + \cdots + \frac{i^n}{n!(2\pi)^n}Tr\ F^n + \cdots, \end{aligned}$$

and $\hat{A}(M)$ is the A-roof sequence of manifold M. Without loss of generality we take $M = S^6$ as an example. Then $\hat{A}(M) = 1$ and $Ind\ \not{D}$ is:

$$Ind\ \not{D} = -\frac{i}{48\pi^3}\int_{S^6} Tr(F^3) \equiv \int_{S^6} \Omega_6^{-1}(F) = Z. \tag{13.12}$$

In which $F = dA + A^2 = \frac{1}{2}F_{\mu\nu}dx^\mu \wedge dx^\nu$, integer Z is the third Chern number, and $\Omega_6^{-1}(F)$ is the third Chern class which is a closed 6-form on base manifold S^6. This third Chern class in general can not be written as an exact form on the whole of S^6 if $Z \neq 0$. However, if we cover $M = S^6$ by two patches $U_i (i = 0, 1)$ such that $U_0 \cup U_1 = S^6$ and each U_i is a contractible 6-dimensional disc, then the closed form Ω_6^{-1} can be written as an exact form on each of U_i respectively. Namely we can find a 5-form $\Omega_5^0(A_i)$ on U_i such that

$$\Omega_6^{-1}(F) = d\Omega_5^0(A_i),$$

where $\Omega_5^0(A_i)$ is the well known Chern-Simons form and can be written as

$$\begin{aligned} \Omega_5^0(A_i) &= 3c_3 \int_0^1 dt\ Tr(AF_t^2), \quad c_3 = \frac{-i}{48\pi^3}, \\ \text{where} \quad F_t &= tF + t(t-1)A^2, \\ \text{so} \quad \Omega_5^0(A_i) &= c_3 Tr(F^2 A - \frac{1}{2} F A^3 - \frac{1}{10} A^5) \\ &= c_3 Tr(A(dA)^2 + \frac{3}{2} A^3 dA + \frac{3}{5} A^5). \end{aligned} \tag{13.13}$$

As $\Omega_5^0(A_i)$ is well defined on U_i only, we have the non-zero difference on the overlap region $U_{01} = U_0 \cap U_1$:

$$\Delta\Omega_5^0(A_0, A_1) \equiv \Omega_5^0(A_1) - \Omega_5^0(A_0).$$

$\Delta\Omega_5^0(A_0, A_1)$ is in general not exact for the simple reason that the overlap region U_{01} is homotopic to S^5, which is not contractible. So $\Omega_5^0(A_i)$ in general are Čech zero-cochains only, but not Čech zero-cocycles.

If we pick a S^5 in U_{01} to divide S^6 into two discs D_i^6, i.e., $D_0^6 \cup D_1^6 = S^6$, then Eq. 13.12 can be written as:

$$\begin{aligned} Z &= \int_{S^6} \Omega_6^{-1}(F) = \sum_{j=0,1} \int_{D_j^6} \Omega_6^{-1}(F) \\ &= \sum_{j=0,1} \int_{D_j^6} d\Omega_5^0(A_j) = \int_{S^5} \Delta\Omega_5^0(A_0, A_1), \end{aligned} \tag{13.14}$$

where the last equality is a result of Stokes formula and the fact that the boundaries of D_j^6 have opposite orientations, $\partial D_j^6 = S_j^5 = (-1)^j S^5$, thus the summation becomes actually a difference.

The two gauge potentials A_i are related by a gauge transformation on the overlap region. Let

$$A_0 = A, \quad \text{then} \quad A_1 = A^g = g^{-1}Ag + g^{-1}dg,$$

and from Eq. 13.13 we have

$$\begin{aligned} &\Delta\Omega_5^0(A, A^g) = \Omega_5^0(A^g) - \Omega_5^0(A) \\ &= \tfrac{c_3}{2} d\ Tr\{v(dA\ A + AdA) + A^3 v - Av^3 + \tfrac{1}{2}(Av)^2\} - \tfrac{c_3}{10} Tr v^5. \end{aligned} \tag{13.15}$$

Where $v = gdg^{-1}$. Substitute this into Eq. 13.14 and we have:

$$Z = -\frac{c_3}{10}\int_{S^5} Tr\ v^5.$$

Namely, the analytic index Z of Dirac operator corresponds to the winding number of gauge transformation between two regions: $\pi_5(G) = Z$. And the procedures above correspond to the following exact sequence:

$$0 \to \Lambda^*(S^6) \xrightarrow{r} \Lambda^*(D_0^6) \oplus \Lambda^*(D_1^6) \xrightarrow{\Delta} \Lambda^*(S^5) \to 0,$$

where r is the restriction map, Δ is the Čech difference map. It is easy to verify that this so called Mayer-Victoris sequence is indeed exact. Note that $\Omega_6^{-1}(A_i) = d\Omega_5^0(A_i)$ is a Čech zero-cocycle, and that the Čech difference operator commutes with the exterior differential operator d, so $\Delta\Omega_5^0(A_0, A_1)$ is a closed form. Combining de Rham cohomology and Čech cohomology, we have the following $\Delta - d$ double complex.

M		U_j		U_{01}
		0		
		$\uparrow d$		
$\Omega_6^{-1}(F)$	$\xrightarrow{r}$	$\Omega_6^{-1}(F)$	$\xrightarrow{\Delta}$	0
		$\uparrow d$		
		$\Omega_5^0(A_j)$	$\xrightarrow{\Delta}$	$\Delta\Omega_5^0(A_0, A_1)$

The overlap region U_{01} is homotopically equivalent to S^5 which is not contractible. $\Delta\Omega_5^0(A_0, A_1)$ is closed but not exact on the whole of S^5. We have to divide the manifold further to overcome this topological obstruction. For instance, we can cover S^6 by three patches U_j $(j = 0, 1, 2)$ as in Fig.13.2. Each U_j is homeomorphic to a 6-dimensional disc D^6, and

$$U_0 \cup U_1 \cup U_2 = S^6,$$

i.e., we can choose the division of S^6 as follows:

$$S^6 = \sum_{i=0}^{2} D_i^6,$$

$$\partial D_i^6 = S_i^5 = \sum_{j \neq i} D_{ij}^5,$$

$$\partial D_{ij}^5 = S_{ij}^4 = (-1)^{i+j} S^4, \quad (i < j).$$

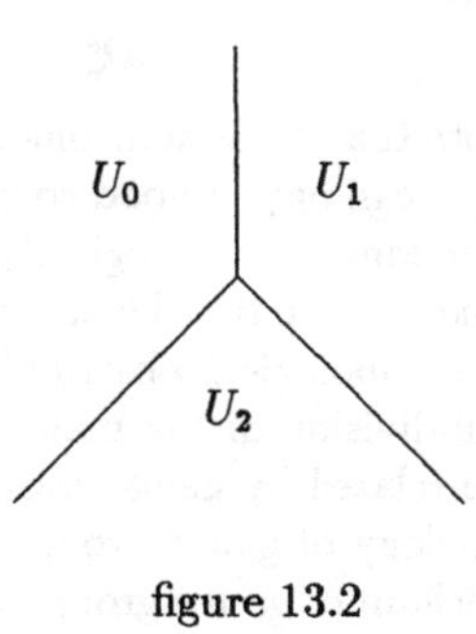

figure 13.2

Now the topologically non-trivial S_i^5 is divided into the the topologically trivial D_{ij}^5, on each of which the Poincare lemma is valid. The closed $\Delta\Omega_5^0$ is exact and there is Ω_4^1 such that

$$\Delta\Omega_5^0(A_i, A_j) = d\Omega_4^1(A_i, A_j),$$

and the index Z can be written as:

$$\begin{aligned} Z &= \int_{S^6} \Omega_6^{-1}(F) = \sum_{i=0}^{2} \int_{D_i^6} \Omega_6^{-1}(A_i) = \sum_{i<j} \int_{D_{ij}^5} \Delta\Omega_5^0(A_i, A_j) \\ &= \sum_{i<j} \int_{D_{ij}^5} d\Omega_4^1(A_i, A_j) \\ &= \int_{S^4} \Delta\Omega_4^1(A_0, A_1, A_2), \\ \text{where} \quad \Delta\Omega_4^1(A_0, A_1, A_2) &= \Omega_4^1(A_1, A_2) - \Omega_4^1(A_0, A_2) + \Omega_4^1(A_0, A_1) \end{aligned} \tag{13.16}$$

is the coboundary of Čech cohomology on the three-fold intersection. It is easy to prove by the commutation of Δ and d that $\Delta\Omega_4^1(A_0, A_1, A_2)$ is a closed form, but it is not an exact form on the whole of S^4.

Note here that the open cover now consists of three patches U_i $(i = 0, 1, 2)$, the intersection U_{ij} $(i \neq j)$ is non-empty and homotopically equivalent to the topologically trivial disc D^5. But $U_0 \cap U_1 \cap U_2$ is homotopically equivalent to the topologically non-trivial sphere S^4. This cover is called an "almost good cover". In general we can cover S^6 by $k+2$ $(k \leq 5)$ patches such that the intersections are all non-empty and homotopically equivalent to the disc D^{6-k}, and the intersection of all the $k+2$ patches is homotopically equivalent to the sphere S^{6-k-1}. Alternatively, we can divide S^6 into $m+2$ discs D^6 and obtain the $\Delta - d$ double sequence:

$$\Delta\Omega_{6-k}^{k-1}(A_0, \cdots, A_k) \equiv \sum_{i=0}^{k} (-1)^i \Omega_{6-k}^{k-1}(A_0, \cdots, \hat{A}_i, \cdots, A_k), \tag{13.17}$$

in which $\hat{A}_i$ stands for the omission of A_i. It is easy to prove from the commutation of Δ and d that this $\Delta\Omega_{6-k}^{k-1}(A_0, \cdots, A_k)$ is closed, and on a topologically trivial region one has

$$\Delta\Omega_{6-k}^{k-1}(A_0, \cdots, A_k) = d\Omega_{6-k-1}^{k}(A_0, \cdots, A_k). \tag{13.18}$$

Note that S^6 is homeomorphic to the boundary of a 7-dimensional standard simplex, we can have a good cover if we divide S^6 into 8 discs D_i^6 $(i = 0, 1 \cdots, 7)$, i.e., all intersections are topologically trivial. Combining the Čech cohomology and de Rham cohomology, we list the complete $\Delta - d$ complex in the table 13.1.

The topological obstruction can be promoted to higher order with the help of the subdivision of the manifold. On the intersection, the different gauge potentials A_i are related by gauge transformations, so Čech cohomology can be related to the cohomology of gauge group, and the subdivision of base manifold can be related to the cochain of gauge group. In what follows we take a closer look at this.

Table 13.1. Čech-de Rham double complex

M	U_j	$\bigcap_2 U$	$\bigcap_3 U$	$\bigcap_4 U$	$\bigcap_5 U$	$\bigcap_6 U$	$\bigcap_7 U$	$\bigcap_8 U$
Λ^7	0							
$\Lambda^6 \xrightarrow{r}$	Ω_6^{-1}	0						
Λ^5	Ω_5^0	$\Delta\Omega_5^0$	0					
Λ^4		Ω_4^1	$\Delta\Omega_4^1$	0				
Λ^3			Ω_3^2	$\Delta\Omega_3^2$	0			
Λ^2				Ω_2^3	$\Delta\Omega_2^3$	0		
Λ^1					Ω_1^4	$\Delta\Omega_1^4$	0	
Λ^0						Ω_0^5	$\Delta\Omega_0^5$	0
$d\uparrow\ \overrightarrow{\Delta}$	C^0	C^1	C^2	C^3	C^4	C^5	$i\uparrow$ C^6	C^7

Here $\bigcap_j$ stands for the intersection of j patches, $\Omega_{5-k}^k \equiv \Omega_{5-k}^k(A_{i0}, \cdots, A_{ik})$, and $C^p \equiv C^p(U, R)$ is the set of locally constant real functions on $\bigcap_{p+1} U$.

We know that $\Omega_6^{-1}(F) = c_3 Tr\ F^3$ is the gauge independent third Chern character on principal bundle $P(M, G)$. It is a closed form defined on the whole of S^6, its integral over S^6 gives the Chern number Z which characterizes the topological properties of $P(M, G)$. Chern-Simons form Ω_5^0 is gauge dependent, and Eq. 13.13 is merely the simplest expression. To remove the gauge dependence, let

$$\alpha^0(A) \equiv 2\pi \int_{S^5} \Omega_5^0(A),$$

then Eq. 13.14 becomes ($A_0 \equiv A, A_1 \equiv A^g = g^{-1}Ag + g^{-1}dg$)

$$\Delta\alpha^0(A, g) \equiv \alpha^0(A^g) - \alpha^0(A) = 2\pi Z,$$

in which the coboundary opeerator Δ can obtained naturally. $\alpha^0(A)$ on S^5 represents a cohomology class and is called the zero-cocycle of gauge group $\mathcal{G}^5$ if it satisfies the equation above. Here $\mathcal{G}^5$ is the set of maps from 5-dimensional base manifold to Lie group G. $\alpha^0(A)$ is the non-trivial zero-cocycle of $\mathcal{G}^5$ if $Z \neq 0$.

If we cover S^6 by three open sets $\{U_i\}$, $i = 0, 1, 2$, the 2-fold intersections are topologically trivial, $\Delta\Omega_5^0(A_i, A_j) = d\Omega_4^1(A_i, A_j)$. From Eq. 13.15 we have:

$$\Omega_4^1(A, A^g) = \frac{c_3}{2} Tr\{v(dAA + AdA) + A^3 v - Av^3 + \frac{1}{2}(Av)^2\} - \frac{c_3}{10} d^{-1}(Tr\ v^3).$$

Here d^{-1} is the Poincare homotopy operator. This equation can be diveded into two parts, the first part is $c_3 Q_4(0, A, v)$, which is known as the generalized Chern-Simons form on connection space (see §10.7). The second part is independent of A and involves pure gauge v only. Namely

$$\Omega_4^1(A, A^g) = c_3 Q_4(0, A, v) + R_4(0, v),$$

$$\text{where} \qquad dR_4(0, v) = c_3 Q_5(0, v) = -\frac{c_3}{10} Tr(gdg^{-1})^5,$$

and Formally $R_4(0,v)$ can be written as

$$R_4(0,v) = -\frac{c_3}{10}d^{-1}Tr(gdg^{-1})^5,$$

$\Omega_4^1(A, A^g)$ is not unique, it could differ by an exact form. This apparent uncertainty disappears if we integrate $\Omega_4^1(A, A^g)$ over the 3-fold intersection S^4, i.e., let

$$\alpha^1(A,g) \equiv 2\pi \int_{S^4} \Omega_4^1(A, A^g),$$

then from Eq. 13.16 we have

$$\begin{aligned} \Delta\alpha^1(A; g_1, g_2) &\equiv \alpha^1(A^{g_1}, g_2) - \alpha^1(A, g_1g_2) + \alpha^1(A, g_1) \\ &= 2\pi \int_{S^4} \Delta\Omega_4^1(A, A^{g_1}, A^{g_1g_2}) = 2\pi Z. \end{aligned}$$

Where $g_1, g_2 \in \mathcal{G}^4$. $\alpha^1(A, g)$ is called the 1-cocycle of gauge group $\mathcal{G}^4$, and $\Omega_4^1(A, A^g)$ is called the 1-cocycle density, if the equation above is satisfied. $\alpha^1(A, g)$ gives the 1-cocycle of the gauge group acting on the manifold of Yang-Mills fields on S^4, which corresponds to the well known Wess-Zumino effective action, whose variation in $\mathcal{G}^4$ gives the anomalous divergence of the chiral current.

Analogously, we can study the higher order topological obstruction of lower dimensional gauge group. For instance, the second order topological obstruction of $\mathcal{G}^3$. As

$$\pi_2(\mathcal{G}^3) = \pi_5(G) = Z,$$

which has the same topological origin as $\pi_1(\mathcal{G}^4)$. Now cover S^6 by four open sets, the common intersection is S^3. Let

$$\alpha^2(A; g_1, g_2) = 2\pi \int_{S^3} \Omega_3^2(A, A^{g_1}, A^{g_1g_2}),$$

then from Eq 13.17 we have

$$\begin{aligned} &\Delta\alpha^2(A; g_1, g_2, g_3) \\ &\equiv \alpha^2(A^{g_1}; g_2, g_3) - \alpha^2(A; g_1g_2, g_3) + \alpha^2(A; g_1, g_2g_3) - \alpha^2(A; g_1, g_2) \\ &= 2\pi \int_{S^3} \Delta\Omega_3^2(A, A^{g_1}, A^{g_1g_2}, A^{g_1g_2g_3}) = 2\pi Z. \end{aligned}$$

$\alpha^2(A; g_1, g_2)$ is the 2-cocycle of gauge group $\mathcal{G}^3$, whose variation in $\mathcal{G}^3$ gives the anomalous Schwinger term in the commutator of 3-dimensional currents.

The procedure above can be generalized and we can obtain the k-cocycle of gauge group $\mathcal{G}^{2n-k-1}$:

$$\alpha^k(A; g_1, \cdots, g_k) = 2\pi \int_M \Omega_{2n-k-1}^k(A, A^{g_1}, \cdots, A^{g_1\cdots g_k}).$$

Keep in mind the duality between the nerve of open covering of base manifold and the simplex of connection space, and making use of the Čech coboundary operator Δ of Ω-sequence, one can obtain the coboundary operator acting on α-sequence of gauge group as follows (assuming the base manifold M to be (2n-k-1)-dimensional)

$$\begin{aligned}\Delta\alpha^k(A;g_1,\cdots,g_{n+1}) &= 2\pi\textstyle\int_M \Delta\Omega^k_{2n-k-1}(A;A^{g_1},\cdots,A^{g_1\cdots g_{n+1}})\\ &= \alpha^k(A^{g_1};g_2,\cdots,g_{n+1})-\alpha^k(A;g_1g_2,g_3,\cdots,g_{n+1})+\cdots\\ &+(-1)^j\alpha^k(A;g_1\cdots g_jg_{j+1},\cdots,g_{n+1})+\cdots+(-1)^{k+1}\alpha^k(A;g_1,\cdots,g_k). \quad (13.19)\end{aligned}$$

It is easy to prove that $\Delta^2=0$. So we can talk about the cohomology of gauge group. α^k is called the k-cocycle if

$$\Delta\alpha^k = 0 \ mod\ 2\pi,$$

and the Ω^k_{2n-k-1} be called the k-cocycle density of gauge group. α_k is called a k-coboundary if there is an α^{k-1} such that

$$\alpha^k = \Delta\alpha^{k-1}.$$

The k-cocycle represents the degree-k topological obstruction, which is related to the variety anomaly. It is easy to prove in terms of Čech-de Rham double complex that the different degrees of topological obstructions of different dimensional gauge groups are closely related. The topologically non-trivial open sets can be divided to obtain the topologically trivial open sets which in turn overcomes the topological obstruction of Poincare homotopy operator d^{-1}. Promote the topological obstruction to a higher order reduces the dimension of the correspobding gauge group (the dimension of base manifold).

13.4 The cocycle density of gauge group and gauge algebra

Let M be a $(2n-k-1)$-dimensional manifold, $\mathcal{G}^{2n-k-1}=\{g(x)\}$ be the set of mappings from M to Lie group G, which is an infinite dimensional group called the gauge group. This gauge group could be topologically non-trivial. Let us look at its k-cocycle

$$\alpha^k(A;g_1,\cdots,g_k) = 2\pi\int_M \Omega^k_{2n-k-1}(A,A^{g_1},\cdots,A^{g_1\cdots g_k}). \quad (13.20)$$

As in the previous section, Ω^k_{2n-k-1} can be divided into two parts, the first part can be written as a known simplex Q-sequence on $\mathcal{U}$, and the second part involves pure gauge v only, i.e.,

$$\begin{aligned}\Omega^k_{2n-k-1}(A,A^{g_1},\cdots,A^{g_1\cdots g_k}) &= (-1)^k c_n Q_{2n-k-1}(0,A,v_1,\cdots,v_n)\\ &\quad +R^k_{2n-k-1}(0,v_1,\cdots,v_n) \quad (13.21)\end{aligned}$$

$$\text{where}\quad c_n=\frac{i^n}{n!(2\pi)^n},\quad v_1 = g_1dg_1^{-1},\quad v_2=(g_1g_2)d(g_1g_2)^{-1},\cdots.$$

The Q-sequence has an explicit expression (see §10.7. Eq.10.15), and we now try to find the R-sequence.

As Eq.13.11: $\pi_{k+1}(\mathcal{U}/\mathcal{G}) = \pi_k(\mathcal{G})$. In this equation, the number of arguments in different terms is different, the Ω-sequence and R-sequence are functionals on gauge group, but the Q-sequence is a functional on orbit space thus it has one more argument than that Ω- and R-sequence. The coboundary operation Δ of gauge group on the equation above needs a peculiar degenerate difference operator $\bar{\Delta}$ which is different from the Δ of orbit space we met in §10.7. From some simple examples, we can obtain its action rules. For instance

$$\Omega_5^0(A) = c_3 Q_5(0, A),$$
$$\Delta\Omega_5^0(A) = c_3 \bar{\Delta} Q_5(0, A).$$

On the other hand, as

$$\begin{aligned} \Delta\Omega_5^0(A) &= \Omega_5^0(A^g) - \Omega_5^0(A) \\ &= c_3(Q_5(0, A^g) - Q_5(0, A)) \\ &= c_3(\Delta Q_5(0, A, v) - Q_5(0, v)), \end{aligned}$$

comparing the two equations above gives

$$\bar{\Delta} Q_5(0, A, v) = \Delta Q_5(0, A, v) - Q_5(0, v),$$

and we can further have the generalized result

$$(-1)^{k-1}\bar{\Delta} Q_{2n-k}(0, A, v_1, \cdots, v_n) = \Delta Q_{2n-k}(0, A, v_1, \cdots, v_n) - Q_{2n-k}(0, v_1, \cdots, v_n).$$

Acting on this equation by the coboundary operator, we have (the degenerate coboundary operator has to be used for the Q-sequence)

$$\begin{aligned} \Delta\Omega_{2n-k}^{k-1}(A, A^{g_1}, \cdots, A^{g_1\cdots g_k}) &= (-1)^{k-1} c_n \bar{\Delta} Q_{2n-k}(0, A, v_1, \cdots, v_k) \\ &\quad + \Delta R_{2n-k}^{k-1}(0, v_1, \cdots, v_k). \end{aligned}$$

Take the exterior differential of Eq. 13.22 and making use of Eq. 13.18 and Eq.10.16, we have

$$\begin{aligned} \Delta\Omega_{2n-k}^{k-1}(A, A_{g_1}, \cdots, A^{g_1\cdots g_k}) &= c_n \Delta Q_{2n-k}(0, A, v_1, \cdots, v_n) \\ &= dR_{2n-k-1}^{k}(0, v_1, \cdots, v_n). \end{aligned}$$

Comparing the three equations above, then we have the recurrence formula for R-sequence

$$dR_{2n-k-1}^{k}(0, v_1, \cdots, v_n) = c_n Q_{2n-k}(0, v_1, \cdots, v_k) + \Delta R_{2n-k}^{k-1}(0, v_1, \cdots, v_k). \quad (13.22)$$

The explicit expression for Q^k_{2n-k} was derived in §10.9 recurrently from Eq. 10.16, as well as the fact that ΔQ^k_{2n-k} is closed and $\mathcal{U}$ is topologically trivial. Now we try to construct the homotopy operator d^{-1} for Eq. 13.22, and get R^k_{2n-k-1} on a topologically trivial region. For instance, as the last example in the previous section, let $n = 3$, we can have the explict expression for Ω^k_{6-k-1} as

$$\Omega_6^{-1} = c_3 Tr\ F^3,$$
$$\Omega_5^0(A) = c_3 Tr(A(dA)^2 + \frac{3}{2}A^3 dA + \frac{3}{5}A^5),$$
$$\Omega_4^1(A, A^g) = -\frac{c_3}{2}Tr(v(dAA + AdA + A^3 v - Av^3 + \frac{1}{2}(Av)^2) - \frac{c_3}{10}d^{-1}Tr(v^3),$$
$$\Omega_3^2(A, g_1, g_2) = \frac{c_3}{2}Tr\{A(dg_1 dg_2 g_2^{-1} g_1^{-1} - g_2^{-1} dg_2 g_2^{-1} g_1^{-1} dg_1 g_1^{-1}$$
$$-d^{-1}[d\ln g_1(d\ln g_1 d\ln g_2 + \frac{1}{2}d\ln g_2 d\ln g_1 + d\ln g_2 d\ln g_2)d\ln g_2]\}.$$

From the analysis above, we can see that the explicit expressions of cocycle densities of gauge group (Ω-sequence) are usually very complicated and in general there are no explicit expressions, thus there are no general expressions for cocycles of gauge group (α-sequence).

In studying the topological obstructions of gauge group, we usually have to consider its infinitesimal form only, i.e., we consider the cohomology of the gauge algebra (BRST cohomology, or the principal part of $\alpha^k(A, g_1, \cdots g_k)$ when $g = I + v$, $v \to 0$):

$$\begin{aligned} \beta^k(A, v) &\equiv \alpha^k(A; I + v, \cdots, I + v) \\ &= 2\pi c_n \int_M \omega^k_{2n-k-1}(A, v), \end{aligned} \tag{13.23}$$

where $\omega^k_{2n-k-1}(A; v)$ is the cocycle density of gauge algebra, which could be obtained as Eq. 13.8 by expanding $\omega_{2n-1}(A + v)$ into the powers of v:

$$\begin{aligned} \omega_{2n-1}(A + v) &= \sum_{k=0}^{2n-1} \omega^k_{2n-k-1}(A, v) \\ &= n\int_0^1 dt Tr(A + v, F_t^{n-1}), \\ \text{where} \qquad F_t &= tF + (t^2 - t)(A + v)^2, \\ \text{so} \qquad \omega^1_{2n-2}(A, v) &= n(n-1)\int_0^1 dt\ (1-t) Str[vd(AF_t^{n-2})], \end{aligned}$$

and

$$\omega^k_{2n-k-1}(A, v) = \frac{n!}{(n-k-1)!k!}\int_0^1 dt\ (1-t)^n Str[(dv)^k, A, (tdA + t^2 A^2)^{n-k-1}],$$
$$0 \le k \le n-1 \tag{13.24}$$
$$\omega^{n+j}_{n-j-1}(A, v) = (-1)^j \frac{(n-1)!n!}{(n-j-1)!(n+j)!} Tr[(dv)^{n-j-1}, v, (v^2)^j],\ 0 \le j \le n-1 \quad .$$

The cocycles of gauge algebra are integrals of cocycle densities over the corresponding base manifold M, so the explicit expressions of ω-sequence (and Ω-sequence) could differ up to an exact form. For instance, let $n = 3$, one may have (up to an exact form):

$$\begin{aligned}
\omega_5^0 &= Tr[A(dA)^2 + \frac{3}{2}A^3dA + \frac{3}{5}A^5],\\
\omega_4^1 &= Tr[vd(AdA + \frac{1}{2}A^2)],\\
\omega_3^2 &= Tr[(dv)^2A],\\
\omega_2^3 &= Tr[(dv)^2v],\\
\omega_1^4 &= -\frac{1}{2}Tr(dvv^3),\\
\omega_0^5 &= \frac{1}{10}Tr(v^5).
\end{aligned}$$

In the following sections, we shall study the anomaly of gauge theory, the variety non-trivial cocycles of gauge algebra induce the variety degree anomalies of the generators of gauge group (currents). For example

$$\begin{aligned}
\alpha^1(A; I+v) &= 2\pi c_3 \int_{S^4} \omega_4^1\\
&= \frac{-i}{24\pi^2}\int_{S^4} Tr[vd(AdA + \frac{1}{2}A^3] \qquad (13.25)
\end{aligned}$$

gives the non-Abelian anomaly for 4-diemnsional divergence of chiral current, and

$$\begin{aligned}
\alpha^2(A; I+v, I+v) &= 2\pi c_3 \int_{S^3} \omega_3^2\\
&= \frac{-i}{24\pi^2}\int_{S^3} Tr[A(dv)^2] \qquad (13.26)
\end{aligned}$$

gives the anomalous Schwinger term in the equal time commutator for the current of 3-dimensional gauge group. Analogously

$$\begin{aligned}
\alpha^3(A; I+v, I+v, I+v) &= 2\pi c_3 \int_{S^2} \omega_2^3\\
&= \frac{-i}{24\pi^2}\int_{S^2} Tr[v(dv)^2] \qquad (13.27)
\end{aligned}$$

gives the anomaly for the Jacobi identity for the current of 2-dimensional gauge group. We shall study this in §13.8.

13.5 Topological properties of 4-dim quantum Yang-Mills theory and θ-vacuum

For a Yang-Mills theory on 4-dimensional Minkowski space,

$$
\begin{aligned}
S &= \int_M d^4x\mathcal{L}, \\
\text{where} \qquad \mathcal{L} &= \frac{1}{2e^2} Tr\ F^{\mu\nu}F_{\mu\nu} = -\frac{1}{4}\mathcal{F}_a^{\mu\nu}\mathcal{F}_{\mu\nu}^a = \frac{1}{2}(E_a^2 - B_a^2).
\end{aligned}
$$

$$E_a^i = \mathcal{F}_a^{i0} \qquad (\vec{E}_a = -\dot{\vec{A}}_a - \nabla\phi_a - ef_{abc}\phi_b\vec{A}_c).$$

$$B_a^i = -\frac{1}{2}\epsilon^{ijk}\mathcal{F}_{jk}^a \qquad (\vec{B}_a = \nabla\times\vec{A}_a - \frac{1}{2}ef_{abc}\vec{A}_b\times\vec{A}_c.$$

One of the standard approaches to quantum field theory is the canonical quantization in a Hamiltonian scheme. As the Lagrangian above is independent of $\dot{\phi}_a$ thus there is no momentum conjugate to ϕ_a, ϕ_a is then the so called auxillary field and one can choose the Weyl gauge $\phi_a = 0$.

The canonical momentum conjugate to $\vec{A}_a$ are

$$\vec{\pi}_a = \frac{\delta\mathcal{L}}{\delta\dot{\vec{A}}_a} = \dot{\vec{A}} = -\vec{E}_a,$$

and the Hamiltonian is

$$H = \frac{1}{2}\int d^3x(\vec{\pi}_a^2 + \vec{B}_a^2) = \frac{1}{2}\int d^3x(\vec{E}_a^2 + \vec{B}_a^2).$$

The dynamical variables $\vec{A}_a$ and $\vec{E}_a$ satisfy the canonical equal time commutation relation:

$$\left[E_a^i(x), A_b^j(y)\right] = i\hbar\delta^{ij}\delta(x-y),$$

and the equations of motion are

$$\begin{cases} \dot{\vec{A}}_a = \frac{i}{\hbar}[H, \vec{A}_a] = -\vec{E}_a, \\ \dot{\vec{E}}_a = \frac{i}{\hbar}[H, \vec{E}_a] = (D\times\vec{B})_a. \end{cases}$$

The second equation is just the spacial component of the classical equation of motion $D_\mu F^{\mu\nu} = 0$, while the temporal component $(D\cdot E)_a = 0$ has not emerged yet. As operator $(D\cdot E)_a$ does not commute with the canonical variables, it is the first class constraint on canonical variables and can not be simple set it to equal to zero.

Under the Weyl gauge, the system is stiil invariant under a time independent gauge transformation:

$$\delta\vec{A}_a = -D\theta_a,$$

where θ_a is a time independent c-number. From Noether's theorem, the conserved charge, corresponding to this symmetry, is

$$\begin{aligned} Q &= \int d^3x \vec{\pi}_a \cdot \delta \vec{A}_a = -\int d^3x \vec{\pi}_a \cdot D\theta_a \\ &= \int d^3x \vec{E}_a \cdot D\theta_a. \end{aligned}$$

And the time independent generator of this gauge transformation is:

$$G_a = (D \cdot \vec{\pi})_a = -(D \cdot \vec{E})_a.$$

It is easy to prove the following commutation relations:

$$\begin{aligned} &\left[G_a(\vec{x}), \vec{A}_b(\vec{y})\right] = -i\hbar\delta_{ab}\nabla\delta(\vec{x}-\vec{y}) - i\hbar f_{abc}\vec{A}_c(\vec{x})\delta(\vec{x}-\vec{y}), \\ &\left[G_a(\vec{x}), \vec{E}_b(\vec{y})\right] = -i\hbar f_{abc}\vec{E}_c(\vec{x})\delta(\vec{x}-\vec{y}), \\ &[G_a(\vec{x}), H] = 0. \end{aligned}$$

Consequently G_a is the generator of infinitesimal gauge transformation, and Hamiltonian H is gauge invariant.

The first class constraint G_a can not be set to zero under quantization, instead we require that the physical states satisfy

$$G_a(x)|\text{physical}>= 0.$$

The physical state in Schrödinger picture can be realized as a functional $\psi(A)$ of canonical variables $\vec{A}_a$, and the equation above can then be written as:

$$(D \cdot \frac{\delta}{\delta A})_a \psi[A] = (\nabla\delta_{ab} - f_{abc}\vec{A}_c) \cdot \frac{\delta}{\delta A_b}\psi[A] = 0. \tag{13.28}$$

From the closedness of gauge algebra

$$[G_a(\vec{x}), G_b(\vec{y})] = -i\hbar f_{abc} G_c(\vec{x})\delta(\vec{x} - y),$$

and the Frobenius theorem, we know that the equation above is integrable. In homotopically trivial case, a finite gauge transformation $U(g_0)$ can be obtained from the infinitesimal ones as

$$U(g_0) = e^{\frac{i}{\hbar}Q}, \tag{13.29}$$

such that Gauss law Eq. 13.28 implies that wave functional $\psi[A]$ is invariant under this "small" finite gauge transformation,

$$U(g_0)\psi[A] = \psi[A^{g_0}] = \psi[A].$$

We can choose $\psi[A]$ as the common eigenvector of H and $U(g)$, as the constraint G_a commute with Hamiltonian H.

Now we look at a Hamiltonian system in which the Yang-Mills fields decay rapidly at infinity. As the equations of motion are conformally invariant, one can compactify the 3-dimensional space to S^3 and study the space $\mathcal{U}^3$ of all static 3-dimensional gauge potentials in Weyl gauge $\phi_a = 0$. As the physical system is gauge invariant, the wave functional $\psi[A]$ is essentially on $\mathcal{U}^3/\mathcal{G}^3$. Since

$$\pi_1(\mathcal{U}^3/\mathcal{G}^3) = \pi_0(\mathcal{G}^3) = \pi_3(G) = Z,$$

topological obstructions show up if $Z \neq 0$ and there are "large" gauge transformations that are not obtainable from iterating the "small" ones. The wave functional is no longer invariant under these "large" gauge transformations and could instead be changed by a phase factor such that the physical result is still invariant:

$$\psi[A^g] = e^{in\theta}\psi[A].$$

In terms of the family index theorem, one can compute the following 1-form (family index) on $\mathcal{U}^3/\mathcal{G}^3$:

$$\begin{aligned} \bar{Q}^1[S^3] &= \int_{S^3} \bar{q}_3^1(\bar{\delta}A, F) \\ &= -\frac{1}{4\pi^2}\int_{S^3} Tr(\bar{\delta}A\, F). \end{aligned}$$

It is $\bar{\delta}$-closed, i.e., $\bar{\delta}Q^1[S^3] = 0$, but it is not exact. We can lift it and obtain a 1-form on $\mathcal{U}^3$ as

$$Q^1[S^3] = -\frac{1}{4\pi^2}\int_{S^3} Tr(\delta A \cdot F),$$

which can be regarded as a connection 1-form on $\mathcal{U}^3$, the corresponding curvature 2-form is

$$\begin{aligned} \delta Q^1[S^3] &= \frac{1}{4\pi}\int_{S^3} Tr(\delta A \delta F) \\ &= \frac{1}{4\pi}\int_{S^3} Tr(\delta A D_A(\delta A)) \\ &= \frac{1}{2\pi^2}\int_{S^3} d(Tr(\delta A \delta A) = 0. \end{aligned}$$

As $\mathcal{U}^3$ has the trivial topology, in terms of Poincare lemma the above equation implies that $Q^1[S^3]$ is also exact on $\mathcal{U}^3$, i.e.,

$$2\pi c_2 Q^1[S^3] = \delta\beta^0[S^3], \tag{13.30}$$

where $\beta^0[S^3]$ is given by Eq. 13.23

$$\begin{aligned} \beta^0[S^3] &= 2\pi c_2 \int_{S^3} \omega_3^0(A, \mathcal{A}) = 2\pi c_2 \int_{S^3} \omega_3^0(A) \\ &= -\frac{1}{4\pi}\int_{S^3} Tr(AdA + \frac{2}{3}A^3). \end{aligned}$$

Thus the connection 1-form $Q^1[S^3]$ is a pure gauge and we look at its integration along an open path Γ^1 in $\mathcal{U}^3$. Let Γ^1 be the path joining two points A^g and A on the same gauge orbit, we have

$$\begin{aligned} c_2 Q^1[\Gamma^1, M] &= c_2 \int_{[}\Gamma^1]Q^1[S^3] = c_2 \int_{\Gamma^1} \delta\beta^0[S^3] = -\int_{\partial\Gamma^1} \beta^0[S^3] \\ &= \frac{1}{8\pi^2}\int_{S^3}[Tr(A^g dA^g + \frac{2}{3}(A^g)^3 - Tr(AdA + \frac{2}{3}A^3)] \\ &= \frac{1}{24\pi^2}\int_{S^3} Tr(g^{-1}dg)^3 = n(g) \ \in Z. \end{aligned}$$

This equation is an explicit manifestation of the relation $\pi_1(\mathcal{U}^3/\mathcal{G}^3) = \pi_0(\mathcal{G}^3) = Z$, which simply says that there are non-contractible loops in $\mathcal{U}^3/\mathcal{G}^3$. Thus we obtain a $U(1)$-vortex field in $\mathcal{U}^3/\mathcal{G}^3$ in agreement with the result by Wu and Zee[75].

When $Z \neq 0$, there are "large gauge transformations" similar to Bohm-Aharonov effect and the corresponding wave functional has a non-integrable phase factor

$$\psi[A^g] = e^{i2\pi c_2 Q^1[\Gamma^1, M]\theta}\psi[A] = e^{-i2\pi n(g)\theta}\psi[A].$$

Then from Eq. 13.30 we know that $c_2 Q^1[\Gamma^1, M]$ is essentially a coboundary

$$\begin{aligned} c_2 Q^1[\Gamma^1, M] &= \frac{1}{2\pi}\Delta\alpha^0(A; g) = \frac{1}{2\pi}(\alpha^0(A^g) - \alpha^0(A)), \\ \text{where } \alpha^0(A) &= \int d^3x\,\omega_3^0(A). \end{aligned}$$

Consequently this phase factor can be removed by a unitary transformation, i.e., let

$$\phi[A] = e^{i\theta\alpha^0(A)}\psi[A].$$

It differs from $\psi[A]$ only by an A dependent phase factor, thus has different transformation under gauge transformation. It is easy to prove that

$$e^{\frac{i}{\hbar}Q}\alpha^0(A)e^{-\frac{i}{\hbar}Q} = \alpha^0(A).$$

Namely $\alpha^0(A)$ is gauge invariant under "small gauge transformation", hence $\phi[A]$ is also gauge invariant

$$U(g_0)\phi(A) = \phi(A^{g_0}) = \phi(A).$$

But under a "large gauge transformation", we have

$$\begin{aligned} \alpha^0(A^g) &= \alpha^0(g^{-1}Ag + g^{-1}\nabla g) \\ &= \alpha^0(g) + \frac{1}{8\pi^2}\int_M d^3x\epsilon^{ijk}\partial_i Tr(\partial_j g \cdot g^{-1}A_k) \\ &\quad + \frac{1}{24\pi^2}\int_M d^3x\epsilon^{ijk}Tr(\partial_i g g^{-1}\partial_j g g^{-1}\partial_k g g^{-1}) \\ &= \alpha^0(A) + n(g). \end{aligned}$$

$$\text{So} \quad U(g)\phi(A) = \phi[A^g] = e^{i\theta(\alpha^0(A) + n(g))}\psi[A^g] = \phi(A).$$

The wave functional $\phi(A)$ is invariant under all gauge transformations, including the topologically non-trivial gauge transformations.

The corresponding transformation of the Hamiltonian is

$$\begin{aligned} H' &= e^{i\theta\alpha_0(A)} H e^{-i\theta\alpha_0(A)} \\ &= e^{i\theta\alpha_0(A)} \frac{1}{2} \int_M d^3x (\pi_a^2 + B_a^2) e^{-i\theta\alpha_0(A)} \\ &= \frac{1}{2} \int d^3x [(\pi_a - \frac{\theta}{8\pi^2} B_a)^2 + B_a^2], \end{aligned}$$

they are dependent on θ. However we know that a phase change in the wave functional correponds to a canonical transformation, which will induce a change in the Lagrangian by a total derivative, i.e.,

$$\begin{aligned} \mathcal{L}' &= \mathcal{L} + \theta C, \\ \text{where} \quad C &= \frac{1}{16\pi^2} Tr({}^*F^{\mu\nu} F_{mu\nu}) = -\frac{1}{8\pi^2} E_a \cdot B_a \\ &= \partial_\mu w^\mu, \\ w^\mu &= \frac{1}{16\pi^2} Tr\ \epsilon^{\mu\alpha\beta\gamma} (A_\alpha F_{\beta\gamma} - \frac{2}{3} A_\alpha A_\beta A_\gamma). \end{aligned}$$

The new Lagrangian depends on the θ parameter, the extra term is the Pontrjagin characteristic class, whose contribution to action I is merely a surface term,

$$\begin{aligned} I &= I + \theta I_c, \\ I_c &= \int d^4x C = \int dS_\mu w^\mu, \end{aligned}$$

and the variation of I_c for A_μ^a is zero

$$\begin{aligned} \delta I_c &= \frac{1}{8\pi^2} \int d^4x Tr({}^*F^{\mu\nu} \delta F_{\mu\nu}) = \frac{1}{4\pi^2} \int d^4x Tr({}^*F^{\mu\nu} D_\mu \delta A_\nu) \\ &= -\frac{1}{4\pi^2} \int d^4x Tr(D_\mu^* F^{\mu\nu}) \delta A_\nu = 0. \end{aligned}$$

So classical equation of motion is unchanged, and consequently classical dynamics is not affected by this extra term.

13.6 3-dim Yang-Mills theory and topological mass term

Now we look at gauge theory at 3-dimensional space-time. Let

$$I = \int d^3x \mathcal{L},$$

$$\mathcal{L} = \frac{1}{2e^2} Tr\ F^{\mu\nu} F_{\mu\nu} - \frac{m}{8\pi e^2} \epsilon^{\mu\nu\lambda} Tr(F_{\mu\nu} A_\lambda - \frac{2}{3} A_\mu A_\nu A_\lambda).$$

Variation $\frac{\delta I}{\delta A^a_\mu} = 0$ gives the equation of motion

$$D_\mu F^{\mu\nu} + \frac{m}{8\pi}\epsilon^{\nu\lambda\rho}F_{\lambda\rho} = 0.$$

Which is gauge invariant and m has the dimensioon of mass.

Now we turn to the quantum theory. Use the canonical quantization in a Hamiltonian scheme, and choose the Weyl gauge $A_0 = 0$. Consider the infinite dimensional affine space $\mathcal{U}^2$ of all static gauge field configurations $\{A^a_i(x)\}$. The wave functional $\psi[A]$ is essentially on $\mathcal{U}^2/\mathcal{G}^2$ as the physical system is gauge invariant.

Since for certain simple Lie group, $\pi_2(G) = 0$, so

$$\pi_1(\mathcal{U}^2/\mathcal{G}^2) = \pi_0(\mathcal{G}^2) = \pi_2(G) = 0,$$

and the orbit space $\mathcal{U}^2/\mathcal{G}^2$ is simply connected. On the other hand, as

$$\pi_2(\mathcal{U}^2/\mathcal{G}^2) = \pi_1(\mathcal{G}^2) = \pi_3(G) = Z,$$

there are topological obstructions if $Z \neq 0$. The family index for Dirac operator on $\mathcal{U}^2/\mathcal{G}^2$ can be obtained from the family index theorem as

$$\bar{Q}^2[M] = \int_M Tr(\mathcal{F}^2) = \int_M Tr(\bar{\delta}A\bar{\delta}A).$$

It can be lifted to a 2-from on $\mathcal{U}^2$

$$Q^2[M] = \int_M Tr(\mathcal{F}^2) = \int_M Tr(\delta A\delta A) + 2\int_M Tr(\mathcal{A}F),$$

which is obviously closed. $\delta Q^2[M] = 0$. Hence $Q^2[M]$ can be identified as a curvature 2-form on $\mathcal{U}^2$. It is clear that $Q^2[M]$ is also exact simply due to $\mathcal{U}^2$ is topologically trivial:

$$2\pi c_2 Q^2[M] = \delta\beta^1[M].$$
$$\beta^1[M] = -\frac{1}{4\pi}\int_{S^2} Tr(A\delta A) - \frac{1}{2\pi}\int_{S^2} Tr(\mathcal{A}F).$$

$\beta^1[M]$ can be identified as a connection 1-from on $\mathcal{U}^2$, and the field velocity operator can be wtitten as:

$$V = i\frac{\delta}{\delta A} + \beta^1[M],$$
$$[V, V] = c_2 Q^2[M],$$

i.e., $c_2Q^2[M]$ is the curvature 2-from corresponding to the gauge field strength. It is easy to prove, from $\delta Q^2[M] = 0$, that the commutator above satisfies the Jacobi identity. Thus the translation (i.e., field velocity operator) on $\mathcal{U}^2$ constitutes a Lie group. The Gauss law fails however, because V satisfies anomalous commutation relations.

The wave functional $\psi[A]$ is changed by a phase factor under a gauge transformation

$$\psi[A^g] = e^{im\int_A^{A^g}\beta^1[M]}\psi[A] = e^{im\alpha^1(A;g)}\psi[A].$$

On the other hand, integrate $\beta^1[M]$ along two different paths from A to A^g, then their difference is an integration of $\beta^1[M]$ along a 1-dimensional loop $S^1 = \partial\Gamma^2$, where Γ^2 is a 2-dimensional disc in $\mathcal{U}^2$, i.e.,

$$\oint_{S^1}\beta^1[M] = \int_{\partial\Gamma^2}\beta^1[M] = \int_{\Gamma^2}\delta\beta^1[M] = 2\pi c_2\int_{\Gamma^2}Q^2[M] = 2\pi Z.$$

Thus wave functional $\psi[A]$ acquires a phase factor $\int_A^{A^g}\beta^1[M] = \alpha^1(A;g)$ under the gauge transformation, it could not be removed by a unitary transformation if $Z \neq 0$ (non-trivial topology), and this topological contribution could affect the equation of motion.

So we see that due to the orbit space $\mathcal{U}^2/\mathcal{G}^2$ is simply connected, $\pi_1(\mathcal{U}^2/\mathcal{G}^2) = 0$, those "large gauge transformations" are not included in it. But the equation

$$c_2\int_{\Gamma^2}Q^2[M] = Z$$

tells us that $\pi^2(\mathcal{U}^2/\mathcal{G}^2) = \pi^1(\mathcal{G}^2) = \pi_3(G) = Z$, if $Z \neq 0$, then this implies that there is a non-contractible 2-dimensional sphere in $\mathcal{U}^2/\mathcal{G}^2$. Thus we have a "monopole" which contributes to the topological mass.

13.7 Fermion interaction and quantum anomalies

In this section we study gauge interaction of fermions. The Lagrangian for a massless fermion interacting with gauge field is

$$\begin{aligned}
\mathcal{L} &= \frac{1}{2e^2}Tr\ F^{\mu\nu}F_{\mu\nu} - \frac{\hbar\theta}{16\pi^2}Tr\ {}^*F^{\mu\nu}F_{\mu\nu} + iTr\bar{\psi}\gamma^\mu(\partial_\mu + A_\mu)\psi \\
&\equiv \mathcal{L}_G + iTr\ \bar{\psi}\not{D}\psi, \\
\text{where}\quad \not{D} &= \gamma^\mu D_\mu = \gamma^\mu(\partial_\mu + A_\mu), \qquad \bar{\psi} = \psi^+\gamma_0.
\end{aligned}$$

Notice that $\{\gamma_5, \not{D}\} = 0$, so the classical action $S = \int d^4x\mathcal{L}$ is invariant under chiral transformation:

$$\begin{aligned}
\psi(x) &\to \psi'(x) = e^{i\alpha\gamma_5}\psi(x), \\
\bar{\psi}(x) &\to \bar{\psi}'(x) = \bar{\psi}(x)e^{i\alpha\gamma_5}.
\end{aligned}$$

Subsequently one can obtain the axial vector current J_μ^5 and the vector current J_μ from the Noether theorem

$$\begin{aligned}
&\partial^\mu J_\mu^5 = 0, \quad \partial^\mu J_\mu = 0, \\
\text{where}\quad &J_\mu^5 = \bar{\psi}\gamma_\mu\gamma_5\psi, \quad J_\mu = \bar{\psi}\gamma_\mu\psi.
\end{aligned}$$

In quantum theory, however, there are anomalies. Let $< >$ denote the expectation value of physical quantities. If one requires the vector current $< J_\mu >$ to be conserved,

$$< \partial_\mu J^\mu >= 0,$$

then the axial vector current $< J^5_\mu >$ is no longer conserved. The gauge potential, in non-Abelian gauge theory, takes value in the matrix representation of gauge group, $A_\mu = A^a_\mu \tau^a$, and the vector current is covariantly conserved: $D_\mu J^\mu = D_\mu D_\nu F^{\mu\nu} = \frac{1}{2}[F_{\mu\nu}, F^{\mu\nu}] = 0$. A pure axial vector interaction with fermions could not be gauge invariant, and gauge transformations could be defined separately for $\pm$ handed spinors only under one of the following two combinations $i\bar{\psi}\gamma^\mu \frac{1}{2}(1 \pm i\gamma_5)\tau_a \psi A^a_\mu$, and

$$\bar{\psi}\gamma^\mu(\partial_\mu + \frac{1}{2}(1 \pm i\gamma_5)A_\mu)\psi = \bar{\psi}_\pm \gamma^\mu(\partial_\mu + A_\mu)\psi_\pm + \bar{\psi}_\mp \gamma^\mu \partial_\mu \psi_\mp.$$

Adler found, through perturbation, the anomalies for axial vector current:

$$\begin{aligned} &< \partial_\mu J^5_\mu >= -\frac{1}{16\pi^2}\epsilon_{\mu\nu\rho\sigma} Tr\ F_{\mu\nu}F_{\rho\sigma}, \\ &< D_\mu J^\mu_{\pm a} >= \frac{\mp i}{24\pi^2} Tr\ \tau^a \epsilon^{\mu\nu\rho\sigma}\partial_\mu[A_\nu \partial_\rho A_\sigma + \frac{1}{2}A_\nu A_\rho A_\sigma], \quad (13.31) \\ \text{where} \quad &J^\mu_{\pm a} = -i\bar{\psi}_\pm \gamma_\mu \tau^a \psi_\pm. \end{aligned}$$

The first equation is called the Abelian anomaly, the second the non-Abelian anomaly. Note that the Abelian anomaly tells us that the axial current is a singlet current and does not involve the gauge group generators.

Next we look at the topological origin of these anomalies.

(1). The Abelian axial anomaly and Atiyah-Singer index theorem.

In the path integral quantization, we extend the 4-dimensional Minkowski space-time to Euclidean space-time,

$$x^0 \to -ix^4, \quad A_0 \to iA_4,$$

and then compactify the 4-dimensional Euclidean space to S^4.

For a massless Dirac fermion ψ coupled to an extended gauge field $A(x)$, we first perform the functional integral over fermion fields and obtain the effective action functional $W[A]$ as follows:

$$e^{iW[A]} = \int \mathcal{D}\bar{\psi}\mathcal{D}\psi e^{i\int d^4x\mathcal{L}}.$$

It is important that the change in the measure of the functional integral has to be considered, i.e., the Jacobian of the change of variables may give rise to an additional contribution. And according to Fujikawa[76], the anomaly is then reduced to the change in measure of the functional integral. We take a closer look at this in the following.

Choose a repesentation such that

$$\gamma_\mu^+ = -\gamma_\mu, \quad \not{D}^+ = \not{D},$$

i.e., $\not{D}$ is Hermitian and its spectrum $\{\lambda_n\}$ is discrete

$$\not{D}\phi_n = \lambda_n \phi_n,$$

where λ_n is real as a result of the Hermiticity of $\not{D}$, and the eigenstates are normalized as

$$\int d^4x \phi_n^+(x)\phi_m(x) = \delta_{mn}.$$

Expand ψ in terms of $\{\phi_n\}$,

$$\psi(x) = \sum_n a_n \phi_n(x), \quad \bar{\psi}(x) = \sum_n \phi_n^+(x)\bar{b}_n.$$

Fermion field anticommute under quantization, so $\{a_n\}$ and $\{b_n\}$ are anticommuting variables.

Under a local chiral transformation

$$\psi(x) \to \psi'(x) = e^{i\alpha(x)\gamma_5}\psi(x),$$

expand the two sides of this equation in terms of $\{\phi_n(x)\}$, and we have:

$$\sum_n a'_n \phi_n(x) = \sum_n a_n e^{i\alpha(x)\gamma_5}\phi_n(x).$$

From the normalization, we finally have

$$a'_m = \sum_n \int d^4x \phi_m^+(x) e^{i\alpha(x)\gamma_5}\phi_n(x) a_n \equiv \sum_n C_{mn} a_n,$$

$$\prod_m da'_m = [det(C_{mn})]^{-1} \prod_n da_n.$$

Here the inverse of $det(C_{mn})$ is due to $\{a_n\}$ are Grassmanians. In case $\alpha(x)$ be an infinitesimal variable, we have

$$det(C_{mn}) \simeq e^{i\sum_n \int d^4x \phi_n^+(x)\alpha(x)\gamma_5\phi_n(x)} = e^{i\int d^4x \alpha(x) I(x)},$$

where

$$I(x) = \sum_n \phi_n^+(x)\gamma_5\phi_n(x).$$

The contribution from the change in $\{b_n\}$ takes the same sign, so under the chiral transformation, the measure of the path integral is changed as

$$d\mu \to d\mu' = d\mu e^{-2i\int d^4x \alpha(x) I(x)}.$$

Thus although the Lagrangian is invariant under chiral transformation, the measure contributes a factor in the path integral quantization and gives the axial current divergence anomaly.

In case $\alpha(x)$ be a constant, we have

$$\int d^4x I(x) = \int d^4x \sum_n \phi_n^+(x)\gamma_5\phi_n(x).$$

The contribution from non-zero modes vanish as a result of $\{\gamma_5, \not{D}\} = 0$. For the zero modes ϕ_0 of Dirac operator $\not{D}$, take ϕ_0 as the eigenstate of γ_5,

$$\gamma_5\, \phi_0^{\pm} = \pm\phi_0^{\pm},$$

$$\text{then} \qquad \int d^4x \sum_n \phi_n^+(x)\gamma_5\phi_n(x) = \nu_+ - \nu_-.$$

Here $\nu_\pm$ are the numbers of left and right zero modes respectively. This equation is just the analytic index formula for Dirac operator $\not{D}$. From Atiyah-Singer index theorem, the corresponding topological index is the second Chern class of gauge potential $A(x)$:

$$\nu_+ - \nu_- = -\frac{1}{32\pi^2}\int d^4x \epsilon^{\mu\nu\rho\sigma} F_{\mu\nu}F_{\rho\sigma}.$$

(2). Non-Abelian anomaly and WZW effective action.

Now we look at non-Abelian anomaly

$$e^{iW[A]} = \int \mathcal{D}\bar{\psi}\mathcal{D}\psi e^{-\int d^4x \bar{\psi}\not{D}\psi} = det\not{D},$$

$$< J_a^\mu > = -\frac{\delta}{\delta A_\mu^a} W[A].$$

The effective action $W[A]$ is no longer gauge invariant although the classical action is, thus emerges the anomaly. Now take the variation along gauge orbit (BRST transformation):

$$\hat{\delta} W[A] = \int_{S^4} (-D_\mu v)_a \frac{\hat{\delta}}{\delta A_\mu^a} W[A] = \int_{S^4} Tr(n\cdot \mathcal{G}) \equiv I[A, v],$$

where $v = g^{-1}dg$, $\mathcal{G}$ is the anomalous divergence of current. As $\hat{\delta}^2 = 0$, the anomaly satisfies the following consistent condition

$$\hat{\delta} I[A, v] = 0.$$

It can be proved that under a gauge transformation, $|det\not{D}|$ is invariant and $det\not{D}$ is changed only by a factor

$$det\not{D} \longrightarrow e^{-i\alpha^1(A,g)} det\not{D},$$

i.e., there is a first order topological obstruction. From the descent equations Eq. 13.25, we can get the following solution for the consistent condition

$$I[A,v] = 2\pi c_3 \int_{S^4} \omega_4^1(A,v),$$

and the anomalous divergence is

$$2\pi \frac{i^3}{3!(2\pi)^3} Tr[\tau^a d(AdA + \frac{1}{2}A^3)] = -\frac{i}{24\pi^2} Tr[\tau^a d(AdA + \frac{1}{2}A^3)],$$

which is the same as Eq. 13.31.

Now we try to find the explicit expressions for $W[A]$, it is not unique and could be determined completely by specific physical requirements.

Integrate v along the gauge group orbit

$$W[A,g] = 2\pi \int_{S^4} \Omega_4^1(\Gamma^1) = \alpha^1(\Gamma^1, S^4) = \alpha^1(A,g),$$

which is the 1-cocycle of gauge group $\mathcal{G}^4$. At the vicinity of identity $g = I + \theta$,

$$\alpha^1(A, I+\theta) = -\frac{i}{24\pi^2} \int Tr\ \theta d(AdA + \frac{1}{2}A^3)$$

gives the anomalous divergence of current, and the 1-cocycle condition

$$\begin{aligned} \Delta\alpha^1(A;h,h^{-1}g) &= \alpha^1(A^h, h^{-1}g) - \alpha^1(A,g) + \alpha^1(A,h) = 0 \quad mod(2\pi) \\ i.e., \quad \alpha^1(A^k, h^{-1}g) &= \alpha^1(A,g) - \alpha^1(A,h) \quad mod(2\pi) \end{aligned}$$

is just the consistent condition for the anomaly. Topologically it comes from the fact that

$$\pi_1(\mathcal{G}^4) = \pi_5(G) = Z.$$

In order to find out explicitly the topological interpretation, one could relate the non-Abelian anomaly with a family index theorem involving a two-parameter family of gauge field configurations. We start from a reference gauge field $A(x) \in \mathcal{U}^4$, and regard it as a connection on a trivial bundle, consider a one-parameter family of gauge transformations, such that one can extend the base manifold from S^4 to a 5-dimensional disc D_5, as follows:

$$\begin{aligned} A(x) \to A(\theta, x) &= A^U \\ &= U^{-1}(\theta,x)A(x)U(\theta,x) + U^{-1}(\theta,x)d_x U(\theta,x). \end{aligned}$$

As $\pi_1(G) = 0$ for semi-simple Lie group G, we can set

$$U(0,x) = U(2\pi,x) = I, \quad U(1,x) = g(x) \ \in \mathcal{G}^4,$$

which induces $S^1 \times S^4 \sim S^5$ and the connection $A^{U(1,x)}$ belongs to the orbit of $\mathcal{G}^4$ through A. Further let us introduce a parameter t, with

$$A \to A(t, \theta, x), \quad 1 \geq t \geq -1.$$

For $t \geq 0$, let $A_+(t, \theta, x)$ be the interpolation between A^U and a background field A_0, which is an arbitrary point in $\mathcal{U}^4$ as follows

$$A_+(t, \theta, x) = tA_0 + (1-t)A^U, \quad 1 \geq t > 0.$$

For $t \leq 0$, let

$$A_-(t, \theta, x) = A + (1+t)d_\theta U\, U^{-1}, \; 0 \geq t \geq -1.$$

Then there exists the following relation for $t = 0$, i.e., on S^5:

$$A_+(0, \theta, x) = U^{-1}(\theta, x)[A_-(0, \theta, x) + d_x + d_\theta]U(\theta, x).$$

The transformtion $U(\theta, x) \in \mathcal{G}^5$ on S^5 determines the topological properties of the bundle on S^6. As

$$\pi_0(\mathcal{G}^5) = \pi_5(G) = Z,$$

the integer Z is the topological index of the bundle on S^6,

$$Z = c_3 \int_{S^6} Tr\mathcal{F}^3,$$
$$\mathcal{F} = (d_x + d_\theta + d_t)A(t, \theta, x) + A^2(t, \theta, x).$$

Under a gauge transformation

$$A \to g^{-1}(A + d)g,$$

assume that U transforms as

$$U \to g^{-1}U, \tag{13.32}$$

then A^U is invariant, namely only the starting point A of the 1-cochain on $\mathcal{U}^4$ is changed, while the other point A^U is unchanged. So

$$\begin{aligned} W[A, U] &= \alpha^1(\Gamma^1, S^4) = \alpha^0(\Sigma^0, D^5) \\ &= 2\pi \int_{D^5} (\Omega_5^0(A) - \Omega_5^0(A^U)), \\ \text{where} \quad & S^4 = \partial D^5, \; \Sigma^0 = \partial \Gamma^1. \end{aligned}$$

This effective functional has the following properties:

1. $\Omega_5^0(A) - \Omega_5^0(A^0)$ is a closed form,

$$d(\Omega_5^0(A) - \Omega_5^0(A^0)) = \Omega_6^{-1}(F) - \Omega_6^{-1}(F) = 0.$$

So it could become an exact form on the 5-dimensional disc D^5. And in terms of Stokes formula one can rewrite the effective action as an integral over 4-dimensional space-time.

2. Since $\alpha^0(\Sigma^0, S^5) = 2\pi Z$, so

$$W[A,U] = \alpha^1(\Gamma^1, S^4) = \alpha^0(\Sigma^0, D^5)$$

can be extended from S^4 to D^5 up to $2\pi Z$. And for the non-trivial topology, $Z \neq 0$, there is the first order topological obstruction to the definition of 4-dimensional Dirac operator and the anomalous divergence for the axial vector current.

3. Under the 4-dimensional gauge transformation Eq. 13.32, $\Omega(A^U)$ is invariant and the anomaly comes from $\Omega_5^0(A)$:

$$\begin{aligned} \hat{\delta} W[A,U] &= 2\pi c_3 \int_{D^5} d\omega_4^1(A,v) \\ &= 2\pi \int_{S^4} \omega_4^1(A,v), \ (v = g^{-1}dg). \end{aligned}$$

The three properties listed above are the necessary conditions for the effective action functional. Namely, it can be written as an integral over a compactified 4-deimensional space-time S^4. Also it can be written as an integral over a 5-dimensional disc and is independent of the way it is extended to the 5-dimensional disc. This effective action generates the anomaly under the 4-dimensional gauge transformation.

13.8 Topological interpretation of quantum anomalies

In the last section we look at the consistent anomaly, whose topological origin is: that the effective action functional $W[A,U]$ is a non-trivial 1-cocycle on the 4-dimensional gauge group $\mathcal{G}^4$,

$$W[A,U] = 2\pi \int_{S^4} \Omega_4^1(\Gamma^1) = \alpha^1(\Gamma^1, S^4).$$

From §13.1-4, we see that the topological obstructions of variety degree on gauge group are related to each other. The higher order topological obstruction of lower dimensional gauge group is related to the lower order topological obstruction of higher dimensional gauge group, all of them are related to the family index theorem.

When there is a first order topological obstruction on the 4-dimensional gauge group, i.e, there is non-trivial 1-cocycle of $\mathcal{G}^4$:

$$\alpha^1(\Sigma^1, S^4) = 2\pi Z \neq 0,$$

then there must be a non-trivial zero-cocycle on $\mathcal{G}^5$, 2-cocycle on $\mathcal{G}^3$, and 3-cocycle on $\mathcal{G}^2$. Here we emphasize again that different orders of topological obstructions have the same topologiacl origin, but have to be realized at different dimensions. In the following we look at various physical problems.

1. Covariant anomaly and the 0-cocycle on $\mathcal{G}^5$.

In quantum field theory, we sometimes have to study the covariant current, which is used to construct gauge invariant coupling with other fields. The quantum anomalous divergence for the covariant current could be preserved by covariance, and is thus called covariant anomaly, i.e., we require the conservation of covariant vector current, $< D_\mu j^\mu >_a = 0$, and the anomaly of the covariant divergence for the covariant axial vector current

$$< D_\mu j_5^\mu >_a = -\frac{i}{16\pi^2} Tr\ \tau^a \epsilon^{\mu\nu\rho\sigma} F_{\mu\nu} F_{\rho\sigma}. \tag{13.33}$$

The consistent anomaly is realized through the gauge group cohomology approach while the covariant anomaly is realized through the orbit space cohomology approach, both have the same topological origin. Covariant anomaly itself is gauge covariant, but it could be constructed from the consistent anomaly, hence the anomaly cancellation conditions are the same for these two types of anomalies.

Covariant anomaly can be obtained by an arbitrary variation δA of the effective action, here we emphasize that the arbitrary variation is not a variation along the orbit of $\mathcal{G}^4$. The 0-cocycle of $\mathcal{G}^5$ can be written as:

$$\begin{aligned}
\alpha^0(\Gamma^0, S^5) &= 2\pi \int_{S^5} \Omega_5^0(A) \\
&= 2\pi c_3 \int_{S^5} Tr(F^2 A - \frac{1}{2} F A^2 + \frac{1}{10} A^5).
\end{aligned}$$

From arbitrary variation δA one has

$$\begin{aligned}
\int \delta A \frac{\delta}{\delta A} \alpha^0(\Gamma^0, S^5) &= 2\pi \int_{S^5} [\frac{i}{16\pi^3} Tr\ F^2 \delta A - \frac{i}{48\pi^3} d\ Tr(ddA\ A + AdA + \frac{3}{2} A^3)\delta A] \\
&= \frac{i}{8\pi^2} \int_{S^5} Tr\ F^2 \delta A.
\end{aligned}$$

So the covariant anomaly of the current interacting with other matter fields can be written as

$$\mathcal{G}^a = -\frac{i}{8\pi^2} Tr\ \tau^a F^2,$$

which is just the right hand side of Eq. 13.33. It is easy to see that this anomaly is gauge covariant.

2. Hamiltonian approach to physical system and the 2-cocycle of 3-dimensional gauge group $\mathcal{G}^3$.

Let us look at a physical system at a fixed time t. In the Hamiltonian approach, the base manifold is a compact 3-dimensional space S^3. Under Weyl gauge $A_0 = 0$, there is still one time independent gauge transformation whose generator is G_a. We require the physical state to satisfy

$$G_a|\text{phys.}>= 0.$$

Under a finite gauge transformation, the Lagrangian $L = \int_{S^3} \mathcal{L}$ is changed as

$$\begin{aligned}\Delta L &= \frac{d}{dt}\int_{-\infty}^{t} \Delta L = \frac{d}{dt}\int_{D^4} \Delta \mathcal{L} \\ &= \frac{d}{dt}\alpha^2(\Sigma^1, D^4) = \frac{d}{dt}\alpha^2(\Gamma^2, S^3).\end{aligned}$$

As the finite variation of the Lagrangian is the time derivative, so the physical wave functional is the projective representation of gauge group $\mathcal{G}^3$.

Let $U(g) = e^{i\int d^3x\theta^a G_a}$ be a finite gauge transformation acting on the wave functional, then

$$U(g_1)U(g_2) = e^{i\alpha^2(A;g_1,g_2)}U(g_1g_2),$$

where the phase factor $\alpha^2(A; g_1, g_2) \equiv \alpha^2(\Gamma^2, S^3)$ satisfies the 2-cocycle condition:

$$\begin{aligned}\Delta\alpha^2(\Gamma^3, S^3) &= \alpha^2(\partial\Gamma^3, S^3) = \alpha^2(\Sigma^2, S^3) \\ &= \alpha^1(\Sigma^1, S^4) = 2\pi Z,\end{aligned} \tag{13.34}$$

and it has the same topological origin as the 1-cocycle on $\mathcal{G}^4$.

Since the wave functional is in the projective representation of gauge transformation, the infinitesmal form of this anomaly is manifested by the anomalous commutation of the generators:

$$[G(u), G(v)] = G([u, v]) + \beta^2(A, u, v).$$

Where $\quad G(U) = i\int_{S^3} d^3x u^a(x) G_a(x),$

and $\quad \beta^2(A; u, v) = 2\pi \int_{S^3} \omega_3^2(A, u, v)$

is the anomalous Schwinger term. The 2-cocyle condition Eq. 13.34 implies that the anomalous extended terms must satisfy the Jacobi identity of Lie algebra:

$$[[G(u), G(v)], G(w)] + \text{cyclic permutation} = 0.$$

The commutator can be re written as:

$$i[G_a(x), G_b(y)] = f_{abc}G_c(x)\delta(x-y) \pm \frac{i}{24\pi^2}\epsilon^{ijk}Tr\{\tau^a, T^b\}\partial_i A_j \partial_k \delta(x-y). \tag{13.35}$$

3. The boundary effect of a hybrid chiral bag model and the 3-cocycle of $\mathcal{G}^2$.

The 3-cocycle of gauge group cohomology implies the non-associativity of the representation of this group and consequently the Jacobi identity of the infinitesimal generator is violated[77,78]. This can be realized in the hybrid chiral bag Skyrme model[79,80]. For a 3-dimensional finite compact space D^3, whose boundary is S^2 of radius r, we look at the topological obstruction of the gauge transformation $\mathcal{G}^2$ on the boundary S^2, and define the charge of the bag as:

$$B(u) \equiv \int_{D^3} d^3x u^a(x) G_a(x), \; u(x) = u^a(x)\tau^a.$$

From Eq. 13.35 one has

$$[B(u), B(v)] = B([u,v]) \pm 2\pi i \int_{D^3} d^3x u^a(x)(\omega_3^2(x))^{ab} v^b(x).$$

The Jacobi identity has an anomaly on the boundary S^2:

$$[[B(u), B(v)], B(w)] + \text{cycltic permutation } = \beta^3(A; u, v, w) \neq 0,$$

where $$\beta^3(A; u, v, w) = 2\pi \int_{S^2} \omega_3^2(u, v, w) \equiv J(u, v, w)$$

is the third order topological obstruction of the 2-dimensional gauge group $\mathcal{G}^2$, it satisfies the 3-cocycle condition

$$\begin{aligned}
&[G(u), J(w,w,z)] - [G(v), J(w,z,u)] + [G(w), J(z,u,v)] - [G(z), J(u,v,w)] \\
&= J([u,v],w,z) - J(u,[v,w],z) + J(u,v,[w,z]) - J([u,w],v,z) \\
&+J([u,z],v,w) - J([v,z],u,w).
\end{aligned}$$

Notations and Formulae

The space of connection: $\mathcal{U}$={A}.
The space of gauge orbit (orbit space): $\mathcal{U}/\mathcal{G}$.
The group of gauge transformation (gauge group): $\mathcal{G}$.

- The connection $\mathcal{A}$ on connection bundle $\mathcal{U}$:

$$\mathcal{A} \equiv -G_A D_A^+ \delta A,$$

here $\delta A = g^{-1}\delta a g - D_A g^{-1}\delta g, \quad G_A \equiv (D_A^+ D_A)^{-1}.$

- The total connection on product bundle $P(M,G) \times \mathcal{U}$:

$$A + \mathcal{A} = A - G_A D_A^+ \delta A,$$

$$\mathcal{F} = F + (1 - D_A G_A D_A^+)\delta A + \delta\mathcal{A} + \mathcal{A}^2,$$

satisfy

$$(d+\delta)Tr(\mathcal{F}^n) = 0,$$

locally $$Tr(\mathcal{F}^n) = (d+\delta)\omega(A+\mathcal{A}).$$

satisfy

$$(d+\delta)Tr(\mathcal{F}^n) = 0,$$
$$\text{locally} \qquad Tr(\mathcal{F}^n) = (d+\delta)\omega(A+\mathcal{A}).$$

- Family index therem:

$$ch^{k+1}(Ind\not{D}_A) = c_n \int_M Tr(\mathcal{F}^n) \in \Lambda^*(\mathcal{A}/\mathcal{G}), \quad c_n = \frac{i^n}{n!(2\pi)^n}.$$

- Cohomology on orbit's space.

  The horizontal derivation $\bar{\delta}$ satisfy:

$$D_A^+ \bar{\delta} A = 0,$$
$$\bar{\delta}\mathcal{A} + \mathcal{A}^2 = 0.$$

Expanding $Tr(\mathcal{F}^n)$ in the powers of $\bar{\delta}A$ yields

$$Tr(\mathcal{F}^n) = \sum_{k=0}^{n} \bar{q}^k_{2n-k}(\bar{\delta}A, F),$$
$$\text{here} \qquad \bar{q}^k_{2n-k}(\bar{\delta}A, F) = \tfrac{n!c_n}{(n-k)!k!} Str[(\bar{\delta}A)^k, F^{n-k}] \in \Lambda^k_{2n-k}.$$

Then from $(d+\bar{\delta})Tr(\mathcal{F}^n) = 0$, and compare terms of the same order, one gets the descent equations:

$$\begin{cases} d\bar{q}^0_{2n} = dTr(F^n) = 0, \\ d\bar{q}^k_{2n-k} + \bar{\delta}\bar{q}^{k-1}_{2n-k+1} = 0 \quad (k = 1, \cdots, n), \\ \bar{\delta}\bar{q}^n_n = 0. \end{cases}$$

- Cohomology on connection space.

  Because the connection space is topological trivial

$$Tr(\mathcal{F}^n) = (d+\delta)\omega_{2n-1}(A+\mathcal{A}),$$

where

$$\begin{aligned} \omega_{2n-1}(A+\mathcal{A}) &= Q_{2n-1}(0, A+\mathcal{A}) \\ &= n\int_0^1 dt Tr(A+\mathcal{A}, \mathcal{F}_t^{n-1}), \end{aligned}$$
$$\mathcal{F}_t = t\mathcal{F} + t(t-1)(A+\mathcal{A})^2.$$

- Cohomology on gauge group.

Now restricting the variation along the gauge orbit,

$$\hat{\delta}A = -D_A g^{-1}\hat{\delta}g = -D_A v, \quad v = g^{-1}\hat{\delta}g.$$

From $Tr(\mathcal{F}^n) = Tr(F^n) = (d+\hat{\delta})\omega_{2n-1}(A+v)$, and expanding

$$\omega_{2n-1}(A+\mathcal{A})|_{\delta A=\hat{\delta}A} = \omega_{2n-1}(A+v),$$

in the powers of v:

$$\omega_{2n-1}(A+v) = \sum_{k=0}^{2n-1} \omega^k_{2n-k-1}(A,v),$$

we get the gauge algebra cohomology hierarchy:

$$d\omega^k_{2n-k-1}(A,v) + \hat{\delta}\omega^{k-1}_{2n-k}(A,v) = 0, \quad (k = 0,\cdots,2n-1).$$

Further integrate $\omega^k_{2n-k-1}(A,v)$ along the gauge orbit (the k-chain Γ^k on fibre $\mathcal{G}_a$), one get the k-cochain density of the gauge group as:

$$\Omega^k_{2n-k-1}[\Gamma^k] \equiv (-1)^{k-1}c_n \int_{\Gamma^k} \omega^k_{2n-k-1}(A,v) \;\in \Lambda^{2n-k-1}(M),$$

where $c_n = \frac{i^n}{n!(2\pi)^n}$. Integrate the k-cochain density $\Omega^k_{2n-k-1}[\Gamma^k]$ over M, then one get the k-cochain of the gauge group,

$$\alpha^k[\Gamma^k, M] = 2\pi \int_M \Omega^k_{2n-k-1}[\Gamma^k].$$

Exercise

1. Find $\omega_5(A^g) - \omega_5(A) = ?$

2. Find the explicite expression of the 1-cocycle $\alpha^1(A,g)$ on $\mathcal{G}^4$, and give the infinitesmal expression when g near identity.

Chapter 14

Noncommutative Geometry, Quantum Group, and q-deformation of Chern-Characters

14.1 Introduction

Symmetry and duality play important roles in theoretical physics and mathematical physics. In differential geometry, one studies manifold M in terms of functions $\mathcal{F}(M)$ and the associated tensor fields on M. Gelfand pointed out that all properties of a topological space M can be studied by studying the commutative algebra $A = F(M)$ of functions on this space, and vice versa. Space M is the spectrum of the algebra A, point $x \in M$ can be regarded as an irreducible representation, i.e., for each $f \in A$,

$$x[f] = f(x).$$

There is a duality between M and $A = \mathcal{F}(M)$ and any thing that is described in terms of M can also be described in terms of A, two algebras are isomorphic if and only if their spectrums are homeomorphism

In quantum physics, the continuum gave way to the discreteness, dynamic variables are described by noncommutative operators in Hilbert space. Meanwhile in modern differential geometry, the global and topological analysis plays an important role. The glocal properties, which are unchanged by continuous variation, are characterized by discreteness, and usually could labelled by integer (or fraction) rather than real number.

Quantum field theory and modern differential geometry are closely interlinked. May be the framework of the quantum group, the q-deformed gauge theories and the non-commutative geometry seems so close to quantum field theory.

Quantum group plays a significant role in quantum integrable systems and solvable statistical models, it provides an insightful example of noncommutative differential geometry. Manin [81] suggested a general construction of quantum groups as linear transformations upon the quantum superplane. Following the general ideas of Connes [82] on the noncommutative geometry, Woronowicz [83] elaborated the framework of the noncommutative differential calculus. He introduced the bimodule over the quantum group, and presented various theorems concerning the differential forms and exterior derivative. The differential calculus on the quantum hyperplane was developed by Wess and Zumino [84]. There have been a lot of papers treating the differential calculus on quantum groups and the q-deformed gauge theories from various viewpoints [85–95].

Since the quantum group is formulated in the language of the Hopf algebra, the gauge transformation will be represented in an abstract language and the term of the transformation parameter becomes obscure. As the alternative formulations of the gauge theory, Bernard [85] firstly raised the idea on q-deformed BRS symmetry [97]. Watamura [95] explicitly constructed a q-deformed BRST algebra, which is the algebra of gauge fields, ghost fields and appropriate matter fields on the space-time point. Now, the gauge transformation of the theory is replaced by the BRST transformation which is represented by a nilpotent "differential operator" δ, and the gauge parameter is replaced by the ghost fields and becomes an object of equal level with the matter and the gauge fields. Watamura [95] extracted appropriate properties from the non-deformed BRST formalism and imposed them as the condition which new algebra should satisfy so that the q-deformed BRST formalism is the q-analogue of the non-deformed one. Watamura proved in detail that the definitions of the covariant commutation relations among the fields and their derivatives are consistent with the operation δ, the space-time derivative d, as well as the *-operation, the antimultiplicative inner involution. Since there are two nilpotent operators δ and d, the double cohomology and cocycle hierarchy can be discussed in this formalism. The q-deformed BRST algebra of Watamura is a good starting point towards developing the deformation of Chern classes and cocycle hierarchy.

The q-deformed Killing form is the key for constructing the q-deformed Chern class and the q-deformed cocycle hierarchy. From the q-gauge covariant condition, the q-deformed Killing form and the second q-deformed Chern class, that contain both components of the identity and adjoint representations, for the quantum group $SU_q(2)$ in the q-deformed BRST formalism, can be defined. Although the components of the two representations have to be mixed in the commutative relations of the q-deformed BRST algebra, we prove that it is possible to define the q-deformed Killing form and the second q-deformed Chern class uniquely up to a common factor such that they only contain the components of the adjoint representation. Then, generalizing Zumino's method [98,99], we introduce a q-deformed homotopy operator to compute the q-deformed Chern-Simons, that contain also only the adjoint components. At last, by making use of the standard method in non-deformed case, we calculate the q-deformed cocycle hierarchy[100].

In §14.2, we look at linear transformations on a quantum plane and obtain quantum group $GL_q(2)$ and $SU_q(2)$. From the $\hat{R}_q$-matrix which satisfies Yang-Baxter equation we obtain both the Hopf algebra A and its dual Hopf algebra A', i.e., the quantum group and its associated quantum enveloping algebra. In order to study the bicovariant differential calculus on quantum group, we look at the adjoint representation and the identity representation of the quantum group. Then in §14.3, we construct the direct product representation of the fundamental representation and its conjugate by using the q-Clebsch-Gordon coefficients (q-Pauli matrices), and proceed to study the bicovariant calculus on quantum group $SU_q(2)$. In §14.4 we construct the q-gauge theory in terms of q-BRST algebra. Finally in the last part of this chapter we look at q-deformed Chern class, Chern-Simons and cocycle hierarchy.

14.2 Linear transformations on the quantum hyperplane, quantum group $GL_q(2)$ and $SU_q(2)$

Let us look at a 2-dimensional quantum hyperplane, its coordinates (x, y) satisfy the following relations:

$$xy = qyx.$$

Correspondingly there exist variables (ξ, η) which satisfy

$$\xi\eta = -\frac{1}{q}\eta\xi, \quad \xi^2 = 0, \quad \eta^2 = 0.$$

For $q = 1$, x, y commute and ξ, η anticommute, and (ξ, η) can be interpreted as the differentials of (x, y)

$$\phi = \begin{pmatrix} x \\ y \end{pmatrix}, \qquad \psi = d\phi = \begin{pmatrix} \xi \\ \eta \end{pmatrix}.$$

Linear transformation on the quantum hyperplane (x, y, ξ, η) is defined as

$$\begin{pmatrix} x' \\ y' \end{pmatrix} = \begin{pmatrix} \alpha & \beta \\ \gamma & \delta \end{pmatrix} \begin{pmatrix} x \\ y \end{pmatrix}, \qquad \begin{pmatrix} \xi' \\ \eta' \end{pmatrix} = \begin{pmatrix} \alpha & \beta \\ \gamma & \delta \end{pmatrix} \begin{pmatrix} \xi \\ \eta \end{pmatrix}.$$

Here we assume that $(\alpha, \beta, \gamma, \delta)$ belong to an algebra A, which commute with (x, y, ξ, η). If (x', y') and (ξ', η') also satisfy the relations of (x, y) and (ξ, η), then we have

$$\begin{cases} \alpha\beta = q\beta\alpha, \quad \alpha\gamma = q\gamma\alpha, \quad \beta\gamma = \gamma\beta, \\ \gamma\delta = q\delta\gamma, \quad \beta\delta = q\delta\beta, \quad \alpha\delta - \delta\alpha = (q - q^{-1})\beta\gamma. \end{cases} \tag{14.1}$$

Let $K < x_1, \cdots, x_n >$ denote a free associative K-algebra generated by $x_1, \cdots, x_n$, then the algebra

$$A = K < \alpha, \beta, \gamma, \delta > .$$

is called the quantum group $GL_q(2)$. Its center is

$$c = det_q T = \alpha\delta - q\beta\gamma = \delta\alpha - q^{-1}\beta\gamma.$$

It is called $SL_q(2)$ if $det_qT = 1$.

Let $(T^i_j) = \begin{pmatrix} \alpha & \beta \\ \gamma & \delta \end{pmatrix}$, Eq. 14.1 can be written as

$$R_{12}T_1T_2 = T_2T_1R_{12} \tag{14.2}$$

where $T_1 = T \otimes 1,\ T_2 = 1 \otimes T$, and

$$R_{12} = \begin{pmatrix} q & & & \\ & 1 & 0 & \\ & \lambda & 1 & \\ & & & q \end{pmatrix}, \quad \lambda = q - q^{-1}.$$

Or let

$$\hat{R}_q = PR = \begin{pmatrix} q & & & \\ & \lambda & 1 & \\ & 1 & 0 & \\ & & & q \end{pmatrix}, \quad (\hat{R}_q)^{ab}_{cd} = (R_{12})^{ba}_{cd},$$

then Eq. 14.2 can be rewritten as

$$\begin{aligned} \hat{R}_q T \otimes T &= T \otimes T \hat{R}_q \\ \text{or} \quad \hat{R}_{q\,ef}^{\ ab} T^e_c T^f_d &= T^a_e T^b_f \hat{R}_{q\,cd}^{\ ef}. \end{aligned}$$

As algebra A is associative, The matrix R ($\hat{R}_q$) satisfies the Yang-Baxter equation

$$R_{12}R_{13}R_{23} = R_{23}R_{13}R_{12}, \tag{14.3}$$

$$\text{or} \quad (1 \otimes \hat{R}_q)(\hat{R}_q \otimes 1)(1 \otimes \hat{R}_q) = (\hat{R}_q \otimes 1)(1 \otimes \hat{R}_q)(\hat{R}_q \otimes 1). \tag{14.4}$$

From which it is easy to see that there is a co-structure in algebra $A = GL_q(2)$, i.e., there are co-product Δ, counit ϵ and antipode κ:

$$\begin{cases} \Delta(T^a_b) = T^a_c \otimes T^c_b, \\ \epsilon(T^a_b) = \delta^a_b, \\ \kappa(T^a_b) = (T^{-1})^a_b = \frac{1}{det_qT}\begin{pmatrix} \delta & -q^{-1}\beta \\ -q\gamma & \alpha \end{pmatrix}. \end{cases}$$

which are compatible with the Yang-Baxter equation, i.e.,

$$\Delta(R_{12}T_1T_2) = \Delta(T_2T_1R_{12}), \quad \Delta(det_qT) = det_qT \otimes det_qT, \quad \text{etc.}$$

Δ and ϵ are homomorphisms and κ is an anti-homomorphism:

$$\begin{aligned} &\Delta(\alpha\beta) = \Delta(\alpha)\Delta(\beta), \\ &\epsilon(\alpha\beta) = \epsilon(\alpha)\epsilon(\beta), \\ &\kappa(\alpha\beta) = \kappa(\beta)\kappa(\alpha), \\ \text{and} \quad &(id \otimes \Delta)\Delta = \Delta \otimes id)\Delta \quad \text{(coassociativity)}, \\ &(id \otimes \epsilon)\Delta(\alpha) = (\epsilon \otimes id)\Delta(\alpha) = \alpha \quad (\text{counit},\ \epsilon * \alpha), \\ &m(id \otimes \kappa)\Delta(\alpha) = m(\kappa \otimes id)\Delta(\alpha) = I \quad (\text{coinverse},\ \kappa * \alpha). \end{aligned}$$

Consequently

$$\Delta\kappa = (\kappa \otimes \kappa)P\Delta,$$
$$\epsilon(\kappa(\alpha)) = \epsilon(\alpha),$$

here P denote the permutation of the two arguments.

An algebra A with unit 1 and endowed with co-structure $\Delta,\ \epsilon,\ \kappa$ satisfying equations above is called a Hopf algebra $A = A(m, 1, \Delta, \epsilon, \kappa)$. Thus both $Gl_q(2)$ and $Sl_q(2)$ are Hopf algebras.

We can further introduce an anti-linear *-operator

$$*:\ A \to A$$

such that $\forall \alpha, \beta \in A,\ c \in \mathbf{C}$

$$(c\alpha\beta)^* = \bar{c}\beta^*\alpha^*,$$
$$\kappa(\kappa(\alpha^*)^*) = \alpha,$$

where $\bar{c}$ is the complex conjugate of c. Co-product Δ and counit ϵ are *-homomorphisms and the conjugate element is written as

$$(T^i_j)^* = T^{*i}_j = T^{+j}_i.$$

We obtain $SU_q(2)$ if we add the unitarity condition $T^+ = T^{-1}$ and require q to be a real number.

In terms of matrix $\hat{R}_q$, which satisfies the Yang-Baxter equation(Eq. 14.4), one can introduce the dual Hopf algebra A' generated by two linear functionals $(L^\pm)^a_b$, defined as

$$(L^+)^a{}_b(T^c{}_d) \ = \ q^{-1/2}\,(\hat{R}_q)^{ac}{}_{db}\,, \quad (L^-)^a{}_b(T^c{}_d) \ = \ q\,(\hat{R}_q^{-1})^{ac}{}_{db},$$

which satisfy the following relations

$$\left(\hat{R}_q\right)^{ab}_{\ \ ef}\left(L^\pm\right)^s_{\ d}\left(L^\pm\right)^r_{\ c} = \left(L^\pm\right)^b_{\ s}\left(L^\pm\right)^a_{\ r}\left(\hat{R}_q\right)^{rs}_{\ \ cd},$$
$$(\hat{R}_q)^{ab}_{ef}\left(L^+\right)^s_{\ d}\left(L^-\right)^r_{\ c} = \left(L^-\right)^b_{\ s}\left(L^+\right)^a_{\ r}\left(\hat{R}_q\right)^{rs}_{\ \ cd}.$$

The co-structure on matrices $L^\pm$ is

$$\Delta(L^{\pm a}_b) = (L^\pm)^a_c \otimes (L^\pm)^c_b,$$
$$\epsilon(L^{\pm a}_b) = \delta^a_b.$$

In the case $A = SU_q(2)$, the dual Hopf algebra $L^\pm \in A'$ can be written as

$$L^+ = \begin{pmatrix} t & F \\ 0 & t^{-1} \end{pmatrix}, \quad L^- = \begin{pmatrix} t^{-1} & 0 \\ -\lambda E & t \end{pmatrix}.$$

And it is easy to check $\{t, E, F\}$ generate the quantum enveloping algebra $U_q(su(2))$:

$$\begin{aligned} &tt^{-1} = t^{-1}t = 1, \\ &tE = q^2 Et, \quad tF = q^{-2} FT, \\ &[E, F] = \frac{1}{\lambda}(t - t^{-1}). \end{aligned}$$

14.3 Bicovariant calculus on quantum group $SU_q(2)$

As is well known, the right or left invariant 1-forms on group manifold belong to the adjoint representation, so in order to study the bicovariant calculus on quantum group one has to consider both the adjoint representation and the identity representation of the quantum group. In the last section we had introduced the fundamental representation $(T^{a_1}_{\ b_1})$ of the quantum group $SU_q(2)$, now let us look at the direct product representation of the fundamental representation and its conjugate

$$M^{a_1\ \ b_2}_{\ \ a_2 b_1} = T^{a_1}_{\ b_1} \kappa\left(T^{b_2}_{\ a_2}\right).$$

Let

$$\begin{aligned} q^{-1}[2]^{1/2}\,(\sigma^0)_{ab} &= q[2]^{1/2}\,(\sigma_0)^{ab} = -\,\epsilon^{ab} = \epsilon_{ab} = \begin{pmatrix} 0 & q^{-1/2} \\ -q^{1/2} & 0 \end{pmatrix}, \\ (\sigma^3)_{ab} &= (\sigma_3)^{ab} = -\,[2]^{-1/2}\begin{pmatrix} 0 & q^{1/2} \\ q^{-1/2} & 0 \end{pmatrix}, \\ (\sigma^+)_{ab} &= (\sigma_+)^{ab} = \begin{pmatrix} 0 & 0 \\ 0 & -1 \end{pmatrix}, \quad (\sigma^-)_{ab} = (\sigma_-)^{ab} = \begin{pmatrix} 1 & 0 \\ 0 & 0 \end{pmatrix}, \\ (\sigma^I)_a^{\ b} &= (\sigma^I)_{ad}\,\epsilon^{db}, \quad (\sigma_I)^a_{\ b} = (\sigma_I)^{ad}\,\epsilon_{db}, \end{aligned}$$

which satisfy following orthogonal retions:

$$\begin{aligned} (\sigma^I)_{ab}\,(\sigma_J)^{ab} &= \delta^I_J, \quad (\sigma^I)_{ab}\,(\sigma_I)^{cd} = \delta^c_a\,\delta^d_b, \\ (\sigma^I)_a^{\ b}\,(\sigma_J)^a_{\ b} &= \delta^I_J, \quad (\sigma^I)_a^{\ c}\,(\sigma_I)^b_{\ d} = \delta^b_a\,\delta^c_d, \end{aligned}$$

where the q-number is defined as usual:

$$[m] = \frac{q^m - q^{-m}}{q - q^{-1}}, \quad [2] = q + q^{-1}. \tag{14.5}$$

Throughout this chapter, the capital Latin, like I, runs over 0, +, 3, and −, the small Latin, like i, runs over +, 3, and −, but the first a few small Latin, like a, runs over 1 and 2.

Now, two kinds of components, the adjoint representation and the identity representation of the quantum group, are separated in M:

$$\begin{aligned} M^I{}_J &= (\sigma^I)_{a_1}^{a_2} \, M^{a_1}{}_{a_2 b_1}{}^{b_2} \, (\sigma_J)^{b_1}_{b_2}, \\ M^i{}_0 &= M^0_i = 0, \quad M^0_0 = \mathbf{1}. \end{aligned}$$

A diagonal matrix D related with the double antipode action is defined:

$$\begin{aligned} \kappa^2(M^I{}_J) &= D^I{}_K \, M^K_L \, (D^{-1})^L{}_J, \\ D^0{}_0 &= D^3{}_3 = 1, \quad D^+_+ = q^2, \quad D^-_- = q^{-2}. \end{aligned}$$

From the q-Clebsch-Gordon coefficients[101] one can construct the projection operators:

$$(\mathcal{P}_A)^{ab}_{cd} = (\sigma_0)^{ab}(\sigma^0)_{cd} = \frac{1}{[2]} \begin{pmatrix} 0 & & & \\ & q^{-1} & -1 & \\ & -1 & q & \\ & & & 0 \end{pmatrix},$$

$$(\mathcal{P}_S)^{ab}_{cd} = (\sigma_i)^{ab}(\sigma^i)_{cd} = \frac{1}{[2]} \begin{pmatrix} [2] & & & \\ & q & 1 & \\ & 1 & q^{-1} & \\ & & & [2] \end{pmatrix}.$$

Now let us consider the space of 1-forms on the quantum group, which can be q-generalization of the space of 1-forms on a Lie group, and denote it by Γ. Then the q-deformed exterior derivative δ is defined as a map from A to bimodule Γ:

$$\delta : A \to \Gamma$$

which satisfies the Leibniz rule

$$\delta(\alpha\beta) = (\delta\alpha)\beta + \alpha(\delta\beta), \quad \forall \alpha, \beta \in A.$$

Differential 1-forms on a Lie group manifold G are sections of the cotangent bundle $T^*(G)$. The space of all sections on the cotangent bundle is a bimodule over the functional $\mathcal{F}(G)$. Now Γ is the q-generalization of the space of 1-forms on a Lie group, which is a bimodule over A, i.e., its elements can be multiplied from either left or right by elements of A. In addition one can introduce the bimodule-bicomodule over A and consider the case where these bimodule-bicomodules are bicovariant as given by Woronowiez[83], i.e., a left action Δ_L and a right action Δ_R of the quantum group on Γ are defined as follows:

$$\begin{aligned} \Delta_L &: \Gamma \to A \otimes \Gamma, \qquad \Delta_L(\alpha\delta\beta) = \Delta(\alpha) \, (id \otimes \delta) \, \Delta(\beta), \\ \Delta_R &: \Gamma \to \Gamma \otimes A, \qquad \Delta_R(\alpha\delta\beta) = \Delta(\alpha) \, (\delta \otimes id) \, \Delta(\beta), \end{aligned}$$

which $\forall \rho \in \Gamma$ satisfies

$$\begin{array}{c}(\epsilon \otimes id)\Delta_L(\rho) = \rho, \quad (\Delta \otimes id)\Delta_L = (id \otimes \Delta_L)\Delta_L \quad , \\ (\epsilon \otimes id)\Delta_R(\rho) = \rho, \quad (\Delta \otimes id)\Delta_R = (\Delta_R \otimes id)\Delta_R \quad , \\ (id \otimes \Delta_R)\Delta_L = (\Delta_L \otimes id)\Delta_R \quad .\end{array}$$

An element $\rho \in \Gamma$ is said to be left invariant if

$$\Delta_L(\rho) = I \otimes \rho,$$

and right invariant if

$$\Delta_R(\rho) = \rho \otimes I.$$

As in the classical case, the right invariant 1-form dgg^{-1} is left covariant. Let the bases of the right-invariant elements of Γ be denoted by η^J, satisfying:

$$\Delta_R(\eta^J) \; = \; \eta^J \; \otimes \; \mathbf{1}, \quad \Delta_L(\eta^J) \; = \; M^J{}_K \; \otimes \; \eta^K.$$

Let ξ, ς be the linear functionals on A, one can define the convolution product $*$ as:

$$\begin{array}{rl} & \xi * \alpha \equiv (id \otimes \xi)\Delta(\alpha) \quad , \forall \alpha \in A, \\ \text{and} & \alpha * \xi \equiv (\xi \otimes id)\Delta(\alpha) \quad , \forall \alpha \in A,\end{array}$$

which satisfy

$$\begin{array}{c}\xi(\alpha * \varsigma) = \varsigma(\xi * \alpha) = (\varsigma * \varsigma)\alpha \equiv (\varsigma \otimes \xi)\Delta(\alpha) \\ \forall \alpha \in A, \quad \xi, \varsigma \in A'\end{array}$$

Then any $\rho \in \Gamma$ can be written as

$$\rho = \alpha_I \eta^I = \eta^I \beta_I.$$

It can be proved that there are linear functionals $L^J_K \in A'$ such that

$$\alpha \, \eta^J \; = \; \eta^K(\alpha \; * \; L^J_K), \quad \alpha \in A, \quad L^J_K \in A',$$

where L^J_K is a special set of elements of A' consistent with the quantum group relations as

$$\Delta'(L^J_K) = L^I_K \otimes L^J_I, \quad \epsilon'(L^J_K) = \delta^J_K, \quad \kappa'(L^J_K) = L^J_K \cdot \kappa,$$

and the bicovariance requires that L^J_K satisfy

$$\begin{array}{c}\Delta(\alpha)\Delta_R(\eta^J) = \Delta_R(\eta^K)\Delta(\alpha * L^J_K), \\ \Delta(\alpha)\Delta_L(\eta^J) = \Delta_L(\eta^K)\Delta(\alpha * L^J_K).\end{array}$$

The first equation is satisfied trivially while the second equation leads to

$$(L^I_K * \alpha)M^J_I = M^I_K(\alpha * L^J_I). \tag{14.6}$$

Accordingly for $SU_q(2)$ one can choose

$$\begin{aligned} L_I{}^J &= (\sigma^J)_a{}^d\, \kappa'((L^+)^a{}_b)\, (L^-)^c{}_d\, (\sigma_I)^b{}_c, \\ L_I{}^J(\alpha\beta) &= L_I{}^K(\alpha)\, L_K^J(\beta), \quad L_I{}^J(\mathbf{1}) = \delta_I^J, \\ (\rho * L_I{}^J) &= (L_I{}^J \otimes id)\, \Delta_L(\rho). \end{aligned}$$

Analogously one can introduced the left-invariant 1-forms. The basis of the left-invariant elements of Γ is easy to be calculated from the right-invariant η^J,i.e., let

$$\omega^J = \kappa(M^J{}_K)\, \eta^K,$$

this elements satisfy:

$$\Delta_L(\omega^J) = \mathbf{1} \otimes \omega^J, \quad \Delta_R(\omega^J) = \omega^K \otimes \kappa(M^J{}_K),$$

which is the basis of the left-invariant elements of Γ.

As the analogue of the ordinary permutation operator, a bimodule automorphism Λ in $\Gamma \otimes \Gamma$ is defined by:

$$\Lambda(\omega^J \otimes \eta^K) = \eta^K \otimes \omega^J,$$

which satisfies

$$\Lambda(a\,\tau) = a\,\Lambda(\tau), \quad \Lambda(\tau a) = \Lambda(\tau)\, a, \quad a \in A, \ \tau \in \Gamma \otimes \Gamma.$$

Λ is invertible and commutes with left and right action of quantum group, so $\Lambda(\eta^J \otimes \eta^K)$ is also right invariant and can be written in terms of the right invariant basis $\eta^J \otimes \eta^K$ as

$$\Lambda(\eta^I \otimes \eta^J) = \Lambda^{IJ}{}_{KL}\, \eta^K \otimes \eta^L, \quad \Lambda^{IJ}{}_{KL} = L_K^J(M^I{}_L).$$

here the matrix $(\Lambda^{IJ}{}_{KL})$ be called the braiding matrix, using which the exterior product of two right invariant 1-forms can be determined as

$$\eta^I \wedge \eta^J = \eta^I \otimes \eta^J - \Lambda^{IJ}_{KL} \eta^K \otimes \eta^L.$$

Taking $\alpha = M^L_M$ in Eq. 14.6 one get

$$\Lambda^{LJ}_{IN} M^I_K M^N_M = M^L_N M^J_I \Lambda^{NI}_{KM}. \tag{14.7}$$

Apply the bimodule automorphism $\Lambda(\alpha\tau) = \alpha\Lambda(\tau)$ to the case $\tau = \eta^I \otimes \eta^J$, then on the left hand side

$$\begin{aligned} \Lambda(\alpha\eta^I \otimes \eta^J) &= \Lambda(\eta^K(\alpha * L^I_K) \otimes \eta^J) = \Lambda(\eta^K \otimes (\alpha * L^I_K)\eta^J) \\ &= \Lambda(\eta^K \otimes \eta^L(\alpha * L^I_K * L^J_L)) \\ &= \Lambda^{KL}_{MN} \eta^M \otimes \eta^N(\alpha * L^I_K * L^J_L), \end{aligned}$$

while on the right hand side one has

$$\alpha \wedge (\eta^I \otimes \eta^J) = \alpha \Lambda^{IJ}_{KL} \eta^K \otimes \eta^L.$$

Comparing the two sides and taking $\alpha = M^P_Q$ one get

$$\Lambda^{KL}_{MN} \Lambda^{PI}_{KR} \Lambda^{RJ}_{LS} = \Lambda^{PK}_{MR} \Lambda^{RL}_{NS} \Lambda^{IJ}_{KL}, \tag{14.8}$$

which is just the Yang-Baxter equation. This is the sufficient condition for the consistency of Eq. 14.7. We can see that M^I_J satisfy a relation identical to the Yang-Baxter relation (RTT relation), only that we are in the adjoint representation.

The eigenvalues of Λ for $SU_q(2)$ are 1, q^2, $-q^{-2}$, and Λ satisfies

$$(\Lambda + q^2)(\Lambda + q^{-2})(\Lambda - 1) = 0. \tag{14.9}$$

The non-vanishing components of ${\Lambda^{IJ}}_{KL}$ are listed as follows:

$$\begin{aligned} {\Lambda^{ij}}_{k\ell} &= {(\Lambda^{-1})^{ij}}_{k\ell} = \delta^i_k\, \delta^j_\ell + \bar{f}^{ij}_n\, f^n_{k\ell}, \quad {\Lambda^{00}}_{00} = {(\Lambda^{-1})^{00}}_{00} = 1, \\ {\Lambda^{i0}}_{jk} &= {(\Lambda^{-1})^{0i}}_{jk} = \lambda\, f^i_{jk}, \quad {\Lambda^{jk}}_{0i} = {(\Lambda^{-1})^{jk}}_{i0} = \lambda\, \bar{f}^{jk}_i, \\ {\Lambda^{0i}}_{j0} &= {(\Lambda^{-1})^{i0}}_{0j} = \delta^i_j, \quad {\Lambda^{i0}}_{0j} = {(\Lambda^{-1})^{0i}}_{j0} = (\lambda^2 + 1)\, \delta^i_j, \end{aligned} \tag{14.10}$$

where the non-vanishing components of $f^i_{jk} = -\bar{f}^{jk}_i$ are [95]:

$$f^+_{+3} = f^-_{3-} = q, \quad f^+_{3+} = f^-_{-3} = -q^{-1}, \quad f^3_{-+} = -f^3_{+-} = 1, \quad f^3_{33} = \lambda. \tag{14.11}$$

Watamura [95] introduced an operator for given tensors ${X^{ab}}_{cd}$ and ${Y^{ab}}_{cd}$:

$$\begin{aligned} {(X,Y)^{IJ}}_{KL} &= (\sigma^I)_{a_1 a_2}\, (\sigma^J)_{b_1 b_2}\, X^{a_1 e_1}{}_{c_1 e_3} \left(\hat{R}_q^{-1}\right)^{a_2 b_1}{}_{e_1 e_2} \\ &\quad \cdot Y^{e_2 b_2}{}_{e_4 d_2} \left(\hat{R}_q\right)^{e_3 e_4}{}_{c_2 d_1} (\sigma_K)^{c_1 c_2}\, (\sigma_L)^{d_1 d_2}. \end{aligned}$$

In terms of this tensors, the matrix ${\Lambda^{IJ}}_{KL}$ can be expressed as combination of four projection operators [95]:

$$\begin{aligned} {\Lambda^{IJ}}_{KL} &= \left(\hat{R}^{-1}, \hat{R}\right)^{IJ}{}_{KL} \\ &= (\mathcal{P}_S, \mathcal{P}_S)^{IJ}{}_{KL} - q^{-2}\, (\mathcal{P}_S, \mathcal{P}_A)^{IJ}{}_{KL} - q^2\, (\mathcal{P}_A, \mathcal{P}_S)^{IJ}{}_{KL} + (\mathcal{P}_A, \mathcal{P}_A)^{IJ}{}_{KL}, \\ {(\Lambda^{-1})^{IJ}}_{KL} &= \left(\hat{R}^{-1}, \hat{R}\right)^{IJ}{}_{KL} \\ &= (\mathcal{P}_S, \mathcal{P}_S)^{IJ}{}_{KL} - q^2\, (\mathcal{P}_S, \mathcal{P}_A)^{IJ}{}_{KL} - q^{-2}\, (\mathcal{P}_A, \mathcal{P}_S)^{IJ}{}_{KL} + (\mathcal{P}_A, \mathcal{P}_A)^{IJ}{}_{KL}. \end{aligned}$$

The exterior product of the elements in Γ is given as follows:

$$\begin{aligned} \rho \wedge \rho' &\equiv \rho \otimes \rho' - \Lambda(\rho \otimes \rho'), \\ \eta^I \wedge \eta^J &= \left(\delta^I_K\, \delta^J_L - {\Lambda^{IJ}}_{KL}\right) \left(\eta^K \otimes \eta^L\right). \end{aligned}$$

$\eta^I \wedge \eta^J$ is annihilated by the projection operator $\mathcal{P}_0$ due to Eq.14.9

$$\begin{aligned}\mathcal{P}_0 &= (\mathcal{P}_S, \mathcal{P}_S)^{IJ}{}_{KL} + (\mathcal{P}_A, \mathcal{P}_A)^{IJ}{}_{KL} \\ &= [2]^{-2} \{\Lambda + \Lambda^{-1} + (\lambda^2 + 2)\, \mathbf{1}\} \\ (\mathcal{P}_0)^{IJ}{}_{KL} \left(\eta^K \wedge \eta^L\right) &= 0.\end{aligned}$$

The projection operator $\mathcal{P}_{SA}$ now can be expressed as follows:

$$\begin{aligned}\mathcal{P}_{SA} &= (\mathcal{P}_S, \mathcal{P}_A)^{IJ}{}_{KL} + (\mathcal{P}_A, \mathcal{P}_S)^{IJ}{}_{KL} \\ &= [2]^{-2} \{\mathbf{2} - \Lambda - \Lambda^{-1}\}.\end{aligned}$$

It is interesting to notice that there is a projection operator $\mathcal{P}_{Adj}$ with only the adjoint components:

$$(\mathcal{P}_{Adj})^{ij}{}_{k\ell} = \frac{[2]^2}{2(\lambda^2+2)} (\mathcal{P}_{SA})^{ij}{}_{k\ell} = -(\lambda^2+2)^{-1} \bar{f}_t^{ij} f_{k\ell}^t,$$

$$\begin{aligned}&(\mathcal{P}_{Adj})^{ij}{}_{rs} (\mathcal{P}_{Adj})^{rs}{}_{k\ell} = (\mathcal{P}_{Adj})^{ij}{}_{k\ell} \\ &(\mathcal{P}_{Adj})^{ij}{}_{k\ell} \left(\eta^k \wedge \eta^\ell\right) = \eta^i \wedge \eta^j \\ &(\mathcal{P}_{Adj})^{ij}{}_{k\ell}\, f_{ij}^t = f_{k\ell}^t, \quad (\mathcal{P}_{Adj})^{ij}{}_{k\ell}\, \bar{f}_t^{k\ell} = \bar{f}_t^{k\ell}.\end{aligned}$$

The high degree differential forms $\Gamma^{\wedge k}$ can be defined as in the classical case but with the quantum permutation Λ.

Next we look at the space dual to the invariant form: the invariant vector fields on quantum group, whose bases elements $\chi_J \in A'$ are defined by

$$\begin{aligned}\delta\alpha &= \eta^J(\alpha * \chi_J), \\ \delta\eta^I &= \eta^J(\eta^I * \chi_J) \\ &= \eta^J(\chi_J \otimes id)\Delta_L \eta^I \\ &= C^I_{JK} \eta^J \otimes \eta^K,\end{aligned} \tag{14.12}$$

where $C^I_{JK} = \chi_J(M^I_K)$, $\chi_J \in A'$ are the quantum analogues of tangent vector fields at the identity element of the group. The commutation relations between the $\chi's$ can be obtained by taking the exterior derivative of Eq. 14.12:

$$\begin{aligned}\delta^2\alpha &= 0 = \delta\eta^J(\alpha * \chi_J) - \eta^J \wedge d(\alpha * \chi_J) \\ &= \eta^K(\eta^J * \chi_K)(\alpha * \chi_J) - \eta^J \wedge \eta^K(\alpha * \chi_J * \chi_K) \\ &= C^I_{JK}\eta^J \otimes \eta^K(\alpha * \chi_I) - (\eta^J \otimes \eta^K - \Lambda^{JK}_{LM}\eta^L \otimes \eta^M)(\alpha * \chi_J * \chi_K) \\ &= \eta^J \otimes \eta^K \alpha * (C^I_{JK}\chi_I - \chi_J\chi_K + \Lambda^{LM}_{JK}\chi_L\chi_M),\end{aligned}$$

$$\text{So} \qquad \chi_I\, \chi_J - \Lambda^{KL}{}_{IJ}\, \chi_K\, \chi_L = \mathbf{C}_{IJ}{}^K\, \chi_K.$$

Apply this equation to M^P_S and one obtains the q-deformed Jacobi identity for the q-deformed structure constants:

$$\mathbf{C}_{IR}{}^P \mathbf{C}_{JS}{}^R - \Lambda^{KL}{}_{IJ}\, \mathbf{C}_{KR}{}^P \mathbf{C}_{LS}{}^R = \mathbf{C}_{IJ}{}^R \mathbf{C}_{RS}{}^P.$$

Since there is a bi-invariant element η^0 which can be used to define a derivative on A and on the whole of $\Gamma^{\wedge}$ as follows:

$$\delta\alpha = \frac{ig}{\lambda}\,(\eta^0\,\alpha\; - \;\alpha\,\eta^0),$$
$$\delta\theta = \frac{ig}{\lambda}\,(\eta^0\wedge\theta\; - \;(-1)^k\theta\wedge\eta^0),$$

where k is the degree of the element $\theta\in\Gamma^{\wedge k}$ and g is a non-zero real constant.

In the case of $SU_q(2)$, comparing with Eq. 14.6

$$\chi_J = \frac{ig}{\lambda}\left(\epsilon\,\delta^0_J\; - \;L_J{}^0\right).$$

The q-deformed structure constants $C^I_{JK} = \chi_J(M^I_K)$ are

$$\mathbf{C}_{JK}{}^0 = \mathbf{C}_{J0}{}^K = 0, \quad \mathbf{C}_{0k}{}^j = -\,ig\lambda\,\delta^j_k, \quad \mathbf{C}_{jk}{}^i = -\,ig\,f^i_{jk}, \tag{14.13}$$

where f^i_{jk} are given in Eq. 14.11. For the adjoint components, from Eq. 14.10 we have

$$(\mathcal{P}_{Adj})^{kl}{}_{ij}\;\mathbf{C}_{kr}{}^p\mathbf{C}_{ls}{}^r = \frac{\lambda^2+1}{\lambda^2+2}\,\mathbf{C}_{ij}{}^r\mathbf{C}_{rs}{}^p.$$

The q-deformed Cartan-Maurer equation can be written as

$$\delta\eta^I = F^I_{JK}\eta^J\wedge\eta^K = F^I_{JK}(\eta^J\otimes\eta^K - \Lambda^{JK}_{LM}\eta^L\otimes\eta^M),$$
$$C^I_{JK} = F^I_{JK} - F^I_{PQ}\Lambda^{PQ}_{JK}.$$

Thus from Eq. 14.10 we can obtain

$$\delta\eta^I = (\lambda^2+2)^{-1}\,\mathbf{C}_{jk}{}^I\,\left(\eta^j\wedge\eta^k\right).$$

14.4 Q-gauge theory in terms of q-BRST algebra

There are a few attempts to construct q-gauge theory by taking the quantum group as a symmetry. As quantum group is formulated as a Hopf algebra, one needs to represent the gauge theory using an appropriate algebraic language.

In what follows we look at the q-deformation of gauge theory in terms of q-BRST algebra developed by Watamura. Here gauge transformation is replaced by BRST transformation which is represented by a nilpotent operator δ and the gauge transformation parameter is replaced by ghost field.

Let Φ be the matter field, its BRST transformation $\delta\Phi$ is an element of the bicovariant bimodule over quantum group A, and one can expand it under the right invariant basis in Γ,

$$\delta\Phi = \eta^J(\Phi * \chi_J) = \frac{ig}{\lambda}[\eta^0,\Phi]_{\mp},$$

where commutator $[\ ,\ ]_-$ is taken for the bosonic field and anticommutator $[\ ,\ ]_+$ is taken for the fermionic field. Here η^J are ghost fields which have the same properties as the right invariant basis in Γ. We must keep in mind that η^J are space-time dependent.

There are two nilpotent operators, the space-time exterior differential operator d and the BRST transformation operator δ. $d\Phi$ is not covariant under the BRST transformation, we can have a covariant differential with the introduction of gauge field A^J:

$$\nabla\Phi = d\Phi + A^J(\Phi * \chi_J) = d\Phi + \frac{ig}{\lambda}[A^0, \Phi]_{\mp}.$$

The gauge coupling between gauge fields A^J with matter field is determined by requiring that the covariant derivative transforms in the same way as the matter field

$$\delta\nabla\Phi = \eta^J(\nabla\Phi * \chi_J) = \frac{ig}{\lambda}[\eta^0, \nabla\Phi]_{\pm}.$$

The gauge potential is introduced in the covariant derivative. The covariance of the covariant derivative under the BRST transformation leads to

$$\delta A^J = d\eta^J + \frac{ig}{\lambda}(\eta^0 A^J + A^J \eta^0).$$

Just as in the classical case, we can get the gauge field strength F by taking the covariant derivative twice:

$$\nabla\nabla\Phi = F^J(\Phi * \chi_J) = d(\nabla\Phi) + A^J(\nabla\Phi * \chi_J).$$

Thus

$$F^J = dA^J + \frac{ig}{\lambda}[A^0, A^J]_+.$$

There are four fundamental kind of fields in the q-BRST algebra $\mathcal{B}$: η_J, $d\eta_J$, A_J and dA_J. In order to describe those covariant commutation relations by unified formulas, we introduce an index n that is equal to the difference between the degree of form and the ghost number. The indices n for η^J, $d\eta^J$, A^J and dA^J are -1, 0, 1, and 2, respectively. Both δ and d operations satisfy the Leibniz rule in the graded sense for the index n, and are nilpotent operators:

$$\begin{aligned} \delta^2 &= 0, \quad d^2 = 0, \quad d\,\delta + \delta\, d = 0, \\ d(XY) &= (dX)Y + (-1)^{n_x} X(dY), \\ \delta(XY) &= (\delta X)Y + (-1)^{n_x} X(\delta Y), \end{aligned}$$

where n_x is the index of X. Both d and δ are covariant for the left and right actions: For any element $\rho \in \mathcal{B}$ they satisfy:

$$\begin{aligned} \Delta_L(\delta\rho) &= (id \otimes \delta)\Delta_L(\rho), \quad \Delta_L(d\rho) = (id \otimes d)\Delta_L(\rho), \\ \Delta_R(\delta\rho) &= (\delta \otimes id)\Delta_R(\rho), \quad \Delta_R(d\rho) = (d \otimes id)\Delta_R(\rho). \end{aligned}$$

A^I are assumed to have following properties, similar to η^I. Hereafter, we neglect the wedge sign $\wedge$ for simplicity.

$$\begin{aligned}
(\mathcal{P}_0)^{IJ}{}_{KL}\left(A^K A^L\right) = 0, \quad (\mathcal{P}_{Adj})^{ij}{}_{k\ell}\left(A^k A^\ell\right) &= A^i A^j, \\
\frac{ig}{\lambda}\left(A^0 A^i + A^i A^0\right) &= (\lambda^2+2)^{-1}\,\mathbf{C}_{jk}{}^i\, A^j A^k.
\end{aligned} \tag{14.14}$$

From the consistent conditions, $d\eta^J$ and dA^J have to satisfy the other conditions:

$$(\mathcal{P}_{SA})^{IJ}{}_{KL}\left(d\eta^K\, d\eta^L\right) = 0, \quad (\mathcal{P}_{SA})^{IJ}{}_{KL}\left(dA^K\, dA^L\right) = 0,$$

i.e., they are q-symmetric fields.

Next we look at the commutation between two different fields X^J and Y^K with indices n_x and n_y, respectively. For definiteness assume $n_x > n_y$. The consistent condition requires the following covariant commutative relations:

$$(-1)^{n_x n_y}\, X^I\, Y^J = \Lambda^{IJ}{}_{KL}\, Y^K\, X^L.$$

From Eq. 14.10 we have

$$\begin{aligned}
(-1)^{n_x n_y}\, X^0\, Y^J &= Y^J\, X^0, \\
\{ig/\lambda\}\left(Y^0 X^I - (-1)^{n_x n_y}\, X^I Y^0\right) &= \mathbf{C}_{JK}^{\ I}\, Y^J X^K.
\end{aligned}$$

At last, the gauge fields F^J can be expressed as:

$$\begin{aligned}
F^J &= dA^J + ig\lambda^{-1}\left(A^0\, A^J + A^J\, A^0\right), \\
\text{i.e.,}\quad F^0 &= dA^0, \quad F^i = dA^i + (\lambda^2+2)^{-1}\,\mathbf{C}_{jk}{}^i\, A^j A^k.
\end{aligned} \tag{14.15}$$

(14.16)

which satisfy the q-Bianchi identities:

$$\begin{aligned}
dF^I &= -\, ig\lambda^{-1}\left(A^0\, F^I - F^I\, A^0\right) \\
&= -\, ig\lambda^{-1}\left(A^0\, dA^I - dA^I\, A^0\right) \\
&= -\, \mathbf{C}_{JK}^{\ I}\, A^J\, dA^K, \\
\text{i.e.,}\quad dF^0 &= 0, \quad dF^i = ig\lambda\, A^0\, dA^i - \mathbf{C}_{jk}{}^i\, A^j\, dA^k.
\end{aligned}$$

It is easy to obtain the following relation

$$F^I \eta^J = \eta^K F^L \Lambda^{IJ}_{KL}.$$

Under the q-BRST transformation

$$\begin{aligned}
\delta F^I &= ig\lambda^{-1}\left(\eta^0\, F^I - F^I\, \eta^0\right) = \mathbf{C}_{JK}^{\ I}\, \eta^J\, F^K, \\
\text{i.e.,}\quad \delta F^0 &= 0, \quad \delta F^i = -\, ig\lambda\, \eta^0\, F^i + \mathbf{C}_{jk}{}^i\, \eta^j\, F^k.
\end{aligned} \tag{14.17}$$

14.5 q-Deformed Chern Class

It is easy to understand that the second q-deformed Chern class has the following form [85,94]:

$$P \equiv F^I F^J g_{IJ},$$

where we omit the possible constant factor in P. The q-deformed Killing form g_{IJ} is chosen from the condition:

$$\delta P = 0, \qquad dP = 0. \tag{14.18}$$

From Eq.14.15 and Eq.14.17 we have:

$$\begin{aligned} \delta P &= \left\{\delta F^R F^S + F^R \delta F^S\right\} g_{RS} \\ &= ig\lambda^{-1} \left\{\eta^0 F^R F^S - F^R F^S \eta^0\right\} g_{RS} \\ &= ig\lambda^{-1} \eta^I F^J F^K \left\{\delta^0_I g_{JK} - \Lambda^{SO}{}_{TK} \Lambda^{RT}{}_{IJ} g_{RS}\right\}, \end{aligned}$$

namely, g_{IJ} has to satisfy:

$$\delta^0_I g_{JK} = \Lambda^{SO}{}_{TK} \Lambda^{RT}{}_{IJ} g_{RS}. \tag{14.19}$$

We are going to show that the condition $dP = 0$ also holds if g_{IJ} satisfy Eq.14.19. From Eq.14.14 a product of four adjoint components of A^j must be vanishing, so that:

$$\begin{aligned} P &= dA^R dA^S g_{RS} \\ &\quad + ig\lambda^{-1} \left\{dA^R (A^0 A^S + A^S A^0) + (A^0 A^R + A^R A^0) dA^S\right\} g_{RS}, \\ dP &= ig\lambda^{-1} \left\{dA^R (dA^0 A^S - A^0 dA^S + dA^S A^0 - A^S dA^0\right. \\ &\quad \left. + (dA^0 A^R - A^0 dA^R + dA^R A^0 - A^R dA^0) dA^S\right\} g_{RS} \\ &= -ig\lambda^{-1} \left(A^0 dA^R dA^S - dA^R dA^S A^0\right) g_{RS} \\ &= -ig\lambda^{-1} A^I dA^J dA^K \left\{\delta^0_I g_{JK} - \Lambda^{SO}{}_{TK} \Lambda^{RT}{}_{IJ} g_{RS}\right\}. \end{aligned}$$

From Eq. 14.10 we can solve Eq.14.19 as follows.
a) When $J = 0$, $I = i \neq 0$ and $K = k \neq 0$, we have:

$$g_{0\ell} = 0.$$

b) When $K = 0$, $I = i \neq 0$ and $J = j \neq 0$, we have:

$$g_{\ell 0} = 0.$$

c) When $I = \pm$ and $J = K = 3$, we have:

$$g_{\pm 3} + g_{3\pm} = 0.$$

However, when $J = \pm$ and $I = K = 3$, we have:

$$\mp q^{\mp 3} g_{\pm 3} + \lambda g_{3\pm} = 0.$$

Thus, we obtain:

$$g_{\pm 3} = g_{3\pm} = 0.$$

d) When $I = J = \pm$ and $K = 3$, we have:

$$g_{++} = g_{--} = 0.$$

e) When $I = -J = \pm$ and $K = 3$, we have:

$$g_{+-} = q\, g_{33}, \qquad g_{-+} = q^{-1}\, g_{33}.$$

For the rest cases there is no new restriction on g_{IJ}. It is worthy to notice that the singlet and adjoint components of g_{IJ} are separated from each other, although those components are mixed in the commutative relations in the $BRST$-algebra $\mathcal{B}$. Furthermore, if one requires that the q-deformed Chern class contains only the adjoint components of gauge fields, the q-deformed Killing form as well as the q-deformed Chern class are determined uniquely up to a common factor.

We are going to choose the q-deformed Killing form g_{IJ} such that g_{IJ} satisfies above equations and is the analogue of the classical Killing form. The following q-deformed Killing form satisfies those conditions:

$$g_{IJ} = D^R{}_S\, \mathbf{C}_{IT}{}^S\, \mathbf{C}_{JR}{}^T = D^r{}_s\, \mathbf{C}_{It}{}^s\, \mathbf{C}_{Jr}{}^t$$

where the summed indices cannot equal to zero owing to Eq.14.13. It is acceptable that the q-deformed Killing form is related to the double antipode action. From Eq.14.13 the non-vanishing components of g_{IJ} are:

$$g_{00} = -g^2\, \lambda^2\, [3], \qquad q^{-1}\, g_{+-} = q\, g_{-+} = g_{33} = -g^2\, (\lambda^2 + 2).$$

Obviously, both g_{IJ} and g_{ij} (without 0 subscript) satisfy Eq.14.19. From the q-deformed Killing form we may define two kinds of the second q-deformed Chern Class:

$$\begin{aligned} P &\equiv \langle\, F\,,\, F\,\rangle = F^i\, F^j\, g_{ij} \\ &= -\, g^2(\lambda^2+2)\, (F^3\, F^3 + q\, F^+\, F^- + q^{-1} F^-\, F^+)\,,\,, \\ \hat{P} &\equiv F^I\, F^J\, g_{IJ} - P = -g^2\, \lambda^2\, [3]\, F^0\, F^0 \end{aligned}$$

where P and $\hat{P}$ contain only the adjoint components and the singlet of F^j, respectively. From now on we will mainly discuss the q-deformed Chern class P. The similar calculation for $\hat{P}$ is much simpler than that for P.

14.6 q-Deformed Chern-Simons

In the classical case Chern[98] and Zumino [99] introduced a homotopy operator k to compute the Chern-Simons. Now we generalize his method to compute the q-deformed Chern-Simons.

Introduce a q-deformed homotopy operator k that is nilpotent and satisfies the Leibniz rule in the graded sense for the index n:

$$k^2 = 0, \quad d\,k + k\,d = \mathbf{1}. \tag{14.20}$$

If k exists, we can compute the q-deformed Chern-Simons $Q(A)$ from q-deformed Chern class:

$$\begin{aligned} P = (d\,k + k\,d)\,P &= d(k\,P) = d\,Q(A), \\ Q(A) &= k\,P, \end{aligned} \tag{14.21}$$

where we used Eq.14.18.

Introduce a real parameter t, $0 \leq t \leq 1$. When t changes from 0 to 1, the gauge potentials A_t^J change from 0 to A^J:

$$\begin{aligned} A_t^J &= t\,A^J, \\ F_t^J &= t\,dA^J + \frac{ig\,t^2}{\lambda}\left(A^0\,A^J + A^J\,A^0\right) \\ &= t\,F^J + \frac{ig\,(t^2 - t)}{\lambda}\left(A^0\,A^J + A^J\,A^0\right). \end{aligned}$$

Due to our definition Eq.14.4 for q-number we have to change a little the usual definition [101] for the q-deformed derivative and the q-deformed integral. Define the q-deformed derivative along t by:

$$\frac{\partial}{\partial_q t}\,f(t) = \frac{f(qt) - f(q^{-1}t)}{t\,(q - q^{-1})}$$

satisfying the q-deformed Leibniz rule:

$$\frac{\partial}{\partial_q t}\,f(t)g(t) = \frac{\partial f(t)}{\partial_q t}\,g(qt) + f(q^{-1}t)\,\frac{\partial g(t)}{\partial_q t}.$$

The q-deformed integral is defined by:

$$\int_0^{t_0} d_q t\,f(t) = t_0(1 - q^2)\sum_{k=0}^{\infty} q^{2k}\,f(q^{2k+1}t_0).$$

At least for a polynomial, the q-deformed integral is the inverse of q-deformed derivative. For example,

$$\frac{\partial}{\partial_q t}\,t^m = [m]\,t^{m-1}, \qquad \int_0^{t_0} d_q t\,t^{m-1} = t_0^m/[m].$$

Now, define the q-deformed Lie derivative $\hat{\delta}_q$ along t in the gauge space:

$$\hat{\delta}_q \equiv d_q t\,\frac{\partial}{\partial_q t}$$

and the q-deformed operator ℓ_t that satisfies the q-deformed Leibniz rule in the graded sense for the index n:

$$\begin{aligned}\ell_t A_t^J &= 0, \quad \ell_t F_t^J = \hat{\delta}_q A_t^J = d_q t\, A^J,\\ \ell_t \{f(t) g(t)\} &= \{\ell_t f(t)\}\, g(qt) + (-1)^n f(q^{-1}t)\, \{\ell_t g(t)\},\end{aligned}$$

where f has index n.

It is easy to check that for all formal polynomials (vanishing at $F_t^J = 0$ and $A_t^J = 0$) we have

$$\begin{aligned}\ell_t\, \ell_t &= 0,\\ \ell_t\, d + d\, \ell_t &= \hat{\delta}_q = d_q t \frac{\partial}{\partial_q t},\\ \hat{\delta}_q\, d = d\, \hat{\delta}_q, &\quad \hat{\delta}_q\, \ell_t = \ell_t\, \hat{\delta}_q\, .\end{aligned}$$

Comparing it with Eq.14.20 we obtain:

$$k = \int_0^1 \ell_t.$$

Now, we are able to compute the q-deformed Chern-Simons by Eq.14.21:

$$\begin{aligned}\ell_t P_t &= \langle \ell_t F_t\, ,\, F_{qt}\rangle + \langle F_{q^{-1}t}\, ,\, \ell_t F_t\rangle\\ &= d_q t\, \{\langle A\, ,\, F_{qt}\rangle + \langle F_{q^{-1}t}\, ,\, A\rangle\}\\ &= d_q t\, \{t[2]\langle A\, ,\, dA\rangle + ig\lambda^{-1}t^2(\lambda^2+2)\, \langle A\, ,\, (A^0 A + A A^0)\rangle\}\\ &= d_q t \left\{ t\, [2]\, \langle A\, ,\, dA\rangle + \frac{t^2(\lambda^2+2)}{\lambda^2+1}\, \langle A\, ,\, A\, ,\, A\rangle \right\},\end{aligned}$$

here

$$\langle A\, ,\, A\, ,\, A\rangle = A^{j_1} A^{j_2} A^{j_3} g_{j_1 j_2 j_3}, \qquad g_{j_1 j_2 j_3} = D^{t_0}_{t_1} C^{t_1}_{j_1 t_2} C^{t_2}_{j_2 t_3} C^{t_3}_{j_3 t_0}.$$

and we have used the relation (S.7) given in appendix S. So one get

$$\begin{aligned}Q(A) &= k\, P_t\\ &= \langle A\, ,\, dA\rangle + \{[4]/[6]\}\, \langle A\, ,\, A\, ,\, A\rangle\\ &= \langle A\, ,\, F\rangle - \{[2]/[6]\}\, \langle A\, ,\, A\, ,\, A\rangle.\end{aligned}$$

Similarly, we can compute $\hat{Q}(A)$ from $\hat{P} = d\hat{Q}(A)$:

$$\hat{Q}(A) = k\, \hat{P}_t = -\, g^2\lambda^2[3]\, A^0\, dA^0.$$

It is easy to prove by direct calculation that:

$$dQ(A) = P, \quad d\hat{Q}(A) = \hat{P}. \tag{14.22}$$

In fact, the second equation is obvious, and the first one can be proved in terms of the formulas given in Appendix S:

$$\begin{aligned} dQ(A) &= \langle dA\,,\, dA\rangle \\ &\quad + \{[4]/[6]\}\ \{\langle dA\,,\, A\,,\, A\rangle\ -\ \langle A\,,\, dA\,,\, A\rangle\ +\ \langle A\,,\, A\,,\, dA\rangle\} \\ &= \langle dA\,,\, dA\rangle\ +\ (\lambda^2+1)^{-1}\ \{\langle dA\,,\, A\,,\, A\rangle\ +\ \langle A\,,\, A\,,\, dA\rangle\} \\ &= \langle dA\,,\, dA\rangle\ +\ \frac{ig}{\lambda}\left\{\langle dA\,,\, (A^0A+AA^0)\rangle\ +\ \langle (A^0A+AA^0)\,,\, dA\rangle\right\} \\ &= \langle F\,,\, F\rangle\ =\ P. \end{aligned}$$

We find that although the components of the identity and the adjoint representations have to be mixed in the commutative relations of the q-deformed BRST algebra, the q-deformed Killing form and the second Chern class can contain only one kind of components: singlet or adjoint components.

14.7 q-Deformed Cocycle Hierarchy

Just like those in the classical case, the gauge fields F^J are invariant under the transformation:

$$A^J\ \rightarrow\ A^J\ -\ \eta^J, \qquad d\ \rightarrow\ d\ +\ \delta.$$

In fact,

$$\begin{aligned} F^J &\rightarrow \mathcal{F}^J \\ &= (d\ +\ \delta)\ (A^J\ -\ \eta^J) \\ &\quad + \frac{ig}{\lambda}\ \left\{(A^0\ -\ \eta^0)(A^J\ -\ \eta^J)\ +\ (A^J\ -\ \eta^J)(A^0\ -\ \eta^0)\right\} \\ &= F^J\ +\ \left\{\delta A^J\ -\ d\eta^J\ -\ \frac{ig}{\lambda}\ (\eta^0A^J\ +\ A^J\eta^0)\right\} \\ &\quad -\ \left\{\delta\eta^J\ -\ \frac{ig}{\lambda}\ \left(\eta^0\eta^J\ +\ \eta^J\eta^0 - A^0\eta^J\ -\ \eta^JA^0\right)\right\} \\ &= F^J. \end{aligned}$$

Now, transforming Eq.14.22 and expanding it by the ghost number, we obtain:

$$\begin{aligned} Q(A-\eta) &= \langle A-\eta\,,\, F\rangle\ -\ \{[2]/[6]\}\ \langle A-\eta\,,\, A-\eta\,,\, A-\eta\rangle \\ &= \omega_3^0\ +\ \omega_2^1\ +\ \omega_1^2\ +\ \omega_0^3, \\ P &= (d\ +\ \delta)\ Q(A-\eta) \\ &= d\omega_3^0\ +\ \{\delta\omega_3^0\ +\ d\omega_2^1\}\ +\ \{\delta\omega_2^1\ +\ d\omega_1^2\} \\ &\quad +\ \{\delta\omega_1^2\ +\ d\omega_0^3\}\ +\ \delta\omega_0^3, \end{aligned}$$

where the subscripts denote the degrees of form of the quantities, and the superscripts denote the ghost numbers. In two sides of above equation, the quantities with the same degree of form and the same ghost number should be equal to each other, respectively:

$$\begin{cases} P = d\omega_3^0, \\ \delta\omega_3^0 + d\omega_2^1 = 0, \\ \delta\omega_2^1 + d\omega_1^2 = 0, \\ \delta\omega_1^2 + d\omega_0^3 = 0, \\ \delta\omega_0^3 = 0. \end{cases}$$

We are going to derive the explicit forms of those ω, and simplify them by the formulas given in Appendix S.

$$\begin{aligned} \omega_3^0 &= Q(A) = \langle A , F\rangle - \{[2]/[6]\} \langle A , A , A\rangle \\ &= \langle A , dA\rangle + \{[4]/[6]\} \langle A , A , A\rangle, \end{aligned}$$

$$\begin{aligned} \omega_2^1 &= - \langle \eta , dA\rangle - \{ig/\lambda\} \langle \eta , (A^0 A + AA^0)\rangle \\ &\quad + \{[2]/[6]\} \{\langle \eta , A , A\rangle + \langle A , \eta , A\rangle + \langle A , A , \eta\rangle\} \\ &= - \langle \eta , dA\rangle, \end{aligned}$$

$$\begin{aligned} \omega_1^2 &= - \{[2]/[6]\} \{\langle \eta , \eta , A\rangle + \langle \eta , A , \eta\rangle + \langle A , \eta , \eta\rangle\} \\ &= - (\lambda^2 + 1)^{-1} \langle \eta , A , \eta\rangle, \end{aligned}$$

$$\omega_0^3 = \{[2]/[6]\} \langle \eta , \eta , \eta\rangle.$$

Similarly, we also have:

$$\begin{aligned} \hat{Q}(A - \eta) &= - g^2\lambda^2[3] \ (A^0 - \eta^0) \ F^0 \\ &= \hat{\omega}_3^0 + \hat{\omega}_2^1, \\ \hat{\omega}_3^0 &= \hat{Q}(A) = - g^2\lambda^2[3] \ A^0 \ F^0, \\ \hat{\omega}_2^1 &= g^2\lambda^2[3] \ \eta^0 \ F^0, \end{aligned}$$

$$\hat{P} = d\hat{\omega}_3^0, \qquad \delta\hat{\omega}_3^0 + d\hat{\omega}_2^1 = 0, \qquad \delta\hat{\omega}_2^1 = 0.$$

Based on the q-deformed $BRST$-algebra, we study the q-deformed Killing form and the q-deformed Chern class for the quantum group $SU_q(2)$. Introducing a q-deformed homotopy operator, that is the quantum analogue of the homotopy operator presented by Chern and Zumino [98,99], we are able to compute the q-deformed Chern-Simons. Then, the q-deformed cocycle hierarchy can be calculated. These method can be generalized to compute the higher q-deformed Chern classes and Chern-Simons, and to the quantum groups $SU_q(N)$ [103]. In terms of two operators d and δ we are able to discuss the double cohomology and cocycle hierarchy.

Notations and Formulae

- Bicovariant calculus on quantum group $SU_q(2)$:
right-invariant elements η^J satisfying:

$$\Delta_R(\eta^J) = \eta^J \otimes \mathbf{1}, \quad \Delta_L(\eta^J) = M^J{}_K \otimes \eta^K.$$

$$\alpha\, \eta^J = \eta^K(\alpha * L^J_K), \quad \alpha \in A, \quad L^J_K \in A',$$

the left-invariant elements ω^J:

$$\omega^J = \kappa(M^J{}_K)\, \eta^K, \quad \Delta_L(\omega^J) = \mathbf{1} \otimes \omega^J, \quad \Delta_R(\omega^J) = \omega^K \otimes \kappa(M^J{}_K).$$

The bimodule automorphism Λ in $\Gamma \otimes \Gamma$ is defined by:

$$\begin{aligned} \Lambda(\omega^J \otimes \eta^K) &= \eta^K \otimes \omega^J, \\ \Lambda(\eta^I \otimes \eta^J) &= \Lambda^{IJ}{}_{KL}\, \eta^K \otimes \eta^L, \quad \Lambda^{IJ}{}_{KL} = L^J_K(M^I{}_L). \end{aligned}$$

$$\begin{aligned} \delta\alpha &= \eta^J(\alpha * \chi_J), \quad \chi_J \in A' \\ \delta\eta^I &= \eta^J(\eta^I * \chi_J) = C^I_{JK}\eta^J \otimes \eta^K, \quad C^I_{JK} = \chi_J(M^I_K), \quad \chi_J \in A' \end{aligned}$$

$$\begin{aligned} \chi_I\, \chi_J - \Lambda^{KL}_{IJ}\, \chi_K\, \chi_L &= \mathbf{C}_{IJ}{}^K\, \chi_K. \\ \mathbf{C}_{IR}{}^P \mathbf{C}_{JS}{}^R - \Lambda^{KL}_{IJ}\, \mathbf{C}_{KR}{}^P \mathbf{C}_{LS}{}^R &= \mathbf{C}_{IJ}{}^R \mathbf{C}_{RS}{}^P. \end{aligned}$$

- q-BRST algebra

$$\delta A^J = d\eta^J + \frac{ig}{\lambda}(\eta^0 A^J + A^J \eta^0).$$

$$\begin{aligned} F^J &= dA^J + ig\lambda^{-1}\left(A^0\, A^J + A^J\, A^0\right), \quad \lambda = q - q^{-1}. \\ dF^I &= -\, ig\lambda^{-1}\left(A^0\, F^I - F^I\, A^0\right) \\ &= -\, ig\lambda^{-1}\left(A^0\, dA^I - dA^I\, A^0\right) = -\, \mathbf{C}_{JK}{}^I\, A^J\, dA^K, \\ F^I\eta^J &= \eta^K F^L \Lambda^{IJ}_{KL}. \\ \delta F^I &= ig\lambda^{-1}\left(\eta^0\, F^I - F^I\, \eta^0\right) = \mathbf{C}_{JK}{}^I\, \eta^J\, F^K, \end{aligned}$$

- q-Deformed Chern Class:

$$P \equiv F^I\, F^J\, g_{IJ}, \quad \delta P = 0, \quad dP = 0.$$

- q-Deformed Chern-Simons:

$$\begin{aligned} Q(A) &= \langle A\,,\, dA\rangle + \{[4]/[6]\}\,\langle A\,,\,A\,,\,A\rangle \\ &= \langle A\,,\,F\rangle - \{[2]/[6]\}\,\langle A\,,\,A\,,\,A\rangle. \\ dQ(A) &= P. \end{aligned}$$

- q-Deformed Cocycle Hierarchy

$$\left\{\begin{aligned} P &= d\omega_3^0, \\ \delta\omega_3^0 + d\omega_2^1 &= 0, \\ \delta\omega_2^1 + d\omega_1^2 &= 0, \\ \delta\omega_1^2 + d\omega_0^3 &= 0, \\ \delta\omega_0^3 &= 0. \end{aligned}\right.$$

Exercise

1. Under linear transformations on the quantum hyperplane,

$$\begin{pmatrix} x' \\ y' \end{pmatrix} = \begin{pmatrix} \alpha & \beta \\ \gamma & \delta \end{pmatrix}\begin{pmatrix} x \\ y \end{pmatrix}, \quad \begin{pmatrix} \xi' \\ \eta' \end{pmatrix} = \begin{pmatrix} \alpha & \beta \\ \gamma & \delta \end{pmatrix}\begin{pmatrix} \xi \\ \eta \end{pmatrix}.$$

If (x',y') and (ξ',η') satisfy the same commutation relations like (x,y) and (ξ,η) respectively, i.e.,

$$xy = qyx. \quad \xi\eta = -\frac{1}{q}\eta\xi, \quad \xi^2 = 0, \quad \eta^2 = 0.$$

And assume that $(\alpha,\beta,\gamma,\delta)$ commute with (x,y,ξ,η). Show that $(\alpha,\beta,\gamma,\delta)$ satisfy (14.1).

Appendix A

Simple Introduction to Set Theory

A.1 Basic definitions and notations

A **set** S consisting of elements $a, b, c, \cdots$ is denoted by

$$S = \{a, b, c, \cdots\}.$$

If a is an element of S, we say a belongs to S or simply

$$a \in S,$$

if not, we say a doesn't belong to S or simply

$$a \notin S.$$

If all elements of a set A belong to S, we say A is a subset of S and write it as

$$A \subset S.$$

The **null set** (empty) is the set which has no element and is denoted by $\emptyset$. As a matter of convention, $\emptyset$ is said to be the subset of any set.

The **complement** of A with respect to S is the following set

$$S \backslash A = \{a | a \in S, \text{ but } a \notin S\}.$$

The **union** of two sets S_1 and S_2 is defined by

$$S_1 \cup S_2 = \{a | a \in S_1, \text{ or } a \in S_2\}.$$

The **intersection** of two sets S_1 and S_2 is defined as

$$S_1 \cap S_2 = \{a | a \in S_1, \text{ and } a \in S_2\}.$$

Two sets S_1 and S_2 are **disjoint** if they have no element in common, i.e.,

$$S_1 \cap S_2 = \emptyset.$$

Cartesian product of S_1 and S_2 is defined as

$$S_1 \times S_2 = \{(a,b)|a \in S_1, b \in S_2\},$$

namely, the set of all ordered pairs (a,b). Note that $S_1 \times S_2 \neq S_2 \times S_1$.

A.2 Equivalence relations and equivalence classes

A relation R between two sets S_1 and S_2 is a subset of $S_1 \times S_2$. a and b are said to be R–related if $(a,b) \in R$.

An **equivalence relation** in S is a relation $R \in S \times S$ which is
(i) reflexive: $(a,a) \in R\ ,\ for\ \ a \in S$;
(ii) symmetric: $(a,b) \in R \Rightarrow\ (b,a) \in R$, for $a,b \in S$;
(iii) transitive: $(a,b) \in R$ and $(b,c) \in R \Rightarrow\ (a,c) \in R$, for $a,b,c \in S$.
We use

$$a = b(\text{mod } R), \qquad \text{or } a \sim b$$

to denote that a is equivalent to b with respect to an equivalence relation R.

The **equivalent class** $[a]$ of $a \in S$ is defined by

$$[a] = \{b \in S|b \sim a\}.$$

Two equivalent classes are identical if they have one common element. Therefore set S can be decomposed into a disjoint union of equivalent classes. The set of equivalent classes is called the **quotient set**, denoted as

$$S/R = \{[a]|a \in S\}.$$

A **Partition** of set S is a collection of classes $\{A_i\}$ of nonempty disjoint subsets of S whose union is S.

An equivalence relation in set S determines a partition of S, the inverse is also true.

A.3 Partial Ordering and Total Ordering

Set S has a **partial ordering** $R \in S \times S$, if the relation R satisfy:
(i) reflexive: $(a,a) \in R,\ \ for\ \ a \in S$;
(ii) antisymmetric: $(a,b) \in R, \text{and} (b,a) \in R\ \Rightarrow\ a = b$;
(iii) transitive: $(a,b) \in R$ and $(b,c) \in R \Rightarrow\ (a,c) \in R$, for $a,b,c \in S$.

then one said that a is related to b by the partial ordering, or simly denoted by

$$a \leq b.$$

The upper bound for a and b are those elements c which satisfy $a \leq c$ and $b \leq c$.

The least upper bound for a and b be called the supremum of a and b, and denote it as sup(a,b).

The lower bound for a and b are those elements c which satisfy $c \leq a$ and $c \leq b$.

The greatest lower bound for a and b be called the infimum of a and b, and denote it as inf(a,b).

A partial ordering set be called totally ordered, if for any two element $a, b \in S$, either $a \leq b$ or $b \leq a$.

A totally ordering set be called also a chain.

A.4 Maps

A **map** f from a set S_1 to a set S_2 associates each element of S_1 a unique element of S_2, written as

$$\begin{aligned} f : S_1 &\rightarrow S_2 \\ a &\longmapsto f(a), \quad a \in S_1, \quad f(a) \in S_2. \end{aligned}$$

$f(a)$ is called the **image** of a under map f.

Set S_1 is the **domain** of f; and the subset of S_2

$$f(S_1) = \{f(a)|a \in S_1\} \subset S_2$$

is the **range** of f.

We say f is **surjective** or map S_1 onto S_2 if $f(S_1) = S_2$; otherwise, we say f map S_1 **into** S_2.

Map f is said to be a one-to-one map or **injective** if for any $b \in f(S_1)$, there is only one $a \in S_1$ satisfying $f(a) = b$. This kind of map can have an inverse map

$$f^{-1} : f(S_1) \rightarrow S_1.$$

A map which is both injective and surjective is called a **bijective** map.

For the following two maps:

$$\begin{aligned} f : S_1 &\rightarrow S_2 \\ g : S_2 &\rightarrow S_3 \end{aligned}$$

we can define a map

$$h : S_1 \rightarrow S_3$$

by $h(a) = g(f(a))$. Map h is called the **composition** of f and g and written as

$$h = g \cdot f.$$

A map g is a restriction of f if the domain U of g belong to the domain V of f and g(a)=f(a) for every $a \in U$, then denote this as

$$g = f|_U.$$

Appendix B

Preliminary Topology

A topological space is a set with a structure allowing the definition of neighborhood, continuity, compactness, connectedness, etc.. These concepts of topological space are abstractions from the corresponding concepts in metric space.

B.1 Metric space

A set M is a **metric space** if for each pair of elements $a, b \in M$, there is a real valued function called **distance function**

$$\begin{aligned} \rho : \quad M \times M &\rightarrow \mathbf{R} \\ (a,b) &\longmapsto \rho(a,b) \end{aligned}$$

satisfying
(*i*) Posivity: $\rho(a,b) \geq 0$, equal iff $a = b$.
(*ii*) Symmetric with respect to a, b: $\rho(a,b) = \rho(b,a)$, $\forall a, b \in M$.
(*iii*) Triangle inequality: $\rho(a,b) + \rho(b,c) \geq \rho(a,c)$, $\forall a, b, c \in M$.
ρ is called **metric** of space M.
An **open ball** $B(a,\epsilon)$ of a point $a \in M$ is a subset of M:

$$B(a,\epsilon) = \{b \in M | \rho(a,b) < \epsilon, \quad \epsilon > 0\}$$

which is also called the **neighborhood** of the point a.

Let U be a subset of M. Then $a \in M$ is said to be an interior point of U if there exists an open ball $B(a,\epsilon) \subset U$. The set of all interior points of U is called the **interior** of U, and denoted by $\dot{U}$.

A point $b \in M$ is said to be a limit point of a subset U if each open ball $B(b,\epsilon)$ has at least one point $a \in U$ different from b. The set of all limit points of U is called the **closure** of U, and denoted by $\bar{U}$.

The **boundary** ∂U of U is defined by

$$\partial U = \bar{U} \setminus \dot{U}.$$

If $U = \dot{U}$, then U is said to be an open set. If $U = \bar{U}$, then U is said to be a closed set.

Example. For two interval U and V of real axis $\mathbf{R}$

$$U = \{t \in \{\mathbf{R}\} | \mathbf{a} < \mathbf{t} < \mathbf{b}\} \equiv (\mathbf{a}, \mathbf{b})$$
$$V = \{t \in \{\mathbf{R}\} | \mathbf{a} \leq \mathbf{t} \leq \mathbf{b}\} \equiv [\mathbf{a}, \mathbf{b}]$$

$$\dot{U} = (a, b), \quad \bar{U} = [a, b], \quad \partial U = a, b.$$
$$\dot{V} = (a, b), \quad \bar{V} = [a, b], \quad \partial V = a, b.$$

So we can see $U = (a, b) = \dot{U}$ is an open set of $\mathbf{R}$. $V = [a, b] = \bar{V}$ is a closed set. Or we can say, an open set U is a subset of space M such that whose any point has neighborhood lies entirely in U. A closed set is a subset of M such that it contains all its limit points.

Note that:

(1.) A limit point b of subset U may or may not be a point of U.

(2.) The complement of an open set is closed. The complement of a closed set is open.

(3.) An open ball is an open set. The union of **finite number** of open balls is still an open set.

(4.) The empty set $\emptyset$ and the whole metric space M are both open and closed.

The **collection of open sets** have the following properties:

(i) The union of **any number** of open sets is an open set.

(ii) The intersection of a **finite number** of open sets is open.

(iii) The empty set $\emptyset$ and the whole space M are open.

Note that in (ii), only finite intersections are allowed. As an example, on the real axis, the intersection of open balls $\bigcap_{n=1}^{\infty} B(b, 1/n)$ is the point b itself only, the set b is not open in $\mathbf{R}$.

B.2 General topological space

A topological space is a generalization of metric space.

Definition:

A collection τ of subsets of a set S is called a topology of S if

(i) The empty set $\emptyset$ and the set S itself are elements of τ.

(ii) The union of any number of elements of τ is also an element of τ.

(iii) The intersection of any finite number of elements of τ is an element of τ.

Elements of τ are called the open sets of τ.

Examples of topological spaces.

(1) Let S be any nonempty set and let the open sets consist of $\emptyset$ and S. This topological space is called trivial.

(2) Let S be any nonempty set and let the open sets consist of all subsets of S, $\emptyset$ and S itself included. This topology is called discrete topology.

(3) Let S be a metric space and let the open sets consist of all open balls in S, $\emptyset$ and S itself included, this topology is called the usual topology.

The coarsest topology is the trivial topology, and the finest topology is the discrete topology.

Topological spaces that arise from metric space are said to be metrizable. There are topological spaces which are not metrizable.

A Hausdorff space is a space in which any two distinct points have distinct neighborhood.

Metric space with the usual topology is **Hausdorff**, and a topological space is metrizable only if it is Hausdorff.

From the topological structure of space, we can study the properties of mappings.

Mapping $f: S_1 \to S_2$ between two topological spaces is continuous if and only if the inverse image of each open set of S_2 is open in S_1.

There is also a localized form of continuity which is useful.

A map $f: M_1 \to M_2$ is continuous at a point $x \in M_1$ if the inverse of each neighborhood of $f(x)$ is a neighborhood of x.

Continuous mapping preserves the neighborhood of topological space. For a mapping f between two topological spaces, if we require the image of f^{-1} to be well defined, then f must be bijective, i.e., one to one and onto.

A Homeomorphism is a continuous bijective mapping with its inverse also continuous. Homeomorphism is an equivalence relation, denoted as $\simeq$.

Properties invariant under homeomorphism are called topological properties, for example, neighborhood, separation properties, compactness, connectedness.

B.3 Connectedness

A topological space (M, τ) is connected if the only subsets of M which are both open and closed are $\emptyset$ and M. An alternative formulation is that (M, τ) is not connected iff there are nonempty open sets A and B such that $A \cap B = \emptyset$, $A \cup B = M$.

A path in M is a continuous map $f: [0,1] \to M$. M is said to be path-connected or arcwise connected if for any two different points $x_1, x_2 \in M$ there exists a path from x_1 to x_2.

It is easy to see if M is path-connected then M is connected. But a connected space is not necessarily path-connected. Here is an example (see Fig. B.1):

$$M = \{(x,y)|y = \sin\frac{1}{x}, 0 < x \leq 1\} \cup (0,0)$$

which is connected but not path-connected.

B.4 Compactness

A family $\{A_i | i \in \mathcal{I}\}$ of subsets of M is a cover of M if

$$M = \bigcup_{i \in \mathcal{I}} A_i.$$

If all A_i are open sets of M then the cover is called an **open cover**.

For a subset $\mathcal{J}$ of $\mathcal{I}$, if the following relation still holds

$$M = \bigcup_{j \in \mathcal{J}} A_j,$$

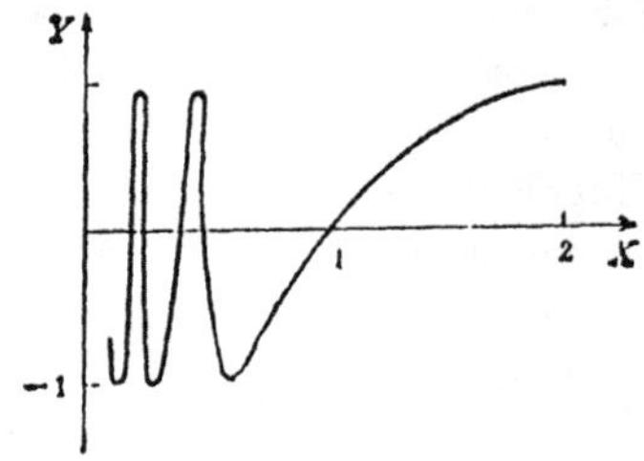

Fig. B.1

then $\{A_j | j \in \mathcal{J}\}$ is called a **subcover**. Particularly, it is a **finite subcover** if the set $\mathcal{J}$ is finite.

On the other hand, if for any point $x \in M$ there exists a neighborhood $U(x)$ such that the index subset $\{i | A_i \cap U(x) \neq \emptyset\}$ is finite, then we say the cover is a **locally finite cover.**

Space M is compact if it is a Hausdorff space and every open cover of M has a finite subcover.

Note that the requirement for compactness is different from simply requiring that the space has a finite open covering. In fact, any space has a finite open covering, it is covered by itself!

Examples:

(i) The closed intervals $[a, b]$ on $\mathbf{R}$ is compact.

(ii) But the open intervals $(0, 1) \subset \mathbf{R}$ is not compact, since the following open cover

$$\{(\frac{1}{n+2}, \frac{1}{n}) | n = 1, 2, 3, \cdots\}$$

does not have a finite subcover.

A topological space M is locally compact if M is Hausdorff and each $x \in M$ has a compact neighborhood.

Examples:

(i) A compact space is always locally compact.

(ii) $\mathbf{R}^n$ is locally compact but not compact.

Let $\{A_i | i \in \mathcal{I}\}$ and $\{B_j | j \in \mathcal{J}\}$ be two covers of M. If for any A_i there is a B_j such that $A_i \subset B_j$, then we say $\{A_i | i \in \mathcal{I}\}$ is a **refinement** of $\{B_j | j \in \mathcal{J}\}$.

We say a topological space is **paracompact** if each of its open cover has a locally finite open subcover, that is any point $p \in M$ has an neighborhood interaction with only finite open set in the cover.

For a space M, the support of a map $f: M \to \mathbf{R}$ is the closure $N = \overline{\{y | f(y) \neq 0\}}$. Now we can introduce the concept of partition of unit.

A family $\{f_i | i \in \mathcal{I}\}$ of continuous maps $f_i : M \to [0,1]$ is a partition of unit of M if

(i) The supports of f_i form a locally finite cover of N;

(ii) $\forall x \in M, \ \sum_{i \in \mathcal{I}} f_i(x) = 1$.

Let $\{U_i | i \in \mathcal{I}\}$ be a cover of M. A partition of unit $\{f_i | i \in \mathcal{I}\}$ of M is said to be subordinate to $\{U_i | i \in \mathcal{I}\}$ if the support of f_i is contained in the corresponding U_i.

It can be proved that if a Hausdorff space M is paracompact then there exists a partition of unit $\{f_i | i \in \mathcal{I}\}$, subordinate to each of its open cover $\{U_i | i \in \mathcal{I}\}$.

An alternative formulation of this statement is that a paracompact space allows integrations on it. So paracompactness is the property which a differentiable manifold must have.

The compact but without boundary manifold is also called closed manifold, which can always be trivially (i.e., with trivial normal bundle) embedded in Euclidean space.

B.5 Product

Given two sets S_1 and S_2, the Cartesian product

$$S_1 \times S_2 = S = \{(a,b) | a \in S_1, \ b \in S_2\}$$

is the set of all ordered pairs (a, b), where $a \in S_1, \ b \in S_2$.

A new topological space (M, τ) can be obtained from the product of two topological spaces (M_1, τ_1) and (M_2, τ_2). As a set, it is the Cartesian product of two sets, while its topology

$$\tau = \tau_1 \times \tau_2 = \{U \times V\} \quad (U \in \tau_1, V \in \tau_2)$$

is given by the Cartesian product of all the open set pairs.

The topological properties (compactness, connectedness etc.) of the product space are determined by the corresponding properties of M_1 and M_2. There is a proposition:

Product space $M_1 \times M_2$ is a Hausdorff space if and only if M_1 and M_2 are Hausdorff spaces.

Product space $M_1 \times M_2$ is compact if and only if M_1 and M_2 are compact.

The product of two connected spaces is a connected space.

The product of two path-connected spaces is path-connected.

Product space $M \times N$ is a differentiable manifold if and only if M and N are differentiable.

Table C.1 The main character of some algebraic systems

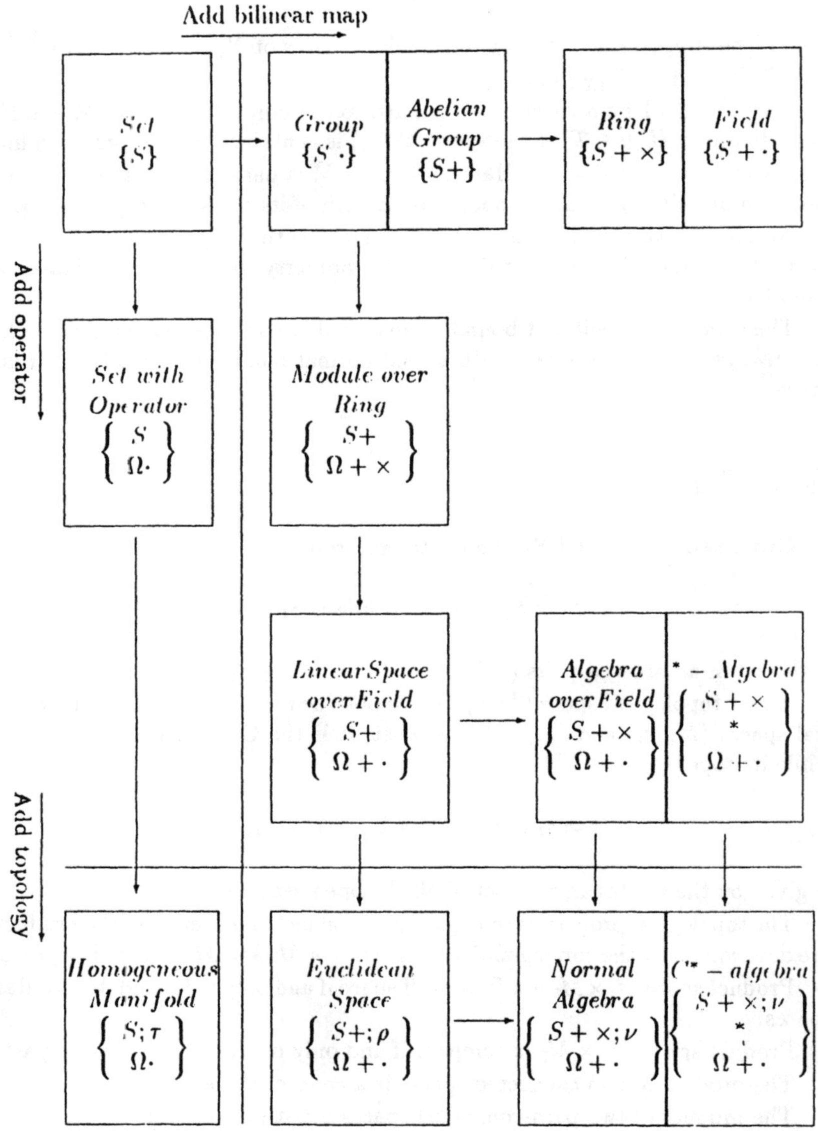

Appendix C

Some Basic Algebraic Structures

Algebra is used extensively in geometry. Here we briefly introduce some basic algebraic structures which are often encountered in this book. Table C.1 shows their characteristics and the relations among them. First we look at them one by one.

C.1 Group, ring and field

An internal operation I on a set M is a map from $S \times S$ to M

$$\begin{aligned} I:\ S \times S &\longrightarrow S \\ (a, b) &\longmapsto I(a, b) \in S \end{aligned}$$

For simplicity, we use $a \cdot b$ or even ab to denote $I(a, b)$. A **group** G is a set with an internal operation satisfying

(*i*) The operation is associative

$$(a \cdot b) \cdot c = a \cdot (b \cdot c) \qquad \forall a, b, c \in G.$$

(*ii*) There is an identity $e \in G$, such that

$$a \cdot e = e \cdot a = a \qquad \forall a \in G.$$

(*iii*) For any $a \in G$, there is an inverse element $a^{-1} \in G$, such that

$$a \cdot a^{-1} = a^{-1} \cdot a = e.$$

This internal operation is often called multiplication.

If this operation is commutative, i.e.,

$$a \cdot b = b \cdot a \qquad \forall a, b \in G,$$

then we say the group is Abelian.

A **ring** R is a set with two internal operations, addition and multiplication, which have the following properties:
(i) R is an Abelian group with respect to the addition operation (denote by $+$).
(ii) The multiplication (denote by $\cdot$) is associative

$$(a\cdot b)\cdot c = a\cdot(b\cdot c) \qquad \forall a,b,c\in R,$$

and is distributive with respect to addition, i.e.,

$$\begin{aligned} a\cdot(b+c) &= a\cdot b+a\cdot c \\ (a+b)\cdot c &= a\cdot c+b\cdot c \qquad \forall a,b,c\in R. \end{aligned}$$

A **field** F is a ring satisfying two additional conditions:
(i) there also exists an identity e with respect to multiplication, i.e.,

$$a\cdot e = e\cdot a = a \qquad \forall a\in F.$$

(ii) All but the zero element have inverse with respect to multiplication.
Examples:

Z, the commutative ring of integers.
R, the field of real numbers.
Q, the field of rational numbers.
C, the field of complex numbers.
H, the noncommutative field of quaternions.

C.2 Vector space, module and algebra

A **vector space (linear space)** V is an Abelian group with an external operation

$$\begin{aligned} \Omega\times V &\longrightarrow V \\ (\alpha,a) &\longmapsto \alpha a\in V \end{aligned}$$

where Ω is a field and satisfy:

$$\begin{aligned} \alpha(a+b) &= \alpha a+\alpha b, \\ (\alpha+\beta)a &= \alpha a+\beta a, \\ (\alpha\beta)a &= \alpha(\beta a). \end{aligned}$$

Elements of V and Ω are called **vectors** and **scalars** respectively.
If

$$\sum_{i=1}^{m}\alpha_i a_i = 0 \Rightarrow \alpha_i = 0\ ,\ i=1,2,\cdots,m,$$

where $\alpha_i\in\Omega$, $a_i\in V$, then we say the m vectors $\{a_1,a_2,\cdots,a_m\}$ are linearly independent. The maximal number of linearly independent vectors in V is called the dimension of V. If the dimension of V is n, we can choose a set of n linearly independent vectors $\{e_1, e_2, \cdots, e_m\}$ as the bases of V. Any element a of V is a linear combination of e_i, i.e.,

$$a = \sum_{i=1}^{n} \alpha_i e_i, \qquad \forall a \in V$$

coefficients α_i are called the components of a with respect to the basis $\{e_1, e_2, \cdots, e_n\}$, and vector a can also be written as

$$a = \{\alpha_1, \alpha_2, \cdots, \alpha_n\}.$$

The n-dimensional vector space over the real field $\mathbf{R}$ is denoted by $\mathbf{R}^n$.

Addition of two vectors in V is defined by adding their corresponding components, i.e.,

$$a = \sum_{i=1}^{n} \alpha_i e_i \in V, \quad b = \sum_{i=1}^{n} \beta_i e_i \in V,$$

$$a + b = \sum_{i=1}^{n} (\alpha_i + \beta_i) e_i \in V.$$

Multiplication of a vector a and a scalar β is defined as

$$\beta a = \sum_{i=1}^{n} (\beta \alpha_i) e_i.$$

For a vector space, we can choose different bases. For example, if $\{e_1, e_2, \cdots, e_n\}$ and $\{e'_1, e'_2, \cdots, e'_n\}$ are two sets of basis of $\mathbf{R}^n$, related by

$$e'_i = \sum_{j=1}^{n} A_{ij} e_j, \qquad i = 1, 2, \cdots, n,$$

where $A_{ij} \in \mathbf{R}$ and matrix (A_{ij}) satisfies

$$det(A_{ij}) \neq 0.$$

Then a vector can have different expressions

$$a = \sum_{i=1}^{n} \alpha_i e_i = \sum_{i=1}^{n} \beta_i e'_i,$$

with the components $\{\alpha_i\}$ and $\{\beta_i\}$ related by

$$\alpha_i = \sum_{j=1}^{n} \beta_j A_{ji}.$$

A **module** over Ω is like a vector space with the difference that Ω is a ring but not a field. When Ω is not Abelian, one has to distinguish between left and right modules.

A **right Ω-module** M_Ω is an Abelian group M together with a ring consisting of external operations

$$\begin{array}{rcl} M \times \Omega & \longrightarrow & M \\ (a, \alpha) & \longmapsto & a\alpha \in M \end{array}$$

such that

$$\begin{array}{rcl} (a+b)\alpha & = & a\alpha + b\alpha, \\ a(\alpha+\beta) & = & a\alpha + a\beta, \\ a(\alpha\beta) & = & (a\alpha)\beta. \end{array}$$

Similarly, we can define a left Ω-module $_\Omega M$.

An **algebra** A over a field Ω is a vector space over Ω with a bilinear map called multiplication, denoted by $\cdot$

$$\begin{array}{rcl} A \times A & \longrightarrow & A \\ (a, b) & \longmapsto & a \cdot b \in A \end{array}$$

which is associative and distributive

$$\begin{array}{rcl} a \cdot (b \cdot c) & = & (a \cdot b) \cdot c, \\ a \cdot (b + c) & = & a \cdot b + a \cdot c, \\ \alpha(a \cdot b) & = & (\alpha a) \cdot b \;=\; a \cdot (\alpha b), \end{array}$$

$$\forall a, b, c \in A, \qquad \alpha \in \Omega.$$

Since A is also a vector space, we can define the dimension of A and choose a bases for it. If $\{e_1, e_2, \cdots, e_n\}$ is a bases of A, then multiplication is defined as

$$e_i \cdot e_j = \sum_{k=1}^{n} f_{ij}^k e_k$$

where coefficients f_{ij}^k are called the **structure constants** which determine the structure of the algebra A.

An ***-algebra** is an algebra with an involution operation.

$$\begin{array}{rcl} * : A & \longrightarrow & A \\ a & \longmapsto & a^* \in A \end{array}$$

which has the following properties:

$$\left\{ \begin{array}{l} (i) \;\; (a^*)^* = a, \\ (ii) \;\; (a \cdot b)^* = b^* \cdot a^*, \\ (iii) \; (\alpha a + \beta b)^* = \bar{\alpha} a^* + \bar{\beta} b^*, \end{array} \right. \qquad \forall a, b \in A, \qquad \alpha, \beta \in \mathbf{C},$$

where $\bar{\alpha}$ and $\bar{\beta}$ are the complex conjugate of α and β respectively.

C.3 Euclidean space

For vector space $\mathbf{R}^n$, we can define many distance functions on it. One of them is

$$\rho(a,b) = \sqrt{\sum_{i=1}^{n}(\alpha_i - \beta_i)^2},$$

where

$$a = \sum_{i=1}^{n} \alpha_i e_i \in \mathbf{R}^n, \qquad b = \sum_{i=1}^{n} \beta_i e_i \in \mathbf{R}^n.$$

Vector space $\mathbf{R}^n$ with distance function thus defined is called a **Euclidean space** and denoted by E^n. It is easy to check that non-negative function ρ satisfies the following:

(*i*) $\rho(a,b) \geq 0$, equal iff a=b.
(*ii*) $\rho(a,b) = \rho(b,a)$.
(*iii*) $\rho(a,b) + \rho(b,c) \geq \rho(a,c)$.

C.4 Normed algebra and C*-algebra

A **normed algebra** A is an algebra to each element $a \in A$ is associated a real non-negative number $\|a\|$, called the norm of a, satisfying

(*i*) $\|a+b\| \leq \|a\| + \|b\|$,
(*ii*) $\|a\| = 0$, iff $a = 0$,
(*iii*) $\|\alpha a\| = |\alpha|\|a\|$, $\qquad a, b \in A, \ \alpha \in \mathbf{R}.$
(*iv*) $\|a \cdot b\| \leq \|a\| \cdot \|b\|$,

The norm defines a **metric topology (uniform topology)** on A, the open neighborhoods of an element a are defined by:

$$U(a,\epsilon) = \{b \in A \mid \|b - a\| < \epsilon\}, \quad \epsilon > 0.$$

A **C*-algebra** A is a complete normed *-algebra with the property

$$\|a^* \cdot a\| = \|a\|^2, \qquad \forall a \in A.$$

Note that it can be proved that $\|a^*\| = \|a\|$.

C.5 Homogeneous space

An external operation on a set S is a map from $\Omega \times S$ to S

$$\begin{array}{rcl} \Omega \times S & \longrightarrow & S \\ (\alpha, a) & \longmapsto & \alpha a \in S \end{array}$$

A internal associative operation between the elements of Ω can be defined as: $(\alpha \cdot \beta)s = \alpha(\beta s)$ where $\Omega = \{\alpha, \beta, \gamma, \cdots\}$ is a set of operators on S. Ω can be a group. If so, we call Ω a transformation group of S, and the set S be called a left Ω-space. If for any two elements $a, b \in S$, there exists an operator $\alpha \in \Omega$ such that $b = \alpha a$, then the operation of Ω on S is transitive, then the Ω-space is said to be homogeneous.

Homogeneous manifolds are differentiable manifolds on which transformation groups are transitive and homeomorphic.

Appendix D

Homomorphism of Algebraic Structure and Tensor Algebra

D.1 Linear functions and dual spaces, dual linear map

For an n-dimensional vector space V over the real field $\mathbf{R}$, we can define a linear function on V

$$\begin{aligned} f \; : \; V &\longrightarrow \mathbf{R} \\ a &\longmapsto f(a) \in \mathbf{R} \end{aligned}$$

which satisfies

$$f(\alpha a + \beta b) \;=\; \alpha f(a) + \beta f(b) \qquad \forall a, b \in V \;\; , \;\; \forall \alpha, \beta \in \mathbf{R},$$

f is a homomorphism from V to $\mathbf{R}$ denoted as $f \in Hom(V, \mathbf{R})$. It is easy to prove, if $f, g \in Hom(V, \mathbf{R})$, $\alpha, \beta \in \mathbf{R}$, then

$$\alpha f + \beta g \in Hom(V, \mathbf{R}),$$

and

$$(\alpha f + \beta g)(a) = \alpha f(a) + \beta g(a).$$

So the set of all real linear functions over V is also a vector space and is denoted by V^* and called the dual space of V. It is easy to see that V^* is also an n-dimensional vector space, and $(V^*)^* = V$. In fact, if $\{e_1, e_2, \cdots, e_n\}$ is a bases of V, then for a linear function $f \in V^*$, we have

$$f(a) \;=\; \sum_{i=1}^{n} a^i f(e_i), \qquad a = \sum_{i=1}^{n} a^i e_i.$$

So we see that f is completely determined by $\{f(e_1), f(e_2), \cdots, f(e_n)\}$, which are its values on $\{e_1, e_2, \cdots, e_n\}$. Let

$$f_i \equiv f(e_i) \in \mathbf{R},$$

then f can be written as

$$f = \sum_{i=1}^{n} f_i \theta^i,$$

where $\{\theta^1, \theta^2, \cdots, \theta^n\}$ are linear functions defined by

$$\theta^i(e_k) = \delta^i_k, \qquad i, k = 1, 2, \cdots, n.$$

It is easy to check that $\{\theta^1, \theta^2, \cdots, \theta^n\}$ are linearly independent, so they form a basis of the dual space V^*. For two vector spaces V and W and their dual spaces V^* and W^*, if $\varphi : V \to W$ is a linear map, then there is a linear map $\varphi^* : W^* \to V^*$ defined by

$$(\varphi^* g)(v) = g(\varphi(v))$$

where $g \in W^*$, $v \in V$.

If we use Dirac bra-ket notation, the equations above are:

$$< f, \alpha a + \beta b >= \alpha < f, a > + \beta < f, b > .$$
$$< \alpha f + \beta g, a >= \alpha < f, a > + \beta < g, a > .$$
$$< f, a >= \sum_i a^i f_i.$$
$$< \theta^i, e_k >= \delta^i_k.$$

D.2 Bilinear functions and tensor products

Let V and W be two vector spaces with bases $\{e_1, e_2, \cdots, e_n\}$ and $\{\tau_1, \tau_2, \cdots, \tau_n\}$ respectively. A map $\Psi : V \times W \to \mathbf{R}$ is a bilinear function denoted as $\Psi \in Hom(V \times W, \mathbf{R})$ with

$$\Psi(\alpha a + \beta b, c) = \alpha \Psi(a, c) + \beta \Psi(b, c),$$
$$\Psi(a, \alpha c + \beta d) = \alpha \Psi(a, c) + \beta \Psi(a, d),$$

where $a, b \in V$; $c, d \in W$; $\alpha, \beta \in \mathbf{R}$.

There is a natural way to construct a bilinear function from two linear functions

$$f : V \to \mathbf{R}, \quad g : W \to \mathbf{R} \qquad f \in V^*, \quad g \in W^*.$$

Function $\Phi = f \otimes g \in Hom(V \times W, \mathbf{R})$ is defined by

$$\begin{aligned} \Phi \ : \ V \times W &\longrightarrow \mathbf{R} \\ (a, b) &\longmapsto \Phi(a, b) = f(a) g(b) \end{aligned}$$

Or simply:

$$< f \otimes g : a, b >=< f, a >< g, b > .$$

Obviously, Φ is a bilinear function and it is called the tensor product of f and g. All such bilinear functions forms a vector space, which be called the tensor product of V^* and W^*, and denote as

$$V^* \otimes W^* \simeq Hom(V \times W, \mathbf{R}).$$

It is easy to verify

$$(\alpha f + \beta f') \otimes g = \alpha f \otimes g + \beta f' \otimes g,$$

i.e.,

$$< (\alpha f + \beta f') \otimes g : a, b >= (\alpha < f, a > +\beta < f', a >) < g, b >,$$

the tensor product is bilinear with respect to each of the factors

$$f \otimes (\alpha g + \beta g') = \alpha f \otimes g + \beta f \otimes g',$$

$$\alpha f \otimes g = f \otimes \alpha g,$$

where $\alpha, \beta \in \mathbf{R}$, $f, f' \in V^*$, $g, g' \in W^*$.

If $\{\theta^1, \theta^2, \cdots, \theta^n\}$ and $\{\varrho^1, \varrho^2, \cdots, \varrho^m\}$ are the bases of V^* and W^* respectively, then the basis of $V^* \otimes W^*$ is

$$\{\theta^i \otimes \varrho^j \ , \ i = 1, 2, \cdots, n; j = 1, 2, \cdots, m\}.$$

So the dimension of $V^* \otimes W^*$ is nm and any element member $\Phi \in V^* \otimes W^*$ can be written as

$$\Phi = \sum_{i=1}^{n}\sum_{j=1}^{m} \Phi_{ij}\theta^i \otimes \varrho^j.$$

Now we take V and W as the dual spaces of V^* and W^* respectively , then we can construct the tensor product $V \otimes W$ of V and W whose basis is

$$\{e_i \otimes \tau_j, \quad i = 1, 2, \cdots, n; j = 1, 2, \cdots, m\} \quad \in Hom(V^* \times W^*, \mathbf{R}).$$

It can be verified that the tensor product $\otimes$ is associative, so we can define n-fold tensor product unambiguously.

Examples:

The (r, s)-type mixed tensor space

$$T^r_s = \underbrace{V \otimes V \otimes \cdots \otimes V}_{r} \otimes \underbrace{V^* \otimes V^* \otimes \cdots \otimes V^*}_{s},$$

whose basis is

$$\{e_{i_1} \otimes e_{i_2} \otimes \cdots \otimes e_{i_r} \otimes \theta^{j_1} \otimes \theta^{j_2} \otimes \cdots \otimes \theta^{j_s}\}.$$

D.3 Direct sum and tensor algebra

The direct sum $V \oplus W$ of V and W is an $(n+m)$-dimensional vector space with basis $\{e_1, e_2, \cdots, e_n, \tau_1, \tau_2, \cdots, \tau_m\}$ for $a = \sum_{i=1}^n a^i e_i \in V$ $b = \sum_{j=1}^m b^j \tau_j \in W$, their direct sum is

$$a \oplus b = \sum_{i=1}^{n} a^i e_i + \sum_{j=1}^{m} b^j \tau_j.$$

It is obvious that for any $a, a' \in V$, $b, b' \in W$, $\alpha \in \mathbf{R}$

$$\begin{cases} a \oplus b + a' \oplus b' = (a + a') \oplus (b + b'), \\ \alpha(a \oplus b) = \alpha a \oplus \alpha b. \end{cases}$$

The direct sum of all types of tensor spaces built on V and V^* is denoted as

$$T = \bigoplus \sum_{r,s} T_s^r.$$

It is an infinite dimensional vector space over $\mathbf{R}$. Since the tensor product of any two elements of T still lies in T, T is an algebra with the multiplication of tensor product $\otimes$.

A map $C : T_s^r \to T_{s-1}^{r-1}$ defined by

$$C(K_{j_1 j_2 \cdots j_s}^{i_1 i_2 \cdots i_r}) = \sum_m K_{j_1 j_2 \cdots j_{s-1} m}^{i_1 i_2 \cdots i_{r-1} m} \in T_{s-1}^{r-1}, \qquad K_{j_1 j_2 \cdots j_s}^{i_1 i_2 \cdots i_r} \in T_s^r$$

is called the contraction of T_s^r.

D.4 Derivator algebras

A map $D : T \to T$ is called a derivator if D satisfies

(*i*) $D : T_s^r \longrightarrow T_s^r$.

(*ii*) $D(a \oplus b) = (Da) \oplus (Db)$.

(*iii*) $D(K \otimes L) = (DK) \otimes L + K \otimes (DL)$.

(*iv*) D commutes with contraction.

For any two derivators D and D', it is easy to check

$$\alpha D + \beta D', \qquad \forall \alpha, \beta \in \mathbf{R},$$

and

$$[D, D'] = DD' - D'D,$$

are still derivators. So, the set of the derivators has a Lie algebra structure.

Appendix E

Exact Sequence of Homomorphism

Let G and G' be two groups. A map

$$\begin{aligned} f: G &\longrightarrow G' \\ g &\longmapsto f(g) \in G' \end{aligned}$$

is called a homomorphism if

$$f(g_1 g_2) = f(g_1)f(g_2), \qquad \forall g_1, g_2 \in G.$$

If f is bijective, it is called an isomorphism.

The ker of the map f is denoted by

$$\mathrm{Ker} f = \{g \in G | f(g) = e'\},$$

where e' is the identity of G'. And the image of f is denoted as

$$\mathrm{Im} f = \{g' \in G' | g' = f(g), \forall g \in G\}.$$

Then we can easily prove

Theorem 1:

(i) $\mathrm{Ker} f$ is a normal subgroup of G.

(ii) The quotient group $G/\mathrm{Ker} f$ and $\mathrm{Im} f$ are isomorphic, i.e.,

$$G/\mathrm{Ker} f \cong \mathrm{Im} f.$$

Next we look at a sequence of homomorphism of groups $\{G_1, G_2, \cdots, G_n\}$

$$G_1 \xrightarrow{f_1} G_2 \xrightarrow{f_2} G_3 \xrightarrow{f_3} \cdots \xrightarrow{f_{n-1}} G_n$$

where all the maps $f_k : G_k \to G_{k+1}$ are homomorphism. If the following relations hold

$$\mathrm{Im} f_k = \mathrm{Ker} f_{k+1},$$

then the sequence is called an exact sequence.

Theorem 2:

$$(i)\ e \xrightarrow{f_0} G_1 \xrightarrow{f} G_2 \ \text{exact}, \quad \Longrightarrow \quad \mathrm{Ker} f = \textit{identity}, \ \text{i.e., f is injective;}$$

$$(ii)\ G_1 \xrightarrow{f} G_2 \xrightarrow{f'} e \ \text{exact}, \quad \Longrightarrow \quad \mathrm{Im} f = G_2, \qquad \text{i.e., f is surjective;}$$

here e denotes the trivial group with only the identity element.

Proof:

(i) As f_0 is a homomorphism, the image $f_0(e)$ must be the identity e_1 of G_1; as the sequence is exact, so

$$\mathrm{Ker} f = \mathrm{Im} f_0 = e_1,$$

so homomorphism f is one to one injective.

(ii) Through the homomorphism f', all elements of G_2 are mapped to e, Kerf' is the whole of G_2, as the sequence is exact, so

$$\mathrm{Im} f = \mathrm{Ker} f' = G_2,$$

and the homomorphism f is surjective.

Theorem 3. If the following sequence is exact

$$e \xrightarrow{f_0} G_1 \xrightarrow{f_1} G_2 \xrightarrow{f_2} G_3 \xrightarrow{f_3} e$$

then G_3 is isomorphic to G_2/G_1.

Proof:

From Theorem 2, we know that

f_1 is injective. $G_1 \cong \mathrm{Im} f_1 = \mathrm{Ker} f_2$.

f_2 is surjective. $G_3 \cong \mathrm{Im} f_2$.

According to Theorem 1,

$$G_2/G_1 \cong G_2/\mathrm{Ker} f_2 \cong \mathrm{Im} f_2 \cong G_3.$$

This completes our proof.

Appendix F

Abelian Group

Abelian groups play a very important role in analyzing the homology and homotopy groups of manifolds.

Conventionally, Abelian groups are denoted additively. Namely, for an Abelian group G, the group operation is denoted by $+$. Thus the commutative property is seen naturally from

$$g_1 + g_2 = g_2 + g_1, \quad g_1, g_2 \in G.$$

And 0 denotes the identity of G; for $g \in G$, its inverse is denoted by $-g$ and its m-th power is denoted by mg (m is an integer).

F.1 Linear independence and rank

Abelian groups is a module over the ring of integers, which is like a vector space with the difference that the underlying number field is replaced by the integer ring, so linear independence and rank of Abelian groups can be considered analogously.

If there exist a set of integers $\{\lambda_1, \lambda_2, \cdots, \lambda_r\}$, at least one of which is nonzero, such that

$$\lambda_1 x_1 + \lambda_2 x_2 + \lambda_3 x_3 + \cdots + \lambda_r x_r = 0, \quad x_i \in G,$$

then $\{x_1, x_2, \cdots, x_r\}$ are said to be linearly dependent. Otherwise, we say that $\{x_1, x_2, \cdots, x_r\}$ are linearly independent.

Given a linearly independent set of $\{y_1, y_2, \cdots, y_m\}$, if for any $g \in G$, $\{g, y_1, y_2, \cdots, y_m\}$ are linearly dependent, then $\{y_1, y_2, \cdots, y_m\}$ are said to be the maximal linearly independent set. We can prove

Proposition: Let $\{x_1, x_2, \cdots, x_m\}$ and $\{y_1, y_2, \cdots, y_n\}$ are two linearly independent sets in an Abelian group G. If both of them are maximal, then $m = n$.

So, we can define such $m = n$ to be the rank $\rho(G)$ of G. If there doesn't exist such an integer, we say the rank of G is infinite, i.e., $\rho(G) = \infty$.

It is easy to check that any subgroup H of an Abelian group G is normal. So there exists a quotient G/H. Obviously, both G/H and H are Abelian. It can be shown that

$$\rho(G) = \rho(H) + \rho(G/H).$$

F.2 Finite generated groups

Let $X = \{x_1, x_2, \cdots, x_m\}$ be a set of elements of an Abelian group G. If each member $x \in G$ can be expressed as

$$x = \lambda_1 x_1 + \lambda_2 x_2 + \cdots + \lambda_m x_m,$$

where $\{\lambda_1, \lambda_2, \cdots, \lambda_m\}$ are integers, then X is called a set of generators of G. If m is a finite positive integer, G is called a finite generated group. If X is also linearly independent, then G is called a free Abelian group of rank m. Any element $g \in G$ can generate a subgroup $< g >$ of G

$$< g > = \{ng | n \in \mathbf{Z}\}$$

which is called a cyclic group. Let p be the number of the elements of $< g >$, then we say g is an element of degree p. If p is infinite, then $< g >$ is a free Abelian group of rank 1.

According to the degrees of the elements of G, G can be decomposed into two subgroups: one is the torsion subgroup consisting of all finite-degreed elements; another is a free Abelian group which includes all infinite-degreed elements. Furthermore, it can be proved

Theorem: Any finite generated group G can be decomposed as

$$G = H_0 \oplus Z_{\tau_1} \oplus Z_{\tau_2} \oplus \cdots \oplus Z_{\tau_n},$$

where H_0 is a rank-r free Abelian group, Z_{τ_i} is an degree-τ_i cyclic group, $\tau_i > 1$ and τ_i/τ_{i+1} are integers. Moreover, $\{r, \tau_1, \tau_2, \cdots, \tau_n\}$ are unique, i.e., they are independent of the decompositions of G.

There is another equivalent statement of the theorem:

Theorem: Any finite generated group G can be decomposed as

$$G = H_0 + H_1 + H_2 + \cdots + H_m,$$

where H_0 is a rank-r free Abelian group, H_i is a prime-degreed cyclic group. And the rank of H_0 and degrees of torsion subgroups $\{H_1, H_2, \cdots, H_m\}$ are invariant.

Appendix G

Stokes' Theorem

The relation between integration over a domain and integration over its boundary is one of the fundamental formulation of calculus. For example, the Ostrogradsky-Green theorem, the Riemann-Ampere-Stokes formulae, etc.. Its generalization to manifold M with boundary ∂M is the Stokes' formula.

$$\int_M \mathrm{d}\omega = \int_{\partial M} \omega,$$

which has a wide range of applications both in physics and in differential geometry. A brief description and proof of the theorem is given as follows.

Stokes' Theorem: Let M be a smooth oriented compact manifold of dimension n with boundary ∂M. Then

$$\int_M \mathrm{d}\omega = \int_{\partial M} \omega,$$

where ω is an $(n-1)$-form on M.

Proof: Let $\{U_i\}_{i\in I}$ be the coordinate covering of M, $\{\rho_i\}_{i\in I}$ be the corresponding partition of unity.

$$\sum_{i\in I} \rho_i(x) = 1; \quad \text{so} \quad \sum_{i\in I} d\,\rho_i(x) = 0,$$

here the set i is finite due to the compactness of M, so

$$\begin{aligned}
\int_M \mathrm{d}\omega &= \sum_{i\in I}\int_M \rho_i\,\mathrm{d}\,\omega \\
&= \sum_{i\in I}\int_M (\rho_i\,\mathrm{d}\,\omega + \mathrm{d}\,\rho_i \wedge \omega) \\
&= \sum_{i\in I}\int_M \mathrm{d}\,(\rho_i\ \omega),
\end{aligned}$$

and

$$\int_{\partial M} \omega = \sum_{i\in I}\int_{\partial M} \rho_i\,\omega,$$

so it is sufficient to prove for any $i \in I$

$$\int_M \mathrm{d}(\rho_i\omega) = \int_{\partial M} \rho_i\omega, \tag{G.1}$$

here $\rho_i(x)\omega(x) = \sigma_i(x) = 0$ if $x \notin U_i$.

Next we look at $\sigma(x)$ for an open set U. (Index i is omitted.)

Let the coordinate expression for $\sigma(x)$ be

$$\sigma(x) = \sum_{j=1}^{n}(-1)^{j-1} f_j(x) dx^1 \wedge \cdots \wedge d\hat{x}^j \wedge \cdots \wedge dx^n \quad \in \Lambda^{n-1}(U),$$

where $d\hat{x}^j$ represents the omission of dx^j.

Then

$$d\sigma(x) = (\sum_{j=1}^{n} \frac{\partial f_j}{\partial x^j})\ \tau,$$

here $\tau = dx^1 \wedge dx^2 \wedge \cdots \wedge dx^n$ is the n-dimensional volume element on U.

Two possibilities now arise

(*i*) $U \cap \partial M = \emptyset$;

(*i*) $U \cap \partial M \neq \emptyset$.

In the first case, the right hand side of (G.1) is zero, and the left hand side

$$\int_M \mathrm{d}\sigma = \sum_{j=1}^{n} \int_U \frac{\partial f_j}{\partial x^j} dx^1 \cdots dx^n.$$

Function $f_j(x)$ which is nonzero on U only can be extrapolated to a cubic in $\mathbf{R}^n$containing U:

$$D = \{x \in \mathbf{R}^n \mid \ |x^i| \leq L,\ i = 1, \cdots, n\} \subset U.$$

Let $f_j(x) = 0,\ \ if\ \ x \notin U$, obviously $f_j(x)$ thus extrapolated is continuous and differentiable in D, so

$$\begin{aligned}
\int_U \frac{\partial f_j}{\partial x^j} dx^1 \cdots dx^n &= \int_D \frac{\partial f_j}{\partial x^j} dx^1 \cdots dx^n \qquad\qquad \text{(G.2)}\\
&= \int_{|x^i| \leq L} dx^1 \cdots d\hat{x}^j \cdots dx^n \{f_j(x^1, \cdots, x^{j-1}, L, x^{j+1}, \cdots, x^n)\\
&\qquad -f_j(x^1, \cdots, x^{j-1}, -L, x^{j+1}, \cdots, x^n)\} = 0.
\end{aligned}$$

Thus both sides vanish, equality (G.1) is true.

In the second case, $U \cap \partial M \neq \emptyset$, the coordinate chart U can be chosen as $x^n = 0$ at points

$$p \in U \cap \partial M$$

and $x^n \geq 0$ at points $p \in U \cap M$. Similar to the first case, $f_j(x)$ can be extrapolated to a cubic D' in $\mathbf{R}^n_+$:

$$D' = \{x \in \mathbf{R}^n_+ \mid \ |x^i| \leq L, \ i = 1, \cdots, n\}.$$

Similar to (G.2), all integrals vanish except the last term, so

$$\int_{M \cap U} d\sigma = - \int_{|x_i| \leq L} f_n(x_1, \cdots, x_{n-1}, 0) dx^1 \cdots dx^{n-1},$$

and the right hand side of (G.1) is

$$\begin{aligned}
\int_{\partial M \cap U} d\sigma &= \sum_{j=1}^{n} (-1)^{j-1} \int_{\partial M \cap U} f_j(x) dx^1 \wedge \cdots \wedge d\hat{x}^j \wedge \cdots \wedge dx^n \\
&= (-1)^{n-1} \int_{\partial M \cap U} f_n(x^1, \cdots, x_{n-1}, 0) dx^1 \wedge \cdots \wedge dx^{n-1} \\
&= (-1)^{n-1}(-1)^n \int_{|x^i| \leq L} f_n(x^1, \cdots, x_{n-1}, 0) dx^1 \cdots dx^{n-1}.
\end{aligned}$$

The last equality is a result of the induced orientation of ∂M. This completes our proof.

Appendix H

Quarternion and Milnor Exotic Spheres

H.1 Division algebras

An algebra over a field Ω is a vector space V over Ω endowed with a bilinear multiplication.

An algebra A is said to have a unit element 1 if

$$1 \cdot a = a \cdot 1 \qquad \forall a \in A.$$

An algebra A with a unit is said to be a division algebra if for each non-zero $a \in A$, there exists its inverse $a^{-1} \in A$, i.e.,

$$a^{-1}a = aa^{-1} = 1.$$

It can be proved that the only possible associated division algebras are the real, complex and quaternion numbers. If associativity is dropped, there is another division algebra: octonion.

Let N denote the norm of this algebra, which satisfy:

$$\begin{aligned} N(ab) &= N(a)N(b), \qquad N(a) > 0 \qquad \forall a, b \in A, \\ N(a) &= 0 \implies a = 0. \end{aligned}$$

It can also been proved that the only division algebras with a norm N satisfying above condition are the algebras of real numbers **R**, complex numbers **C**, quarternions **H** and octonions O. Division algebras lie at the very core of the classification of possible symmetries in nature. We know that for the four categories of semi-simple Lie groups: orthogonal, unitary, symplectic and exceptional groups, they are the compact automorphism group of the corresponding algebras: **R**, **C**, **H**, **O**.

Topologically unit norm real numbers are $S^0 \simeq \{1, -1\} \simeq Z_2$;

Unit norm complex numbers are $S^1 \simeq U(1)$;
Unit norm quarternions are $S^3 \simeq SU(2)$;
Unit norm octonions are S^7.

S^1, S^3, S^7 are paralellizable manifolds (its tangent bundle is trivial), of which S^1 and S^3 are Lie group manifold, but S^7 is not, as the octonion is not associative.

H.2 Quarternion H

The quarternion **H** is a 4−dimensional algebra over **R** with the basis $\{e_0 = 1, e_1, e_2, e_3\}$ satisfying

$$e_i e_j = -\delta_{ij} + \epsilon_{ijk} e_k.$$

Any quarternion u can be written as

$$u = \sum_{i=0}^{3} u^i e_i,$$

and its conjugate $\bar{u}$ is defined by

$$\bar{u} = u^0 - \sum_{i=1}^{3} u^i e_i.$$

The norm $\|u\|$ of u is defined by

$$\|u\| = u\bar{u} = \bar{u}u = \sum_{i=0}^{3} (u^i)^2,$$

which has the following property

$$\|u\|\|v\| = \|uv\|, \qquad \forall u, v \in \mathbf{H}.$$

Now $u^{-1} = \bar{u}/\|u\|$ can be taken as the inverse of u. So, algebra **H** is a division algebra. One important realization of its basis is given by $e_i = -i\sigma_i$, where σ_i's are the Pauli matrices. Thus a quarternion u can be expressed as

$$u = u^0 - i\sum_{i=1}^{3} u^i \sigma_i = \begin{pmatrix} u^0 - iu^3 & -u^2 - iu^1 \\ u^2 - iu^1 & u^0 + iu^3 \end{pmatrix},$$

here we can see a clear correspondence between the unit quarternions and the $SU(2)$ group, and its Lie algebra su(2) consists of the three imaginary quarternions.

H.3 Milnor exotic spheres

In 1956[11], J. Milnor found an example of manifolds which are homeomorphic, but not diffeomorphic to S^7. This very surprising result stimulated intensive research. Here we give a very simple introduction.

Let us consider the fiber bundle over S^4 with S^3 as the fiber. $SO(4)$ is its structure group. It is well known that S^4 can be covered by two open sets U_1 and U_2 which are homeomorphic to $\mathbf{R}^4$, and $U_1 \cap U_2$ is homeomorphic to $\mathbf{R}^4 - \{0\}$. We use non-zero quaternion u to denote element of $\mathbf{R}^4 - \{0\}$ and unit quaternion v to denote element of fiber S^3. A family of nontrivial fiber bundles Σ_k^7 can be constructed by gluing the two trivial fiber bundles $U_1 \times S^3$ and $U_2 \times S^3$ with the following transition function

$$\begin{array}{rlcl} \Psi: & (\mathbf{R}^4 - \{0\}) \times S^3 & \longrightarrow & (\mathbf{R}^4 - \{0\}) \times S^3 \\ & (u, v) & \longrightarrow & (u', v') \\ & & = & (\frac{u}{\|u\|^k}, \frac{u^{m+1} v u^{-m}}{\|u\|}) \end{array}$$

where $k = 2m + 1$ is an odd integer. The transition function is smooth so one can obtain smooth 7-dimensional manifold denoted Σ_k^7.

By Morse theory, it can be proved that Σ_k^7 is homeomorphic to S^7. Furthermore, J. Milnor proved

$$\lambda(\Sigma_k^7) = k^2 - 1(\text{mod } 7)$$

is invariant under diffeomorphism. Note that $k = \pm 1$ correspond to the standard 7-dimensional sphere S^7. When $\lambda = k^2 - 1 \neq 0(\text{mod}7)$, we get some peculiar spheres Σ_k^7 which are homeomorphic to S^7, but have different differential structures. These spheres are called Milnor exotic spheres.

Appendix I

Generalized Kronecker δ notation

I.1 Definition

The generalized order-p Kronecker δ notation is defined by

$$\delta^{j_1 j_2 \cdots j_p}_{k_1 k_2 \cdots k_p} = \begin{vmatrix} \delta^{j_1}_{k_1} & \delta^{j_2}_{k_1} & \cdots & \delta^{j_p}_{k_1} \\ \delta^{j_1}_{k_2} & \delta^{j_2}_{k_2} & \cdots & \delta^{j_p}_{k_2} \\ \vdots & \vdots & & \vdots \\ \delta^{j_1}_{k_p} & \delta^{j_2}_{k_p} & \cdots & \delta^{j_p}_{k_p} \end{vmatrix} \qquad (j_i, k_i = 1, 2, \cdots, n),$$

where $\delta^{j_n}_{k_m}$'s are just the ordinary δ notation, i.e.,

$$\delta^{j_n}_{k_m} = \begin{cases} 1 & j_n = k_m, \\ 0 & j_n \neq k_m. \end{cases}$$

It is easy to see from the definition that

$$\delta^{j_1 j_2 \cdots j_p}_{k_1 k_2 \cdots k_p} = \begin{cases} 1 & \text{if } j_i\text{'s is an even permutation of } k_m\text{'s}, \\ -1 & \text{if } j_i\text{'s is an odd permutation of } k_m\text{'s}, \\ 0 & \text{otherwise.} \end{cases}$$

and

$$\delta^{j_1 j_2 \cdots j_l \cdots j_m \cdots j_p}_{k_1 k_2 \cdots k_l \cdots k_m \cdots k_p} = -\delta^{j_1 j_2 \cdots j_m \cdots j_l \cdots j_p}_{k_1 k_2 \cdots\cdots\cdots\cdots k_p} = -\delta^{j_1 j_2 \cdots\cdots\cdots\cdots j_p}_{k_1 k_2 \cdots k_m \cdots k_l \cdots k_p}.$$

From here on, we take the convention of summation over repeated indices.

I.2 Contractions of order-p δ notation ($p \leq n$)

$$\delta^{i_1 i_2 \cdots i_p}_{i_1 i_2 \cdots i_p} = \frac{n!}{(n-p)!}, \qquad \delta^{i_1 i_2 \cdots i_{p-1} j}_{i_1 i_2 \cdots i_{p-1} k} = \frac{(n-1)!}{(n-p)!}\delta^j_k,$$

$$\delta^{i_1 i_2 \cdots i_{p-s} j_1 \cdots j_s}_{i_1 i_2 \cdots i_{p-s} k_1 \cdots k_s} = \frac{(n-s)!}{(n-p)!}\delta^{j_1 j_2 \cdots j_s}_{k_1 k_2 \cdots k_s} \quad (s < p).$$

If $p = n$, we have

$$\delta^{i_1 i_2 \cdots i_n}_{i_1 i_2 \cdots i_n} = n!, \qquad \delta^{i_1 i_2 \cdots i_{n-1} j}_{i_1 i_2 \cdots i_{n-1} k} = (n-1)!\delta^j_k,$$

$$\delta^{i_1 i_2 \cdots i_{n-s} j_1 \cdots j_s}_{i_1 i_2 \cdots i_{n-s} k_1 \cdots k_s} = (n-s)!\delta^{j_1 j_2 \cdots j_s}_{k_1 k_2 \cdots k_s}.$$

I.3 Contraction of δ and tensors

It is easy to check $\delta^{j_1 j_2 \cdots j_p}_{k_1 k_2 \cdots k_p}$ is a (p,p)-typed tensor. So we can get a completely skewsymmetric tensor from the contraction of δ and tensors. For example:

$$\begin{aligned}
\delta^{j_1 j_2}_{k_1 k_2} dx^{k_1} \otimes dx^{k_2} &= dx^{j_1} \otimes dx^{j_2} - dx^{j_2} \otimes dx^{j_1} \\
&\equiv dx^{j_1} \wedge dx^{j_2}. \\
\delta^{j_1 j_2 j_3}_{k_1 k_2 k_3} dx^{k_1} \otimes dx^{k_2} \otimes dx^{k_3} &= dx^{j_1} \wedge dx^{j_2} \wedge dx^{j_3}. \\
\delta^{j_1 j_2 \cdots j_p}_{k_1 k_2 \cdots k_p} A_{j_1 j_2 \cdots j_p} &= A_{[k_1 k_2 \cdots k_p]},
\end{aligned}$$

where $[k_1 k_2 \cdots k_p]$ has the following property:

$$[k_1 \cdots k_l \cdots k_m \cdots k_p] = -[k_1 \cdots k_m \cdots k_l \cdots k_p].$$

For an $(n \times n)$ matrix $M = (a^i_k)$, it is easy to check that

$$\det M = \frac{1}{n!}\delta^{i_1 \cdots i_n}_{k_1 \cdots k_n} a^{k_1}_{i_1} \cdots a^{k_n}_{i_n}.$$

$$\det(1+M) = 1 + \delta^i_k a^k_i + \frac{1}{2!}\delta^{i_1 i_2}_{k_1 k_2} a^{k_1}_{i_1} a^{k_2}_{i_2} + \cdots + \frac{1}{m!}\delta^{i_1 i_2 \cdots i_m}_{k_1 k_2 \cdots k_m} a^{k_1}_{i_1} a^{k_2}_{i_2} \cdots a^{k_m}_{i_m} + \cdots.$$

I.4 Generalized Levi-Civita notation

Let $\{g_{ij}\ ,\ i,j = 1,2,\cdots,n\}$ be a metric tensor and $g = \det(g_{ij})$. The generalized Levi-Civita notation is defined by

$$\epsilon_{12\cdots n} \equiv |g|^{\frac{1}{2}},$$

and

$$\epsilon_{i_1 i_2 \cdots i_n} \equiv \delta^{12\cdots\cdots n}_{i_1 i_2 \cdots i_n} \epsilon_{12\cdots n} = |g|^{\frac{1}{2}} \delta^{12\cdots\cdots n}_{i_1 i_2 \cdots i_n}.$$

On the other hand,

$$\begin{aligned}
\epsilon^{i_1 i_2 \cdots i_n} &= g^{i_1 j_1} \cdots g^{i_n j_n} \epsilon_{j_1 j_2 \cdots j_n} \\
&= g^{i_1 j_1} \cdots g^{i_n j_n} \delta^{12\cdots\cdots n}_{j_1 j_2 \cdots j_n} \epsilon_{12\cdots n} \\
&= \det(g^{ij}) \delta^{i_1 i_2 \cdots i_n}_{12\cdots\cdots n} |g|^{\frac{1}{2}} \\
&= \mathrm{sgn}(g) |g|^{-\frac{1}{2}} \delta^{i_1 i_2 \cdots i_n}_{12\cdots\cdots n}.
\end{aligned}$$

So we have

$$\epsilon^{12\cdots n} = \operatorname{sgn}(g)|g|^{-\frac{1}{2}}, \qquad \epsilon^{i_1 i_2 \cdots i_n} = \delta^{i_1 i_2 \cdots i_n}_{12\cdots\cdots n}\epsilon^{12\cdots n}.$$

Under a coordinate transformation, we have

$$g'_{ij} = \frac{\partial x^k}{\partial x'^i}\frac{\partial x^l}{\partial x'^j}g_{kl},$$

$$g' = \det(g'_{ij}) = J^2 g, \qquad |g'|^{\frac{1}{2}} = J|g|^{\frac{1}{2}},$$

where J is the Jacobi determinant

$$J = \frac{\partial(x^1, \cdots, x^n)}{\partial(x'^1, \cdots, x'^n)}.$$

It can be verified that $\epsilon_{i_1 i_2 \cdots i_n}$ is a order-n tensor. In fact

$$\begin{aligned}
& \frac{\partial x^{j_1}}{\partial x'^{i_1}}\frac{\partial x^{j_2}}{\partial x'^{i_2}}\cdots\frac{\partial x^{j_n}}{\partial x'^{i_n}}\epsilon_{j_1 j_2 \cdots j_n} \\
= & \frac{\partial x^{j_1}}{\partial x'^{i_1}}\frac{\partial x^{j_2}}{\partial x'^{i_2}}\cdots\frac{\partial x^{j_n}}{\partial x'^{i_n}}\delta^{12\cdots n}_{j_1 j_2 \cdots j_n}\epsilon_{12\cdots n} \\
= & J\delta^{12\cdots n}_{i_1 i_2 \cdots i_n}|g|^{\frac{1}{2}} \\
= & \delta^{12\cdots n}_{i_1 i_2 \cdots i_n}\epsilon'_{12\cdots n} \\
= & \epsilon'_{i_1 i_2 \cdots i_n}.
\end{aligned}$$

But the ordinary Levi-Civita notation

$$\varepsilon_{12\cdots n} = 1, \qquad \varepsilon_{i_1 i_2 \cdots i_n} = \delta^{12\cdots n}_{i_1 i_2 \cdots i_n},$$

is not a tensor under the transformation of coordinates,

$$\varepsilon'_{i_1 i_2 \cdots i_n} = J\,\varepsilon_{i_1 i_2 \cdots i_n},$$

so it is a tensor density with weight 1 only.

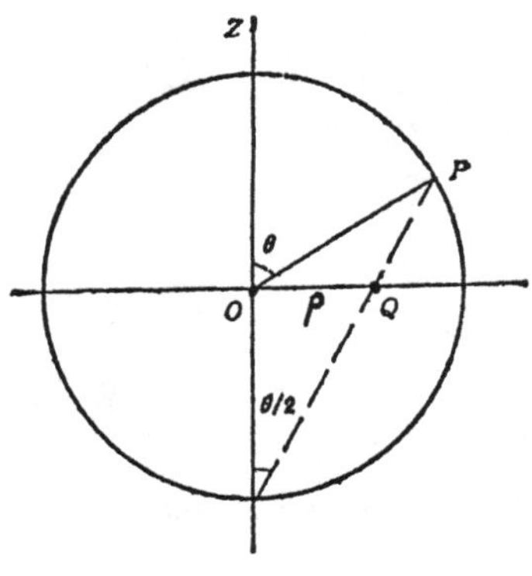

Fig. J.1

Appendix J

n-dimensional Sphere S^n

S^n is a hyperplane in an $(n+1)$-dimensional Euclidean space E^{n+1} given by

$$\sum_{i=1}^{n} y_i^2 + z^2 = \mathbf{R}^2.$$

The y_i's and z can be written in terms of the spherical coordinates $\theta_1, \theta_2, \cdots, \theta_{n-1}$ and θ as

$$\begin{aligned}
y_1 &= R\sin\theta\sin\theta_1\cdots\sin\theta_{n-2}\cos\theta^{n-1},\\
y_2 &= R\sin\theta\sin\theta_1\cdots\sin\theta_{n-2}\sin\theta^{n-1},\\
y_3 &= R\sin\theta\sin\theta_1\cdots\cos\theta^{n-2},\\
&\cdots\cdots\cdots\cdots\cdots\cdots\\
y_n &= R\sin\theta\cos\theta_1,\\
z &= R\cos\theta,
\end{aligned}$$

where $0 \le \theta < \pi$, $0 \le \theta_i < \pi$, $0 \le \theta_{n-1} < 2\pi$, $i = 1, \cdots, n-1$.

As illustrated in Fig. J.1, point P of S^n is projected to Q, a point on the hyperplane $z = 0$. The coordinates of Q is

$$\begin{aligned}
x_1 &= \rho\sin\theta_1\cdots\sin\theta_{n-2}\cos\theta_{n-1},\\
x_2 &= \rho\sin\theta_1\cdots\sin\theta_{n-2}\sin\theta_{n-1},\\
x_3 &= \rho\sin\theta_1\cdots\cos\theta_{n-2},\\
&\cdots\cdots\cdots\cdots\cdots\cdots\\
x_n &= \rho\cos\theta_1,
\end{aligned}$$

where

$$\rho = R\,\mathrm{tg}\frac{\theta}{2} = R\frac{\sin\theta}{1+\cos\theta}.$$

After some calculations, we obtain

$$\begin{aligned}
&y_i = \tfrac{R}{\rho}x_i\sin\theta = \tfrac{2R^2}{R^2+\rho^2}x_i = \tfrac{R+z}{R}x_i, \quad && \mathrm{d}y_i = \tfrac{2R^2}{R^2+\rho^2}\mathrm{d}x_i + \tfrac{x_i}{R}\mathrm{d}z,\\
&z = R\cos\theta = R\tfrac{R^2-\rho^2}{R^2+\rho^2}, && \mathrm{d}z = -\tfrac{4R^3}{(R^2+\rho^2)^2}\rho\,\mathrm{d}\rho.
\end{aligned}$$

Therefore

$$\begin{aligned} ds^2 &= \sum_{i=1}^{n}(dy_i)^2 + (dz)^2 \\ &= \sum_{i=1}^{n}\frac{4R^4}{(R^2+\rho^2)^2}(dx_i)^2 + \frac{4R}{R^2+\rho^2}\rho d\rho dz + \frac{\rho^2}{R^2}(dz)^2 + (dz)^2 \\ &= 4\sum_{i=1}^{n}(dx_i)^2/(1+\frac{\rho^2}{R^2})^2. \end{aligned}$$

If we project P to the hyperplane $z = R$, the coordinates are given by

$$\xi_i = 2x_i, \qquad r = 2\rho.$$

So

$$ds^2 = \frac{\sum_{i=1}^{n}(d\xi_i)^2}{(1+(\frac{r}{2R})^2)^2} = \sum_{i=1}^{n}(\vartheta^i)^2,$$

where $\vartheta^i = \frac{d\xi_i}{1+(\frac{r}{2R})^2}$. These ϑ^i are just the frame 1-forms on S^n. As a result of the zero torsion of S^n, from the first Structure Equations of Cartan $\tau^i = d\vartheta^i + \omega^i_j \wedge \vartheta^j$, we get the connection 1-form

$$\omega^i_j = \frac{\xi^i d\xi^j - \xi^j d\xi^i}{2R^2(1+(\frac{r}{2R})^2)} = \frac{\xi^i\vartheta^j}{2R^2} - \frac{\xi^j\vartheta^i}{2R^2}.$$

Then from the second Structure Equations of Cartan $\Omega^i_j = d\omega^i_j + \omega^i_k \wedge \omega^k_j$, we obtain

curvature 2-form:	$\Omega^i_j = \frac{1}{R^2}\vartheta^i\vartheta^j$,
Riemannian curvature tensor:	$R_{ijkl} = \frac{1}{R^2}(\delta_{ik}\delta_{jl} - \delta_{il}\delta_{jk})$,
Ricci tensor:	$\Re_{ij} = \frac{n-1}{R^2}\delta_{ij}$,
Ricci scalar curvature:	$\Re = \frac{n(n-1)}{R^2}$,
Weyl curvature tensor:	$W_{ijkl} = 0$.

The area element of S^n is $dA = \vartheta^1 \wedge \vartheta^2 \wedge \cdots \wedge \vartheta^n/(1+(\frac{r}{2R}^2)^n$, and the total area of S^n is

$$\begin{aligned} A &= \int d\theta d\theta_1 d\theta_2 \cdots d\theta^{n-1} R^n \sin\theta_1 \sin^2\theta_2 \cdots \sin^{n-1}\theta_{n-1} \\ &= 2\pi R^n \prod_{k=1}^{n-1}\int_0^\pi d\theta_k \sin^k\theta_k = 2\pi^{\frac{n+1}{2}} R^n/\Gamma(\frac{n+1}{2}). \end{aligned}$$

Here we had used the well known identity:

$$\int_0^\pi d\theta \sin^k\theta = \sqrt{\pi}\Gamma(\frac{n+1}{2})/\Gamma(\frac{n+2}{2}).$$

Appendix K

$SU(2)$ Group Manifold

The fundamental representation of $SU(2)$ is

$$g = \begin{pmatrix} z_1 & -\bar{z}_2 \\ z_2 & \bar{z}_1 \end{pmatrix},$$

where the complex variables z_1 and z_2 satisfy

$$|z_1|^2 + |z_2|^2 = 1.$$

Let

$$z_1 = t - iz, \qquad z_2 = y - ix,$$

then

$$t^2 + z^2 + y^2 + x^2 = 1.$$

Therefore $SU(2)$ is equivalent to the 3-dimensional sphere S^3. The length of infinitesimal arc is

$$\begin{aligned} ds^2 &= dx^2 + dy^2 + dz^2 + dt^2 \\ &= dz_1 d\bar{z}_1 + dz_2 d\bar{z}_2 \\ &= \tfrac{1}{2}\mathrm{Tr}(dg^{-1}dg) \\ &= \tfrac{1}{2}\mathrm{Tr}[(ig^{-1}dg)(ig^{-1}dg)]. \end{aligned}$$

The generators of Lie algebra $su(2)$ are just the Pauli matrices $\vec{\tau} = (\tau_1, \tau_2, \tau_3)$, the left invariant Maurer-Cartan form can be written as

$$g^{-1}dg = i\vec{\tau}\cdot\vec{\vartheta},$$

where $\vec{\vartheta} = (\vartheta_1, \vartheta_2, \vartheta_3)$ are the left invariant 1-forms satisfying the Maurer-Cartan Equations

$$d\vartheta_i = \epsilon_{ijk}\vartheta_j \wedge \vartheta_k.$$

As a result of the identity

$$(\vec{\tau}\cdot\vec{a})(\vec{\tau}\cdot\vec{b}) = (\vec{a}\cdot\vec{b}) + i\vec{\tau}\cdot[\vec{a}\times\vec{b}],$$

it is easy to see that

$$ds^2 = \vartheta_1^2 + \vartheta_2^2 + \vartheta_3^2.$$

Noticing

$$g = \begin{pmatrix} t - iz & -y - ix \\ y - ix & t + iz \end{pmatrix} = t - i\vec{\tau} \cdot \vec{x},$$

we get

$$\begin{aligned} g^{-1}dg &= (t + i\vec{\tau} \cdot \vec{x})(dt - i\vec{\tau}d\vec{x}) \\ &= i\vec{\tau} \cdot (\vec{x}dt - td\vec{x} + \vec{x} \times d\vec{x}). \end{aligned}$$

Now we can write $\vec{\vartheta}$ as

$$\begin{aligned} \vartheta_1 &= xdt - tdx + ydz - zdy, \\ \vartheta_2 &= ydt - tdy + zdx - xdz, \\ \vartheta_3 &= zdt - tdz + xdy - ydx. \end{aligned}$$

Using the Structure Equations of Cartan, we have the connection 1-forms

$$\omega_j^i = \epsilon_{ijk}\vartheta_k,$$

and the curvature two forms

$$\Omega_j^i = \vartheta_i \wedge \vartheta_j.$$

Appendix L

S^2 Manifold and Hopf Maps

As is known in Appendix J, the unit sphere in E^3

$$S^2 = \{(x, y, z) \in E^3 | x^2 + y^2 + z^2 = 1\}$$

can be expressed in spherical coordinates as

$$x = \sin\theta\cos\phi,$$

$$y = \sin\theta\sin\phi,$$

$$z = \cos\theta \; .$$

or after projective mapping, as

$$x_1 = \rho\cos\phi = \frac{x}{1+z},$$

$$x_2 = \rho\sin\phi = \frac{y}{1+z},$$

where

$$\rho = \mathrm{tg}\frac{\theta}{2} = \frac{\sin\theta}{1+\cos\theta}.$$

The length of a line element on the sphere is

$$ds^2 = (d\theta)^2 + \sin^2\theta(d\phi)^2 = (\vartheta^1)^2 + (\vartheta^2)^2,$$

where

$$\vartheta^1 = d\theta, \qquad \vartheta^2 = \sin\theta d\phi.$$

It is easy to obtain the connection 1−form

$$\omega_2^1 = -\cos\theta d\phi = -\frac{\cos\theta}{\sin\theta}\vartheta^2,$$

and the curvature 2−form

$$\Omega^1_2 = \sin\theta \mathrm{d}\theta \wedge \mathrm{d}\phi = \vartheta^1 \wedge \vartheta^2.$$

If we denote the connection 1−form by

$$A = A_\theta \vartheta^1 + A_\phi \vartheta^2,$$

then

$$A_\theta = 0, \qquad A_\phi = -\frac{\cos\theta}{\sin\theta}.$$

We see that A_ϕ is singular at both $\theta = 0$ and $\theta = \pi$. After a gauge transformation,

$$iA^\pm_\phi = e^{\pm i\phi}(iA_\phi + \frac{1}{\sin\theta}\frac{\partial}{\partial\phi})e^{\mp i\phi},$$

we get

$$A^\pm_\phi = \frac{\mp 1 + \cos\theta}{\sin\theta}.$$

Each of $A^\pm_\phi$ has only one singular point.

Remember that

$$S_2 = SU(2)/U(1),$$

we can express the unit vector N in terms of matrices

$$N = \sum_{i=1}^{3} N^i \tau_i = g\tau_3 g^{-1}$$

where τ^i's are Pauli matrices and $g \in SU(2)$.

Obviously, N has the following properties:

(*i*) $N^2 = 1$.

(*ii*) If we take the transformation $g \to g e^{i\gamma\tau_3}$, N is invariant. So $N \in SU(2)/U(1)$.

The unit vector $N = (\sin\theta\cos\phi, \sin\theta\sin\phi, \cos\phi)$ can be obtained by a rotation g from z axis along the vector $\vec{a} = (-\sin\phi, \cos\phi, 0)$, see Fig. L.1, where

$$g = \begin{pmatrix} \cos\frac{\theta}{2} & -e^{-i\phi}\sin\frac{\theta}{2} \\ e^{i\phi}\sin\frac{\theta}{2} & \cos\frac{\theta}{2} \end{pmatrix} = \frac{1}{\sqrt{1+|\zeta|^2}}\begin{pmatrix} 1 & -\bar{\zeta} \\ \zeta & 1 \end{pmatrix},$$

$$\zeta = \mathrm{tg}\frac{\theta}{2}e^{i\phi} = x_1 + ix_2.$$

Therefore we have

$$N^i = \frac{1}{2}\mathrm{Tr}(N\tau_i) = \frac{1}{2}\mathrm{Tr}(g(1+\tau_3)g^{-1}\tau_i) = z^\dagger\tau_i z,$$

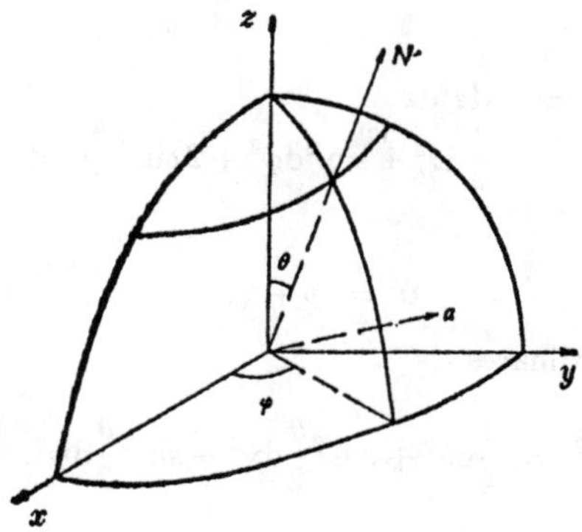

Fig. L.1

where

$$z = \begin{pmatrix} z_1 \\ z_2 \end{pmatrix} = \begin{pmatrix} \cos\frac{\theta}{2} \\ e^{i\phi}\sin\frac{\theta}{2} \end{pmatrix} = (1+|\zeta|^2)^{-\frac{1}{2}} \begin{pmatrix} 1 \\ \zeta \end{pmatrix}.$$

Thus

$$\begin{aligned} \mathrm{d}s^2 &= \tfrac{1}{2}\mathrm{Tr}(\mathrm{d}N\mathrm{d}N) = \mathrm{d}z^\dagger \mathrm{d}z = (\vartheta^1)^2 + (\vartheta^2)^2 \\ &= \tfrac{4\mathrm{d}\zeta\mathrm{d}\bar{\zeta}}{(1+|\zeta|^2)^2} = 4\tfrac{\mathrm{d}x_1^2+\mathrm{d}x_2^2}{(1+x_1^2+x_2^2)^2} \\ &= 4\tfrac{\partial^2 K}{\partial\zeta\partial\bar{\zeta}}\mathrm{d}\zeta\mathrm{d}\bar{\zeta}, \end{aligned}$$

where $K = \ln(1+|\zeta|^2)$ is the Kählar potential.

$$\vartheta^a = i\mathrm{Tr}[(g^{-1}\mathrm{d}g)\tau_a] \qquad (a = 1, 2).$$

Moreover, we have Kählar 2−form

$$\begin{aligned} \Omega &= \mathrm{d}z^\dagger \mathrm{d}z \\ &= 2i\tfrac{\mathrm{d}\zeta\mathrm{d}\bar{\zeta}}{(1+|\zeta|^2)^2} = 2i\tfrac{\partial^2 K}{\partial\zeta\partial\bar{\zeta}}\mathrm{d}\zeta\mathrm{d}\bar{\zeta} \\ &= 4\tfrac{\mathrm{d}x_1\wedge \mathrm{d}x_2}{(1+x_1^2+x_2^2)^2}. \end{aligned}$$

Let

$$z = \begin{pmatrix} \cos\frac{\theta}{2} \\ e^{i\phi}\sin\frac{\theta}{2} \end{pmatrix} e^{i\chi}.$$

Now z is a point on 3-dimensional sphere S^3. Therefore, we see that S^3 can be expressed as a non-trivial $U(1)$ bundle over S^2. Such a bundle is called a Hopf bundle. The projection map

$$\pi : S^3 \longrightarrow S^2$$

is called a Hopf map. Thus

$$\begin{aligned} ds^2 &= dz^\dagger dz \\ &= \frac{1}{4}d\theta^2 + \sin^2 d\phi^2 + 2\sin^2\frac{\theta}{2}d\phi d\chi. \end{aligned}$$

Introducing

$$\psi = \phi + \chi,$$

we get an orthogonal coordinates

$$ds^2 = \frac{1}{4}d\theta^2 + \cos^2\frac{\theta}{2}d\chi^2 + \sin^2\frac{\theta}{2}d\psi^2.$$

Also the connection

$$A = -iz^\dagger dz = \cos^2\frac{\theta}{2}(d\chi)^2 + \sin^2\frac{\theta}{2}(d\psi)^2.$$

So we arrive at the following non-singular connection

$$A_\theta = 0, \qquad A_\chi = \cos\frac{\theta}{2}, \qquad A_\psi = \sin\frac{\theta}{2}.$$

This could be understood by noticing the topological fact that

$$H^2(S^3) = 0,$$

Thus the closed curvature 2−form F must be exact

$$F = dA,$$

and A can be defined globally on S^3.

Appendix M

Clifford Algebra, Spin Group and its Representations

M.1 Clifford Algebra

Cliford algebra Cl_n is a real vector space endowed with an associative product, distributive with respect to addition, generated by n elements e_i and a unit 1. Their product satisfies:

$$e_i e_j + e_j e_i = -2\delta_{ij}\, 1. \tag{M.1}$$

As a vector space, its basis are consisted of these polynomials

$$e_{i_1} e_{i_2} \cdots e_{i_k}, \qquad i_1 < i_2 < \cdots < i_k, \quad 0 \le k \le n.$$

Here the monomial $k = 0$ corresponds to the unit 1, and $k = n$ is corresponds to chirality operator, and is denoted as

$$e_f = e_1 e_2 \cdots e_n.$$

There are $\sum_{k=0}^{n} \begin{pmatrix} n \\ k \end{pmatrix} = 2^n$ elements, so this algebra Cl_n is of 2^n dimensional.

The Clifford algebra can be decomposed into a direct sum:

$$Cl_n = Cl_n^+ \oplus Cl_n^-,$$

where Cl_n^+ is generated by the product of even number of elements and is a 2^{n-1}-dimensional subalgebra of Cl_n. It is easy to prove that Cl_n^+ is isomorphic to Cl_{n-1}.

1. $n = 2k$ case

The center of Cl_{2k} is trivial, namely multiples of the unit element $\mathbf{R}\, 1$.

The complex extension $Cl_{2k}^c = Cl_{2k} \otimes \mathbf{C}$ is a simple algebra over $\mathbf{C}$. So it is isomorphic to the algebra $\mathcal{M}_{2^k}(\mathbf{C})$ of $2^k \times 2^k$ complex matrices. There exist a set of $2^k \times 2^k$ matrices γ_i satisfying

$$\gamma_i \gamma_j + \gamma_j \gamma_i = -2\delta_{ij} 1_{2^k}. \tag{M.2}$$

The algebra over the real numbers generated by the matrices $\{\gamma_i\}_1^{2k}$ is a faithful representation of Cl_{2k}.

The space ψ of representation can also be called the Cl_{2k}-module, it itself is a 2^k-dimensional vector space.

2. $n = 2k+1$ case

The center of Cl_{2k+1} is $\mathbf{R}1 \otimes \mathbf{R}\, e_f$.

The complex extension Cl^c_{2k+1} is the direct sum of two simple algebras over $\mathbf{C}$. So it is isomorphic to $M_{2^k}(\mathbf{C}) \oplus M_{2^k}(\mathbf{C})$. The representation of Cl^c_{2k+1} on a 2^k-dimensional vector space is not faithful. The corresponding representation of Cl_{2k+1} may not be faithful either.

M.2 Spin Group

The vector subspace of Clifford algebra Cl_n spanned by $\{e_i\}^n$ is an n-dimensional subspace of Cl_n, denoted as Cl_n^V, which can be endowed with a Euclidean metric as follows:

$$(e_i,\ e_j) = \delta_{ij}.$$

So Cl_n^V is isomorphic to an n-dimensional Euclidean space, its isometry group is $O(n)$.

If $u, v \in Cl_n^V$, then

$$uv + vu = -2(u, v),$$

i.e., u and v are anticommutative if and only if $u \perp v$.

The subset of elements of Cl_n, which is generated by the norm 1 elements u_i of Cl_n^V, constitutes a group $Pin(n)$, i.e.,

$$Pin(n) = \{a \in Cl_n | a = u_1 \cdots u_l, \quad u_i \in Cl_n^V \quad \text{and } (u_i, u_i) = 1\}.$$

The **spin group** $Spin(n)$ is the subgroup of $Pin(n)$ with index 2, which can be defined as

$$\begin{aligned} Spin(n) &= Pin(n) \cap Cl_n^+ \\ &= \{a \in Pin(n) | aa^* = 1\}, \end{aligned}$$

here $*: \quad a \to a^*$ is an involution (antiautomorphism) of Cl_n, in terms of basis e_i

$$(e_{i_1} e_{i_2} \cdots e_{i_k})^* = e_{i_k} \cdots e_{i_2} e_{i_1}.$$

So for $a, b \in Cl_n$, we have $(ab)^* = b^* a^*$.

Group $Pin(n)$ is not connected, group $Spin(n)$ is the connected component containing identity in $Pin(n)$.

Theorem M.1 *There exists a homomorphism ρ from $Pin(n)$ group into the isometry group $O(n)$ of Cl_n^V defined by*

$$\rho(a)u = aua^* \qquad \forall\ a \in Pin(n), \quad u \in Cl_n^V.$$

and this homomorphism is surjective.

Proof: First one can show $\rho(a)(a \in Pin(n))$ is an orthogonal transformation of Cl_n^V.
We know that for $u, v \in Cl_n^V$, $u^* \equiv u$, $uu = -(u,u)$, so

$$uvu* = uvu = (-vu - 2(u,v))u = (v - 2\frac{(u,v)}{(u,u)}u)(u,u).$$

Now suppose $(u,u) = 1$, i.e., u is a generator of elements of $Pin(n)$, then

$$\rho(u)v = uvu^* = v - 2(u,v)u.$$

So the action of $\rho(u)$ on Cl_n^V is a reflection on the hyperplane perpendicular to u. For $a = u_1 u_2 \cdots u_l \in Pin(n)$, the action of $\rho(a)$ on Cl_n^V corresponds to multiple reflection, so $\rho(a) \in O(n)$, the isometry group of Cl_n^V.

Next it is easy to show that map ρ is surjective. Since every orthogonal transformation of Cl_n^V is a multiple product of reflections. And this completes our proof.

Spin(n) is defined as $Spin(n) = Pin(n) \cap Cl_n^+$, each element $a \in Spin(n)$ satisfies

$$\begin{cases} (i) \quad aa^* = 1, \qquad \text{so } a^* = a^{-1}. \\ (ii) \quad ae_ia^* \in Cl_n^V. \end{cases} \tag{M.3}$$

These can also be taken as the definition of *Spin*(n).

As each $a \in Spin(n)$, $\rho(a)$ is a product of even reflections, the linear mapping $v \to ava^*$ is in $SO(n)$, thus we get a homomorphism:

$$\rho: \quad Spin(n) \to SO(n)$$

homomorphism ρ is surjective and its kernel is exact: $\{\pm 1\} \subset Spin(n)$, so we obtain the exact sequence

$$1 \to \mathbf{Z}_2 \to Spin(n) \to SO(n) \to 1.$$

Thus ρ is a double covering of $SO(n)$. For $n \geq 3, Spin(n)$ is simply connected, which is the universal covering of $SO(n)$.

M.3 The representation of *Spin*(n)

Now we consider the fundamental representation of *Spin*(n).

Every $(Cl_n^+ \otimes \mathbf{C}) - module$ Ψ can be given a representation τ of *Spin*(n)

$$\tau: \quad Spin(n) \to End_{\mathbf{C}}(\Psi)$$

1. $n = 2k$.

From (M.2), there exists a faithful representation of Cl_{2k}, the representation space Ψ is 2^k-dimensional.

From (M.3), matrix Λ satisfying

$$\Lambda\gamma_i\Lambda^{-1} = L_i^j\gamma_j,$$

can represent elements of $Spin(2k)$. Here L_i^j are real numbers satisfying

$$\delta_{ik} = L_i^j L_k^l \delta_{jl}.$$

So transformation $L \in SO(n)$. And

$$\Lambda \subset Cl_{2k} \otimes \mathbf{C}, \quad det\Lambda = 1.$$

It is easy to prove that the map $\rho: \quad \Lambda \to L$ is a surjective homomorphism, its kernel Ker$\rho = \mathbf{Z}_2$. So Λ is a spin representation τ of $Spin(n)$, i.e., there exists a 2^k-dimensional $(Cl_n^+ \otimes \mathbf{C})$-module Ψ, from which we can obtain a representation

$$\tau: \quad Spin(2k) \to End_C(\Psi)$$

such that for $a \in Spin(2k), \quad s \in \Psi$

$$\tau(a)s = \Lambda(a)s.$$

Now the representation of $Spin(n)$ is obtained from $(Cl_{2k}^+ \otimes \mathbf{C})$-module, but it is not irreducible. Let

$$\gamma_f = i^k \gamma_1 \gamma_2 \cdots \gamma_{2k},$$

which commutes with each element of Cl_{2k}^+. Since

$$Spin(2k) \subset Cl_{2k}^+$$

it follows that Ψ can be decomposed into the ± 1 eigenspaces of γ_f

$$\Psi = \Psi^+ \otimes \Psi^-,$$

here

$$\Psi^+ = \{\phi \in \Psi \quad |\gamma_f \phi = \phi\},$$
$$\Psi^- = \{\phi \in \Psi | \gamma_f \phi = -\phi\}.$$

So $Spin(2k)$ has two spin representations

$$\tau^+: \quad Spin(2k) \to End_C(\Psi^+)$$

$$\tau^-: \quad Spin(2k) \to End_C(\Psi^-)$$

These two complex representations are both 2^{k-1}-dimensional and irreducible. As $Spin(2k)$ is a connected compact Lie group, so $\tau^\pm$ both are unitary representations of $Spin(2k)$.

2. $n = 2k + 1$

As $Spin(2k+1) \subset Cl_{2k+1}^+ \simeq Cl_{2k}$. Let Ψ be the 2^k-dimensional irreducible $Cl_{2k} \otimes \mathbf{C}$-module. Through the isomorphism between $Cl_{2k+1}^+ \otimes \mathbf{C}$ and $Cl_{2k} \otimes \mathbf{C}$, we can obtain a 2^k-dimensional irreducible representation τ of $Spin(2k+1)$.

$$\tau: \quad Spin(2k+1) \to End_C \Psi$$

M.4 spin(n) algebra module

$Spin(n)$ is a $\frac{1}{2}n(n-1)$-dimensional compact simply connected Lie group, it is the 2-fold covering of $SO(n)$ by a homomorphism ρ, the derived mapping ρ_* is an isomorphism between tangent spaces at the unit elements of these groups. So they have the same Lie algebra. We have

$$\begin{array}{rccc} \rho: & Spin(n) & \to & SO(n) \\ & \Lambda & \to & L \end{array}$$

with

$$\Lambda\gamma_a\Lambda^{-1} = L^b_a\gamma_b.$$

Let

$$\lambda = d\Lambda\Lambda^{-1} \in spin(n), \quad l = dLL^{-1} \in so(n).$$

Then

$$\lambda\gamma_a - \gamma_a\lambda = l^b_a\gamma_b, \quad l^{ab} = -l^{ba}, \quad \mathrm{tr}\lambda = 0.$$

The general solution satisfying these conditions can be obtained as:

$$\lambda = \frac{1}{2}l^{ab}\gamma_a\gamma_b = \frac{1}{4}l^{ab}[\gamma_a, \gamma_b].$$

Thus we obtain the isomorphism between algebras $so(n)$ and $spin(n)$.

The Lie algebra valued connection 1-form on the spin bundle can be expressed as

$$dx^i\omega_i^{ab}\frac{1}{4}[\gamma_a, \gamma_b],$$

so the covariant derivative on spin bundle can be expressed as

$$\nabla_i = \partial_i + \frac{1}{4}[\gamma_a, \gamma_b]\omega_i^{ab}.$$

$Sp(n)$ is a group generated by $exp(te_ie_j)$, and its algebraic bases is $e_ie_j, \quad i < j$.

$spin(n)$ algebraic bases		$so(n)$ algebraic bases
e_ie_j	$\leftrightarrow$	$2\theta^i \wedge \theta^j$

here

$$\theta^i \wedge \theta^j(e) = \theta^i(e)\theta^j - \theta^j(e)\theta^i.$$

For example $Sp(4)$-module

$$\begin{aligned} Cl_4 \otimes \mathbf{C} &= \mathbf{C}\{W_1, W_2, \bar{W}_1, \bar{W}_2\} \\ &= V \oplus \bar{V}, \\ W_1 &= \frac{1}{\sqrt{2}}(e_1 - ie_2), \\ W_2 &= \frac{1}{\sqrt{2}}(e_3 - ie_4). \end{aligned}$$

$$\psi = \wedge^* V = \{1, W_1, W_2, W_1 \wedge W_2\} = \underset{\text{even}}{\psi^+} \oplus \underset{\text{odd}}{\psi^-}.$$

$$\begin{aligned} V(\psi) &= V \wedge \psi. \\ \bar{V}(\psi) &= i_V(\psi). \end{aligned}$$

The operators of V and $\bar{V}$ on ψ are as creation and annihilation operators respectively in quantum field theory.

Appendix N

Good Cover and the Nerve of a Cover. Čech Cohomology

Given a manifold of dimension n, it is always possible to choose an open cover U= $\{U_\alpha\}_{\alpha\in J}$ where the index set J is a countable ordered set, such that each open set and each nonempty finite intersection is diffeomorphic to an open ball in $\mathbf{R}^n$, then these covers are called a **good cover**, each of its intersection is topologically trivial. If the manifold is compact, then the cover may be chosen to be finite, i.e., set J is finite.

For example, S^2 can be covered by two open sets:

$$S^2 = U_0 \cup U_1,$$

here $U_0(U_1)$ is the upper (lower) more hemisphere, the overlapping region $U_0 \cap U_1$ is topologically nontrivial, so this open cover is not a good cover.

If we cover the lower hemisphere with three open sets, together with the upper hemisphere U_0, as shown in Fig N.1, this gives a good cover of the entire sphere S^2, i.e., we can cover S^2 by four open sets and obtain a good cover.

In generally, in order to obtain a good cover of $S^n(n \geq 1)$, the number of open sets in the open cover has to be $\geq n+2$ because the unit sphere S^n in $\mathbf{R}^{n+1}$ is homeomorphic to an $(n+1)$-simplex, which has $n+2$ vertices.

The topology of the manifold determines its permissible triangulations. The good cover of a manifold M defines a triangulation of M, i.e., we choose a point α in the interior of each open set U_α, these points define the vertices of the triangulation. If $U_\alpha \cap U_\beta$ is nonempty, we join vertices α and β by an edge, yielding a 1-simplex, denoted by $U_{\alpha\beta}$. If $U_{\alpha\beta\gamma} = U_\alpha \cap U_\beta \cap U_\gamma$ is nonempty, we associate with it a 2-simplex. Repeating this procedure for all finite intersections, we obtain the **nerve** of this open cover, denoted as $N(\mathcal{U})$. Nerve $N(\mathcal{U})$ is the complex whose vertices are denoted by J, and a set $\alpha_0, \alpha_1, \cdots, \alpha_k$ of distinct elements of J is a k-simplex of $N(\mathcal{U})$ if and only if

$$U_{\alpha_0\alpha_1\cdots\alpha_k} \equiv U_{\alpha_0} \cap U_{\alpha_1} \cap \cdots \cap U_{\alpha_k} \neq 0,$$

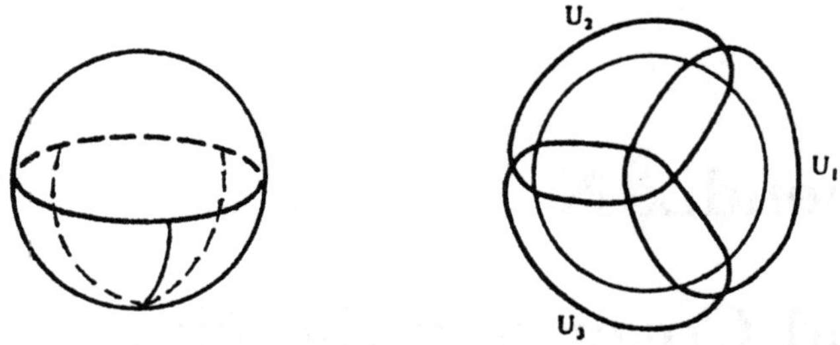

a b

Fig. N.1 The good cover of S^2

The nerve N of a cover U is the building block of Čech homology $H_*(N, Z)$.

Next we look at Čech cohomology. A k-chain ω taking values in l-forms is an assignment of a nonsingular l-form to each k-chain,

$$\omega \in \prod \wedge^l(U_{\alpha_0\cdots\alpha_k}) = C^k(\mathcal{U}, \wedge^l),$$

or it can be written so that ω has "components" as follow: for example,

0-cocycle $\omega_j \in \Lambda^l(U_j)$

1-cocycle $\omega_{jk} \in \Lambda^l(U_j \cap U_k)$, $(\omega_{jk} = -\omega_{kj})$.

k-cocycle $\omega_{\alpha_0\cdots\alpha_k} \in \Lambda^l(U_{\alpha_0\cdots\alpha_k})$, $(\omega_{\cdots i\cdots j\cdots} = -\omega_{\cdots j\cdots i\cdots})$.

The coboundary operator δ can be defined as an alternating difference operator as follows

$$(\delta\omega)_{\alpha_0\cdots\alpha_{k+1}} = \sum_{i=0}^{k+1}(-1)^j \omega_{\alpha_0\cdots\hat{\alpha}_k\cdots\alpha_{k+1}},$$

here $\hat{\alpha}$ represents omission.

It is easy to prove that $\delta^2 = 0$, so one obtains Čech cohomology.

On a vector bundle, the collection of transition functions $g_{\alpha\beta}$ defines a 1-cochain taking values in the Lie algebra of the structure group, the consistency condition is the 1-cocycle condition as follows:

$$(\delta g)_{\alpha\beta\gamma} = g_{\beta\gamma} g_{\alpha\gamma}^{-1} g_{\alpha\beta} = g_{\beta\gamma} g_{\gamma\alpha} g_{\alpha\beta} = 1.$$

Appendix O

Symmetric Trace

Let Σ be Lie algebra g-valued k-forms

$$\Sigma = \sigma \otimes \lambda \quad \in \Omega^k(M, \mathrm{g}),$$

here λ is the matrix representation of Lie algebra g, σ is the k-form on manifold M, $\sigma \in \Lambda^k(M)$. Assume that under gauge transformation Σ transforms gauge covariant

$$\Sigma \to \Sigma^g = g^{-1}\Sigma g, \quad g \in \mathrm{G}.$$

The symmetric trace of matrix-valued forms Σ_i is defined as

$$\mathrm{Str}(\Sigma_1, \cdots, \Sigma_l) = \sigma_1 \wedge \cdots \wedge \sigma_l \otimes \frac{1}{l!}\sum_P \mathrm{Tr}(\lambda_{P_1}, \cdots, \lambda_{P_l}),$$

here P denotes permutation, i.e., pull out the form part and symmetries in the matrix part. The symmetries trace defined as above has the following properties:
(1) it is a symmetric invariant polynomial, i.e.,

$$\begin{aligned} \mathrm{Str}(\Sigma_{P_1}, \cdots, \Sigma_{P_l}) &= \mathrm{Str}(\Sigma_1, \cdots, \Sigma_l), \\ \mathrm{Str}(\Sigma_1^g, \cdots, \Sigma_l^g) &= \mathrm{Str}(\Sigma_1, \cdots, \Sigma_l). \end{aligned}$$

(2) for any Lie algebra g-valued 1-form ω, the following equation exists:

$$\sum_{i=1}^{l}(-1)^{k_1+\cdots+k_{i-1}}\mathrm{Str}(\Sigma_1, \cdots, [\omega, \Sigma_i], \cdots, \Sigma_l) = 0.$$

By using the above properties, it is easy to prove

$$d\,\mathrm{Str}(\Sigma_1, \cdots, \Sigma_l) = \sum_{i=1}^{l}(-1)^{k_1+\cdots+k_{i-1}}\mathrm{Str}(\Sigma_1, \cdots, D\Sigma_i, \cdots, \Sigma_l),$$

here $D\Sigma_i = d\Sigma_i + [A, \Sigma_i]$ is the covariant exterior differential. For example, take $\Sigma_i = F = dA + A^2$, since F satisfies the Bianchi identity $DF = dF + [A, F] = 0$, so one get

$$d\,\mathrm{Str}\,(\,F^l\,) \;=\; 0.$$

Appendix P

Cohomology with Compact Support and Cohomology on Fibre Bundle, Thom Class

The support of a continuous function f on a topological space M is the smallest closed set on which f is not zero, i.e.,

$$Supp\, f = \{p \in M | f(p) \neq 0\}.$$

If in the definition of de Rham complex one uses the C^∞ function with compact support only, i.e., by restricting to differential forms with compact support, the resulting complex is called de Rham complex with compact support and the cohomology of this complex is called compact cohomology, denoted by $H^*_C(M)$. $H^*_C(M)$ on compact manifold coincides with the usual de Rham cohomology $H^*(M)$, but they are different for non-compact manifold. For instance, as $(ker\ d) \cup \Lambda^0(\mathbf{R}^1)$ are constant functions, so

$$H^0(\mathbf{R}^1) = \mathbf{R},$$

but there is no constant function on $\mathbf{R}^1$ with compact support, so

$$H^0_C\mathbf{R}^1) = 0.$$

All 1-forms on $\mathbf{R}^1$ are closed, but it is easy to show that every 1-form on $\mathbf{R}^1$ is also exact. For instance, let

$$\omega = f(x)dx \ \in \Lambda^1(\mathbf{R}^1),$$

then by taking the integral:

$$\int_0^x f(y)dy = g(x),$$
$$dg = f(x)\, dx = \omega \quad \Longrightarrow \quad H^1(\mathbf{R}^1) = 0.$$

Now we look at 1-forms with compact support, $\omega \in \Lambda^1_C(\mathbf{R}^1)$, they are all closed so

$$Z^1_C(\mathbf{R}^1) = \Lambda^1_C(\mathbf{R}^1).$$

Take the integration map:

$$\int_R : \ \Lambda^1_C(\mathbf{R}^1) \to \mathbf{R}^1.$$

this map is clearly surjective:

$$\Lambda^1_C(\mathbf{R}^1) \xrightarrow{\int_R} \mathbf{R}^1 \to 0.$$

It vanishes on exact 1-form df where f has compact support so we have the following exact sequence:

$$0 \to H^1_C(\mathbf{R}^1) = \Lambda^1_C(\mathbf{R}^1)/B^1_C(\mathbf{R}^1) \xrightarrow{\int_R} \mathbf{R}^1 \to 0$$

so

$$H^1_C(\mathbf{R}^1) = \mathbf{R}^1.$$

More generally, $\mathbf{R}^n$ is a trivial manifold which is homotopically equivalent to a single point,

$$H^*(\mathbf{R}^n) = H^*(\text{point}) = \begin{cases} \mathbf{R} & \text{in dimension } 0, \\ 0 & \text{otherwise.} \end{cases}$$

The de Rham cohomology is invariant under homotopy equivalence. For compact cohomology,

$$H^*_C(\mathbf{R}^n) = H^*(\text{point}) = \begin{cases} \mathbf{R} & \text{in dimension } n, \\ 0 & \text{otherwise,} \end{cases}$$

where the isomorphism $H^n_C(\mathbf{R}^n) \simeq \mathbf{R}$ is given by the integration over $\mathbf{R}^n$,

$$0 \to H^n_c(\mathbf{R}^n) \xrightarrow{\int_{\mathbf{R}^n}} \mathbf{R} \to 0.$$

A note of reminder here, cohomology with compact support is not invariant under homotopy equivalence, although it is invariant under diffeomorphism.

Next we look at cohomology on vector bundle E.

Let E be a rank-k vector bundle over n-dimensional manifold M. The zero section of E embeds M diffeomorphically in E, i.e., $M \times \{0\}$ is a deformation of E, so E and M have the same homotopy type, and they have the same de Rham cohomology:

$$H^*(E) \simeq H^*(M).$$

Now we look at compact cohomology. First look at the case of the trivial bundle:

$$E = M \times \mathbf{R}^k. \quad \pi : \ M \times \mathbf{R}^k \to M$$

is a projection map. The pullback of a form on M to a form on E may not have compact support, the pullback map π^* does not send $\Lambda_C^*(M)$ to $\Lambda_C^*(M \times \mathbf{R}^k)$, but there is a push-forward map π_*, called integration along the fibre which could send $\Lambda_C^*(M \times \mathbf{R}^k)$ to $\Lambda^{*-k}_C(M)$. Let $\{t_i\}_1^k$ be the coordinate on the fibre $\mathbf{R}^k$, a form on E is a linear combination of the following two types of forms:

$$\begin{array}{l}(1).\ (\pi^*\omega)f(x,t_1,\cdots,t_n)dt_{i_1}\cdots dt_{i_l}, \quad l<k;\\ (2).\ (\pi^*\omega)f(x,t_1,\cdots,t_n)dt_1\cdots dt_k,\end{array} \tag{P.1}$$

where $f(x,t_1,\cdots,t_n)$ is a function with compact, ω is a form on the base manifold, map π_* is defined by:

$$\begin{array}{l}(\pi^*\omega)f(x,t_1,\cdots,t_n)dt_{i_1}\cdots dt_{i_l}=0, \quad l<k,\\ (\pi^*\omega)f(x,t_1,\cdots,t_n)dt_1\cdots dt_k=\omega\int_{\mathbf{R}^k}f(x,t_1,\cdots,t_k)dt_1\cdots dt_k.\end{array} \tag{P.2}$$

It easy to show that the integration along the fibre π_* commutes with exterior differential d:

$$\pi_* d = d\pi_*.$$

Map π_* induces a map in compact cohomology

$$\pi_*:\quad H_C^*(M\times\mathbf{R}^k)\to H_C^*(M).$$

For cohomology on vector bundle E, there is another C^∞-invariance: the cohomology with compact support in the vertical direction $H_{CV}^*(E)$, simply called compact vertical cohomology.

Let E be a rank-k oriented vector bundle. A form on E is a real linear combination of two types of forms in Eq. P.1 where $f(x,t_1,\cdots t_n)$ is a function with compact support for each point $x\in M$. Making use of the integration along fibre π_* like Eq. P.2, this map defines an isomorphism:

$$\pi_*:\quad H_{CV}^*(E)\to H^{*-k}(M)$$

which is called the Thom isomorphism.

Theorem P.1 (**Thom isomorphism**) *Let E be a rank-k orientable vector bundle over M, then there is an isomorphism:*

$$H_{CV}^*(E)\simeq H^{*-k}(M).$$

The isomorphism image of $1\in H^0(M)$ under π_*^{-1} determines a cohomology class and is called the Thom class of E. In terms of which the Thom isomorphism is given by:

$$\begin{array}{rcl}\mathcal{J}:\ H^*(M) & \to & H_{CV}^{*+n}(E)\\ \omega & \to & \pi^*(\omega)\wedge\phi(E)\end{array}$$

We know that the Euler class $\chi(E) \in H^k(M)$ measures the twisting of the oriented vector bundle, its Poincare dual with respect to the zero section of E is just the Thom class, i.e., let

$$s: \ M \to E$$

be the zero section of E, then the pull back of Thom class to M by the zero section is the Euler class:

$$\chi = s^*\phi.$$

And its integration along the fibre equals to 1

$$\pi_*\phi = 1.$$

Appendix Q

K-theory, Topological Invariance of Elliptic Operator

Let M be a compact Hausdorff space, $V(M)$ be the set of all isomorphism classes of complex vector bundle over M. One can define an addition in $V(M)$ by forming the direct sum pointwise (Whitney sum). This operation is associative, but $V(M)$ is only an Abelian semigroup, i.e., the relation

$$E \oplus F = E' \oplus F, \quad E, E', F \in V(M)$$

does not imply $E = E'$. For example, let $M = S^2$ be embedded in $\mathbf{R}^3$, and $T(S^2)$ be the tangent bundle, which is a non-trivial rank-2 bundle over S^2, $N(S^2) = I$ be the normal bundle which is a trivial line bundle over S^2. Now let $I^2(S^2)$ be the rank-2 trivial bundle over S^2, $I^2(S^2)$ and $T(S^2)$ are not isomorphic, but the addition of a trivial line bundle to each of them yields an isomorphic rank-3 trivial bundle over S^2,

$$T(S^2) \oplus N(S^2) = T(\mathbf{R}^3) = I^3.$$
$$I^2 \oplus I = I^3.$$

One is tempted to construct a group by defining negative elements. Let us introduce the virtual bundle $(E, F) \sim E \ominus F$ to represent the formal difference between E and F, the virtual rank of which is (rank E - rank F). The ordinary vector bundle can be written as $(E, 0) \sim E$ and $(0, E) \sim -E$. So one can obtain an Abelian group $K(M)$ whose elements are virtual bundles. Thus $T(S^2)$ and I^2 represent the same element of $K(S^2)$,

$$T(S^2) \sim I^2.$$

Relation $\sim$ in $K(M)$ is an equivalence relation, called stably equivalent. Thus $T(S^2)$ is stably trivial.

Two bundles E and F are stably equivalent ($E \sim F$) if

$$E \oplus I^k = F \oplus I^l.$$

This relation $\sim$ is an equivalence relation. The equivalence class $[E]$ of a bundle E is regarded as a stable vector bundle. In the set of all such equivalence classes one can also introduce the tensor product pointwise. the tensor product is distributive with respect to Whitney sum. Thus one obtains a ring structure on $K(M)$. For high rank vector bundle, stably equivalent could also be isomorphic. Let k denote the rank of vector bundle over M, the stable ranges are:

$$k > dim\ M, \quad \text{for real vector bundle.}$$
$$k > \frac{1}{2} dim\ M, \quad \text{for complex vector bundle.}$$

Stably equivalence implies isomorphism if k is in the stable range.

The K-theory $K(M)$ of vector bundle plays an essential role in index theory. One knows that the topological index (topological invariance) of differential elliptic operator on a compact manifold depends on the homotopy class of its leading symbols only, which defines an equivalence class of compact K-theory of tangent bundle

$$\Sigma(D) \in K_{\text{cpt}}(TM).$$

Now look at the following sequence:

$$K_{\text{cpt}}(TM) \xrightarrow{f_\uparrow} K_{\text{cpt}}(T\mathbf{R}^N) \xrightarrow{\phi} K(\text{pt}) \simeq Z.$$

First we choose a smooth embedding:

$$f:\ M \hookrightarrow \mathbf{R}^N,$$

and there is the induced map

$$f_*:\ TM \to T\mathbf{R}^N.$$

The normal bundle can be obtained by pullback to TM, which has a complex structure. Therefore one can have the associative map

$$f_\uparrow:\ K_{\text{cpt}}(TM) \to K_{\text{cpt}}(T\mathbf{R}^N).$$

Then we define a contracted map taking $T\mathbf{R}^N$ to a point, which is just the inverse of Thom isomorphism. So one obtains the topological index of the symbol bundle, which is an integer. Applying the homotopy invariance of K_{cpt} one can obtain the A-S index theorem. Furthermore, the Chern class of a complex vector bundle depends on the equivalence class only, and Chern character extends it to a ring homomorphism. The cohomology formulation of the topological index can be derived from the Chern character ring homomorphism.

Appendix R

Cohomology of Lie group and Lie algebra

The cohomology of Lie group and Lie algebra has wide application in topology, algebra and physics.

A chain complex over **R** is a pair (C, ∂), where $C = \{C_k\}_{k \in \mathbf{Z}}$ is a graded **R**-module and the mappings $\partial = \{\partial_k\}_{k \in \mathbf{Z}}$:

$$\partial_k : \quad C_k \to C_{k-1}$$

is a **R**-module homomorphism of degree -1 such that $\partial^2 = 0$, then the homology of this chain complex can be defined as:

$$H_k(C, \partial) = \frac{Ker\partial_k}{Im\partial_{k+1}}.$$

Analogously, the cochain complex and the corresponding cohomology group can be defined: when there exists an endomorphism δ satisfying $\delta^2 = 0$, gut its degree is $+1$ instead of -1.

$$\delta = \delta_k : C^k \to C^{k+1},$$

so,

$$H^k = \frac{Ker\delta_k}{Im\delta_{k-1}}.$$

For example:

1.**Cohomology of Lie group**

Let G be a Lie group, (ρ, V) be its linear representation, i.e., ρ is a homomorphism map from G to $End(V)$, V is a module. Now one can defined k-cochain f^k of G with values in V, which is the k-linear smooth map

$$f^k : G^k \to V$$

and satisfy

$$f^k(gg_0, gg_1, ;gg_{k-1}) = \rho(g) f^k(g_0, g_1, \cdots, g_{k-1}).$$

The space of f^k is denoted by $C^k(G,\rho)$, and $C(G,\rho) = \oplus_{k\in \mathbf{Z}} C^k(G,\rho)$. Let δ be the graded linear map of degree 1 defined as

$$\delta f^k(g_0,\cdots,G_k) = \sum_{i=0}^{k}(-1)^i f^k(g_0,\cdots,\hat{g}_i,\cdots,g_k),$$

where $\hat{g}_i$ denotes the omission of g_i. It is easy to prove that $\delta^2 = 0$.

There are several equivalent definition of cohomology of G. For example, one can introduce the non-homogeneous k-cochain of G with value in V as follow

$$F^k(g_1,g_2,\cdots,g_k) = f^{k+1}(e,g_1,g_1g_2,\cdots,g_1g_2\cdots g_k),$$

and

$$\begin{aligned} f^{k+1}(g_0,g_1,\cdots,g_k) &= \rho(g_0)f^{k+1}(e,g_0^{-1}g_1,\cdots,g_0^{-1}g_k) \\ &= \rho(g_0)F^k(g_0^{-1}g_1,g_1^{-1}g_2,\cdots,g_{k-1}^{-1}g_k). \end{aligned}$$

The corresponding coboundary operator Δ can be expressed as

$$\begin{aligned} \Delta F^k(g_1,g_2,\cdots,g_{k+1}) &= \rho(g_1)F^k(g_2,\cdots,g_k) \\ &+ \textstyle\sum_{i=1}^{k}(-1)^k F^k(g_1,\cdots,g_ig_{i+1},\cdots,g_{k+1}) + (-1)^{k+1}F^k(g_1,\cdots,g_k), \end{aligned}$$

which also satisfy $\Delta^2 = 0$.

2.Cohomology of Lie algebra

Let g be the Lie algebra of G, by using differential, the representation (ρ,V) of G where one can obtain a representation (λ,V) of g is as follow:

$$\lambda(X)(P) = fracddt\rho(e^{tX})(P)|_{t=0},\ X\in \mathbf{g},\ \ \mathrm{P}\in \mathrm{V}$$

Let ϕ^k be a k-linear skewsymmetric mapping from $\mathbf{g}^k$ to V

$$\phi^k(X_0,\cdots,X_{k-1}) = (-1)\phi^k(X_{\sigma_0},\cdots,X_{\sigma_{k-1}})\in V,$$

here $X_i\in \mathbf{g}$ and σ belong to the permutation group S_k. The cohomology operator can be defined as

$$\begin{aligned} \Delta\phi^k(X_0,\cdots,X_k) &= \textstyle\sum_{i=0}^{k}(-1)^i\lambda(X_i)\phi^k(X_0,\cdots,\hat{X}_i,\cdots,X_k) \\ &+ \textstyle\sum_{i<j}(-1)^{i+j}\phi^k([X_i,X_j)],X_0,\cdots,\hat{X}_i,\cdots,\hat{X}_j,\cdots,X_k), \end{aligned}$$

which also satisfy $\delta^2 = 0$.

Note that the equation of de Rham cohomology d acting on difference forms is just like this one

3. Cohomology of gauge group

The gauge group cochain complex (α^k,Δ) can be defined as

$$\begin{aligned} \Delta\alpha^k(A,g_1,\cdots,g_k) &= \alpha^k(A^{g_1},g_2,\cdots,g_{k+1}) \\ &+ \textstyle\sum_{i=1}^{k}(-1)^i\alpha^k(A,g_1,\cdots,g_ig_{i+1},\cdots,g_{k+1}) + (-1)^{k+1}\alpha^k(A,g_1,\cdots,g_k). \end{aligned}$$

The element of gauge group is smooth mapping g $:M\to$ G. In general, the gauge group is topologically non-trivial and cocycles, α^k denotes the topological obstruction.

Appendix S

Some Recursion Relations of Q-BRST Algebra

Now we consider some recursion relations of Q-BRST algebra, let: (some notations see chapter 14)

$$\langle X_1 , X_2, \cdots , X_m\rangle = X_1^{j_1} X_2^{j_2} \cdots X_m^{j_m} g_{j_1 j_2 \cdots j_m} \tag{S.1}$$

where

$$g_{j_1 j_2 \cdots j_n} \equiv D^{t_0}_{t_1} \mathbf{C}_{j_1 t_2}^{t_1} \mathbf{C}_{j_2 t_3}^{t_2} \cdots \mathbf{C}_{j_m t_0}^{t_m}, \tag{S.2}$$

and X_n are fields η^J, $d\eta^J$, A^J or dA^J in the BRST algebra $\mathcal{B}$. They can also be replaced by, for example, XY^0, Y^0X or F. The following relations are easy to prove by direct calculation:

$$\begin{array}{ll} \Lambda^{RS}{}_{IJ}\, g_{RS} = g_{IJ}, & \Lambda^{rs}{}_{ij}\, g_{rs} = g_{ij}, \\ f^i_{rs}\, g_{rs} = 0, & (\mathcal{P}_{Adj})^{rs}{}_{ij}\, g_{rs} = 0, \\ (\mathcal{P}_{Adj})^{rs}{}_{ij}\, g_{rsk} = g_{ijk}, & (\mathcal{P}_{Adj})^{rs}{}_{jk}\, g_{irs} = g_{ijk}. \end{array} \tag{S.3}$$

Now, we are going to derive four important recursive relations(see S.6-S.9). Assume X^J and Y^J are two different fields in $\mathcal{B}$ with the indices $n_x > n_y$. Z^J denotes a field in $\mathcal{B}$, or a field multiplying by the zero component of this or another field. Since the following relations are linear for the field Z^J, Z^J can also be replaced by their linear combination, for example, F^J.

In chap.14 we known:(see P.460)

$$(-1)^{n_x n_y} X^I Y^J = \Lambda^{IJ}{}_{KL} Y^K X^L. \tag{S.4}$$

By using (14.10) one get

$$\begin{aligned} (-1)^{n_x n_y} X^i Y^0 &= Y^J X^K \Lambda^{i0}{}_{JK} \\ &= (\lambda^2+1) Y^0 X^i - (ig)^{-1}\lambda\, Y^j X^k \mathbf{C}_{jk}^{\ \ i}. \end{aligned}$$

Due to the adjoint components of the q-structure constents satisfy: (see P.458)

$$(\mathcal{P}_{Adj})^{k\ell}{}_{ij}\ \mathrm{C}_{kr}{}^{p}\mathrm{C}_{\ell s}{}^{r} \;=\; \frac{\lambda^2+1}{\lambda^2+2}\ \mathrm{C}_{ij}{}^{r}\mathrm{C}_{rs}{}^{p}, \tag{S.5}$$

so one get

$$(-1)^{n_x n_y}\ X^iY^0\ \mathrm{C}_{is}{}^{r} \;=\; (\lambda^2+1)\ Y^0X^i\ \mathrm{C}_{is}{}^{r} \;-\; \frac{\lambda(\lambda^2+2)}{ig(\lambda^2+1)}\ Y^uX^v\ (\mathcal{P}_{Adj})^{jk}{}_{uv}\ \mathrm{C}_{jt}{}^{r}\mathrm{C}_{ks}{}^{t}.$$

Now, according to the definition (S.1) and (S.3), we have:

$$\begin{aligned}(-1)^{n_x n_y}\ \langle Z\ ,\ X\rangle Y^0 &= (\lambda^2+1)\ \langle Z\ ,\ Y^0X\rangle \;-\; \frac{\lambda(\lambda^2+2)}{ig(\lambda^2+1)}\ \langle Z\ ,\ Y\ ,\ X\rangle;\\ (-1)^{n_x n_y}\ \langle XY^0\ ,\ Z\rangle &= (\lambda^2+1)\ Y^0\langle X\ ,\ Z\rangle \;-\; \frac{\lambda(\lambda^2+2)}{ig(\lambda^2+1)}\ \langle Y\ ,\ X\ ,\ Z\rangle.\end{aligned} \tag{S.6}$$

Similarly, from (S.4) and (14.14) one get

$$\begin{aligned}\{ig/\lambda\}\ (A^0A^i\ +\ A^iA^0)\ \mathrm{C}_{is}{}^{r} &= (\lambda^2+2)^{-1}\ A^jA^k\ \mathrm{C}_{jk}{}^{i}\mathrm{C}_{is}{}^{r}\\ &= (\lambda^2+1)^{-1}\ A^jA^k\ \mathrm{C}_{jt}{}^{r}\mathrm{C}_{ks}{}^{t}.\end{aligned}$$

The equation holds for η^J, too. Thus, we have:

$$\begin{aligned}&\{ig/\lambda\}\ \langle\cdots\ ,\ Z_1\ ,\ (A^0A+AA^0)\ ,\ Z_2\ ,\ \cdots\rangle\\ &\quad= (\lambda^2+1)^{-1}\ \langle\cdots\ ,\ Z_1\ ,\ A\ ,\ A\ ,\ Z_2\ ,\ \cdots\rangle;\\ &\{ig/\lambda\}\ \langle\cdots\ ,\ Z_1\ ,\ (\eta^0\eta+\eta\eta^0)\ ,\ Z_2\ ,\ \cdots\rangle\\ &\quad= (\lambda^2+1)^{-1}\ \langle\cdots\ ,\ Z_1\ ,\ \eta\ ,\ \eta\ ,\ Z_2\ ,\ \cdots\rangle.\end{aligned} \tag{S.7}$$

In terms of the q-deformed Jacobi identity for the q-deformed structure constants:

$$\mathrm{C}_{IR}{}^{P}\mathrm{C}_{JS}{}^{R} \;-\; \Lambda^{KL}{}_{IJ}\ \mathrm{C}_{KR}{}^{P}\mathrm{C}_{LS}{}^{R} \;=\; \mathrm{C}_{IJ}{}^{R}\mathrm{C}_{RS}{}^{P},$$

and Eq. (S.6) one can obtain:

$$\begin{aligned}(-1)^{n_x n_y}\ X^IY^J\ \mathrm{C}_{It}{}^{r}\mathrm{C}_{Js}{}^{t} &= Y^KX^L\ \Lambda^{IJ}{}_{KL}\ \mathrm{C}_{It}{}^{r}\mathrm{C}_{Js}{}^{t}\\ &= Y^KX^L\ \{\mathrm{C}_{Kt}{}^{r}\mathrm{C}_{Ls}{}^{t}\ -\ \mathrm{C}_{KL}{}^{t}\mathrm{C}_{ts}{}^{r}\}\\ &= Y^KX^L\ \mathrm{C}_{Kt}{}^{r}\mathrm{C}_{Ls}{}^{t}\ -\ \{ig/\lambda\}\ (Y^0X^j\ -\ (-1)^{n_x n_y}\ X^jY^0)\ \mathrm{C}_{js}{}^{r}.\end{aligned}$$

Then,

$$\begin{aligned}&\{(-1)^{n_x n_y}\ X^iY^j\ -\ Y^iX^j\}\ \mathrm{C}_{it}{}^{r}\mathrm{C}_{js}{}^{t}\\ &\quad= \left\{(-1)^{n_x n_y}\ X^IY^J\ -\ Y^IX^J\right\}\ \mathrm{C}_{It}{}^{r}\mathrm{C}_{Js}{}^{t}\\ &\qquad+ ig\lambda\ \{(-1)^{n_x n_y}\ X^jY^0 - Y^0X^j\}\ \mathrm{C}_{js}{}^{r}\\ &\quad= -\ ig\lambda^{-1}(\lambda^2+1)\ \{Y^0X^j\ -\ (-1)^{n_x n_y}\ X^jY^0\}\ \mathrm{C}_{js}{}^{r}.\end{aligned}$$

So we have,

$$\begin{aligned}(-1)^{n_x n_y}\,&\langle\cdots\,,\,Z_1\,,\,X\,,\,Y\,,\,Z_2\,,\,\cdots\rangle\\ &=\langle\cdots\,,\,Z_1\,,\,Y\,,\,X\,,\,Z_2\,,\,\cdots\rangle\\ &\quad-ig\lambda^{-1}(\lambda^2+1)\,\langle\cdots\,,\,Z_1\,,\,(Y^0X-(-1)^{n_x n_y}XY^0)\,,\,Z_2\,,\,\cdots\rangle.\end{aligned}\tag{S.8}$$

By making use of (S.6) we have:

$$\begin{aligned}(-1)^{n_x n_y}\,\langle Z\,,\,X\,,\,Y\rangle &= ig\lambda(\lambda^2+1)\,\langle Z\,,\,Y^0X\rangle-(\lambda^2+1)\,\langle Z\,,\,Y\,,\,X\rangle;\\ (-1)^{n_x n_y}\,\langle X\,,\,Y\,,\,Z\rangle &= ig\lambda(\lambda^2+1)\,\langle Y^0X\,,\,Z\rangle-(\lambda^2+1)\,\langle Y\,,\,X\,,\,Z\rangle.\end{aligned}\tag{S.9}$$

From these four recursive relations and (S.3), we can derive the following useful formulas. First of all, from (S.3) we have:

$$\langle\eta\,,\,\eta\rangle=\langle A\,,\,A\rangle=0,\qquad(-1)^{n_x n_y}\,\langle X\,,\,Y\rangle=\langle Y\,,\,X\rangle.\tag{S.10}$$

From (S.7) we have:

$$\begin{aligned}\langle F\,,\,A\rangle &= \langle A\,,\,F\rangle=\langle A\,,\,dA\rangle+\frac{ig}{\lambda}\langle A\,,\,(A^0A+AA^0)\rangle\\ &= \langle A\,,\,dA\rangle+(\lambda^2+1)^{-1}\,\langle A\,,\,A\,,\,A\rangle.\end{aligned}\tag{S.11}$$

If the fields Y^J and Z^J are not the field η^J, we have:

$$\langle Y\,,\,Z\rangle\,\eta^0=(-1)^{n_y+n_z}\,\eta^0\,\langle Y\,,\,Z\rangle.\tag{S.12}$$

Let Z^J be a field in $\mathcal{B}$ with the index n_z, and X^J denote the field η^J or A^J with the index $n_x=-1$ or 1, respectively. From the recursive relations we obtain that:
(1) If $n_z<n_x$:

$$\begin{aligned}&\langle X\,,\,X\,,\,Z\rangle=\langle Z\,,\,X\,,\,X\rangle;\\ &(-1)^{n_x n_z}\,\langle X\,,\,Z\,,\,X\rangle=-(\lambda^2+1)\,\langle Z\,,\,X\,,\,X\rangle;\\ &\langle X\,,\,X\,,\,Z\rangle-(-1)^{n_x n_z}\langle X\,,\,Z\,,\,X\rangle+\langle Z\,,\,X\,,\,X\rangle\\ &\quad=[3]\,\langle Z\,,\,X\,,\,X\rangle.\end{aligned}\tag{S.13}$$

(2) If $n_z>n_x$:

$$\begin{aligned}&\langle Z\,,\,X\,,\,X\rangle=ig\lambda(\lambda^2+1)(\lambda^2+2)\,X^0\langle X\,,\,Z\rangle+(\lambda^2+1)\,\langle X\,,\,X\,,\,Z\rangle\\ &(-1)^{n_x n_z}\,\langle X\,,\,Z\,,\,X\rangle=-ig\lambda(\lambda^2+1)\,X^0\langle X\,,\,Z\rangle-\langle X\,,\,X\,,\,Z\rangle\\ &\quad=-(\lambda^2+2)^{-1}\,\{\langle X\,,\,X\,,\,Z\rangle+\langle Z\,,\,X\,,\,X\rangle\}\\ &\langle X\,,\,X\,,\,Z\rangle-(-1)^{n_x n_z}\langle X\,,\,Z\,,\,X\rangle+\langle Z\,,\,X\,,\,X\rangle\\ &\quad=-(-1)^{n_x n_z}\,[3]\,\langle X\,,\,Z\,,\,X\rangle.\end{aligned}\tag{S.14}$$

At last, through direct calculation we have:

$$\langle A\,,\,A\,,\,A\rangle\,\eta^0=-\eta^0\,\langle A\,,\,A\,,\,A\rangle=ig^3(\lambda^2+1)(\lambda^2+2)\,\eta^0A^3A^+A^-.\tag{S.15}$$

The general reference texts

The general refrence texts

I. S.Kobayashi and K.Nomizu, Fundations of Differential Geometry, Vol. I(1963), Vol. II(1969), Interscience.

II. Y.C.Bruhat, C.D.Morette and M.D.Bleick, Analysis, Manifolds and Physics(1977), North-Holland.

III. M.Spivak, Differential Geometry, Vol. 1-5(1979), Publish or perish Inc.

IV. S.Helgason, Differential Geometry, Lie Group and Symmetric Space(1978), Academic Press.

V. T.Eguchi, P.B.Gilkey and A.J.Hanson, *Phys.Rep.*, **66**(1980), 213.

VI. C.Nash and S.Sen, Topology and Geometry for Physicists(1983), Academic Press.

VII. B.A.Dubrovin, A.T.Fomenko, S.P.Novikov, Modern Geometry, Part I(1984), Part II(1985), Part III(1990), Springer-Verlag.

VIII. Z.H.Jiang, Introduction to topology, (1979) (in Chinese).

References

[1] L.Alvarez Gaumé, Comm.Math.Phys. **90** (1983) 161.

[2] S.K.Donaldson, J.Diff.Geom. **18** (1983) 269.

[3] N.Seiberg and E.Witten, Nucl Phys. **B426** (1994) 19.

[4] A.Einstein, Annalen der Phys. **49** (1916) 769.

[5] V.I.Arnol'd, Mathematical Methods of Classical Mechanics. Springer-Verlag (1978).

[6] M.B.Green, J.H.Schwarz, E.Witten. Superstring theory. Cambridge University Press (1987).

[7] C.N.Yang and R.L.Mills, *Phys.Rev.*, **96**(1954), 191.

[8] B.Zumine, Y.S.Wu and A.Zee, *Nucl.Phys.*, **B239**(1984), 477.

[9] T.T.Wu and C,N.Yang. Phys.Rev., **D12** (1975), 3845.

[10] M.V.Berry, *Proc.R.Soc.Lond.*, **A392**(1984), 45.

[11] J.Milnor, Ann. Math. **64** 1956, 399.

[12] P.A.M.Dirac, *Proc.Roy.Soc.London, A*, **133**(1931), 60.

[13] F. B. Estabrook and H. D. Wahlquist, *J.Math.Phys.*, **16**(1975), 1.

[14] I. Robinson and A. Trautman, *Proc.Roy.Lond.*, **A265**(1962), 463.

[15] K,Schwarzschild, Uber das Gravitationsfeld eines Massenpunktes nach der Einsteinschen Theorie, Sitz. Preuss. Akad. Wiss.(1916) 189.

[16] T. Tguchi and J. Hanson. *Phys.Lett.*, **74B**(1975), 249.

[17] K. Pohlmeyer, *Com.Math.Phys.*, **46**(1976), 207, **72**(1980), 317.

[18] R. Sasaki, *Nucl.Phys.*, **B154**(1979), 343.

[19] B. Y. Hou, B. Y. Hou and P. Wang, *J.Phys.A : Math.Gen.*, **18**(1985), 165.

[20] B. Y. Hou, B. Y. Hou and P. Wang, Proc. 3-rd Grassmann Meeting Shanghai(1983), 1033.

[21] B. Y. Hou, B. Y. Hou and P. Wang, *Int.J.Morden.Phys.* **1** (1986), 193.

[22] B.Y.Hou, M.L.Ge, Y.S.Wu, Phys. Rev. **D24** (1981),2238.

[23] L.L.Chau, B.Y.Hou, X.C.Song, Phys. Lett. **151B** (1985), 421.

[24] S. S. Chern and C. L. Terng, *RockyMountainJour.Math.*, **10**(1980), 105.

[25] A. Adler, *J.Math.*, **91**(1969), 657.

[26] A.Nijenhuis, Indag. Math. $\overset{13}{-}$ (1951) 200-212.

[27] G.Fubini, Atti. Instit. Veneto 6 (1903), 501-513.

[28] E.Study, Math. Ann. 60 (1905), 321-377.

[29] E. Calabi, Algebraic Geometry and Topology. A Symposium in Honor of S. Lefschtz, Princeton Univ. Press(1957).

[30] S. T. Yau, *Proc.Natl.Acad.Sci.* U. S. A., **74**(1977), 1798.

[31] B. Y. Hou, B. Y. Hou and P. Wang, Advances in Science of China Phys., **2**, Science Press, China.(1987) 1-26.

[32] J.M.Maillet, Nucl. Phys. B269 (1986) 54-76.

[33] M.Forger, J.Larry, U.Schaper, Comm. Math. Phys. **146** (1992) 397. M.Bordemann, M.Forger, J.Larry, U.Schaper, Comm. Math. Phys. **152** (1993) 167.

[34] B.Y.Hou, B.Y.Hou, Y.W.Li, B.Wu, J. Phys. A. Math. Gen. **27** (1994) 7209-7216.

[35] W. V. D. Hodge, Theory and Application of Harmonic Integrals, Cambridge Univ. Press.

[36] H.Toda, Composition methods in homotopy groups of spheres (1962).

[37] E. Witten, *InternationalJ.Mod.Phys,A.*, **1**(1986), 39.

[38] B.Y.Hou, B.Y.Hou, Phys. Energ. Fortis Phys. Nucl.**3** (1979),255-265. Chin. Phys. Vol. 1, No. 3, July-Sept (1981) 624-634.

[39] Y. Aharonov and B. Bohm, *Phys.Rev.*, **115**(1954), 485.

[40] A.Belavin, A.Polyakov, A,Schwartz, Y,Tyupkin, Phys. Lett. **59B**. (1975) 85-87.

[41] M.K.Prasad, C.M.Sommerfield, Phys. Rev. Lett. **35**, (1975) 760.

[42] E.B.Bogomolny, Sov. J. Nuel. Phys. **24**, (1976) 861-870.

[43] T.S.Wu, C.N.Yang, Phys.Rev. **D13** (1976) 3233.

[44] N.Seiberg, E.Witten, Nucl. Phys. **B431** (1994) 581-640

[45] S. S. Chern and J. Simons, *Ann.Math.*, **99**(1974), 48.

[46] H. Y. Guo, K. Wu and S. K. Wang, *Com.Theor.Phys.*(Beijing), **4**(1985), 113.

[47] H. Y. Guo, B. Y. Hou, S. K. Wang and K. Wu, *Com.Theor.Phys.*(Beijing), **4**(1985), 145,233.

[48] M. F. Atiyah, Comm. Pure Appl. Math. **20** (1967), 237-249.

[49] M. F. Atiyah, Sem. Bourbaki, Exp. **253**, (1963),

[50] M. F. Atiyah, I.M.Singer, Bull. Amer. Math. Soc. **69** (1963), 422-433.

[51] E. Brieskorn. et al., Seminarvorträge, Bonn, (1963).

[52] R. S. Palais ed., Ann. of Math. Studies **57** (1965), Princeton.

[53] M. F. Atiyah, I.M.Singer, Ann. Math. **87** (1968), 484-530. 5466-604. **93** (1971), 119-138. 139-149.

[54] R.S.Patodi., J. Diff. Geom. **5** (1971), 233–249, 251-283.

[55] P.B.Gilkey, Adv. in Math. **11** (1973), 311-325.

[56] B.Y.Hou, B.Y.Hou, Phys. Energ. Fortis Phys. Nucl.**3** (1979),697-707.

[57] M.F.Atiyah, N.J.Hitchin, I.M.Singer, Proc. Nat. Acad. Sci. **74** (1977), 2662.

[58] M.F.Atiyah, N.J.Hitchin, I.M.Singer, Proc. R. Soc. Lond. **A362** (1978), 425.

[59] S.W. Hawking, Phys. Lett. **60A** (1977), 81.

[60] M. F. Atiyah and R. Bott., *Ann.ofMath.*, **86**(1967), 374.

[61] M. F. Atiyah, V. K. Patodi, I. M. Singer, Math. Proc. Camb. Phil. Soc. **77**(1975), 43-69; **78**(1975), 405; **79**(1976), 71.

[62] N.Hitchin, Advances in Math. **14** (1974), 1.

[63] A. J. Niemi and G. W. Semenoff, *Phys.Rep.*, **135**(1986), 99.

[64] A. P. Polychronaros, *Nucl.Phys.*, **B278**(1986), 297.

[65] M. F. Atiyah and I. M. Singer, *Proc.Nat.Acad.Sci.*, U. S. A., **81**(1984), 2597-2600.

[66] B. Y. Hou, Y. Z. Zhang, J. Math. Phys. **28** (1987), 1700. J. Phys. **A21** (1988), 345.

[67] B. Y. Hou, B. Y. Hou and P. Wang, *Lett.Math.Phys.*, **11**(1986), 179.

[68] R.Jackiw, "Topological Investigations of Quantized Gauge Theories". In: "Relativity, Groups and Topology II" B. DeWitt and R.Stora, ed. North-Holland, Amsterdam (1984).

[69] B.Zumino,"Chiral anomalies and diferential geometry." In: "Relativity, Groups and Topology II" B. DeWitt and R.Stora, ed. North-Holland, Amsterdam (1984).

[70] R. Jackiw and C. Rebbi, *Phys.Rev.Lett.*, **37**(1976), 172.

[71] S. Deser, R. Jackiw and S. Templeton, *Phys.Rev.Lett.*, **48**(1983), 975. *Ann.Phys.NY.*, **140**(1984), 372.

[72] J. Wess and B.Zumino, *Phys.Lett.*, **B87**(1971), 95.

[73] E. Witten, *Nucl.Phys.*, **B223**(1983), 422.

[74] K.C.Chou, H.Y.Guo, X.C.Song and K.Wu, *Phys.Lett.*, **B134**(1984), 67.

[75] Y.S.Wu and A.Zee, Nucl.Phys. B258 (1985) 157.

[76] K. Fujikawa, *Phys.Rev.*, **D21**(1980), 2848.

[77] R. Jackiw, *Phys.Rev.Lett.*, **54**(1985), 159.

[78] B. Y. Hou, B. Y. Hou, *ChinesePhys.Lett.*, **2**(1985), 149.

[79] B. Y. Hou, B. Y. Hou and P. Wang, *AnnPhys.*, **171**(1986), 172.

[80] B.Y.Hou, P.Wang, Comm. Theor. Phys. **12** (1989), 109.

[81] Yu. I. Manin, *Quantum Groups and Non-Commutative Geometry*, Montreal Univ. preprint, CRM-1561, 1988.

[82] A. Connes, Publ. Math. IHES. **62**(1986)41.

[83] S. L. Woronowicz, *Commun. Math. Phys.* **111**(1987)613; **122**(1989)125.

[84] J. Wess and B. Zumino, *Nucl. Phys.* (Proc. Suppl.) **B18**(1990)302.

[85] D. Bernard, Quantum Lie Algebras and Differential Calculus on Quantum Groups – Proc. 1990 Yukawa Int. Seminar (kyoto); *Phys. Lett.* **B260**(1991)389.

[86] B. Jurčo, *Lett. Math. Phys.* **22**(1991)177.

[87] B. Zumino, *Introduction to the differential geometry of quantum groups*, LBL-31432 and UCB-PTH-62-91, Notes of a plenary talk given at the 10th IAMP Conf., Leipzig, 1991.

[88] U. Carow-Watamura, M. Schlieker, S. Watamura and W. Weich, *Commun. Math. Phys.* **142**(1991)605.

[89] A. P. Isaev and Z. Popowicz, *Phys. Lett.* **B281**(1992)271; K. Wu and R. J. Zhang, *Commun. Theor. Phys.* **17**(1992)175.

[90] I. Ya. Aref'eva and I. V. Volovich, *Phys. Lett.* **B264**(1991)62; *Mod. Phys. Lett.* **A6**(1991)893.

[91] T. Brzeziński and S. Majid, *Commun. Math. Phys.* **157**(1993)591.

[92] X. D. Sun and S. K. Wang, *Bicovariant differential calculus on quantum group* $GL_q(n)$, Worldlab-Beijing preprint CCAST-92-04, 1992.

[93] P. Aschieri, L. Castellani, *Inter. J. Mod. Phys.* **A8**(1993)1667.

[94] L. Castellani, *Phys. Lett.* **B292**(1992)93; $U_q(N)$ *Gauge Theories*, Preprint, DFTT-74/92, 1992.

[95] S. Watamura, *Commun. Math. Phys.* **158**(1993)67.

[96] T. Brzeziński and S. Majid, *Phys. Lett.* **B298**(1993)339.

[97] C. Becchi, A. Rouet, and R. Stora, *Ann. Phys.* **98**(1976) 287; I. V. Tyuitin, Lebedev preprint FIAN **39**(1975), unpublished; T. Kugo, I. Ojima, *Prog. Theor. Phys. Suppl.* **66**(1979)1.

[98] S. S. Chern, *Complex manifolds without potential theory*, (especially the Appendix: *Geometry of characteristic classes*), Springer, New York (1979).

[99] B. Zumino, *Chiral Anomalies and Differential Geometry*, Lectures given at Les Houches, Preprint, LBL-16747, 1983; J. Mañes, R. Stora and B. Zumino, *Commun. Math. Phys.* **102**(1985)157.

[100] B.Y.Hou, B.Y.Hou, Z.Q.Ma, J. Phys.A. Gen. **28** (1995), 543-558.

[101] Zhong-Qi Ma, *Yang-Baxter Equation and Quantum Enveloping algebras*, World Scientific, Singapore, 1993.

[102] G. Gasper and M. Rahman, Basic Hypergeometric Series, in *Encyclopedia of Mathematics and its Applications 35*, Cambridge, 1990; F. H. Jackson, *Quart. J. Pure Appl. Math.* **41**(1910)193.

[103] B.Y.Hou, B.Y.Hou, Z.Q.Ma, J. Math. Phys. **36** (1995), 5110.

Index

D

E

F

G

H

M

N

O

P

Q

R

S

T

U

V

W

Y

www.ingramcontent.com/pod-product-compliance
Lightning Source LLC
Chambersburg PA
CBHW061516060426
42492CB00005B/187
9789810231057